Structural Life Assessment Methods

Structural Life Assessment Methods

A.F. Liu

The Materials Information Society

First printing, July 1998

Library of Congress Cataloging-in-Publication Data

Alan F. Liu
Structural Life Assessment Methods
Includes bibliographical references and index.
1. Metals—Fracture
2. Fracture mechanics
3. Accelerated life testing
I. Title
TA460.L59 1998 620.1'66 98-14143
ISBN: 0-87170-653-9
SAN: 204-7586

ASM International®
Materials Park, OH 44073-0002

Printed in the United States of America

To my greatest supports:

my wife, Iris, and our three sons,

Kent, Willy and Henry.

Contents

Preface

In any profession, it is tempting to try to identify the origin and development of one's specialized field. In the case of fracture mechanics, the Griffith theory of fracture, which was introduced at the turn of the century, is known to many scientists and engineers. However, it was not until World War II, when many Liberty ships experienced brittle fracture (suddenly separated into two halves) in the ocean, that engineers began studying and treating "fracture" seriously. Since then, many fracture theories have been developed. However, good material fracture properties characterization and structural residual strength prediction were not possible until the birth of the Irwin theory of fracture in the mid-1950s.

In a symposium held at the College of Aeronautics, Cranfield, England in 1961, the delegation from Boeing (D.R. Donaldson and W.E. Anderson) reported that many pre-existing cracks (both short and long cracks) had been found in brand-new Boeing airplanes. No matter how bizarre it might seem, this did point to the fact that pre-existing cracks in structural components are unavoidable. Engineers just need to learn how to deal with them. They need to know the smallest flaw that can be found (or the largest flaw that might have been missed) during inspection, and they need to know whether the vehicle can be operated with such a flaw. This led to the need for a method of calculating fatigue crack propagation life. The Paris fatigue crack growth law, which suggests that the Irwin theory of fracture is applicable to incremental fatigue crack damage, has paved the way for engineers to apply fracture mechanics to structural life assessment.

Over the years, the U.S. Federal Aviation Administration has adopted the fail-safe philosophy in the design of commercial aircraft, and the U.S. Air Force has set up an Aircraft Structural Integrity Program to monitor the performance and safety of military aircraft. However, it was only the failure of the wing carrythrough structure of the F-111 aircraft in the early 1970s that triggered the requirement for designing aircraft structures with assumed pre-existing flaws. Today, virtually every industry, including defense, aerospace, electric power, nuclear, pressure vessel, shipbuilding, turbine, pipeline, and offshore drilling imposes stringent fracture control procedures regulating the design/development and maintenance of their products.

Owing to the need and interests generated, the state-of-the-art in damage-tolerant life assessment technique has changed at a rapid pace during the last four decades. More than 20,000 technical papers and reports on various aspects of fracture and fatigue crack growth of metals have been published. It is impossible to include all the methods, even from a fraction of these publications, in a single book. Only those relevant to the specific themes intended are included in this book. Many excellent fracture mechanics textbooks exist, and some are listed in Chapter 1 as selected references. However, no single book covers the full spectrum of the methodologies currently used in industry. It is desirable to have a treatise that includes the updated materials, such as the state-of-the-art techniques for fatigue crack growth rate representation, linear-elastic and elastic-plastic crack tip stress field solutions, and the new parameters for characterizing creep-fatigue crack growth and fracture at high temperature, with illustrations showing how to apply them.

This book covers a wide range of subjects relating to application of fracture mechanics for assessing the safe-life and fracture strength of metallic structures, and the methods for improving structural longevity. Most of the discussions, methods, and examples in this book concern aircraft structures, but these methods are equally applicable to other types of metallic structures and machinery. This book provides a practical, useful source of information on fracture-mechanics-based analytical methods for life assessment and damage tolerance design/analysis. Fundamental and metallurgical aspects of fatigue crack growth and fracture are discussed throughout this book to improve the reader's understanding. Strong emphasis is placed on the stress analysis aspects of problem solving. The book also maintains a balance of theory and practice, the "how and why."

This book is intended for use as a reference book by engineers and engineering students who are involved in design/analysis of structures or machinery, or evaluation of material properties. It also can be used as a textbook for a course in fracture mechanics, to be taken by undergraduate and graduate students of aeronautical, civil, mechanical, or metallurgical engineering.

A.F. Liu
West Hills, California
September, 1997

Chapter 1

Fundamentals of Fracture Mechanics

Energy Considerations

The quantitative relationship that engineers and scientists use today in determining the fracture strength of a cracked solid is based on the Griffith theory of fracture (Ref 1-1). Griffith recognized that the driving force for crack extension was the difference between the energy that could be released if the crack was extended and the energy needed to create new surfaces. For a through-the-thickness crack of length $2a$ in an infinitely wide sheet (Fig. 1-1), Griffith estimated the strain energy stored in the system (per unit thickness) to be:

$$U_e = -\frac{\pi a^2 S^2}{E} \qquad \text{(Eq 1-1)}$$

where S is the applied stress and E is the modulus of elasticity of the material. The minus sign indicates that this quantity would be released from the system. He then estimated the second term, i.e., the energy associated with the total area of the crack surface (per unit thickness), to be:

$$U_s = 2 \cdot (2a) \cdot \gamma \qquad \text{(Eq 1-2)}$$

where γ is the specific surface energy (surface tension). Here, $2 \cdot 2a$ denotes that the crack (of total length $2a$) had two surfaces, top and bottom.

Griffith's criterion of fracture states that the crack will propagate under constant applied stress S if an incremental increase in crack length produces no change in the total energy of the system, i.e., the derivative of the sum of Eq 1-1 and 1-2 is zero. That is:

$$\frac{\partial \bar{U}}{\partial a} = 0 = \frac{\partial}{\partial a}\left[4a\gamma - \frac{\pi a^2 S^2}{E}\right] = 4\gamma - \frac{2\pi a S^2}{E} \quad \text{(Eq 1-3)}$$

which gives:

$$S_{\text{cr}} = \left(\frac{2E\gamma}{\pi a}\right)^{1/2} \quad \text{(Eq 1-4)}$$

or:

$$2\gamma = \frac{\pi a S_{\text{cr}}^2}{E} \quad \text{(Eq 1-5)}$$

where the subscript "cr" stands for "critical."

The failure criterion of Eq 1-3 requires the change of surface energy to be greater than the change of strain energy in order to maintain the integrity of a structural member. We will elucidate this concept through a graphical illustration that is presented in the following section. Before we do that, we like to demonstrate how Eq 1-1 is derived. Because Griffith's method of calculation was rather complicated, instead, we will examine this problem from an engineer's point of view. A simple method of derivation is presented below.

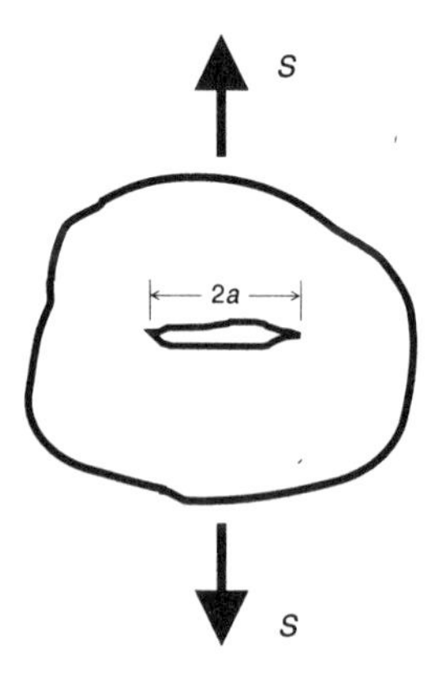

Fig. 1-1 A cracked body subjected to uniaxial tension

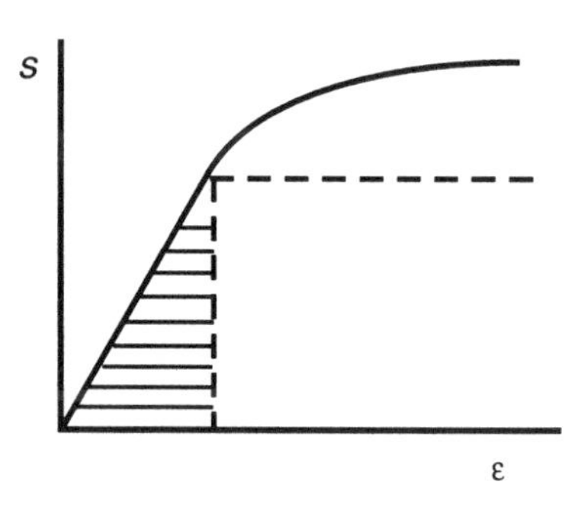

Fig. 1-2 Representation of strain energy per unit volume by the area under a tensile stress-strain curve

Engineering Estimate of Energy Release Rate

To understand what makes a crack propagate, first consider the case of a plate in the absence of a crack. When a body is loaded, e.g., under uniaxial tension, the movement of the applied loads does work on the body and is stored as strain energy. While the loading condition within the system is within the elastic limit, the strain energy per unit volume is represented by the area under the stress-strain curve of a tensile test (the shaded area of Fig. 1-2):

$$W = \frac{1}{2}S\varepsilon = \frac{1}{2E}S^2 \qquad \text{(Eq 1-6)}$$

This relationship is valid whether the material in question is a strain hardening type or an elastic/ideally plastic solid (the solid line and the dotted line, respectively, in Fig. 1-2).

Now consider a plate containing a crack of length $2a$. The applied load will have to go around the crack and be supported by the uncracked area. Thus, there will be a general relaxation of the material above and below the crack, and some strain energy will be released. A sketch of such a relaxed zone is given in Fig. 1-3(a), shown as shaded areas above and below the crack. In order to obtain an approximate solution, let us assume that the

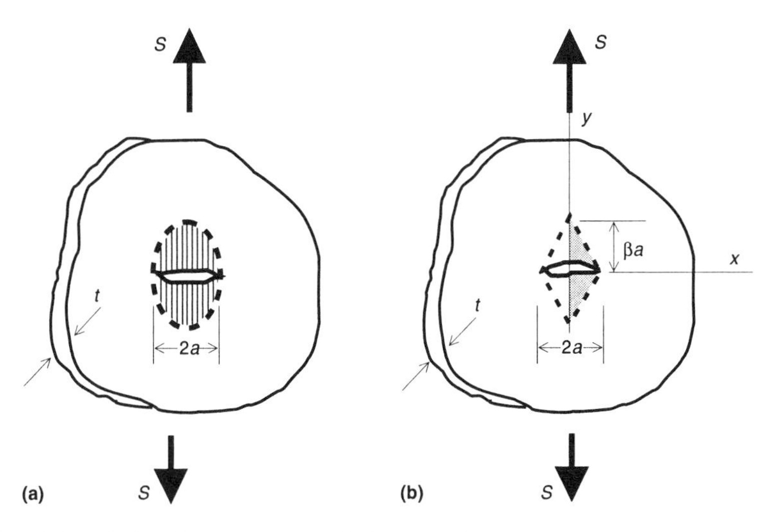

Fig. 1-3 Load relaxed zones above and below a crack

stress-free region (i.e., the relaxed zone on either side of the crack) has the shape of a triangle (Fig. 1-3b). The height of the triangle is somewhat proportional to the crack length, i.e., equal to βa. The relaxed volume of the rhombus to the right of the y-axis is given by:

$$V = 2\left(\frac{1}{2}\beta a \cdot a\right)t \qquad \text{(Eq 1-7)}$$

Because the Griffith crack is a straight separation of the two surfaces, the thickness across is infinitesimal, and no volume is occupied. Therefore the energy per unit thickness U is the energy per unit volume times the volume and divided by the thickness:

$$U = \frac{W \cdot V}{t} = \left(\frac{S^2}{2E}\right) \cdot \beta a^2 \qquad \text{(Eq 1-8)}$$

By letting $\beta = \pi$, we can see that we have just derived a relationship that is equivalent to one-half of the Griffith solution (i.e., quantitatively equal to $U_e/2$). In other words, Eq 1-8 can be regarded as the energy per unit volume for one crack tip. Therefore, the strain energy release rate for each crack tip for plane stress becomes:

$$\frac{\partial U}{\partial a} = \frac{S^2 \pi a}{E} \qquad \text{(Eq 1-9)}$$

While for plane strain:

$$\varepsilon = (1 - \nu^2)\, S/E \qquad \text{(Eq 1-10)}$$

$$W = \frac{1}{2}SE = \frac{1}{2E}S^2(1 - \nu^2) \qquad \text{(Eq 1-11)}$$

$$U = \frac{W \cdot V}{t} = \left(\frac{S^2}{2E}\right) \cdot \pi a^2 (1 - \nu^2) \qquad \text{(Eq 1-12)}$$

$$\frac{\partial U}{\partial a} = \frac{S^2 \pi a}{E}(1 - \nu^2) \qquad \text{(Eq 1-13)}$$

Failure Criteria

As stated above, the driving force for crack extension is the difference between the energy that could be released if the crack was extended and the energy needed to create new surfaces. Here the energy that could be released is U and the rate of release as the crack grows is $\partial U/\partial a$ (Eq 1-9, 1-13). From now on we will call the energy release rate as $\mathcal{G}$ (Ref 1-2 to 1-4). The variations of U and $\partial U/\partial a$ with crack length are schematically shown in Fig. 1-4. Because U represents a release of energy, it is plotted as a negative quantity. The energy input required to create new crack surfaces will be called $\overline{W}$. For one crack tip, the quantity for $\overline{W}$ is equal to one-half of the surface energy ($U_s/2$). For our purposes it is sufficient to assume that the energy required for each equal increment of crack length is constant. This is equivalent to saying that the energy required to break an atomic bond ahead of the crack is the same as that required to break the next, and so on. Hence $\overline{W}$ increases linearly with increasing crack length (top part of Fig. 1-4), and $\partial \overline{W}/\partial a$ is a constant (bottom part of Fig. 1-4).

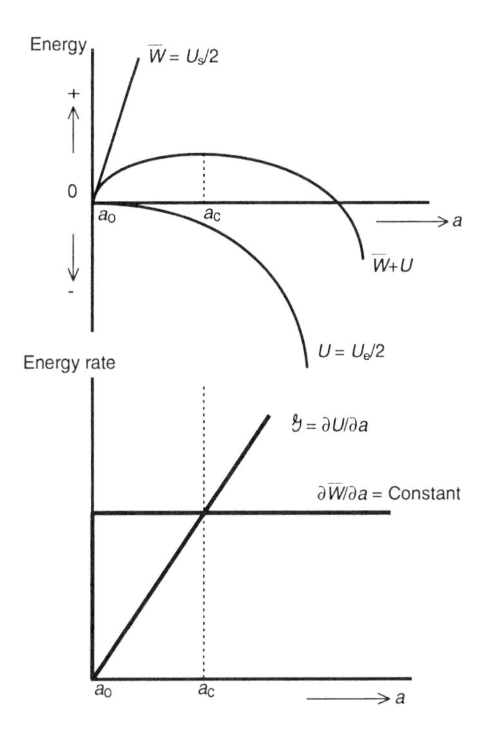

Fig. 1-4 Failure criteria

The failure criterion based on the top part of Fig. 1-4 is that unstable crack extension occurs at a_c where the total energy $(\overline{W} + U)$ reaches a maximum. At this location the change of total energy with respect to a is zero. The failure criterion based on the bottom part of Fig. 1-4 would be that the strain energy release rate, $\mathcal{G}$, exceeds the rate of energy input to the system (i.e., $\mathcal{G} > \partial \overline{W} / \partial a$ for creating new crack surfaces). Hence, a_c is called the critical crack length. The corresponding value of $\mathcal{G}$ (at a_c) is called $\mathcal{G}_c$ (the plane stress fracture toughness) or $\mathcal{G}_{Ic}$ (the plane strain fracture toughness). The unit for U is work done (energy) per unit thickness. Therefore, the physical unit for $\mathcal{G}$, i.e., $\partial U / \partial a$, is work done per unit area (inch-pound per square inch). Both $\mathcal{G}_c$ and $\mathcal{G}_{Ic}$ are regarded as material properties.

Stress Analysis of Cracks

So far we have proved, qualitatively, that during crack propagation (in an infinite solid) there is something called the strain energy release rate $\mathcal{G}$ operating in the system; $\mathcal{G}$ increases as the crack grows. The plate (or solid) will fail (rapid crack propagation) when $\mathcal{G}$ reaches its critical value $\mathcal{G}_c$, or $\mathcal{G}_{Ic}$. It is possible to determine $\mathcal{G}_c$, or $\mathcal{G}_{Ic}$, by testing a specimen containing a crack (its initial length is $2a_o$) and noting the stress level and crack length at failure; i.e., S_{cr} and $2a_c$. According to Eq 1-9 and 1-13, we have

$$\mathcal{G} = S_c^2 \pi a_c / E \qquad \text{(Eq 1-14)}$$

for plane stress, and

$$\mathcal{G}_{Ic} = S_c^2 \pi a_c (1 - \nu^2) / E \qquad \text{(Eq 1-14a)}$$

for plane strain. Rearranging these equations, we have

$$\mathcal{G}_c \cdot E = S_c^2 \pi a_c \qquad \text{(Eq 1-14b)}$$

for plane stress, and

$$\mathcal{G}_{Ic} \cdot E / (1 - \nu^2) = S_c^2 \pi a_c \qquad \text{(Eq 1-14c)}$$

for plane strain.

Note that Eq 1-14(b) and (c) contain only material properties on the left-hand side, and geometrical and loading parameters on the right-hand side. Using an experimentally generated $\mathcal{G}_c$, or $\mathcal{G}_{Ic}$, one may determine the fracture stress, S_{cr}, for a desired crack length, or the critical crack length for a predetermined operating stress level. Due to the fact that most engineering problems involve complex loading conditions and geometric details, the determination will require mathematical solutions to the right-hand side of Eq 1-14(b) and (c). Therefore, a mathematically analyzable parameter called K is used.

Stresses and Displacements at Crack Tip

It is well known that a notch acts as a stress raiser; i.e., the stress at the notch root is actually higher (sometimes much higher) than the remotely applied stress. For an infinite plate with remote uniform stress S, the theoretical stress concentration factor for an elliptical hole of semimajor axis a (on the x-axis) and end radius ρ is equal to:

$$K_t = 1 + 2(a/\rho)^{1/2} \qquad \text{(Eq 1-15)}$$

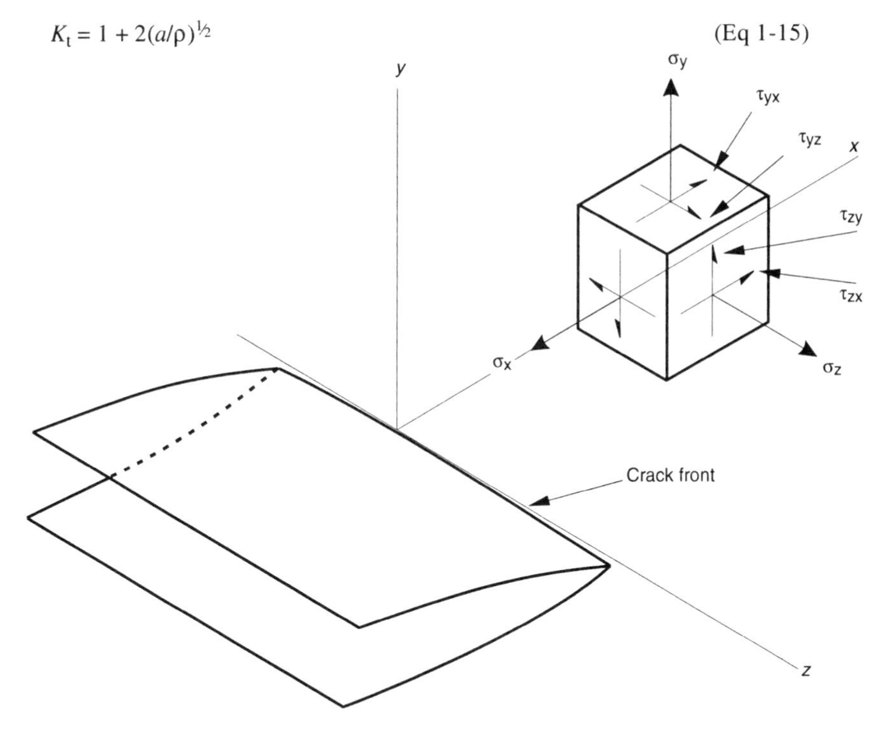

Fig. 1-5 Coordinates used to describe stresses near a crack tip

or

$$K_t = 1 + (2a/b) \qquad \text{(Eq 1-15a)}$$

where b is the semiminor axis (on the y-axis) of the ellipse. For $a = b$, i.e. a circle, Eq 1-15(a) yields $K_t = 3$, the well-known stress concentration factor for a circular hole in an infinite sheet. As b approaches zero, the ellipse may be regarded as a crack having an infinite stress concentration at the crack tip. Referring to the coordinate system shown in Fig. 1-5, the local stress and displacement distributions near the crack tip are given by:

$$\sigma_y = \frac{K_1}{\sqrt{2\pi r}} \cos\frac{\theta}{2}\left[1 + \sin\frac{\theta}{2}\sin\frac{3\theta}{2}\right]$$

$$\sigma_x = \frac{K_1}{\sqrt{2\pi r}} \cos\frac{\theta}{2}\left[1 - \sin\frac{\theta}{2}\sin\frac{3\theta}{2}\right]$$

$$\tau_{xy} = \frac{K_1}{\sqrt{2\pi r}} \sin\frac{\theta}{2}\cos\frac{\theta}{2}\cos\frac{3\theta}{2} \qquad \text{(Eq 1-16)}$$

For plane strain:

$$\sigma_z = \nu(\sigma_x + \sigma_y)$$

$$\tau_{xz} = \tau_{yz} = 0$$

$$v = \frac{K_1}{G}\left(\frac{r}{2\pi}\right)^{1/2} \sin\frac{\theta}{2}\left[2 - 2\nu - \cos^2\frac{\theta}{2}\right]$$

$$u = \frac{K_1}{G}\left(\frac{r}{2\pi}\right)^{1/2} \cos\frac{\theta}{2}\left[1 - 2\nu + \sin^2\frac{\theta}{2}\right]$$

$$w = 0 \qquad \text{(Eq 1-17)}$$

where r is the absolute distance from the crack tip (i.e., for a point ahead or behind the crack tip), v, u, and w are the displacements corresponding to the directions of y, x, and z, respectively. The constant G is the shear modulus of elasticity (also called the modulus of rigidity). It relates to Young's modulus (modulus of elasticity in tension, E) and Poisson's ratio (ν) by:

$$G = \frac{E}{2(1+\nu)} \qquad \text{(Eq 1-18)}$$

For plane stress:

$$\sigma_z = \tau_{xz} = \tau_{yz} = 0$$

$$v = \frac{K_1}{G}\left(\frac{r}{2\pi}\right)^{1/2} \cdot \sin\frac{\theta}{2}\left[2 - 2\left(\frac{\nu}{1+\nu}\right) - \cos^2\frac{\theta}{2}\right]$$

$$u = \frac{K_1}{G}\left(\frac{r}{2\pi}\right)^{1/2} \cdot \cos\frac{\theta}{2}\left[1 - 2\left(\frac{\nu}{1+\nu}\right) + \sin^2\frac{\theta}{2}\right]$$

$$w = -\frac{K_1}{G} \cdot \frac{z}{\sqrt{2\pi r}} \cdot \left(\frac{\nu}{1+\nu}\right) \cdot \cos\frac{\theta}{2} \qquad \text{(Eq 1-19)}$$

where K_1 is the stress intensity factor. The subscript 1 refers to "crack opening mode," which means that the crack is subjected to, and normal to, tension load. The materials covered in the first six chapters of this book are primarily related to mode 1 only. In Chapter 7 we will discuss other types of crack tip stress and displacement distributions, and cracks subjected to loading conditions other than uniaxial tension. The terminology for those crack tip displacement modes will be K_2 for the shear mode, and K_3 for the tearing mode. In any event, K without a subscript is often used to mean K_1.

It should be emphasized that K is in no way related to plane stress or plane strain, because in either case the crack tip stress distributions are unchanged. The difference between the hypothetical plane-stress and plane-strain loci is not due to the stress components in the plane, because they are identical. Referring to Eq 1-16, K is not a function of the coordinates r and θ, but rather depends on the configuration of the body, including the crack, i.e., crack size, geometry, location of the crack, and loading condition. As will be discussed in Chapter 2, the difference between plane stress and plane strain is hinged on the presence or absence of transverse constraint in material deformation in the vicinity of the crack tip.

Some higher-order non-singular terms have been omitted in Eq 1-16, 1-17, and 1-19. As r becomes small compared to planar dimensions (in the x-y plane), the magnitudes of the non-singular terms become negligible compared to the leading $1/\sqrt{r}$ term. Therefore, under ordinary circumstances inclusion of these terms is unnecessary.

Speaking in terms of solid mechanics, the state of stress in a plate, whether it is plane stress or plane strain, is determined by the stress and displacements in the z-direction (i.e. the thickness direction) of the plate, i.e., whether or not σ_z, τ_{xz}, τ_{yz}, and w are equal to zero. Translating into plain layman language, when a plate is subjected to in-plane loading, there is no stress acting normal to the free surfaces of the plate. When the plate is sufficiently thin, the z-directional stress across the thickness of the plate can be considered negligible. Therefore, the stress state in a thin sheet is usually plane stress (i.e. stresses are only acting on the x-y plane). Conversely, the z-directional stress in the inside in a very thick plate is non-zero, whereas the z-directional strain is zero. Therefore, it is considered a plane-strain condition.

In the cases of material fracture toughness, which will be discussed later in this book, terms such as K_c (or K_{1c}), K_{Ic}, K_{2c}, K_{IIc}, and K_{3c} are used in the literature. The Roman numerals I and II are reserved for the plane strain cases, whereas the Arabic numerals 1, 2, and 3 are used for an unspecified state of stress (anywhere from plane stress to any degree of mixture in between plane stress and plane strain). The tearing mode is neither plane stress nor plane strain. These subscripts in Roman or Arabic are assigned by the test engineer/metallurgist to indicate whether a given fracture toughness value was obtained from a very thick section (i.e., plane strain), or for a plate thickness that exhibits a behavior of generalized plane stress. They have nothing to do with calculating the K value. In some literature (especially older publications), only the Roman numerals are used.

Stress Intensity Factor

The stress intensity factor K is actually a physical quantity, not a factor. By definition, a factor (e.g., the stress concentration factor) is unitless (or dimensionless). However, it is clear in Eq 1-16 that K has a unit of stress times the square root of the crack length (a quantity needed to balance the stress on the left, and the $1/\sqrt{r}$ on the right, of Eq 1-16). K has been called the stress intensity factor because it appears as a factor in Eq 1-16. As shown in Chapter 5 and in other sections of this book, K is often expressed as a closed-form solution that includes a conglomeration of many dimensionless factors to account for the unique geometry under consideration. In the remainder of this book we will refer to K as stress intensity or the stress intensity factor whenever the situation is called for. In any event, K is mathematically analyzable, and it characterizes the stress and displacement distributions at the crack tip, so it also characterizes the behavior and the criticality of the crack. The solution for K consists of terms representative of stress (or load), crack length, and geometry. It fully accounts for the geometry of a local area in a structure in question, the crack morphology, and how the load is applied. The general expression for K can be written as:

$$K = S\sqrt{\pi a} \cdot \phi_1 \cdot \phi_2 \cdot \phi_3 \cdot \ldots \quad \text{(Eq 1-20)}$$

where ϕ is normally expressed as a dimensionless geometric parameter, e.g., crack length to specimen width ratio, specimen width to length ratio, etc. For example, for a crack of length $2a$ located at the center of an infinitely long strip subjected to farfield uniaxial tension:

$$K_1 = S\sqrt{\pi a} \cdot \phi_w \quad \text{(Eq 1-21)}$$

with

$$\phi_w = \left(\sec\frac{\pi a}{W}\right)^{1/2} \quad \text{(Eq 1-22)}$$

where, W is the width of the strip and ϕ_w is the width correction factor for a center-cracked panel. In a straightforward sense, a K-solution is normally obtained for a stationary configuration, i.e., a fixed crack length in a geometry having fixed dimensions. A close-form solution is almost nonexistent. The so-called correction factor in the above case is obtained by fitting a curve through many K-factors corresponding to different crack sizes. Each K-factor is

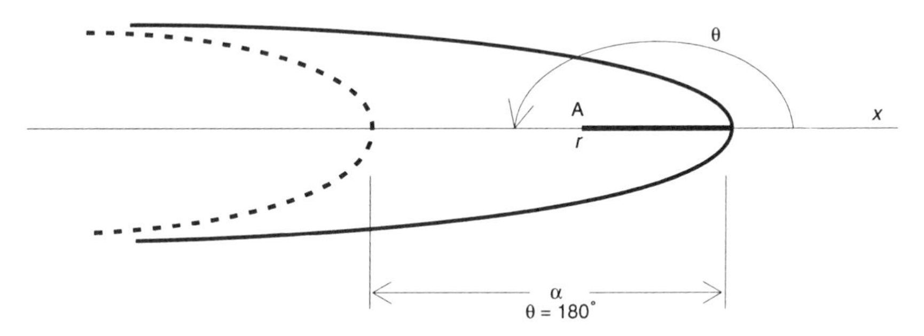

Fig. 1-6 A crack closing concept for relating strain energy release rate and crack tip stress intensity

obtained by solving one problem at a time. We will discuss Eq 1–22 and the K-solutions for the other geometries and loading conditions in Chapters 4 and 5. For now, we ought to remember that the stress term in a K-expression is the applied remote gross area stress unless it is otherwise specified.

In some earlier literature the $\sqrt{\pi}$ term was omitted in the crack tip stress field equation. For example, Eq 1-16 was written as $\sigma_y = K/\sqrt{2r} \cdot F(\theta)$ instead of $\sigma_y = K/\sqrt{2\pi r} \cdot F(\theta)$. Accordingly, the equation for K was expressed in terms of $\sqrt{a}$ instead of $\sqrt{\pi a}$.

Relationship Between $\mathcal{G}$ and K

When the width of a sheet is infinitely wide, Eq 1-22 gives $\phi_w = 1$ and $K = S\sqrt{\pi a}$. Relating these to Eq 1-14, we have:

$$\mathcal{G} = K^2 \cdot E' \qquad \text{(Eq 1-23)}$$

where $E' = 1/E$ for plane stress and $E' = (1 - \nu_2)/E$ for plane strain. Therefore, the failure criterion is that rapid fracture occurs when $\mathcal{G}$ reaches $\mathcal{G}_c$ (or $\mathcal{G}_{Ic}$), or K reaches K_c (or K_{Ic}). Now we can see that there are two approaches to the relationships among stresses, crack sizes, and material properties: by a physical criterion of fracture process, $\mathcal{G}$, or by stress analysis, K. In the following, we will prove the relationship between K and $\mathcal{G}$ analytically (Ref 1-4).

Consider the energy (work done) required to re-close part of a crack that had formed in a solid. In reverse manner, then, once this energy was removed the crack should reopen; i.e., the energy required to extend an increment α ahead of a crack is the same that is required to close the crack of the same amount (α). In Fig. 1-6, we treat the solid lines of the elliptical contour as the old crack, the dotted line as the new crack. The work done to close the tip of the old crack by an amount α is equal to:

$$\alpha\mathcal{G}_1 = \alpha\left(\frac{\partial U}{\partial a}\right) = \lim_{\alpha \to 0} 2\int_0^{\alpha} \frac{1}{2}(v \cdot \sigma_y)\, dr \qquad \text{(Eq 1-24)}$$

where v is the instantaneous height (displacement) along α. The quantity inside the integral is the elastic strain energy (depicted by the shaded area in Fig. 1-2). The constant "2" outside the integral accounts for the total distance of closure because v is only one-half of the total crack opening displacement (i.e., v is equal to the distance from 0 to $+y$, or from 0 to $-y$ in a x-y coordinate system).

To quantify v and σ_y, first we recall that in Eq 1-19, the vertical displacement was given as:

$$v = \frac{K_1}{G}\left(\frac{r}{2\pi}\right)^{1/2} \cdot \sin\frac{\theta}{2}\left[2 - 2\left(\frac{\nu}{1+\nu}\right) - \cos^2\frac{\theta}{2}\right]$$

Because we are doing this proof by means of a reverse process, the displacement v (to be compressed) at a given point along α (e.g., point A in Fig. 1-6) is behind the old crack. Thus, the grid location for point A (relating to the old crack) is r and $\theta = 180°$. Substitute $\theta = 180°$ into Eq 1-19, we have:

$$v = \frac{K_1}{G}\left(\frac{2r}{\pi}\right)^{1/2} \cdot \left(1 - \frac{\nu}{1+\nu}\right) = \frac{K_1}{(1+\nu)\,G}\left(\frac{2r}{\pi}\right)^{1/2} = \frac{2K_1}{E}\left(\frac{2r}{\pi}\right)^{1/2} \qquad \text{(Eq 1-25)}$$

To obtain σ_y, we recall Eq 1-16, realizing that we need to treat point A as being in front of the new crack. Substituting $(\alpha - r)$ for r; and 0 for θ, we have:

$$\sigma_y = K_1/\sqrt{2\pi(\alpha - r)} \qquad \text{(Eq 1-26)}$$

Substituting v and σ_y into Eq 1-24, we have:

$$\alpha\mathcal{G} = \frac{2K_1^2}{E}\int_0^{\alpha}\left(\frac{2r}{\pi} \cdot \frac{1}{2\pi(\alpha - r)}\right)^{1/2} \cdot dr$$

$$= \frac{2K_1^2}{\pi E}\int_0^{\alpha}\sqrt{\frac{r}{\alpha - r}}\,dr = \frac{2K_1^2}{\pi E}\left(\frac{\alpha\pi}{2}\right)$$

Therefore:

$$\mathcal{G} = K_1^2/E \qquad \text{(Eq 1-27)}$$

Now we have proved that $\mathcal{G}$ is related to K. By analogy, $\mathcal{G}_c$ is related to K_c, and $\mathcal{G}_{Ic}$ is related to K_{Ic}. In structural analysis and fracture testing, K is conveniently computed as a function of damage size (crack length) and the applied stress level. Most importantly, K can account for the influence of the geometric details. Therefore, the critical value of K (whether it is K_c, K_{Ic},

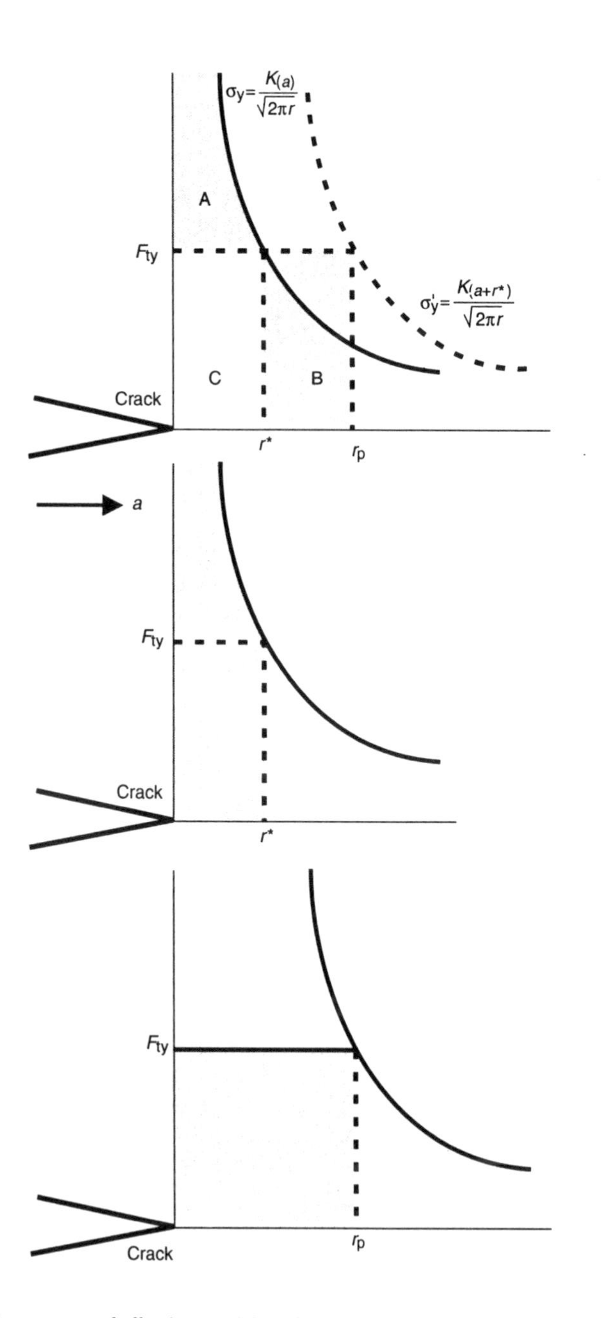

Fig. 1-7 The concept of effective crack length

K_{2c}, K_{IIc}, or K_{3c}) is regarded as the residual strength of a structural member or the fracture toughness of the material, whichever case is applicable. The existence of a relationship between $\mathcal{G}$ and K is most important in today's fracture mechanics technology, because it allows engineers to use K, which is mathematically analyzable, to perform structural sizing and failure analysis.

The Role of Crack Tip Plasticity

Consider the distribution of σ_y along $y = 0$ for a remotely loaded cracked sheet in a state of plane stress. The solid line in Fig. 1-7 was obtained by using Eq 1-16 with $\theta = 0$. Examining Eq 1-16, it is noticed that $\sigma_y \rightarrow \infty$ when $r \rightarrow 0$. This is called the stress singularity. Therefore, the material at the crack tip would have undergone plastic yielding at any applied stress level. It seems logical to assume that the material in between the crack tip ($r = 0$) and some distance r^* is yielded because σ_y over this area has exceeded the tensile yield strength of the material. Recalling Eq 1-16

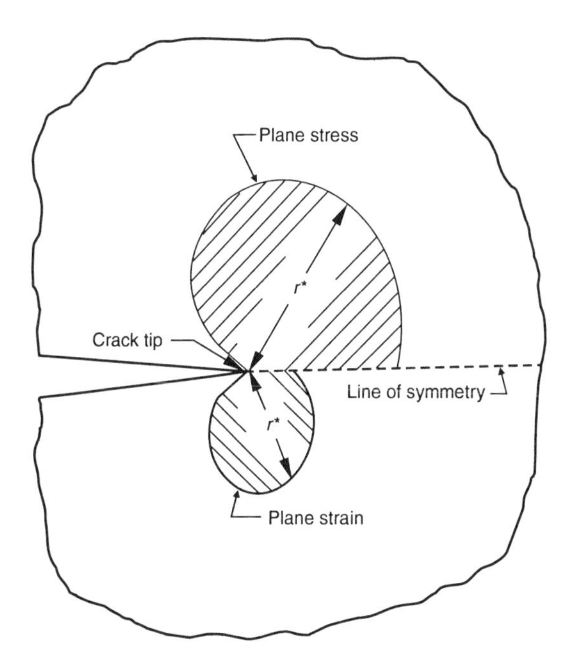

Fig. 1-8 Comparison of analytical mode 1 plastic zone sizes (plane stress versus plane strain)

and substituting F_{ty} (the material tensile yield strength) for σ_y, and r^* for r, we have:

$$F_{ty} = K_1/\sqrt{2\pi r^*}$$

or

$$r^* = \frac{1}{2\pi}(K_1/F_{ty})^2 \qquad \text{(Eq 1-28)}$$

If we take r^* as the first estimate of the extent of the plastic zone, clearly the force produced by the stress acting over the length r^* will produce further yielding. For the purpose of maintaining equilibrium within the system, the whole curve for σ_y must be shifted (shown as the dotted line in Fig. 1-7). In other words, area A must be balanced by area B. The area under the distribution curve of σ_y, i.e., area A plus area C, can be obtained by integrating σ_y from 0 to r^*. Because the size of area A is equal to that of area B, the area to be integrated is algebraically equal to area B plus area C:

$$\int_0^{r^*} \sigma_y dr = F_{ty}\left(r_p - r^*\right) + F_{ty} \cdot r^*$$

or

$$\int_0^{r^*} \frac{K_1}{\sqrt{2\pi r}} dr = F_{ty} \cdot r_p \qquad \text{(Eq 1-29)}$$

After integration, we have:

$$\left(\frac{2}{\pi}\right)^{1/2} \cdot K_1\sqrt{r^*} = F_{ty} \cdot r_p$$

$$\frac{2}{\pi}\left(\frac{K_1}{F_{ty}}\right)^2 \cdot r^* = r_p^2$$

$$\frac{2}{\pi}\left(\frac{K_1}{F_{ty}}\right)^2 \cdot \frac{1}{2\pi}\left(\frac{K_1}{F_{ty}}\right)^2 = r_p^2$$

Therefore:

$$r_p = \frac{1}{\pi}\left(\frac{K_I}{F_{ty}}\right)^2 \quad \text{(Eq 1-30)}$$

Comparing Eq 1-30 with Eq 1-28, we have:

$$r_p = 2r^* \quad \text{(Eq 1-31)}$$

From now on we will call r_p the full plastic zone, or plastic zone width. If we imagine the crack tip plastic zone as a circle, the radius of the circle will be $r_p/2$ (= r^*). The plastic zone radius is designated by r_y.

The solution for r_y under plane strain has been derived by Irwin (Ref 1-5) as:

$$r_y = \frac{1}{4\sqrt{2\pi}}\left(\frac{K_I}{F_{ty}}\right)^2 \cong \frac{1}{6\pi}\left(\frac{K_I}{F_{ty}}\right)^2 \quad \text{(Eq 1-32)}$$

Therefore, the plane-strain plastic zone radius is equal to one-third of the plane-stress plastic zone radius. Similarly, the plane-strain plastic full zone width is equal to one-third of the plane-stress full zone.

The Effective Crack Length

The shifting of σ_y to σ_y' (the dotted line in Fig. 1-7) implies that the K-factor that is associated with this crack (of length $2a$) has become a function of $(a + r_y)$. Thus, it can be said that the crack behaves as if it were

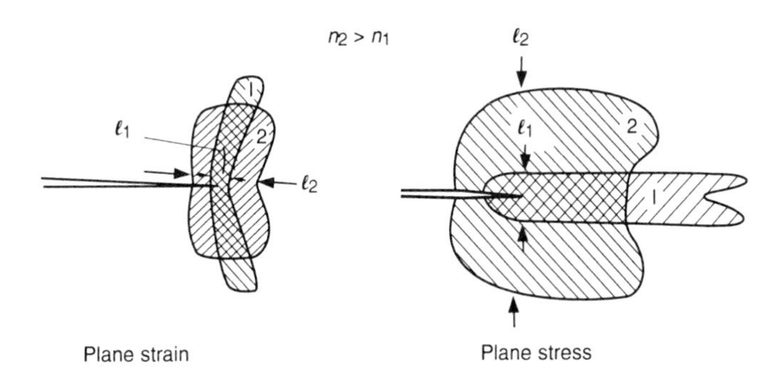

Fig. 1-9 Schematic of the effect of strain hardening exponent on crack tip plastic zone shape. Source: Ref 1-12

of length $(a + r_y)$. The quantity $(a + r_y)$ is called the effective crack length. It is academically correct to use the effective crack length $(a + r_y)$ in place of the physical crack length a, to compute K. However, in the above equations, the crack tip plastic zone size (r_p or r_y) has been given as a function of K and F_{ty}. Now, we are also saying that K is a function of the effective crack length $(a + r_y)$. Clearly, implementation of the effective crack length approach to fatigue crack growth prediction is difficult and cumbersome.

Other than making an adjustment to the effective crack length (from a physical crack length), there are two additional uses for the crack tip plastic zone. The first one is its contribution to the special meaning of plane stress

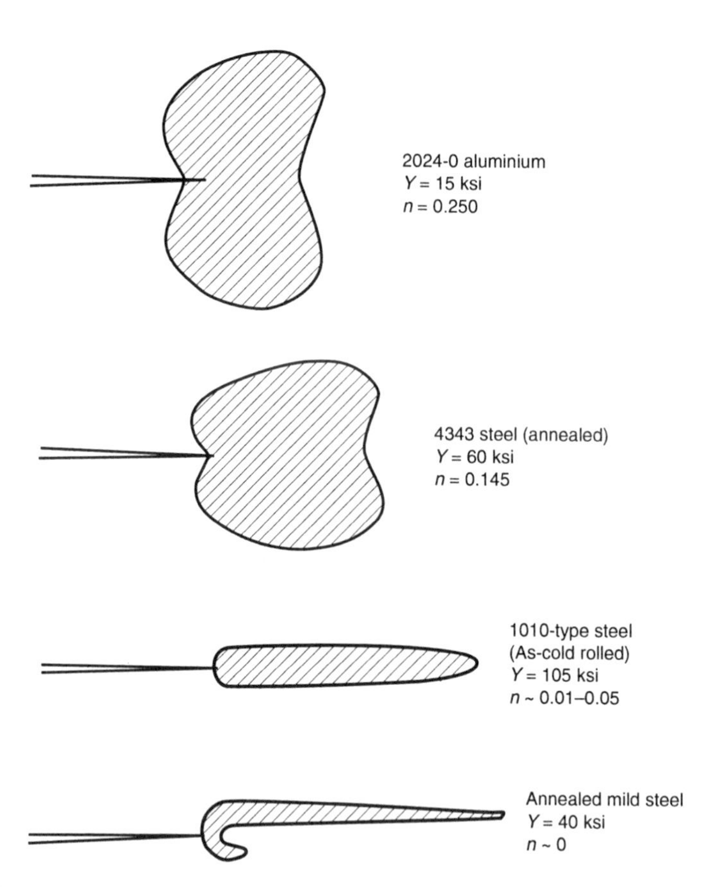

Fig. 1-10 Effect of strain hardening exponent on crack tip plastic zone shape of three low-alloy steels and one aluminum. Source: Ref 1-12

and plane strain in fracture mechanics terminology. The second one is its role in governing the behavior of spectrum (variable-amplitude loading) crack growth. We will defer discussion of these to Chapters 2 and 3, respectively.

Other Analytical and Physical Aspects of the Crack Tip Plastic Zone

It is often shown in the literature that the crack tip plastic zone r^* is shaped like a butterfly (Fig. 1-8). These are stress (or strain) contour maps computed on the basis of the Von Mises yield criterion:

$$\sigma_0 = \sqrt{\frac{1}{2}\left[(\sigma_x - \sigma_y)^2 + (\sigma_y - \sigma_z)^2 + (\sigma_z - \sigma_x)^2 + 6\tau_{xy}\right]} \qquad \text{(Eq 1-33)}$$

where σ_0 is the yield stress under uniaxial tension. It can be regarded as the material tensile yield strength F_{ty}. Combining Eq 1-16 with Eq 1-33, the resulting elastic-plastic boundary is given by:

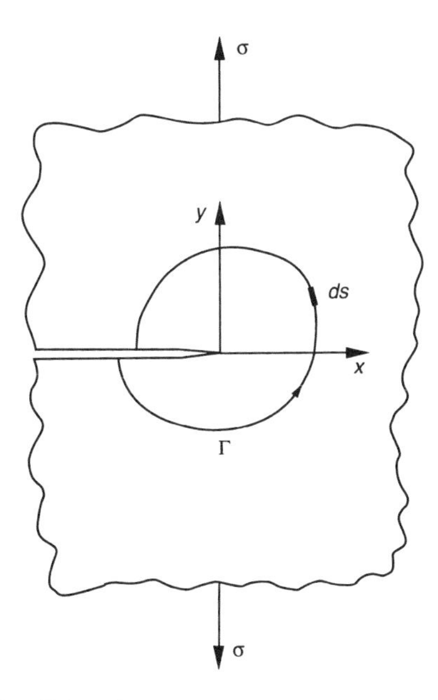

Fig. 1-11 Definition of the *J*-integral

$$r^* = \frac{1}{2\pi}\left(\frac{K_1}{F_{ty}}\right)^2 \cos^2\left(\frac{\theta}{2}\right)\left[1 + 3\sin^2\left(\frac{\theta}{2}\right)\right] \quad \text{(Eq 1-34a)}$$

for plane-stress, and by

$$r^* = \frac{1}{2\pi}\left(\frac{K_1}{F_{ty}}\right)^2 \cos^2\left(\frac{\theta}{2}\right)\left[(1 - 2\nu)^2 + 3\sin^2\left(\frac{\theta}{2}\right)\right] \quad \text{(Eq 1-34b)}$$

for plane strain. At $\theta = 0$, Eq 1-34(a) reduces to Eq 1-32. Therefore, the plastic zone contour maps are actually plotting r_y, not r_p.

The plastic zones described in the previous section, which are derived from the local stress distribution at the crack tip, are commonly referred to as the Irwin plastic zones. Other approaches have also been used to derive crack tip plastic zone expressions. The most notable one is the Dugdale plastic zone (Ref 1-6), and for plane stress its width equals:

$$\omega = a\left[\sec\left(\frac{\pi S_y}{2F_{ty}}\right) - 1\right] \quad \text{(Eq 1-35)}$$

For small-scale yielding, this leads to:

$$r_p = \frac{\pi}{8}\left(\frac{K_1}{F_{ty}}\right)^2 \quad \text{(Eq 1-36)}$$

and is known as the Rice plastic zone (Ref 1-7). The Dugdale zone represents a product of large-scale yielding that results from high applied stress levels. Physically the Rice plastic zone is another form of the Irwin plane-stress plastic zone, just approximately 23% larger. The Dugdale zone width has a shape like a long and thin strip in front of the crack tip, whereas the Irwin or Rice plastic zone has a round circular shape. Many other investigators have also made estimates of crack tip plastic zone size for the plane-stress condition. Their results are within the two boundaries represented by Eq 1-30 and 1-35 (Ref 1-8, 1-9).

Because crack tip plastic zone is a function of material mechanical properties, it is logical to relate it to material strain hardening exponent and loading rate. Newman calculated the plastic zone sizes for a number of metals having different combinations of strain hardening exponent, tensile yield strength, and loading rate. His results show that the plastic zone sizes for these metals also are within the range bounded by Eq 1-30 and 1-35 (Ref 1-10).

Experimental investigations on crack tip plastic zone shapes have been carried out by Hahn et al. (Ref 1-11 to 1-13). Their results show that there is a basic difference between the shapes of the plane-stress and plane-strain plastic zones. The strain hardening exponent of the material further influences the shape of the plastic zone of a given type (Fig. 1-9). For a nearly elastic/perfectly plastic material (i.e., the strain hardening exponent $n \approx 0$) such as annealed mild steel, the plane-stress plastic zone becomes a very thin strip, resembling the shape of the Dugdale plastic zone (Fig. 1-10).

The foregoing discussion has undoubtedly complicated the general subject of the crack tip plastic zone. However, it helps to explain why it is often difficult to match up field service life or experimental results with analytical prediction.

Other Fracture Indices

The concept of linear elastic fracture mechanics (LEFM) assumes that the stress intensity factor is a valid fracture index and that the material behaves in a "brittle" manner. As a rule of thumb, the applicability of LEFM is limited to a crack tip plastic zone radius (r^* or r_y) smaller than one-fifth of the current crack length. The linear elastic stress intensity factor approach to fracture analysis is not applicable to the situation in which nonlinear behavior is encountered in the vicinity of the crack tip. Large-scale yielding at the crack tip and time-dependent crack growth behavior, such as stress relaxation due to creep, belong to the category of nonlinear behavior. Alternate approaches are needed and will be presented in the following paragraphs.

The J-Integral

For a stationary crack, the HRR crack tip stress field, named after Hutchinson, Rice, and Rosengren, can be used in place of Eq 1-16 as a generalized crack tip stress distribution (Ref 1-14, 1-15). For mode 1 loading, the HRR field is defined as:

$$\sigma_{ij} = \sigma_0 \tilde{\sigma}_{ij}(\theta, n)\left[\frac{EJ}{\sigma_0^2 I_n r}\right]^{1/(n+1)} \qquad \text{(Eq 1-37)}$$

or

$$\sigma_{ij} = \sigma_0 \tilde{\sigma}_{ij}(\theta, n) \left[\frac{J}{\alpha \sigma_0 \varepsilon_0 I_n r} \right]^{1/(n+1)} \qquad \text{(Eq 1-37a)}$$

where J is the J-integral (in. · lb/in.^{2}), θ and r are the polar coordinates centered at the crack tip, $\tilde{\sigma}_{ij}(\theta, n)$ is a dimensionless function that is a function of θ and n, I_n is a dimensionless constant that is a function of n, and σ_0 is yield stress, which relates to yield strain ε_0.

In the pure power stress-strain law, the uniaxial plastic strain ε is related to the uniaxial stress σ by:

$$\frac{\varepsilon}{\varepsilon_0} = \alpha \left(\frac{\sigma}{\sigma_0} \right)^n \qquad \text{(Eq 1-38)}$$

where α is a material constant and n is the strain hardening exponent. The numerical values of $\tilde{\sigma}_{ij}(\theta, n)$ and I_n, for $2 \le n \le 100$, are given in Ref 1-16. For an elastic body (i.e., small-scale yielding) under plane stress, with $n = 1$, $I_n = 6.28$ (i.e., 2π), and $J = K^2/E$, Eq 1-37 reduces to the familiar K field distribution:

$$\sigma_{ij} = \tilde{\sigma}_{ij}(\theta) \left[\frac{K_2}{2\pi r} \right]^{1/2} \qquad \text{(Eq 1-39)}$$

By comparison, it is clear that $\tilde{\sigma}_{ij}\theta$ is the same as the θ-functions in Eq 1-16.

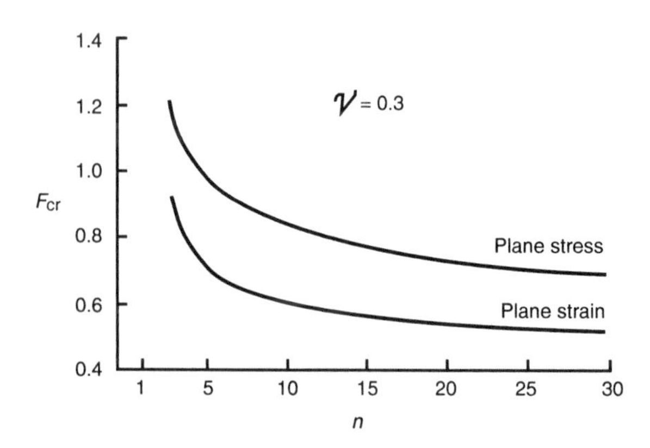

Fig. 1–12 Creep zone size dimensionless coefficient F_{cr}

The J-integral may be regarded as a change of potential energy of the body with an increment of crack extension. The general expression for J, as defined by Rice (Ref 1-17), is given by:

$$J = \int_{\Gamma} \left(W d_y - \vec{T} \cdot \frac{\partial \vec{u}}{\partial x} ds \right) \qquad \text{(Eq 1-40)}$$

where $\vec{T}$ is the traction vector acting on a segment ds of contour Γ, and $\vec{u}$ is a displacement vector (the displacement on an element along arc s), and Γ is any contour in the x-y plane that encircles the crack tip. Γ is taken in a counterclockwise direction, starting from one crack face, ending on the opposite face, and closing the crack tip (see Fig. 1-11). W is the strain energy density, given by:

$$W = \int_{0}^{\varepsilon_{ij}} \sigma_{ij} \cdot d\left(\varepsilon_{ij}\right) \qquad \text{(Eq 1-41)}$$

That is,

$$W = \int [\sigma_x + d\varepsilon_x + \tau_{xy} + d\gamma_{xy} + \tau_{xz} + d\gamma_{xz} + \sigma_y + d\varepsilon_y + \tau_{yz} + d\gamma_{yz} + \sigma_z + d\varepsilon_z] \qquad \text{(Eq 1-41a)}$$

For generalized plane stress, Eq 1-41(a) is reduced to:

$$W = \int [\sigma_x + d\varepsilon_x + \tau_{xy} + d\gamma_{xy} + \sigma_y + d\varepsilon_y] \qquad \text{(Eq 1-41b)}$$

The total magnitude of J consists of two parts: an elastic part and a fully plastic part. For small-scale yielding or linear elastic material behavior, J is equivalent to Irwin's strain energy release rate $\mathcal{G}$. Therefore, the portion for "elastic-J" is directly related to K, and the plastic part vanishes. However, unlike $\mathcal{G}$ or K, J can be used as a generalized fracture parameter for small- or large-scale yielding. Rice also noted that J is path-independent. That is, the values of J would be identical whether the contour Γ is taken far away or very close to the crack tip. In addition, there is no restriction as to what specific course Γ should follow. In Chapter 4, the methods for determining J will be presented.

The Characterizing Parameters for Creep Crack Growth

A crack growth parameter C^* has been introduced by Landes and Begley (Ref 1-18) as a characterizing parameter for materials under the steady-state creep range, i.e., in which the material follows a creep law in the form:

$$\dot{\varepsilon} = \alpha\dot{\varepsilon}_0\left(\frac{\sigma}{\sigma_0}\right)^n \qquad \text{(Eq 1-42)}$$

The constant α is usually absorbed in the constant $\dot{\varepsilon}_0$. Therefore, frequently in the literature the creep law (i.e., Eq 1-42) is written as:

$$\dot{\varepsilon} = \dot{\varepsilon}_0\left(\frac{\sigma}{\sigma_0}\right)^n \qquad \text{(Eq 1-42a)}$$

For a material that is primarily influenced by second-stage creep, where the creep strain rate is:

$$\dot{\varepsilon} = A \cdot \sigma^n \qquad \text{(Eq 1-43)}$$

the steady-state creep crack growth parameter C^* is analogous to the J-integral in the fully plastic condition. It is simply a modification of the J-integral where strain and displacement, ε and u, are replaced by their rates, $\dot{\varepsilon}$ and $\dot{u}$. Therefore, the close-form solution for C^* takes the same form as that for J. An expression for C^* is derived by substituting the material creep strain rate coefficients of Eq 1-43 for the material stress-strain coefficients of Eq 1-38. Comparing Eq 1-42 and 1-43, one has:

$$A = \alpha\dot{\varepsilon}_0\varepsilon_0^{-n} \qquad \text{(Eq 1-44)}$$

Substituting Eq 1-44 into Eq 1-37(a) for α and ε_0, the HRR field for elastic-plastic deformation can be expressed for creep as:

$$\sigma_{ij} = \tilde{\sigma}_{ij}(\theta, n)\left[\frac{C^*}{AI_n r}\right]^{1/(n+1)} \qquad \text{(Eq 1-45)}$$

where the physical unit for C^* is (in. · lb/in.^{2})/hour, or (MPa · m)/hour. For small-scale creep, Eq 1-45 can be written as:

$$\sigma_{ij} = \tilde{\sigma}_{ij}(\theta, n)\left[\frac{C(t)}{AI_n r}\right]^{1/(n+1)} \qquad \text{(Eq 1-46)}$$

with

$$C(t)_{ssc} = \frac{(1-\nu^2)K^2}{(n+1)Et} \qquad \text{(Eq 1-47)}$$

for plane strain; for plane stress the factor $(1 - \nu^2)$ must be omitted. Equation 1-46 is called the RR field, named after Riedel and Rice (Ref 1-19). At time zero, i.e., $t = 0$, $C(t)$ can be considered as a crack growth driving force similar to K, or elastic-J. Under steady-state creep, $C(t)$ becomes time-independent, and its value approaches C^*:

$$C(t \to \infty) = C^* \qquad \text{(Eq 1-48)}$$

The characteristic transition time t_1 between small-scale creep and steady-state creep is estimated by setting the value of $C(t)$ of Eq 1-47 equal to C^*, i.e., for plane strain:

$$t_1 = \frac{(1-\nu^2)K^2}{(n+1)EC^*} \qquad \text{(Eq 1-49)}$$

Again, the factor $(1 - \nu^2)$ in Eq 1-49 should be deleted for the plane-stress condition. The characteristic time can be used to decide whether or not the test data set in question is correlated with K. In other words, K can be used as a controlling parameter if $t << t_1$. Otherwise, C^* should be used for the steady state creep. In the transient regime (Ref 1-20):

$$C(t) \approx C^*\left[\frac{t_1}{t} + 1\right] \qquad \text{(Eq 1-50)}$$

or

$$C(t) \approx \frac{(1-\nu^2)K^2}{(n+1)Et} + C^* \qquad \text{(Eq 1-50a)}$$

Similar to r_p (the crack tip plastic zone), which is a result of crack tip plastic yielding, a creep zone r_{cr} will be formed at the crack tip during creep deformation. Much like the crack tip stress distribution associated with r_p is the HRR field, the crack tip stress distribution associated with r_{cr} is the RR field. In small-scale creep, the creep zone is defined as the region around the crack tip where the creep strains exceed the elastic strains. For a stationary crack, the creep zone size is given by Ref 1-19 as:

$$r_{cr}(\theta, t) = \frac{K^2}{2\pi}(EAt)^{2/(n-1)} \cdot F_{cr}(\theta, n) \qquad \text{(Eq 1-51)}$$

where

$$F_{cr}(\theta, n) = \tilde{F}_{cr}(\theta, n)\left[\frac{(n+1)I_n}{2\pi\eta}\right]^{2/(n-1)} \qquad \text{(Eq 1-52)}$$

where $\eta = (1 - \nu^2)$ for plane strain (= 1 for plane stress). The term $\tilde{F}_{cr}(\theta, n)$ is roughly maximum at $\theta = 90°$ for plane strain and at $\theta = 0°$ for plane stress. The values of $\tilde{F}_{cr}(\theta, n)$ are given in Ref 1-19 (equal to 0.2 to 0.5 for plane strain and 0.5 to 0.6 for plane stress, depending on n). For readers' convenience, a plot for Eq 1-52, i.e., a plot of $F_{cr}(\theta, n)$ versus n (with ν = 0.3), is shown in Fig. 1-12.

Equation 1-51 implies that the degree of creep deformation is a function of stress level and time at load. For a stress cycle of which the hold time is relatively short, the size of a creep zone will be relatively small (or nonexistent for no hold time). Equation 1-51 also implies that for a given pair of K and t, r_{cr} is a function of the material properties E, A, and n. Even for a stationary crack (i.e., the crack is not moving but accumulates damage inside the creep zone), r_{cr} is in proportion to $t^{2/(n-1)}$. The rate of creep zone expansion, $\dot{r}_{cr}$, i.e., the derivative of r_{cr} with respect to t, is in proportion to $t^{-(n-3)/(n-1)}$. For $n \leq 3$, $\dot{r}_{cr}$ increases with time. Thus, steady-state creep is not possible; the crack is dominated by the K field. For $n > 3$, $\dot{r}_{cr}$ decreases with time so that steady-state creep is possible (because $\dot{r}_{cr}$ has to approach zero to reach the time-independent state). Therefore, to capture the history of deformation in the crack tip region, the parameter that controls crack growth should be uniquely related to the rate of creep zone expansion. Such a parameter has been defined by Saxena (Ref 1-21) as:

$$C_t = (C_t)_{ssc} + C^* \qquad \text{(Eq 1-53)}$$

In this equation, the first term denotes the contribution from small-scale creep and the second term denotes the contribution from steady-state, large-scale creep. The first term is time-variant whereas the second term is time-invariant. Methods for determining the value for $(C_t)_{ssc}$ and C^* will be presented in Chapter 6.

All of the above equations have been formulated assuming the following conditions: (1) the crack is stationary; (2) the applied load (either sustained or cyclic) has a step loading profile, i.e., the load rising time is very short; and (3) the material at the crack tip region is dominated by the second-stage creep deformation mechanism.

The crack tip stress field changes with time during crack propagation. On the subject of a moving crack, readers are referred to Ref 1-22 and 1-23. For a ramp-type waveform (i.e., it takes time for the load to rise), the magnitude of $C(t)$ is increased by a factor of $(1 + 2n)/(n + 1)$, and the creep zone size will thereby change accordingly (Ref 1-24, 1-25). If the degree of primary creep is significant, a whole new set of analytical solutions would be required. Discussions on this matter are given in Ref 1-21 and 1-26.

REFERENCES

1-1. A.A. Griffith, The Theory of Rupture, *Proc., First Int. Congress of Applied Mechanics,* 1924, p 55–63

1-2. G.R. Irwin, Relation of Stresses Near a Crack to the Crack Extension Force, *Ninth International Congress of Applied Mechanics,* Vol 8, 1957, p 245

1-3. G.R. Irwin, Analysis of Stresses and Strains Near the End of a Crack Transversing a Plate, *J. Appl. Mech. (Trans. ASME),* Series E, Vol 24, 1957, p 361

1-4. G.R. Irwin, Fracture, *Hanbuch der Physik,* Vol VI, Springer-Verlag, Berlin, Germany, 1958, p 551–590

1-5. G.R. Irwin, Plastic Zone Near a Crack and Fracture Toughness, *Mechanical and Metallurgical Behavior of Sheet Materials,* Proc. of Seventh Sagamore Ordnance Materials Conf., Syracuse University Research Institute, 1960, p IV-63 to IV-78

1-6. D.S. Dugdale, Yielding of Steel Sheets Containing Slits, *J. Mech. Phys. Solids,* Vol 8, 1960, p 100–104

1-7. J.R. Rice, Mechanics of Crack Tip Deformation and Extension by Fatigue, *Fatigue Crack Propagation,* STP 415, American Society for Testing and Materials, 1967, p 247–311

1-8. M.A. Hussain and S.L. Pu, Variational Method for Crack Intensity Factors and Plastic Regions of Dugdale Model, *Eng. Fract. Mech.,* Vol 4, 1972, p 119–128

1-9. P.T. Head, G.M. Spink, and P.J. Worthington, Post Yield Fracture Mechanics, *Mater. Sci. Eng.,* Vol 10, 1972, p 129–138

1-10. J.C. Newman, Jr., Fracture of Cracked Plates Under Plane Stress, *Eng. Frac. Mech.,* Vol 1, 1968, p 137–154

1-11. A.R. Rosenfield, P.K. Dai, and G.T. Hahn, Crack Extension and Propagation Under Plane Stress, *Proc., First Int. Conf. on Fracture,* 1966, p 223–258
1-12. G.T. Hahn and A.R. Rosenfield, Sources of Fracture Toughness: The Relation Between K_{Ic} and the Ordinary Tensile Properties of Metals, *Symposium on Applications Related Phenomena in Titanium and Its Alloys,* STP 432, American Society for Testing and Materials, 1968, p 5
1-13. W.W. Gerberich, Plastic Strains and Energy and Density in Cracked Plates, Part I: Experimental Techniques and Results, *Exp. Mech.,* Vol 4, 1964, p 335
1-14. J.W. Hutchinson, Singular Behavior at the End of a Tensile Crack in a Hardening Material, *J. Mech. Phy. Solids,* Vol 16, 1968, p 13–31
1-15. J.R. Rice and G.F. Rosengren, Plane strain Deformation Near a Crack Tip in a Power-Law Hardening Material, *J. Mech. Phys. Solids,* Vol 16, 1968, p 1–12
1-16. C.F. Shih, "Table of Hutchinson-Rice-Rosengren Singular Field Quantities," Report MRL E-147, Brown University, June 1983
1-17. J.R. Rice, A Path Independent Integral and the Approximate Analysis of Strain Concentration by Cracks and Notches, *J. Appl. Mech. (Trans. ASME),* Series E, Vol 35, 1968, p 379–386
1-18. J.D. Landes and J.A. Begley, A Fracture Mechanics Approach to Creep Crack Growth, *Mechanics of Crack Growth,* STP 590, American Society for Testing and Materials, 1976, p 128–148
1-19. H. Riedel and J.R. Rice, Tensile Cracks in Creeping Solids, *Fracture Mechanics, Twelfth Conference,* STP 700, American Society for Testing and Materials, 1980, p 112–130
1-20. R. Ehlers and H. Riedel, A Finite Element Analysis of Creep Deformation in a Specimen Containing a Macroscopic Crack, *Advances in Fracture Research,* Vol 2, Pergamon Press, 1981, p 691–698
1-21. A. Saxena, Mechanics and Mechanism of Creep Crack Growth, *Fracture Mechanics: Microstructure and Micromechanisms,* ASM International, 1989, p 283–334
1-22. C.Y. Hui and H. Riedel, The Asymptotic Stress and Strain Field Near the Tip of a Growing Crack Under Creep Conditions, *Int. J. Frac.,* Vol 17, 1981, p 409–425
1-23. D.E. Hawk and J.L. Bassani, Transient Crack Growth Under Creep Conditions, *J. Mech. Phys. Solids,* Vol 34, 1986, p 191–212
1-24. H. Riedel, Crack-Tip Stress Fields and Crack Growth Under Creep-Fatigue Conditions, *Elastic-Plastic Fracture: Second Symposium,* Vol I, *Inelastic Crack Analysis,* STP 803, American Society for Testing and Materials, 1983, p I-505 to I-520
1-25. A. Saxena, Limits of Linear Elastic Fracture Mechanics in the Characterization of High-Temperature Fatigue Crack Growth, *Basic Questions in Fatigue,* Vol II, STP 924, American Society for Testing and Materials, 1988, p 27–40
1-26. H. Riedel, Creep Crack Growth, *Flow and Fracture at Elevated Temperatures,* American Society for Metals, 1983, p 149–177

SELECTED REFERENCES

- T.L. Anderson, *Fracture Mechanics—Fundamentals and Applications,* 2nd ed., CRC Press, 1995

- D. Broek, *Elementary Engineering Fracture Mechanics,* Noordhoff, The Netherlands, 1974
- K. Hellan, *Introduction to Fracture Mechanics,* McGraw-Hill, 1984
- R.W. Hertzberg, *Deformation and Fracture Mechanics of Engineering Materials,* 3rd ed., John Wiley & Sons, 1983
- J.W. Hutchinson, *A Course on Nonlinear Fracture Mechanics,* Department of Solid Mechanics, The Technical University of Denmark, 1979
- J.F. Knott, *Fundamentals of Fracture Mechanics,* John Wiley & Sons, 1973
- A.P. Parker, *The Mechanics of Fracture and Fatigue,* E. & F.N. Spon, Ltd., London, 1981
- S.T. Rolfe and J.M. Barsom, *Fracture and Fatigue Control in Structures—Application of Fracture Mechanics,* Prentice-Hall, 1977

Chapter 2

Fracture Phenomena

Modes of Failure: Plane Stress versus Plane Strain

The state of stress throughout most bodies falls in between the limits defined by the states of plane stress and plane strain. In the theory of elasticity, a state of plane stress exists when $\sigma_z = \tau_{xz} = \tau_{yz} = 0$ (refer to Fig. 1-5 for notation). This two-dimensional state of stress is frequently assumed in practice when one of the dimensions of the body is small relative to the others. In a thin sheet loaded in the plane of the sheet (the x-y plane), there will be virtually no stress acting perpendicular to the sheet. The remaining stress system will consist of two normal stresses, σ_x and σ_y, and a shear stress, τ_{xy}. The term *generalized plane stress* is applied to cases of deformation where the above definitions apply on the average through the thickness of a thin plate subjected to extensional forces. Often, when the term *plane stress* is used, generalized plane stress is actually implied. Similarly, plane strain is mathematically defined as $\varepsilon_z = \gamma_{xz} = \gamma_{yz} = 0$, which ensures that $\tau_{xz} = \tau_{yz} = 0$ and $\sigma_z = \nu(\sigma_x + \sigma_y)$. In such a system, all displacements can be considered to be limited to the x-y plane (i.e., $w = 0$), so that strains in the z-direction can be neglected in the analysis. Neither total plane-stress conditions nor total plane-strain conditions are found in real structural configurations. However, in stress field analysis, employment of either of these constraints allows for two-dimensional solutions of three-dimensional problems, which provides the foundation for linear elastic fracture mechanics theory as it is applied today.

In fracture mechanics terminology, the terms *plane stress* and *plane strain* take on special, more restricted meanings. Instead of characterizing stress and strain states throughout a body, in fracture mechanics special concern is given to the crack tip and surrounding region. Because a bulk of material tends to deform in all directions, to develop a plane-strain condition it is necessary to constrain the flow in one direction. In a region near the leading edge of a crack, the magnitudes of stresses and strains are very high compared to those at some relatively greater distances away. Because of this high degree of tensile deformation, the material near the crack front tends to shrink (a Poisson's ratio effect) in a direction parallel to the leading edge. However, this shrinkage is constrained by the surrounding material that is less deformed. If the leading edge length is long compared to other dimensions of the plastically deformed zone, then it is highly constrained against shrinkage parallel to the leading edge. This is considered to be in a state of localized plane strain. Conversely, a plane-stress state exists if the plastic zone dimensions are large compared to the leading edge length of the crack. By viewing the plastic zone as a short cylinder with free ends, it is bound to be relatively unconstrained or in a state of plane stress. Naturally, plane-stress conditions would still exist at free surfaces of a thick plate containing a flaw, while localized plane-strain conditions would prevail in the interior of the plate (Fig. 2-1). The above considerations are equally relevant for three-dimensional cracks with curved crack fronts.

In order to understand the reasons for this apparently complex failure mechanism, consider two small rectangular elements of material in a moderately thick plate, close to the crack tip. One of them (element A in Fig. 2-2)

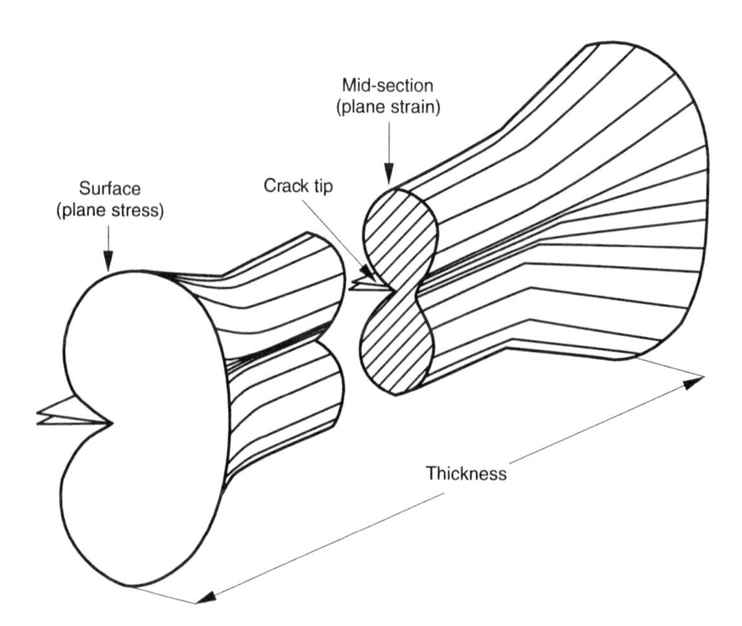

Fig. 2-1 Variation of crack tip plastic zone through the thickness of a plate

is on the free surface of the plate normal to the crack front. The other (element B) is at a similar position relative to the crack, but in the mid-thickness of the plate. If the test panel is thick enough to produce sufficient constraint, a plane-strain condition exists in the interior. However, the condition of plane stress will still prevail at the surface, because it is free of stresses. As the remote loading is increased, each of these elements will fail at some particular load level, either by shear or by cleavage. A shear failure means one atomic plane is sliding over another and the mechanism of failure is governed by the Von Mises criterion or some other yield criterion. Cleavage failure is direct separation of one plane of atoms from another that is caused by loading normal to the eventual fracture surface. Because a hydrostatic state of stress ($\sigma_x = \sigma_y = \sigma_z$) cannot produce a ductile fracture, element B (which is near-hydrostatic) may fail by cleavage in the plane of the crack before it is able to achieve a critical shear stress level (a stress level that will cause ductile failure of element A). Thus, we associate the slant

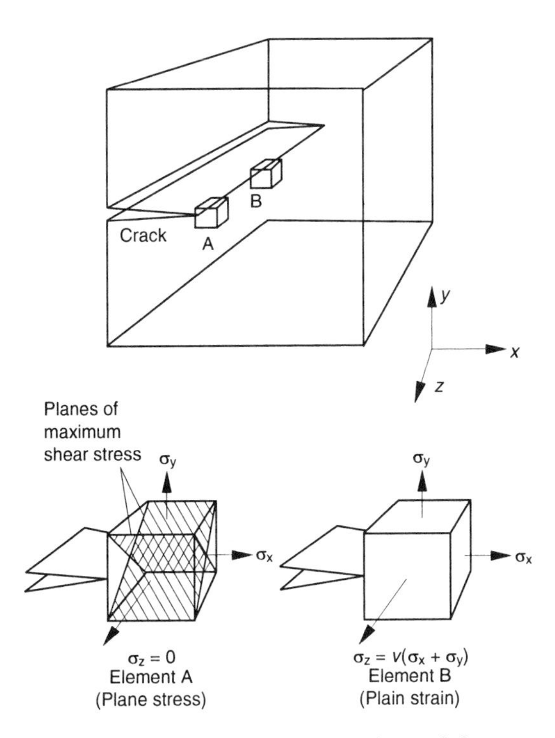

Fig. 2-2 Plane-stress and plane-strain elements near the crack tip

fracture with a ductile failure on inclined planes, and that flat fracture (brittle failure) with cleavage separation.

Fracture appearances are indicated by the percentage of shear lips that appear on the rapid growth portion of the fracture surfaces. A plane-strain state causes a flat fracture appearance. The fracture plane is perpendicular to the loading direction and is called the tensile mode failure. The plane-stress state causes a slant fracture appearance (having single or double shear lips 45° to the loading direction) and is called the shear mode failure. A combined state of plane strain and plane stress results in a mixed failure mode (see Fig. 2-3). It is also evident that the tensile mode of failure is usually associated with a shorter crack or a lower K-level, whereas the shear mode occurs at higher K-levels or is associated with a longer crack. Refer to the scenario for cleavage and shear fracture discussed earlier. Should the applied load continue to increase, the plastic zone ahead of the crack tip will also become larger as the crack grows, allowing less through-the-thickness restraint on internal elements. Under these conditions the proportion of flat fracture surface reduces as the crack extends, and the proportion of shear lip

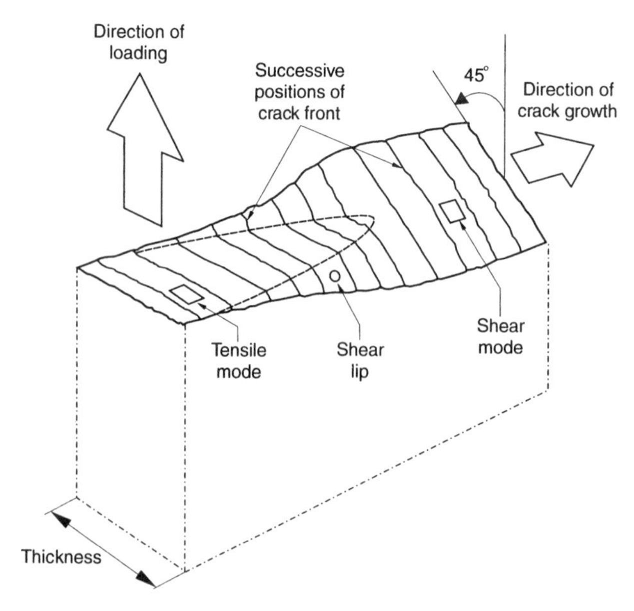

Fig. 2-3 The surface of a fatigue crack in a sheet (or plate), showing the transition from a tensile mode to a shear mode

increases. It might even approach plane stress right through the thickness. The resulting failure is illustrated in Fig. 2-3.

The brittleness of a given material in a given atmosphere is affected by the specimen geometry (e.g., thickness, panel width, crack size), testing temperature, and testing speed (i.e., loading rate). All these variables contribute to the final failure mode of a given test (or failure of a structural member). Elaboration of these is given below.

Fracture Toughness Data Representation

Fracture Indices: K_c versus K_{Ic}

Although there is no difference in the K-solution for the conditions of plane stress and plane strain, the critical stress intensity factor (i.e., the fracture index) for a material does vary with test specimen configuration and crack morphology. The two types of fracture indices most generally used are the plane-stress fracture toughness, K_c, and the plane-strain fracture toughness, K_{Ic}. As will be shown later in this chapter, K_{Ic} is independent of structural geometry but is normally associated with thick sections. However, K_c is a function of plate thickness and other geometric variables. Its values are obtained from testing of plate or sheet specimens of various thicknesses.

Table 2-1 ASTM standard specimens

Specimen configuration(a)	Usage	ASTM standard(b)
Compact, C(T) (c)	K_{Ic}, K_Q	E 399–90
Compact, C(T)	J_{Ic}	E 813–89
Compact, C(T) (c)	R-curve	E 561–94
Compact, C(T)	J-R-curve	E 1152–87
Compact, C(T) (c)	da/dN vs. ΔK	E 647–95
Compact, C(T)	da/dt vs. C^*(f)	E 1457–92
Compact, C(T) (c)	K_{ISCC}	E 1681–95
Bend, SE(B)	K_{Ic},K_Q	E 399–90
Bend, SE(B)	J_{Ic}	E 813–89
Bend, SE(B)	J-R curve	E 1152–87
Arc-shaped, A(T)	K_{Ic}, K_Q	E 399–90
Arc-shaped bend, A(B)	K_{Ic}, K_Q	E 399–90
Disk-shaped, DC(T)	K_{Ic}, K_Q	E 399–90
Crack line wedge loaded, C(W)	R-curve	E 561–94
Center-cracked, M(T) (c)	R-curve	E 561–94
Center-cracked, M(T) (c)	da/dN vs. ΔK	E 647–95

(a) The alphabetic code is a ASTM designation code. (b) The number after the hyphen is the year for the latest revision. (c) The K-expression for the specimen is also given in Chapter 5. Source: Ref 2-1

Several test specimen configurations are available for obtaining plane-stress and plane-strain fracture toughness. Table 2-1 lists those standardized by the American Society for Testing and Materials (ASTM). As noted, some of these specimens are also used for fatigue crack growth testing and for determination of J_c, J_{Ic}, and C^* values. The ASTM standards specify the specimen configuration and size, test setup and loading requirements, and data interpretation techniques (Ref 2-1, 2-2). Theoretically, a K-solution is available for each type of these specimens. If these K-solutions were correctly derived, and the fracture stress for a test is below 80% of the material tensile yield strength (the ASTM criterion for validation), the K_c value for a given thickness and product form should be the same, regardless of which type of specimen was used.

The problem in K_{Ic} testing is to meet the valid K_{Ic} requirement imposed by the ASTM. To ensure a plane-strain condition, ASTM E 399 specifies that both specimen thickness and crack length must meet a minimum size requirement, i.e., the conditions:

$$B > 2.5 \cdot (K_{Ic}/F_{ty})^2 \qquad \text{(Eq 2-1a)}$$

and

$$a > 2.5 \cdot (K_{Ic}/F_{ty})^2 \qquad \text{(Eq 2-1b)}$$

must be met in order to obtain consistent K_{Ic} values. Here B is the ASTM designation for specimen thickness. In this book both t and B are used for thickness. The term $(K_{Ic}/F_{ty})^2$ suggests that the required plate thickness B and crack length a are related to some measure of crack tip plastic zone size. Taking $r_p = (K_{Ic}/F_{ty})^2/3\pi$, it is seen that the criteria for plane strain reflect a condition where B and a are both greater than, or equal to, $24r_p$.

To arrive at a valid K_{Ic}, it is first necessary to compute a tentative value called K_Q based on a graphical construction on the load-displacement test record. If K_Q satisfies the conditions of Eq 2-1(a) and 2-1(b), i.e., if the quantity $(K_Q/F_{ty})^2$ is less than $0.4B$ (and also less than $0.4a$), then $K_Q = K_{Ic}$. The graphical procedure for obtaining a K_Q value will not be discussed here because it is specified in ASTM E 399 (Ref 2-1). If the test result fails to satisfy Eq 2-1 and/or other requirements specified in ASTM E 399, it will be necessary to use a larger specimen to determine K_{Ic}. For some relatively ductile materials it will be very difficult to establish a valid K_{Ic} value. One may have to compromise and use a K_Q value instead.

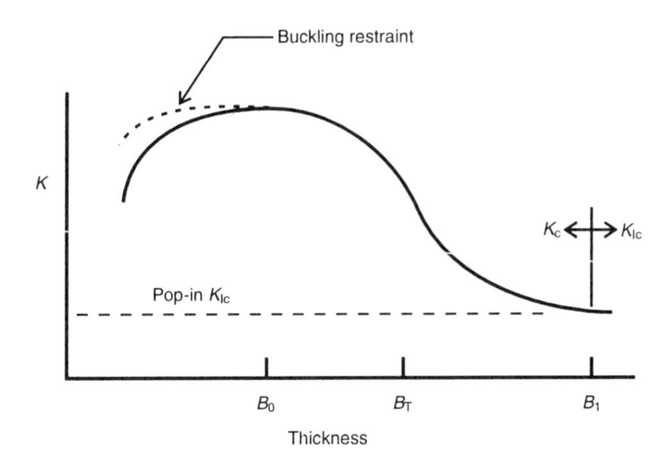

Fig. 2-4 Schematic representation of fracture toughness as a function of plate thickness

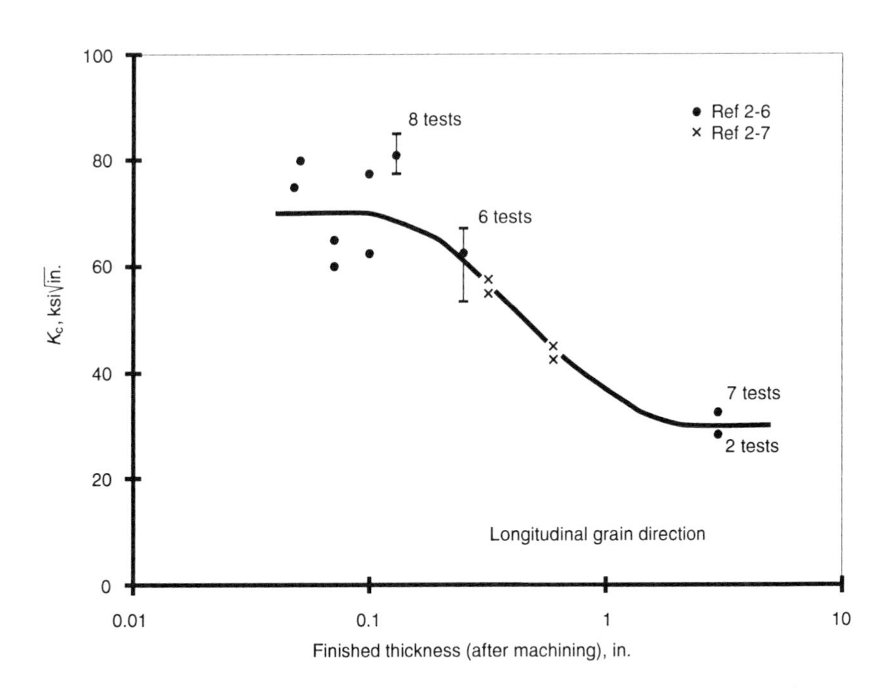

Fig. 2-5 Fracture toughness properties of 7075-T6 aluminum sheet and plate. Source: Ref 2-6, 2-7

The Pop-in K_{Ic}

As will be explained in the next section, during monotonic loading some amount of crack extension usually takes place prior to rapid fracture. For some materials the onset of slow crack growth occurs in a discontinuous manner (i.e., the so-called *pop-in* failure). The pop-in load is identifiable in the load-displacement record of a fracture test. In addition, an audible sound can be heard during pop-in. A sudden crack extension associated with a slight load-drop is usually observed during pop-in. After pop-in, the applied load is further increased until final fracture occurs at the expected fracture load. Occasionally, multiple pop-in's occur in a single test. Pop-in can occur in sheets or plates of any thickness, thin or thick. The crack growth process initiates from a local area in the mid-thickness region of the plate, and thus it is plane strain in nature. Experimental tests show that the pop-in K_c values are relatively constant, irrespective of panel thicknesses, and are equivalent to K_{Ic} values obtained from thick panels. Or, more appropriately, the pop-in value is more or less equivalent to K_Q. However, there is one problem: we do not know under what conditions pop-in will occur. Pop-in just happens during a test; its occurrence can not be prearranged. In other words, we

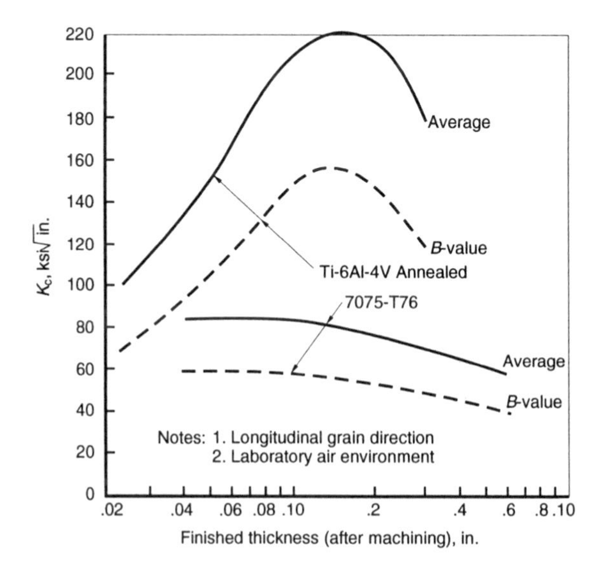

Fig. 2-6 Fracture toughness properties of Ti-6Al-4V and 7075-T76 aluminum sheet and plate. Source: Ref 2-8

cannot determine K_{Ic} by relying on pop-in. We will discuss more about pop-in in the section "The Crack Growth Resistance Curve" later in this chapter.

The Effect of Plate Thickness on K_c

The dependency of K_c on plate thickness is one of the better-known phenomena in fracture toughness testing. Figure 2-4 shows that K_{Ic} and pop-in K_{Ic} are independent of panel thickness. Beyond a certain thickness B_1, a state of plane strain prevails and fracture toughness reaches the plane-strain value K_{Ic}, independent of thickness as long as $B > B_1$. According to Irwin (Ref 2-3), K_{Ic} will be reached at a thickness $B \geq 2.5\ (K_{Ic}/F_{ty})^2$. Incorporating this thickness requirement into Eq 1-32, one concludes that the size of a plane-strain crack tip plastic zone width has to be smaller than 4% of the plate thickness. There is an optimum thickness B_0 where the toughness reaches its highest level. This level is usually considered to be the real plane-stress fracture toughness. According to Broek (Ref 2-5), r_p has to reach the size of the sheet thickness in order for the full plane stress to

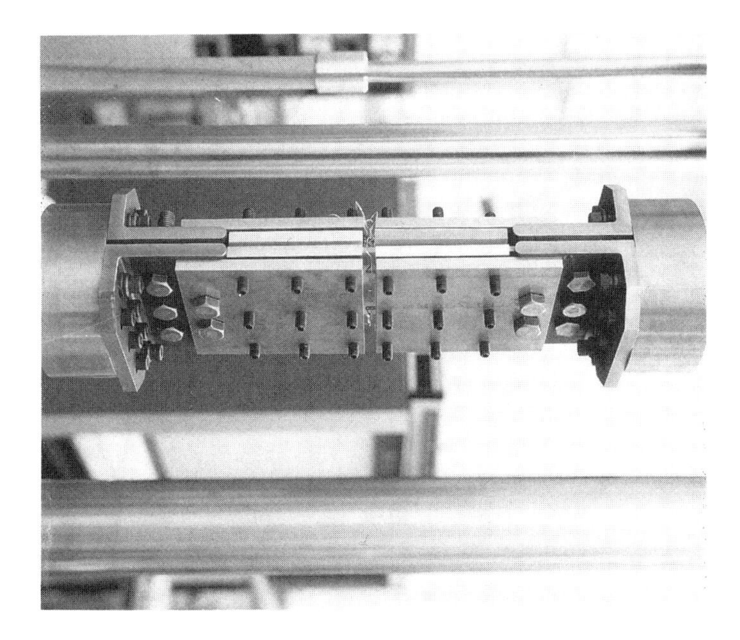

Fig. 2-7 Special setup of a buckling guide device. Source: Ref 2-4

develop. In the transition region, $B_0 < B < B_1$, the toughness has intermediate values. Actual examples showing the effect of (geometric) thickness on K_c for two aluminum alloys and a titanium alloy are given in Fig. 2-5 and 2-6. Again according to Irwin (Ref 2-3), the inflection point B_T (in the transition region) occurs at approximately $B = (K_{Ic}/F_{ty})^2$. Therefore, from the standpoints of design and material selection, it is desirable to design a part with $B < (K_{Ic}/F_{ty})^2$, or select a material that has a K_{Ic} value higher than $F_{ty}\sqrt{B}$.

For thicknesses below B_0 there is uncertainty about the toughness. In some cases a horizontal level is found; in other cases a decreasing K_c value is observed. Recall the K_c versus B curves in Fig. 2-5 and 2-6; thin sheets of the 7075 aluminum (in either overaged or peak-aged conditions) are insensitive to thickness variation. In contrast, the Ti-6Al-4V alloy exhibits a significant drop in K_c values for thicknesses below B_0. Such a phenomenon might be attributable to progress of the extremely large plastic zone during slow stable crack growth, leading to very high local strains at the crack tip, and possibly yielding across the net section of the specimen. Sheet buckling is also responsible for the drop in K_c value. Sheet buckling occurs above and below the crack line and induces lateral compressive stress ($-\sigma_x$) at the crack tip. During sheet buckling some additional normal stress, σ_y, is induced at the crack tip in order to maintain a balance of the stress system. The K level might actually have been higher than $S\sqrt{\pi a} \cdot \phi$.

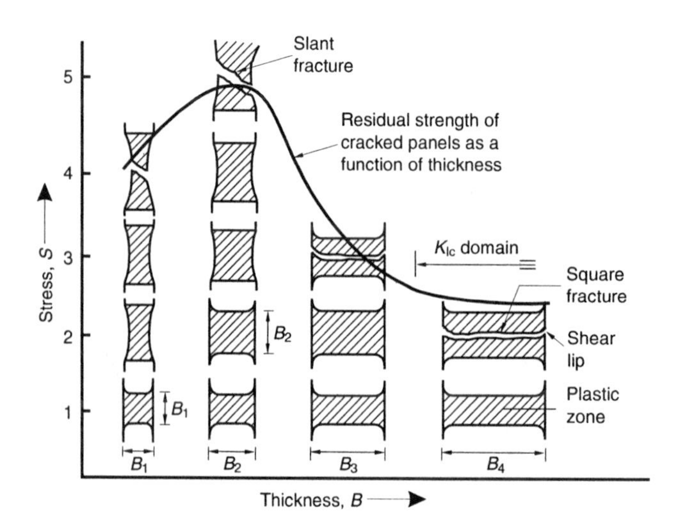

Fig. 2-8 Residual strength as a function of plate thickness. Source: Ref 2-5

Using an anti-buckling device in a test setup could help to minimize the amount of this additional σ_y at the crack tip, thereby allowing the test panel to sustain a higher applied load level before failure, so that the computed K_c value will be higher. An anti-buckling device can be as simple as four horizontal bars loosely hanging on both surfaces of the sheet, resting above and below the crack line. It also can be as sophisticated as the one shown in Fig. 2-7, which consists of fixed plates with adjustable angle plates that provide for friction gripping over a large area of the specimen surface. A fixed outside plate is attached to the angle plate to constrain the inner motion to a vertical direction. Set screws provide a means for adjusting the gap between the specimen and the inside constraint plate, which is covered with a sheet of Teflon to inhibit friction between surfaces.

The foregoing discussion leads to the conclusion that large plastic zones tend to allow plane-stress conditions to develop. An enlarging plastic zone under increasing load would tend to reduce the through-the-thickness constraint, allowing more of the plastic zone to deviate from the plane-strain condition. Thus, the ratio of plastic zone size to specimen thickness must be much less than unity for plane strain to predominate. According to ASTM criteria, this ratio should be 0.04 or smaller. Because it is the plane-strain condition that produces the lower limiting value of fracture toughness, very

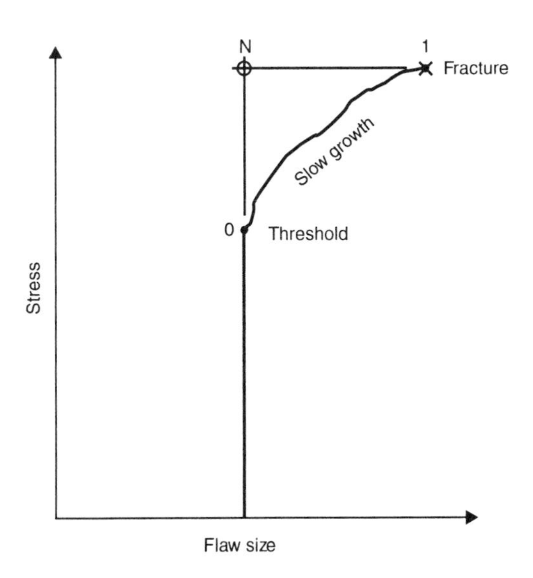

Fig. 2-9 Crack extension under monotonic increasing load

thick plates may be required for plane-strain fracture toughness testing. This is particularly true for a material that has high K_{Ic} value and a low F_{ty} value, because it would allow the specimen to reach a high K/F_{ty} ratio (and a high ratio of plastic zone size to specimen thickness) before failure.

To better understand the thickness dependence of fracture toughness, consider a scenario that was originated by Broek (Ref 2-5). Consider four panels of thicknesses B_1, B_2, B_3, and B_4, having $B_1 < B_2 < B_3 < B_4$, as shown in Fig. 2-8. All panels have the same crack length and all panels are loaded to the same stress, S_1; therefore, all panels have the same K level. Consequently, the plastic zones in these panels are of equal size. This is depicted in the lower part of Fig. 2-8, which shows through-the-thickness sections of the four panels. The plastic zone sizes for these panels at each applied stress level are indicated by the shaded areas.

Assume that at S_1 the plastic zone in panel B_1 is as big as the plate thickness. Therefore, the plastic zones in the other panels are smaller than the thickness of the plate. Yielding cannot take place freely in the thickness direction, because it is restrained by the surrounding elastic material. Thus, the state of stress at the mid-thickness of panel B_2, B_3, or B_4 is predominantly plane strain. In panel B_1 the plastic zone is equal to the thickness, and yielding in the thickness direction is unconstrained. That means that the plane-stress plastic zone has grown from the surfaces through the mid-thickness of the panel. Thereby a state of plane stress has been fully developed. Possibly the plastic zone size in B_1 may enlarge further due to slow stable crack growth. From now on the plastic zone across the thickness in B_1 will be larger than the plastic zones in the other panels.

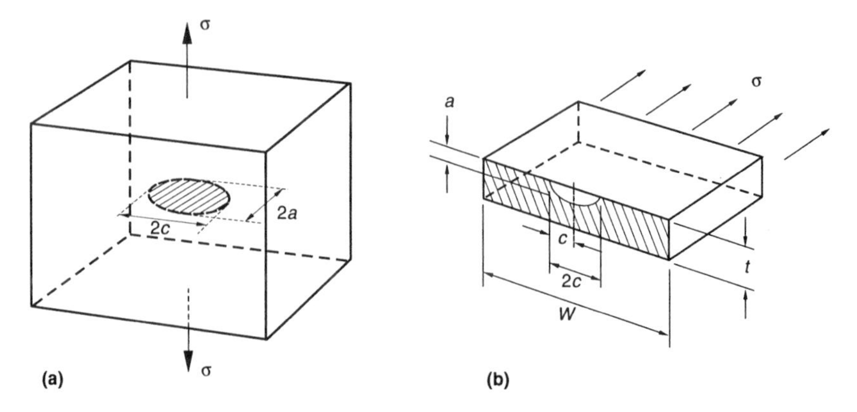

Fig. 2-10 Two types of 2-D crack: fully embedded vs. partially exposed

For a given K-level, i.e., for a given crack length and stress level, the distribution of σ_y is unchanged whether it is plane stress or plane strain. However, the plastic zone radius for plane stress is three times larger than that for plane strain. Therefore, as far as the effective crack length is concerned, at $(a + r_y)$ the crack undergoes a higher σ_y for plane strain than the one for plane stress (see Fig. 1-7). Consequently, the panel that exhibits a higher degree of plane strain will fail first. While further increase of the stress to S_2 causes failure in panel B_4, panel B_3 does not fail, due to the effect of the regions of plane stress (lower crack tip stress) that exist near the specimen surfaces, which are relatively influential in this thinner panel.

Now further assume that at stress level S_2 the plastic zone size in panel B_2 has reached a dimension equal to its thickness, which implies that plane stress now develops in panel B_2. Consequently, further increase of stress will cause failure of B_3 first, and then B_1, and B_2. A rationale has been offered by Broek (Ref 2-5) attempting to explain why B_1 should fail first instead of B_2. It considers that the local strain in the much thinner sheet B_1 is much higher than the strain in B_2. As shown in Fig. 1-2, both stress and strain contribute to final fracture of a panel. At stress level S_4, the strains in B_1 are sufficiently high to cause failure, while B_2 can withstand more load and finally fails at S_5. Here the thickness B_2 can be taken as equivalent to B_0 in Fig. 2-4.

The Phenomenological Aspects of Fracture under Monotonic Loading

Consider the case of a sheet (or a plate) containing a through-the-thickness crack. If loads are applied perpendicular to the crack so that tensile stresses act to open the crack, the level of K increases linearly with the level of the tension stress component normal to the crack. As the level of K increases, some point will be reached at which the crack will start to increase in length. Then the crack will grow to a critical size and onset of rapid crack propagation (fracture) will occur. Schematic illustration of this is given in Fig. 2-9. In practice, the stress at which slow crack growth starts (point 0 in Fig. 2-9) is usually not very well defined. For practical purposes, the stress at the onset of rapid crack propagation (point 1 in Fig. 2-9) may be taken as the maximum stress reached in a test. The critical crack length at rapid fracture is not sharply defined, because the crack length is increasing rapidly up to the length at failure. However, it may be measured to a useful degree of accuracy by observing the fracture appearance of the specimen, or by taking a high-speed motion picture during the test.

Strictly speaking, fracture toughness K_c is computed using critical crack length and fracture load. For engineering purposes, however, a critical K-value, i.e., K_c, can be computed based on initial crack length and maximum load (point N in Fig. 2-9). This engineering value is often called the K_{app} (K-apparent). Therefore, the terminology for critical crack length of a structural part is not clearly defined. Most likely it refers to the criticality of the structural part under consideration. Any critical crack length that is computed using a single value of K_c (which is probably a K_{app} value) is actually the crack length at onset of fracture. However, if desired, the real critical crack length can be determined analytically. We will discuss the technique for doing that in the section "The Crack Growth Resistance Curve" in this chapter.

If the material is ductile and/or the test specimen is in a state of generalized plane stress, slow crack growth is expected to occur. On the other hand,

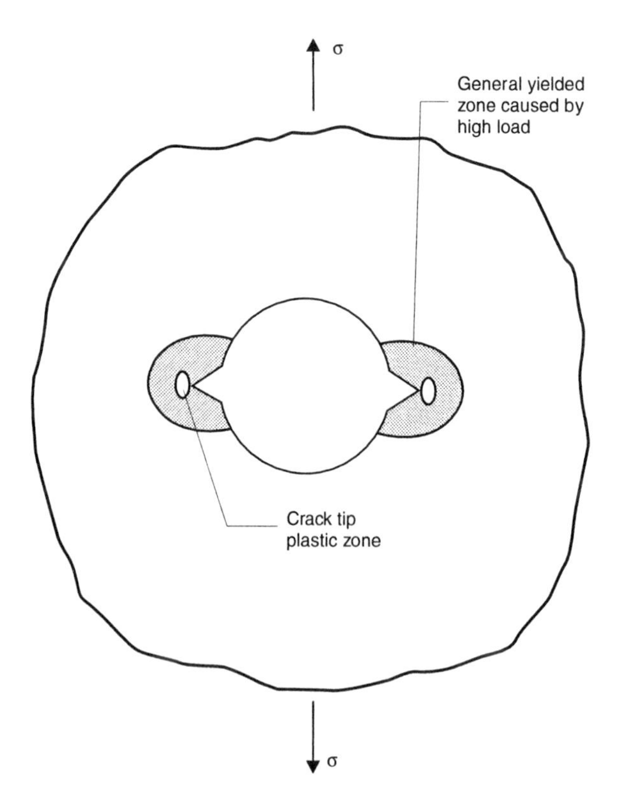

Fig. 2-11 Small cracks embedded inside a locally deformed zone

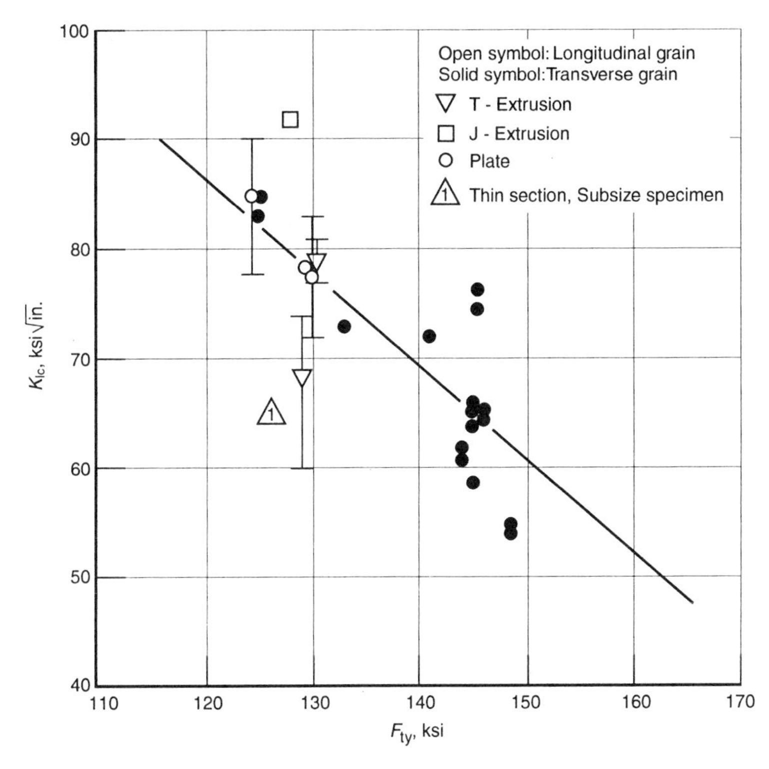

Fig. 2-12 Plane-strain fracture toughness as a function of material tensile yield strength. Four-point notch-bend specimens. Ti-6Al-4V, mill annealed. Source: Ref 2-6

Table 2-2 Heat treating methods vs. heat treating operations for D6AC steel

Heat treating operation	Heat treating methods							
	A	B	C	D	E	F	G	H
Austenitizing	1700 ± 25 °F		1650 ± 25 °F			1650 ± 25 °F		
Ausbay (interrupted) quenching	Cooled from austenitizing temperature to 975 ± 25 °F in austenitizing furnace and held at 975 ± 25 °F until material is stabilized at this temperature. (Note: Cooling rate between 1350 and 1150 °F must not be less than 6 °F per minute)							
Quenching	140 °F oil		Salt			Salt		
			325 °F	325 °F	400 °F	400 °F	400 °F	375 °F
Tempering	Double tempered at 1025 °F; held at temperature for 2 h per cycle							

for the case in which the environment is inert, and temperature, thickness, and other characteristics of the material are such that it is quite brittle, the start of slow crack growth will be followed immediately by the onset of rapid fracture and the obtained fracture toughness will be equivalently plane

Table 2-3 Summary of D6AC Steel K_{Ic} Data

Thickness, in.	Product form	Heat treatment(a)	Avg. K_{Ic}, ksi$\sqrt{in.}$	Number of specimens
0.8	Forging	E, 400 °F salt	65.3	60
0.8	Plate	E, 400 °F salt	64.5	100
0.8	Plate	C, 325 ° F salt + agitation	81.8	4
0.8	Plate	F, 400 ° F salt	53.8	12
0.8	Plate	A, 140 °F oil	94.6	25
0.8	Forging	A, 140 ° F oil	96.9	26
1.5–1.8	Forging	G, 400 °F salt	43.8	6
1.5–1.8	Forging	H, 375 °F salt + agitation	49.4	14
1.5–1.8	Plate	H, 375 °F salt + agitation	61.3	20
1.5–1.8	Plate	D, 325 °F salt + agitation	47.0	3
1.5–1.8	Plate	B, 140 °F oil	79.1	5
1.5–1.8	Forging	B, 140 °F oil	89.6	8

(a) See Table 2-2 for heat treatment designations

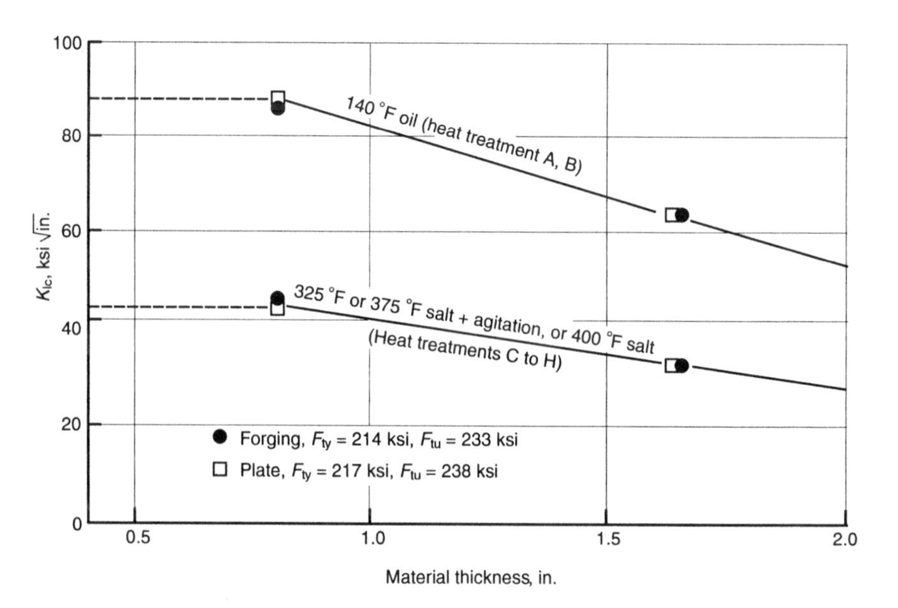

Fig. 2-13 MIL-HDBK-5 B-scale plane strain fracture toughness values for D6AC steel as a function of heat treatment. Heat treatment designations refer to Table 2-2. Source: Ref 2-10

strain (in magnitude). Thus, the fracture indices are, respectively, K_c or K_{app} for the plane-stress (or mixed-mode) fracture and K_{Ic} for the plane-strain fracture. In reality, it is too complicated to use a real critical crack length to calculate the residual strength (or failure load) by running a computer program for crack growth life prediction. It actually makes more sense to use a K_{app} value, because the crack length at onset of structural failure (i.e., initial crack length) is what is important in structural life/structural damage tolerance prediction.

For a fully embedded flaw (Fig. 2-10a) and the partially exposed flaw (the so-called surface flaw or part-through crack, Fig. 2-10b), the constraint at the leading border of the crack is very high and the mechanism of crack propagation is controlled by a plane-strain condition. Therefore, it is customary to use the K_{Ic} value to predict the residual strength for a surface flaw. Incidentally, the National Aeronautic and Space Administration (NASA) is using an alternate fracture toughness value for routine engineering tasks in the space shuttle program. This fracture toughness value (being called K_{IE}) is specifically developed from fracture testing of a surface flaw specimen, because NASA workers found that the fracture behavior of the surface flaw is not entirely the same as K_{Ic}. After testing a number of engineering alloys, including aluminum, titanium, steel, Inconel, magnesium, and beryllium-copper, Forman (Ref 2-9) has found that K_{IE} is related to K_{Ic} in the following manner:

$$K_{IE} = K_{Ic} \cdot (1 + C_k \cdot K_{Ic}/F_{ty}) \qquad \text{(Eq 2-2)}$$

where C_k is an empirical constant whose value equals 1.0 $(\text{in.})^{-1/2}$, or 0.1984 $(\text{mm})^{-1/2}$, or 6.275 $(\text{m})^{-1/2}$. However, the relationship of Eq 2-2 is not applicable to very ductile materials (which have a very high K_{Ic}/F_{ty} ratio). The limitation specified by Forman is $K_{IE} \leq 1.4\ K_{Ic}$.

In a real ductile material, if the critical crack depth for a surface flaw is large in comparison with the thickness of the specimen, or if the operating stress level is quite low (for the case of fatigue cycling), then the crack might grow through the thickness before rapid fracture. Once the crack has grown through the thickness, the crack propagation behavior is the same as that of a through-the-thickness crack. Whether it should be called plane-strain or plane-stress failure still depends on the conditions described in the foregoing paragraph.

In summary, K_c failure is associated with a slant fracture appearance with shear lips and slow stable crack growth prior to rapid fracture. The amount of slow stable crack growth depends on the ductility of the material and the extent of plastic constraint at the crack tip. However, if the material is extremely brittle, rapid fracture may occur without an appreciable amount of slow stable crack growth. This may happen even in a very thin sheet (i.e., in the state of plane stress). An example for 7178-T6 thin aluminum sheet will be shown later in this chapter in the section "The Crack Growth Resistance Curve." The plane-strain fracture index K_{Ic} is associated with a flat fracture appearance without an appreciable amount of crack growth prior to fracture. K_{Ic} is regarded as the minimum value of K_c and is an intrinsic material mechanical property.

Structural details often play an important role in contributing to the residual strength of a structural member. The local area in the immediate vicinity of a stress riser, or a cutout, is known to be susceptible to stress concentration. Taking a circular hole, for example, the local tangential stress at the

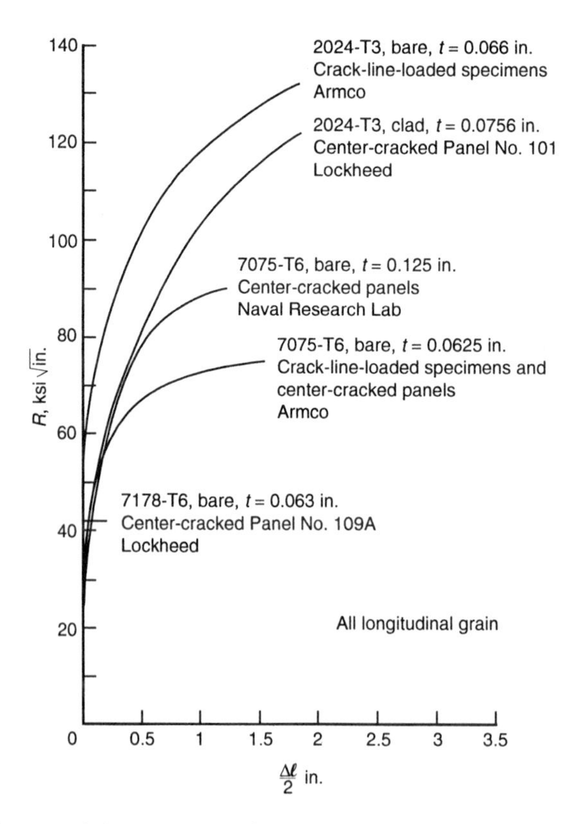

Fig. 2-14 *R*-curves of aluminum alloys. Source: Ref 2-15 to 2-18

hole edge is at least three times the applied far-field stress. Because the material cannot forever follow Hook's law under monotonic increasing load, in reality it follows the stress-strain relationship of the tensile stress-strain curve. Therefore the magnified stress at the hole edge eventually causes gross-scale yielding around the hole. Depending on the applied stress level,

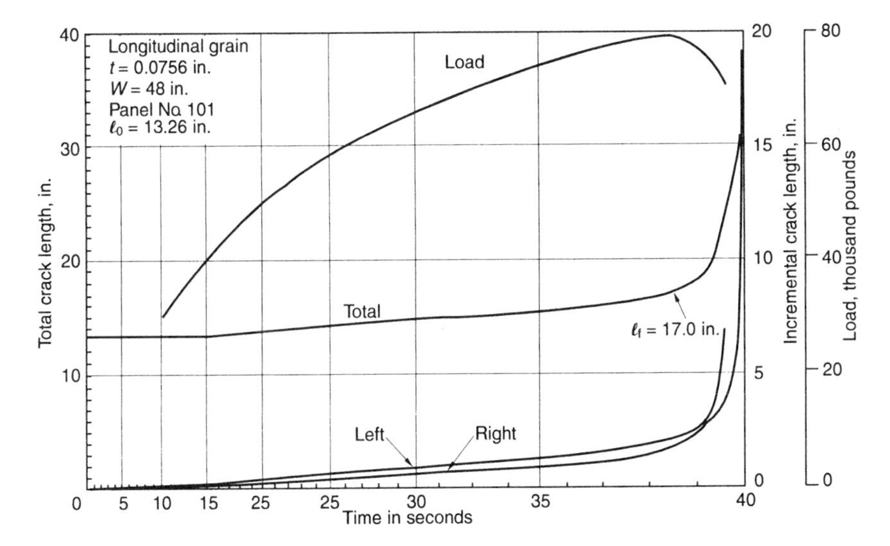

Fig. 2-15 Crack growth behavior of 2023-T3 aluminum sheet. Source: Ref 2-15

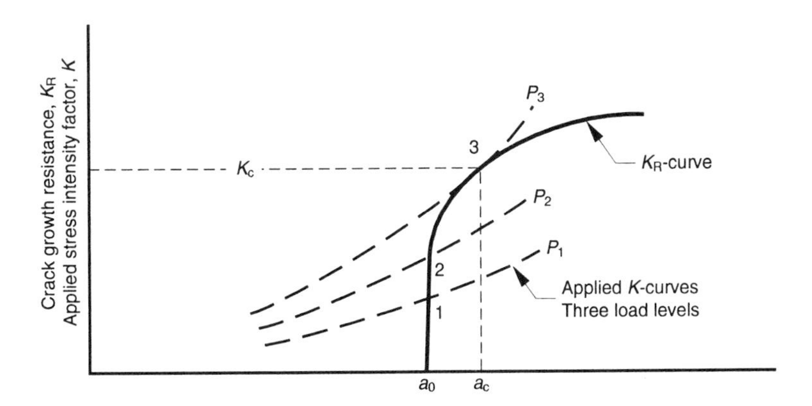

Fig. 2-16 Schematic representation of superposition of *R*-curve and applied *K*-curves to predict instability. $P_1 < P_2 < P_3$.

the yielded zone adjacent to the hole can be very large, to the extent that a very short crack would be totally embedded inside this zone. A schematic illustration of this situation is shown in Fig. 2-11. It shows that the entire effective crack length (i.e., the physical crack plus the crack tip plastic zone) is embedded inside the locally deformed area. Although this type of fracture behavior has long been recognized, a simple engineering solution to the problem does not exist. Direct application of the current linear fracture mechanics technology to predict the residual strength of this configuration is inappropriate. So far, the J-integral is used as an alternate fracture index to characterize fracture behavior involving large-scale yielding at the crack tip. However, Shows et al. have shown that the J-integral can also be used to determine the fracture strength of the yielded hole configuration (Ref 2-4). We will demonstrate how this is done in Chapter 8.

Heat treatment and other mechanical (or thermal mechanical) means are known to cause strength variations in a given alloy. The fracture toughness K_c and K_{Ic} of a given alloy are generally found to be inversely proportional to its tensile ultimate or tensile yield strength. A typical example showing the K_{Ic} as a function of tensile yield strength for a titanium alloy is given in Fig. 2-12. For no other reason, heat treatment alone might affect the fracture toughness values of an alloy even if the heat treatments had made no change in tensile strength. To explain what this means, some data for D6AC steel are presented in Tables 2-2 and 2-3 and in Fig. 2-13. These data indicate that the alloy had gone through several different heat treatments but yielded similar tensile strengths. However, depending on the heat treatment, its fracture toughness could have been different by a factor of 2. These data were reported as K_{Ic}, however, their values still depend on thickness.

Normally, three basic factors contribute to a brittle cleavage type of fracture: triaxial state of stress, low temperature, and fast loading rate. Although not all three of these factors have to be present at the same time, the conventional dynamic test methods used for determining the ductile-to-brittle transition behavior do involve all these elements. The Charpy and Izod tests (Ref 2-11, 2-12) and the drop-weight test being used by the Naval Research Laboratory (Ref 2-13, 2-14) are typical examples. In fracture mechanics testing, however, it has been shown that loading rate does not affect the fracture toughness of aluminum alloys. An increase in loading rate results in lower K_c for low-carbon steels, but higher K_c for titanium alloys (Ref 2-6).

Aside from a few exceptions reported in the literature, fracture toughness K_c and K_{Ic} are in proportion to test temperature. An increase in test tempera-

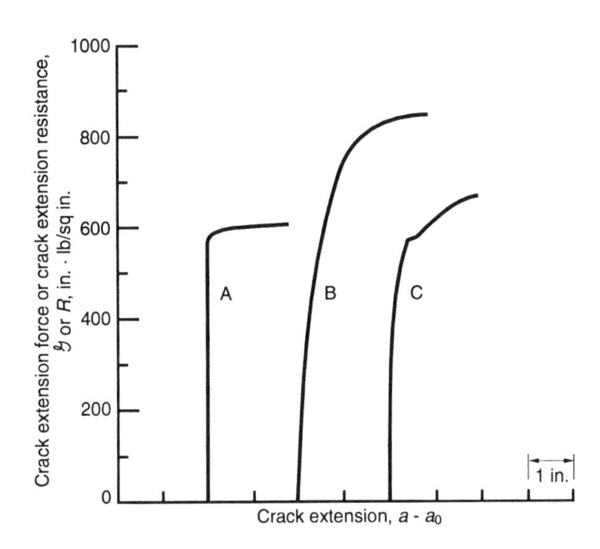

Fig. 2-17 Some conceivable types of *R*-curves. Source: Ref 2-19

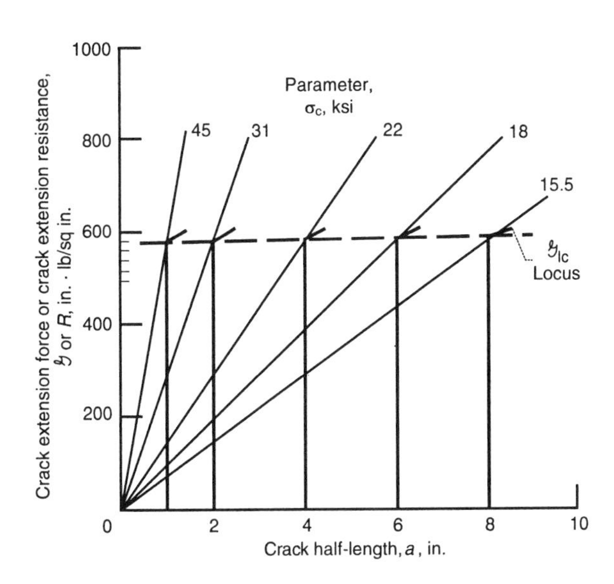

Fig. 2-18 Crack extension instability condition for various crack lengths in a brittle material. Source: Ref 2-19

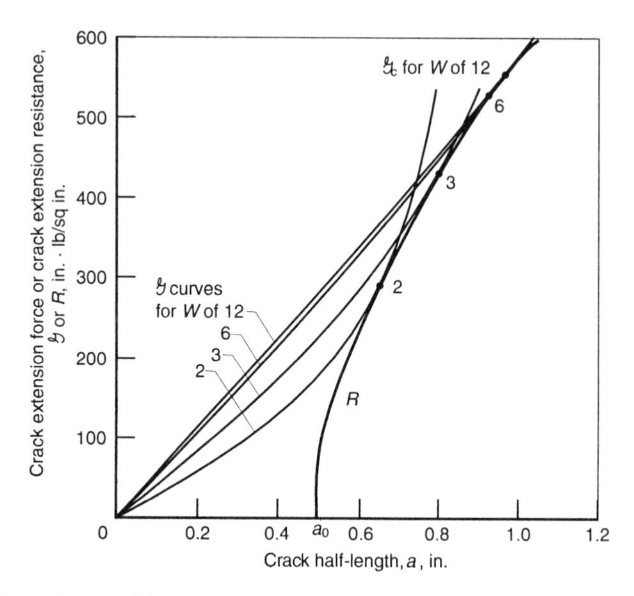

Fig. 2-19 Dependence of fracture toughness on specimen width, W, for center-cracked plate specimens having the same initial half crack length. Source: Ref 2-19

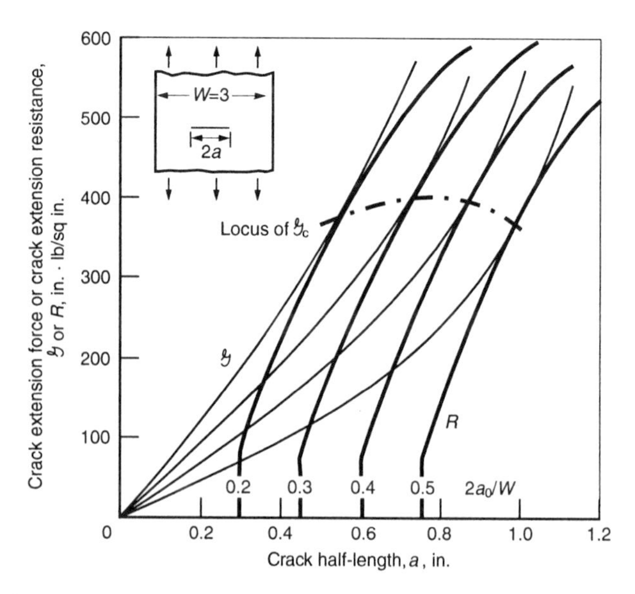

Fig. 2-20 Dependence of fracture toughness on relative initial crack length for a finite width specimen having an R-curve identical to that of Fig. 2-19. Source: Ref 2-19

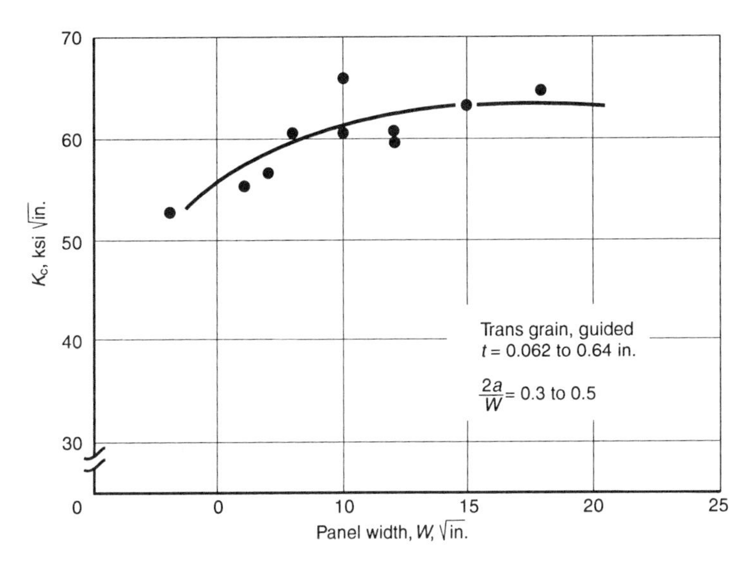

Fig. 2-21 Effect of panel width on fracture toughness for bare 7075-T6 sheets. Source: Ref 2-20

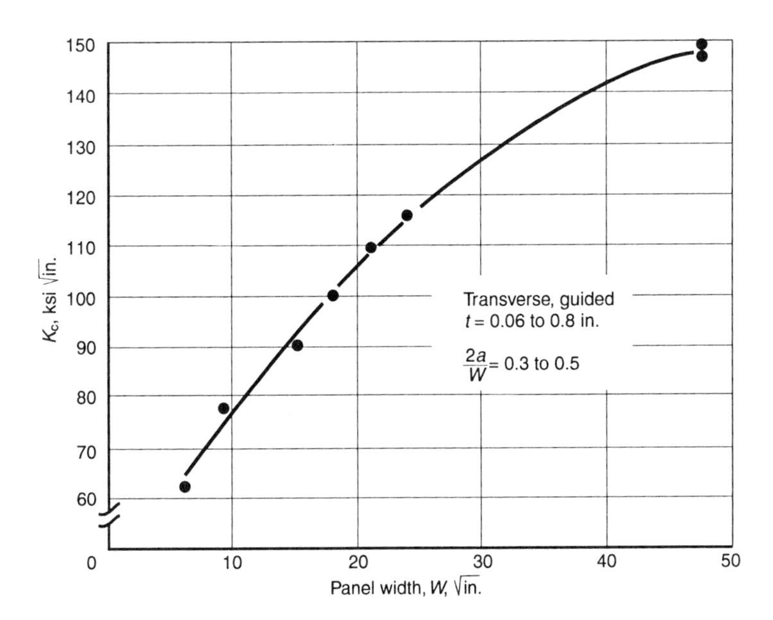

Fig. 2-22 Effect of panel width on fracture toughness for bare 2024-T3 sheets. Source: Ref 2-20

ture from room temperature normally results in increased fracture toughness associated with decreasing material tensile yield strength. In general, lowering the test temperature leads to an increase in tensile strength coupled with a decrease in crack tip plastic zone size. Consequently, subzero temperatures will cause a normally ductile material to become brittle and thereby exhibit a lower material fracture toughness. This behavior is normally associated with a change from slant fracture appearance to flat fracture appearance. The sensitivity of material tensile strength to low-temperature exposure also depends on crystal structure. Among the three basic types of crystal structures, the body-centered cubic structure, such as in carbon steels, is most susceptible to low temperatures. The face-centered cubic metals, such as aluminum, nickel, copper, and austenitic stainless steels, are much less susceptible to low temperatures.

The effect of corrosive atmosphere probably is not applicable to monotonic load fracture testing, but it applies to fracture testing with a sustained load. We will discuss such behavior in the section "Stress Corrosion Cracking" in Chapter 6.

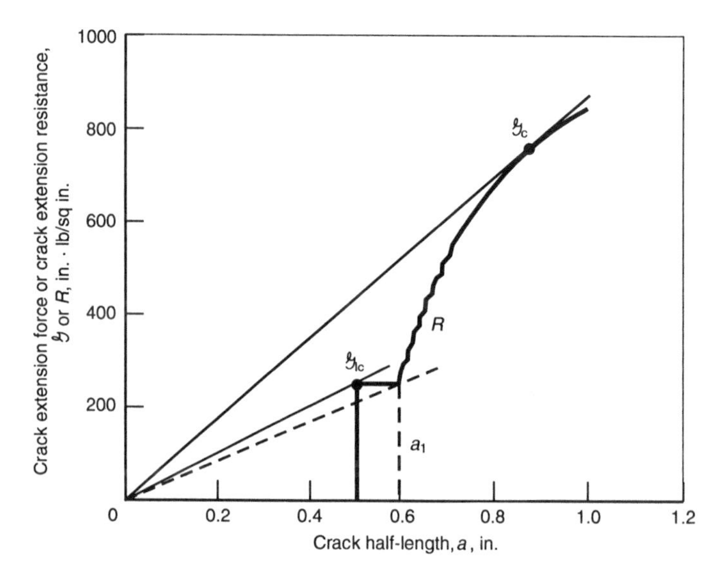

Fig. 2-23 Meta-instability (K_{Ic}) and ultimate instability (K_c) for wide plate specimen exhibiting pronounced pop-in behavior. Source: Ref 2-19

The Crack Growth Resistance Curve

As discussed earlier, the fracture process of a metal plate can be characterized by the type of curve shown in Fig. 2-9, in which load (or stress) is plotted against crack length. For some materials, considerable slow stable crack growth takes place (along with increasing load) prior to catastrophic failure. The stress intensity factor, K, which is determined from linear elastic stress analysis of a cracked plate, uniquely defines the stresses in the crack tip vicinity, so tests on different panels may be correlated by the use of this parameter. The S versus a curve can be converted to a K versus a curve. By subtracting the initial crack length from a, the K versus a curve becomes a K versus Δa curve. Actual examples of this type of curve for several aluminum alloys are shown in Fig. 2-14. Each of these curves is called a crack growth resistance curve, or just an R-curve, or K_R-curve, because the K-parameter here is specifically designated as R or K_R. Once an

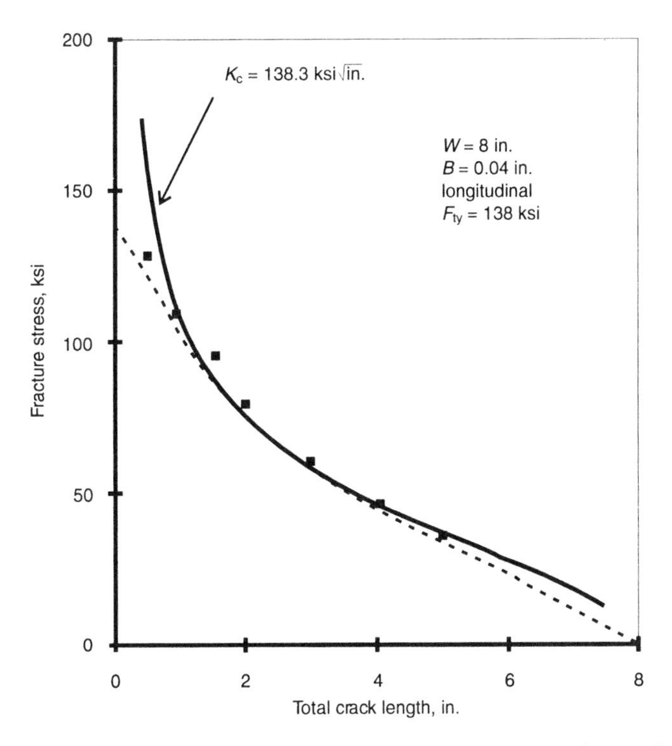

Fig. 2-24 Display of residual strength data for Ti-6Ai-4V, mill annealed. Source: Ref 2-6

R-curve representing the fracture behavior of a given material is determined, it is regarded as a material property.

Everything else being equal, i.e., specimen geometry, initial crack length, loading condition, temperature, etc., a more ductile material will have a longer period for slow stable growth (a longer R-curve). In reality, the amount of slow stable tear also is dependent on structural configuration (or the geometry of the test specimen). However, as will be demonstrated later, a material R-curve of any origin is equally applicable for determining the residual strength of a structural member of any configuration and loading combination.

An R-curve can be obtained by conducting a fracture test of a center-precracked panel, or using any other type of specimen such as those recommended by ASTM (Table 2-1), provided that the load and crack length increments can be correlated on a time basis. An example of such data is shown in Fig. 2-15. The data are obtained from a test of a precracked

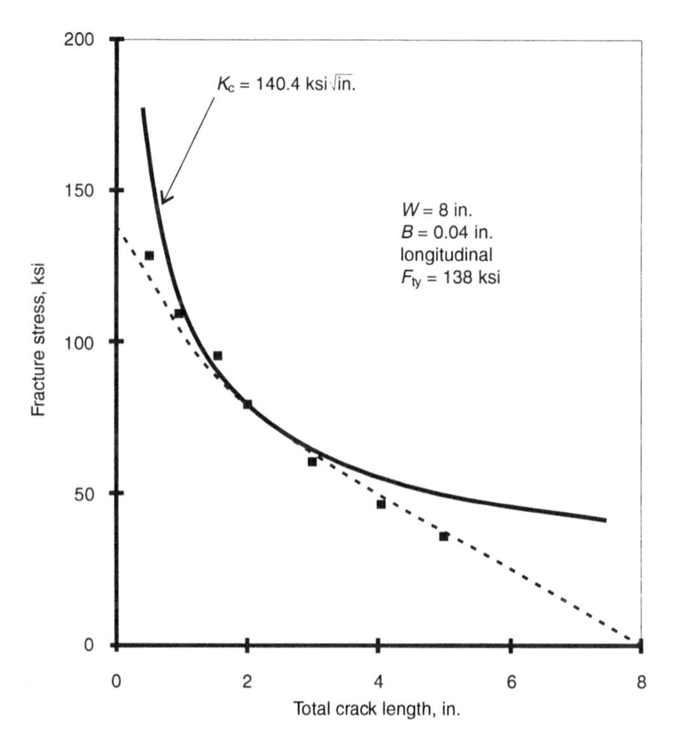

Fig. 2-25 Replot of Fig. 2-24

2024-T3 Alclad aluminum sheet, 0.0756 in. thick, 48 in. wide and 100 in. long. First a 10 in. saw cut was made at the center of the panel. Very low tension-tension cyclic loads were applied to the panel to produce a simulated fatigue crack approximately 13 in. long, symmetric to the vertical centerline of the panel. During static fracture testing, close-up motion pictures (film speed was 200 frames per second) and electronically recorded load data were taken. Superposition of the load versus time and the crack length versus time curves yielded a load versus crack length plot similar to that shown in Fig. 2-9. From these data, K-values were computed for each

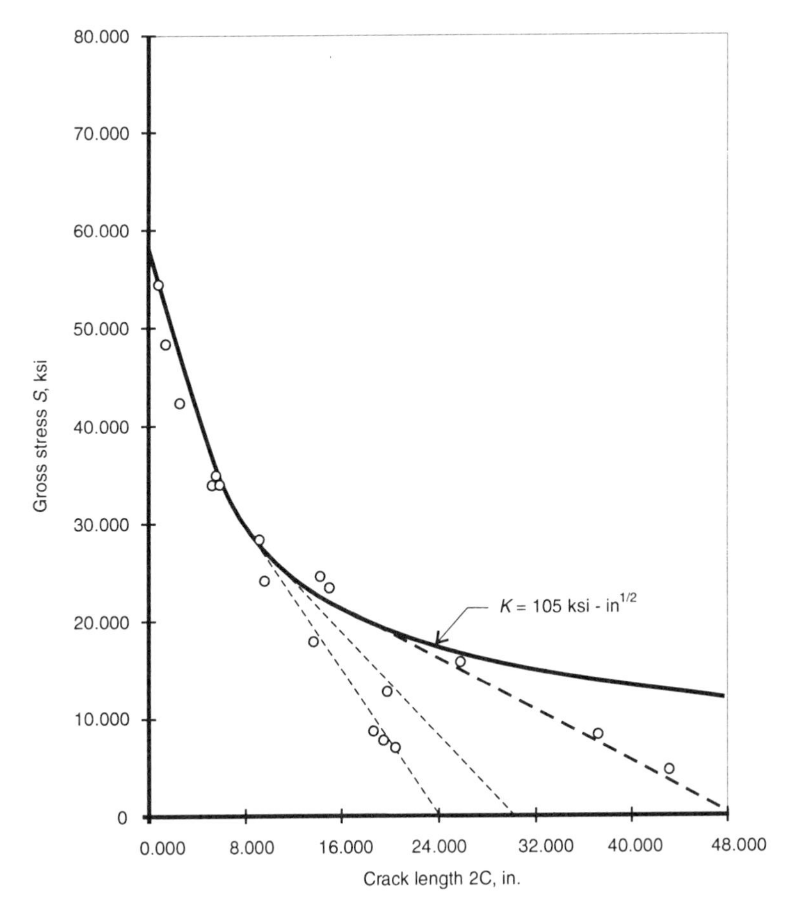

Fig. 2-26 Display of residual strength data for 2219-T87 aluminum, 0.1 in. thick, having multiple combinations of specimen width and crack sizes. Source: Ref 2-21

consecutive crack length and the corresponding stress levels using an appropriate stress intensity factor equation (i.e., Eq 1-21 for the center cracked panel). By setting K equal to K_R, an R-curve was constructed (with K_R versus a or K_R versus Δa). The result is shown in Fig. 2-14 (see the curve that is labeled as Panel No. 101).

Perhaps the best way to explain an R-curve is by going through a hypothetical exercise that demonstrates how to use an R-curve to determine the residual strength of a piece of structure. For simplicity, we consider a center-cracked panel having an initial crack length $2a_0$. To predict the failure load, we need to construct a diagram such that the R-curve of the *same material* is placed at $a = a_0$, as shown in Fig. 2-16. The material R-curve need not be developed from a specimen configuration identical to the one that is being analyzed. That is, the R-curve could have come from a compact specimen, or other types of specimens, as long as the thickness of the specimen and the part in question are the same.

The three dotted lines in Fig. 2-16, labeled as P_1, P_2, and P_3, are K-curves hypothetically computed for four stress levels (as a function of a). Each K-curve corresponds to a stress level applied to the structure (with $P_1 < P_2 < P_3$). Thus, these are the applied K-curves. Let's focus on the points on P_1 and P_2; the applied K-levels on the right of the R-curve are lower than the K_R of the R-curve. This means that at stress levels P_1 and P_2 the material has more energy ($K_R \cong \mathcal{G}_R$) available for release against the input (the applied K). When the crack starts to extend at point 1 (stress level P_1) we have $K_R > K$. That is, the material is capable of taking a higher load. The same is true for point 2. But as the stress level reaches P_3, the applied K-levels corresponding to subsequent crack extension (i.e., beyond point 3) are higher than K_R. This means that the material can no longer sustain more load. Therefore, the failure criterion based on the crack growth resistance curve concept is that crack growth will be stable as long as the increase in resistance K_R, as the crack grows is greater than the increase in applied K. Otherwise, unstable fast fracture will occur. That is, failure will occur when:

$$K = K_R$$

and

$$\delta K/\delta a \geq \delta K_R/\delta a \qquad \text{(Eq 2-3)}$$

The failure point can be graphically determined when the applied K-curve for a given load is tangent to the R-curve. This tangential point defines the

K_c value, or the fracture load (P_3 in this case) of the structure (or the test panel). Recalling that for Panel No. 101 (Fig. 2-14 and 2-15), the failure load was 79,000 lb. The tangential point where the applied K-curve merges with the R-curve gives a K_c value of 122 ksi$\sqrt{\text{in.}}$. However, on the basis of initial crack length and final load, a K_{app} value of 104 ksi$\sqrt{\text{in.}}$ is obtained.

Now we can use an R-curve to explain the phenomena of K_{Ic} and pop-in K_{Ic} and the effect of geometry on K_c. We begin this by showing several possible types of R-curves (see Fig. 2-17). Obviously, the one that has a sharp point (i.e., type A) is representative of the plane-strain K_{Ic} behavior. The smooth curve (type B) represents most of the thin-sheet (or mixed-mode K_c) behavior. As shown in Fig. 2-18, the value of K_{Ic} for a given material is independent of initial crack lengths. In contrast to the K_{Ic} behavior, the type B material exhibits a geometry dependency, as schematically shown in Fig. 2-19 and 2-20. The former figure illustrates K_c as a function of panel width, whereas the latter shows K_c as a function of the $2a/W$ ratio for a fixed specimen width. It is interesting to note that for a given panel width there is an optimum K_c value that can be obtained at $2a \cong W/3$. Although the general trend for the panel width effect is that K_c increases with panel width, the degree of susceptibility depends on the ductility of the material. As shown in Fig. 2-21 and 2-22, of the K_c values for two aluminum alloys (7075-T6 and 2024-T3), the one that is more ductile (i.e., the 2024-T3 alloy) exhibits a higher degree of panel width susceptibility. Therefore, using an R-curve to determine structural residual strength is perhaps the most efficient and economical method in damage tolerance assessment. One can depend on a single R-curve to analytically determine residual strengths for multiple combinations of panel widths and ratios of crack length to panel width, as well as any structural detail for which a K-solution is available.

In a given condition and at a given temperature and rate of loading, the plane-strain fracture toughness K_{Ic} represents a practical lower limit to the fracture toughness of a material. From the point of view of having a single value representing the fracture toughness of a material, K_{Ic} is conceptually independent of the dimensions of specimen. However, this is only true provided that the size of the specimen is sufficiently large for a proper K_{Ic} measurement. Of course, materials exhibit nonuniformity and anisotropy with respect to K_{Ic}, just as they do for other properties, and this has to be taken into consideration in materials evaluation. The most obvious way to measure K_{Ic} would be to test a sufficiently thick plate specimen, but this may not always be convenient, or even possible, and certainly it is not economical.

For those materials that exhibit metal-instability (pop-in) during a test, K_{Ic} can be determined by using a thin specimen. As explained above, pop-in is usually associated with an audible ping (or click) at the pop-in load and an abrupt extension of the crack front, while the load remains constant (or even drops slightly). It is then followed by a stage of gradual crack extension as the load further increases (see Fig. 2-23). While more than one pop-in may occur in a given test, a distinct pop-in is not always observed. Therefore, it cannot be depended upon for K_{Ic} measurement in all cases. When pop-in does occur, it satisfies the instability condition $\delta K/\delta a \geq \delta K_R/\delta a$, but the instability is only temporary so that it is referred to as meta-instability.

The fracture mechanism for the hypothetical R-curve shown in Fig. 2-23 (with pop-in) is just the same as those described earlier. During the early stage of a test, the K-curve intercepts the R-curve so that equilibrium is maintained until the step in the R-curve is reached. At this point, the value of K_R over a certain interval of a-a_0 is lower than the value of K corresponding to the stress at the point, so the condition of K_{Ic} is reached. Thus, the balance between K and K_R is temporarily upset until the crack has extended to the point a_1, or somewhere beyond a_1. The extent to which the load drops in this interval is a function of several variables. However, K and K_R will again become balanced at some value of a slightly greater than a_1. It will remain so, on the average, until the final point for K_c is reached. Beyond this point the load cannot increase further, even though extension of the specimen is halted at this point. The excess of K over K_R will continue to increase with increasing a. Crack extension is therefore accelerated under the driving force of the excess elastic strain energy of the system. This may be referred to as the ultimate instability point of the test, as distinguished from the meta-instability that occurs at pop-in.

According to Srawley and Brown (Ref 2-19), maintaining the balance between K and K_R after pop-in is not always possible. This means that some specimen may reach its ultimate instability at pop-in. Consequently, some believe that K_{Ic} should always be used in structural sizing in order to avoid the possibility of catastrophic failure owing to a single pop-in. However, using a K_{Ic} value as an across-the-board criterion in characterizing structural component failure may not be practical. So far, pop-in only occurs in laboratory tests. The chance of having this type of failure in a real structure would be slim. In addition, an R-curve with pop-in (type C) is required for making prediction of pop-in failure in a real structure. Such an R-curve for a given material is not always obtained from mechanical testing; it is rather a byproduct of a test. The controlling variable for having (or not having) pop-in during a test is not known. Normally a smooth *R*-curve (type B) is used for structural residual strength prediction. In any event, the policy of using K_{Ic} as a single fracture index is not generally implemented by fracture mechanics analysts today.

The Data Display Technique

It has been shown that the fracture strength for a center-cracked plate is not always represented by a single fracture toughness value. Although a crack growth resistance curve is a useful tool for predicting residual strength, some limitations of using K_c as a fracture index still exist. Taking center-cracked panels as an example, the residual strengths for seven panels of the same material and width but different crack lengths are displayed in Fig. 2-24. The average K_c value is 138.3 ksi$\sqrt{\text{in.}}$ (computed by using Eq 1-21). The corresponding fracture map is plotted in the figure as a solid line. It is seen that this curve fits through most of the individual data points, except one (or maybe two) of the very short crack length. However, the residual strength of a given panel cannot exceed the net section ultimate stress on one hand (for short cracks) and should be zero as $2a/W$ approaches unity. The analytical residual strengths at both ends of the curve (the solid line in Fig. 2-24), which are functions of K_c, do not meet these criteria.

To better represent the realistic behavior, Feddersen has proposed a technique that is based on displaying the test data (Ref 2-21). The technique calls for dividing the data set into three groups. The middle group, which covers the intermediate $2a/W$ ratios, is represented by K_c. Another group correlates with a straight line that connects the F_{ty} on the ordinate and another point that is tangent to the curve of the mid-region. The third group is represented by another straight line that connects the end of the panel (where $2a = W$) to another tangential point on the curve of the middle group. Two straight lines are added to Fig. 2-24 as dotted lines. Some improvement in data correlation is apparent.

Feddersen actually constructed the fracture map in the following way. The curve for the middle group was computed using Eq 1-21 without the width correction factor. The *y*-component of the tangential point for the short crack region was set at two-thirds of F_{ty}. The *x*-component of the tangential point for the long crack region was set at $2a/W = 1/3$. Follow these guidelines; the fracture map of Fig. 2-24 has been reconstructed and replotted (Fig. 2-25). This time, the K_c value is taken to be 140.4 ksi$\sqrt{\text{in.}}$, which is the average of three data points in the middle region that exhibit the linear elastic K_c behavior. The result is quite dramatic. Upon examining a large quantity of test data for various aluminum alloys, Feddersen proclaimed that the three-segment data display technique is a useful engineering tool for residual strength representation and prediction. One of his data plots covering three panel widths (of center-cracked panels made of the 2219 aluminum alloy) is shown in Fig. 2-26 in support of this concept.

REFERENCES

2-1. Metals Test Methods and Analytical Procedures, *Annual Book of ASTM Standards,* Vol 03.01, *Metals—Mechanical Testing; Elevated and Low Temperature Tests; Metallography,* American Society for Testing and Materials, 1995

2-2. J.D. Landes, Fracture Toughness Testing, *ASM Handbook,* Vol 19, *Fatigue and Fracture,* ASM International, 1996, p 393–409

2-3. G.R. Irwin, Fracture Mode Transition of a Crack Traversing a Plate, *J. Basic Eng. (Trans. ASME),* Series D, Vol 82, 1960, p 417–425

2-4. D. Shows, A.F. Liu, and J.H. FitzGerald, Application of Resistance Curves to Crack at a Hole, *Fracture Mechanics: Fourteenth Symposium,* Vol II, *Testing and Applications,* STP 791, American Society for Testing and Materials, 1983, p II-87 to II-100

2-5. D. Broek, *Elementary Engineering Fracture Mechanics,* Noordhoff Publishing, Layden, The Netherlands, 1974, p 101–111

2-6. J.C. Ekvall, T.R. Brussat, A.F. Liu, and M. Creager, "Engineering Criteria and Analysis Methodology for the Appraisal of Fracture Resistant Primary Aircraft Structure," Report AFFDL-TR-72-80, Air Force Flight Dynamics Laboratory, Wright-Patterson Air Force Base, Dayton, OH, 1972

2-7. F.C. Allen, Effect of Thickness on the Fracture Toughness of 7075 Aluminum in the T6 and T73 Conditions, *Damage Tolerance in Aircraft Structures,* STP 486, American Society for Testing and Materials, 1971, p 16–38

2-8. J.C. Ekvall, T.R. Brussat, A.F. Liu, and M. Creager, Preliminary Design of Aircraft Structures to Meet Structural Integrity Requirements, *J. Aircr.,* Vol 11, 1974, p 136–143

2-9. R.G. Forman, V. Shivakumar, and J.C. Newman, Jr., "Fatigue Crack Growth Computer Program NASA/FLAGO Version 2.0," Report JSC-22267A, National Aeronautics and Space Administration, Washington, D.C., May 1994

2-10. C.E. Feddersen and D.P. Moon, A Compilation and Evaluation of Crack Behavior Information on D6AC Steel Plate and Forging Materials for the F-111 Aircraft, Defense Metals Information Center, Battelle Columbus Laboratories, Columbus, OH, 25 June 1971

2-11. G.E. Dieter, Jr., Chapter 14, *Mechanical Metallurgy,* McGraw-Hill, 1961

2-12. Standard Test Method for Notched Bar Impact Testing of Metallic Materials, E 23, *Annual Book of ASTM Standards,* American Society for Testing and Materials, 1995

2-13. Standard Test Method for Conducting Drop-Weight Tests to Determine Nil-Ductility Transition Temperature of Ferritic Steels, E 208, *Annual Book of ASTM Standards,* American Society for Testing and Materials, 1995

2-14. Standard Test Method for Drop-Weight Tear Tests of Ferritic Steels, E 436, *Annual Book of ASTM Standards,* American Society for Testing and Materials, 1995
2-15. A.F. Liu and M. Creager, On the Slow Stable Crack Growth Behavior of Thin Aluminum Sheet, *Mechanical Behavior of Materials,* Vol I, *Deformation and Fracture of Metals,* The Society of Material Science, Japan, 1972, p 558–568
2-16. J.M. Krafft, A.M. Sullivan, and R.W. Boyle, Effect of Dimensions on Fast Fracture Instability of Notched Sheets, *Proc. of the Crack Propagation Symposium,* Vol 1, College of Aeronautics, Cranfield, England, Vol 1, 1961, p 8–26
2-17. R.H. Hyer and D.E. McCabe, Plane-Stress Fracture Toughness Testing Using a Crack-Line-Loaded Specimen, *Eng. Fract. Mech.,* Vol 4, 1972, p 393–412
2-18. R.H. Hyer and D.E. McCabe, Crack Growth Resistance in Plane-Stress Fracture Testing, *Eng. Fract. Mech.,* Vol 4, 1972, p 413–430
2-19. J.E. Srawley and W.F. Brown, Fracture Toughness Testing, *Fracture Toughness Testing and Its Applications,* STP 381, American Society for Testing and Materials, 1965, p 133–198
2-20. A.F. Liu, "Statistical Variation in Fracture Toughness Data of Airframe Materials," *Pro. of the Air Force Conf. on Fatigue and Fracture of Aircraft Structures and Materials,* AFFDL-TR-70-144, Air Force Flight Dynamics Laboratory, Wright-Patterson Air Force Base, Dayton, OH, 1970, p 323–341
2-21. C.E. Feddersen, Evaluation and Prediction of the Residual Strength of Center Cracked Tension Panels, *Damage Tolerance in Aircraft Structures,* STP 486, American Society for Testing and Materials, 1971, p 50–78

SELECTED REFERENCES

- *Damage Tolerance Design Handbook,* University of Dayton Research Institute, 1993
- J.A. Joyce, *Elastic-Plastic Fracture Laboratory Test Procedures,* Manual 27, American Society for Testing and Materials, 1995
- W.E. Krupp and D.W. Hoeppner, Fracture Mechanics Applications in Materials Selection, Fabrication Sequencing and Inspection, *J. Aircr.,* Vol 10, 1973, p 682–688
- *Fatigue and Fracture,* Vol 19, *ASM Handbook,* ASM International, 1996
- *Materials Selection and Design,* Vol 20, *ASM Handbook,* ASM International, 1997
- H. Liebowitz, Ed., *Fracture,* Academia Press, 1971
- A. Nadai, *Theory of Flow and Fracture of Solids,* Vol 1, 2nd ed., McGraw-Hill, 1950
- E.R. Parker, *Brittle Behavior of Engineering Structures,* John Wiley & Sons, 1957
- N. Perrone, H. Liebowitz, D. Mulville, and W. Pilkey, Ed., *Fracture Mechanics,* University Press of Virginia, 1978
- R.L. Tobler and H.I. McHenry, Fracture Mechanics, *Materials at Low Temperatures,* American Society for Metals, 1983, p 269–293

Chapter 3

Fatigue Crack Propagation

Many structural failures are the result of the growth of pre-existing subcritical flaws or cracks to a critical size under fatigue loading. The growth of these flaws will occur at load levels well below the ultimate load that can be sustained by the structure. A quantitative understanding of this behavior is required before the performance of the structure can be evaluated. Information on the crack propagation behavior of metals is needed for selecting the best-performing material, evaluating the safe-life capability of a design, and establishing inspection periods.

Constant-Amplitude Loading

Fatigue crack propagation is a phenomenon in which the crack extends at every applied stress cycle. A clear illustration of this phenomenon is shown in a fractograph (Fig. 3-1) that was taken from the fracture surface of a specimen after the specimen was terminated from a cyclic crack growth test (Ref 3-1). The amount of crack extension due to a stress cycle can be calculated by counting the number of striations and correlating the width of each striation with the cyclic stress profile shown in Fig. 3-1.

In this example, a loading block contained 24 constant-amplitude stress cycles in which three of the cycles had a lower magnitude than the others. A crack growth test was conducted by repeatedly applying this loading block until the crack had grown to a preselected length (or the specimen failed). In each loading block, the biggest jump of crack length, i.e., the largest striation width, corresponded to the stress cycle for which the load was increased from point 4 (the lowest valley) to point 1 (the highest peak). This widest striation is followed by 20 medium-size striations and 3 tiny striations, corresponding to 20 (remaining) higher stress cycles and 3 lower stress cycles, respectively. Therefore, the amount of crack growth (Δa) per stress cycle is simply denoted as da/dN (or $\Delta a/\Delta N$).

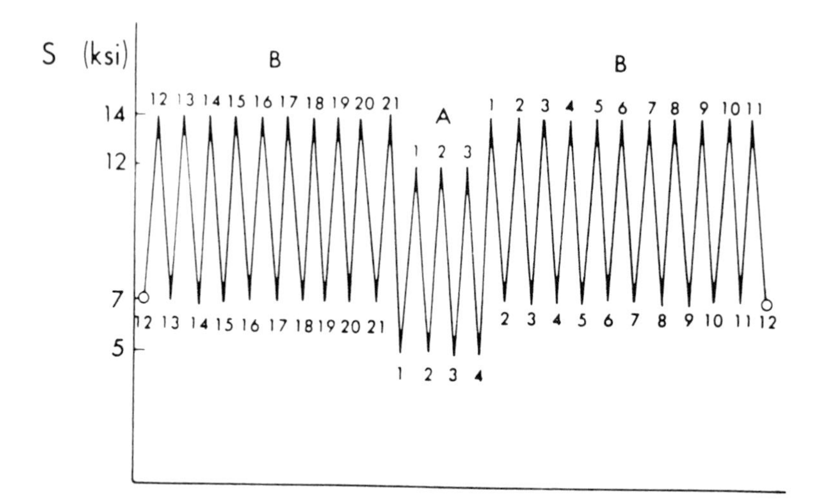

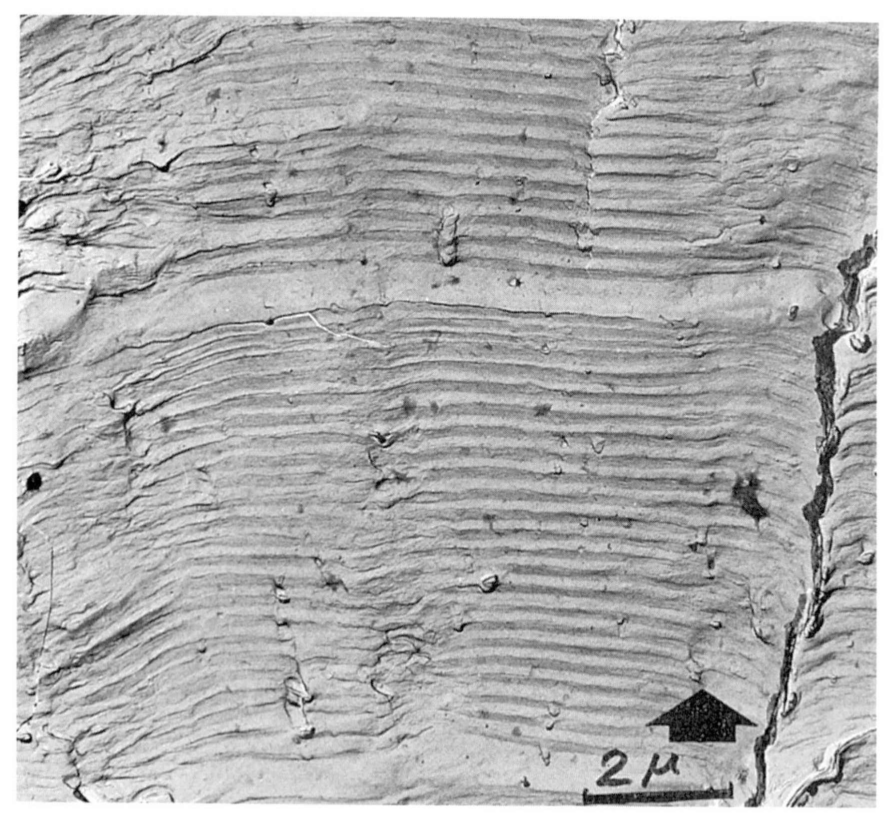

Fig. 3-1 Typical fractograph showing striation markings corresponding to fatigue load cycles. Note the large striation spacing due to the load amplitude A4-B1, preceded by three smaller striations corresponding to the load cycles of block A. The arrow indicates the crack propagation direction. Source: Ref 3-1

In the early 1960s, Paris was the first to demonstrate that each small increment of crack extension is actually governed by the stress intensity at the crack tip (Ref 3-2, 3-3). In other words, he showed that the Irwin theory of fracture is also applicable to incremental subcritical crack growth. The following paragraphs will show that da/dN (for a given loading cycle) is a function of crack tip stress intensity, K.

Data Presentation

The methods for conducting fatigue crack growth testing as well as data reduction procedures are specified in ASTM E 647 (Ref 3-4). An excellent article that provides a concise summary of the techniques is available in *ASM Handbook,* Volume 19 (Ref 3-5). The following presents a simple description of a general procedure for obtaining material crack growth rate data, serving as a stepping stone to the discussion of the fatigue crack growth mechanism.

Fatigue crack growth testing can be conducted on any type of test specimen or structural component. As indicated in Table 2-1, either compact specimens or center-cracked specimens can be used for generating material da/dN data. The test is run under constant-amplitude loading, i.e., the maximum and minimum load levels are kept constant for the duration of the test. Either a sinusoidal or saw-tooth wave form is used for fluctuating the input loads. The data recorded from the test include the testing parameters as well as the crack length as a function of the number of cycles. The crack growth data are usually obtained by visually measuring the crack length and noting the number of load cycles on a counter. Crack gages or compliance techniques can be used in lieu of visual measurement to monitor the progress of crack propagation (Ref 3-4, 3-5). This gives a series of points that describe the crack length as a function of load cycles for a given test.

Crack growth rates are computed on every two consecutive data points. The last point of each pair of the a versus N data points is used as the first point for the following pair of data points. The amount of each crack growth increment is Δa. The difference in the number of fatigue cycles between two consecutive data points is ΔN. Dividing Δa by ΔN is the crack growth rate per cycle.

Stress intensity factors, corresponding to each increment of crack growth, are computed using the average crack length of each pair of consecutive data points. Alternatively, one may prefer to draw a smooth curve through the a versus N data points first, then determine the slope for each selected point on the curve. The slope of a given point on the smooth curve is the *da*/*dN* for that point. The K_{max} and K_{min} values corresponding to each of these crack lengths also can be determined.

It will be explained below that *da*/*dN* is a function of K_{max} and K_{min}. Having this in mind, we let $\Delta K = K_{max} - K_{min}$ and $R = K_{min}/K_{max}$. Then the relation between *da*/*dN* and *K* is expressed as:

$$\frac{da}{dN} = f(\Delta K, R) \qquad \text{(Eq 3-1)}$$

or

$$\frac{da}{dN} = f(K_{max}, R) \qquad \text{(Eq 3-1a)}$$

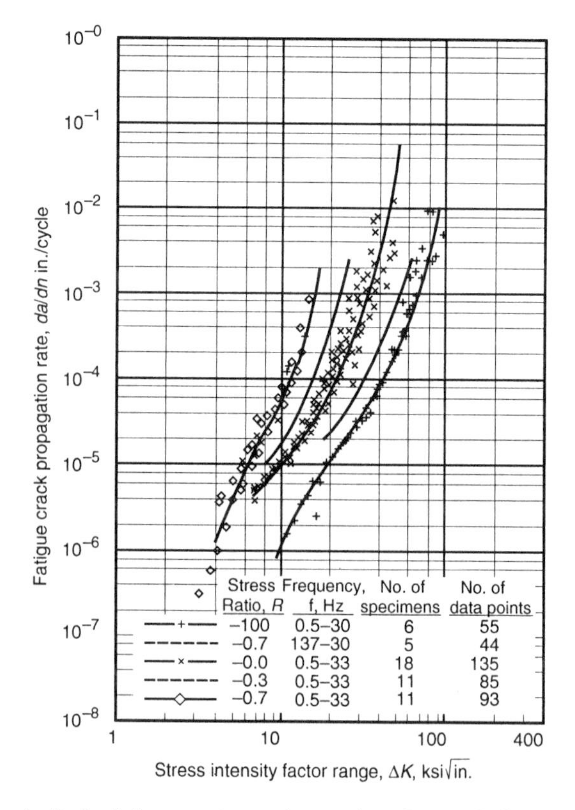

Fig. 3-2 Sample display fatigue crack growth rates data. Source: Ref 3-6

For zero-to-tension loading, i.e., $R = 0$, these two equations are identical. Therefore, da/dN can be plotted as a function of ΔK, or K_{max}. However, it has become a standard practice to use ΔK as the independent variable for data presentation. A set of typical da/dN versus ΔK data (plotted in a log-log scale) for the 7075-T6 alloy, is shown in Fig. 3-2.

To put da/dN and ΔK in an equation, Paris (Ref 3-2, 3-3) described the fatigue crack growth rate data by a power law equation:

$$\frac{da}{dN} = C(\Delta K)^n \qquad \text{(Eq 3-2)}$$

where n is the slope and C is the coefficient at the intercept of a log-log plot. In other words, there is a linear relationship (in log-log scale) between da/dN and ΔK. Since then, no less than fifty similar equations have been suggested by various investigators. A survey paper published in 1981 covers only a small fraction of the population (Ref 3-7). The diversification in crack growth rate description is deemed necessary because each individual investigator faces a situation that is unique to the material and application associated with the product of a particular industry.

Now, define mean stress as the average of the maximum and minimum stress of a given fatigue cycle. When two tests are conducted at different

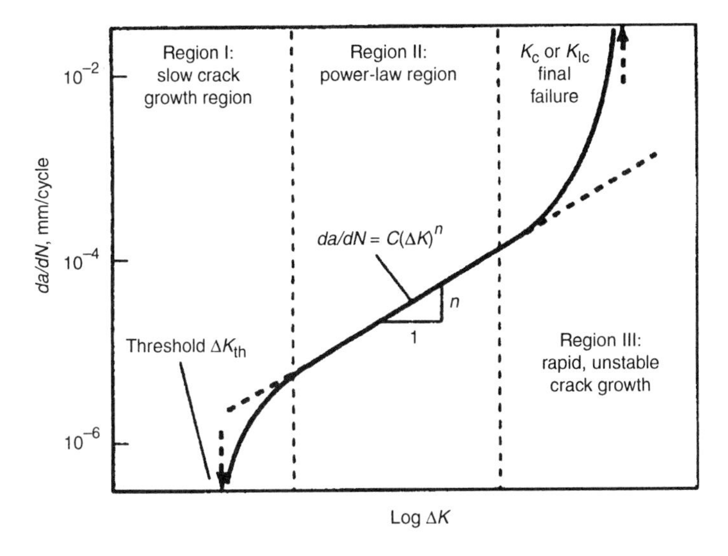

Fig. 3-3 Typical shape of a fatigue crack growth rate curve for a given R-value

mean stress levels (e.g., the tests have the same ΔS but different R values), their cyclic crack growth rates will not be the same. Examining the fractograph of Fig. 3-1, evidently the striation band corresponding to the cycle of lower mean stress is narrower than the striation band for the cycle that has a higher mean stress level. That is, the fatigue crack growth rate for the former is slower than for the latter. This phenomenon is called the mean stress effect, or stress ratio (R-ratio) effect. Using Eq 3-2, one can only plot a da/dN curve for a given R-ratio. Eventually a series of equations is needed to fully describe the da/dN behavior for a wide range of R-ratios. This means that a different pair of C and n values is needed for each equation (for a particular R-ratio). This type of data presentation is called correlating the data *individually* (see Fig. 3-2).

In addition to the mean stress effect, there is another element that inherently associates with the da/dN data and needs to be dealt with. In general, a da/dN curve appears to have three regions: a slow-growing region (the so-called threshold), a linear region (the middle section of the curve), and a terminal region (toward the end of the curve where ΔK approaches K_c). A smooth connection of all three regions forms a sigmoidal curve representative of the entire da/dN versus ΔK curve for a given R (see Fig. 3-3). Although some alloys may not exhibit a clearly defined threshold, i.e., the entire da/dN curve is apparently linear up to the termination point, the existence of these regions has been well recognized.

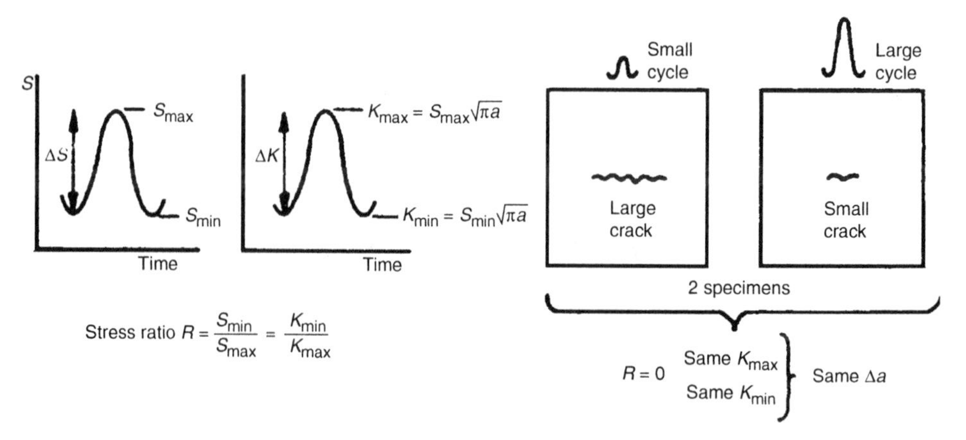

Fig. 3-4 Schematic illustration of K being the driving force of fatigue crack propagation. Source: Ref 3-8

The physical significance of the slow-growing and terminal regions is that there are two obvious limits of ΔK in a da/dN curve. The lower limit (the threshold value ΔK_{th}) implies that cracks will not grow if $\Delta K < \Delta K_{th}$, i.e., $da/dN \to 0$ when $(\Delta K - \Delta K_{th}) \to 0$. On the other hand, if ΔK becomes too high, which implies that K_{max} exceeds the fracture toughness K_c, static failure will follow immediately. This is equivalent to ΔK exceeding $(1 - R)K_c$, i.e., $da/dN \to \infty$ when $[(1 - R)K_c - \Delta K] \to 0$. Consequently, Eq 3-2 should be modified to have a general form:

$$\frac{da}{dN} = C(\Delta K)^n \frac{\Delta K - \Delta K_{th}}{(1 - R)K_c - \Delta K} \qquad \text{(Eq 3-3)}$$

This equation gives a sigmoidal relation on a log-log plot with two vertical asymptotes. The values C and n are the intercept and slope of the line that fit through the linear region of the da/dN curve. One should keep in mind that the $(1 - R)$ term in Eq 3-3 is solely for providing the termination point of the da/dN curve. It does not form any connection between da/dN at different R-ratios. Therefore Eq 3-3 can only be used for correlating the data individually (i.e., with one R-ratio at a time).

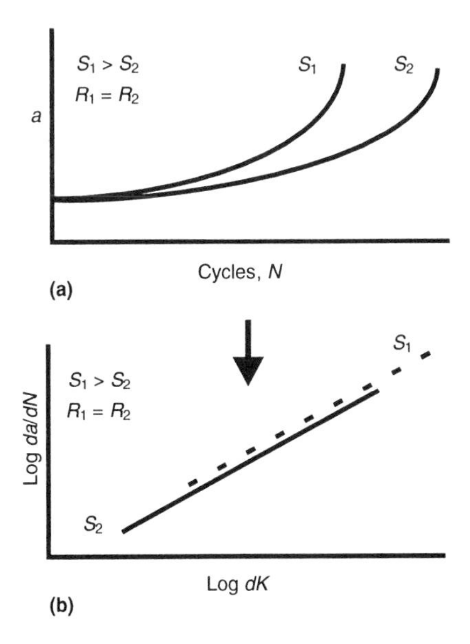

Fig. 3-5 Fatigue crack growth rate correlation technique

In summary, we face two problems in crack growth rate data description. One of them is the need to establish a relationship between all the R-ratios in obtaining a mathematical function that can collapse all the da/dN data points into a common scale. In so doing, only one pair of C and n values is needed for the entire data set. This approach is called correlating the data *collectively*. Another problem is the need to select a curve-fitting equation capable of fitting through all three regions of the da/dN data. Among the many crack growth rate equations published in the literature, some deal with the mathematical formulation of a sigmoidal curve, or normalizing the R-ratio effect, or both. Some even divide the da/dN curve into multiple segments, attempting to obtain a closer fit between experimental data and a set of equations. In the following paragraphs, we will discuss a few techniques that are commonly used today. However, prior to doing that, we ought to demonstrate how and why linear elastic stress intensity, K, is the controlling parameter for fatigue crack propagation.

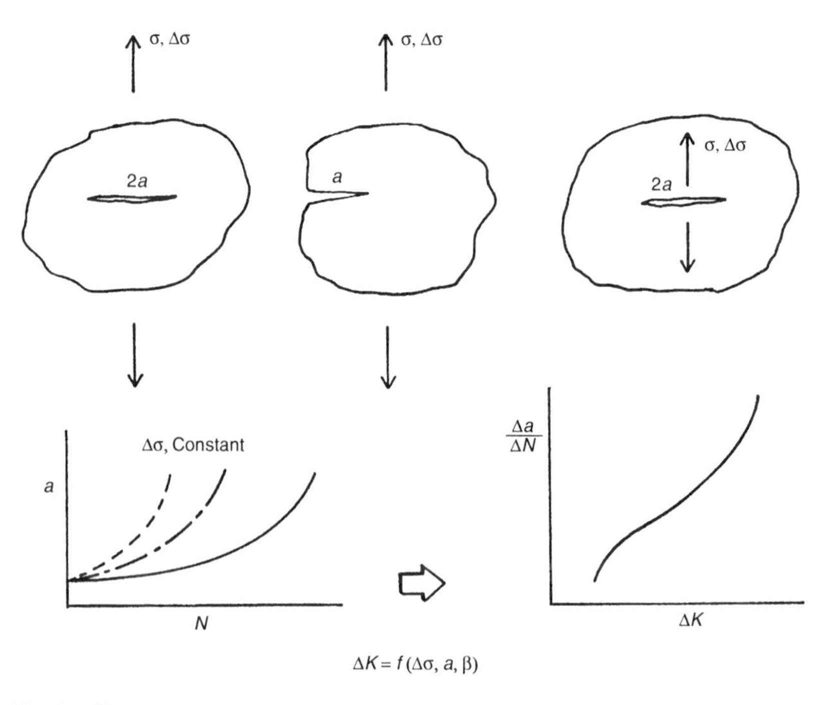

Fig. 3-6 The expanded crack driving force concept

The Driving Force

The mechanics of fatigue crack propagation is similar to that of rapid fracture. From the definition of $K = S\sqrt{\pi a}$ (see Eq 1-22), it follows that a cyclic variation of S will cause a similar cyclic variation of K, i.e., K_{max} and K_{min} (see Fig. 3-4). Following the similarity rule, i.e., comparing two specimens having different crack lengths and applied stresses, one may have a short crack subjected to high stress, whereas another may have a longer crack subjected to a lower stress. If these combinations lead to the same stress intensity level, an equal amount of crack growth should result. In other

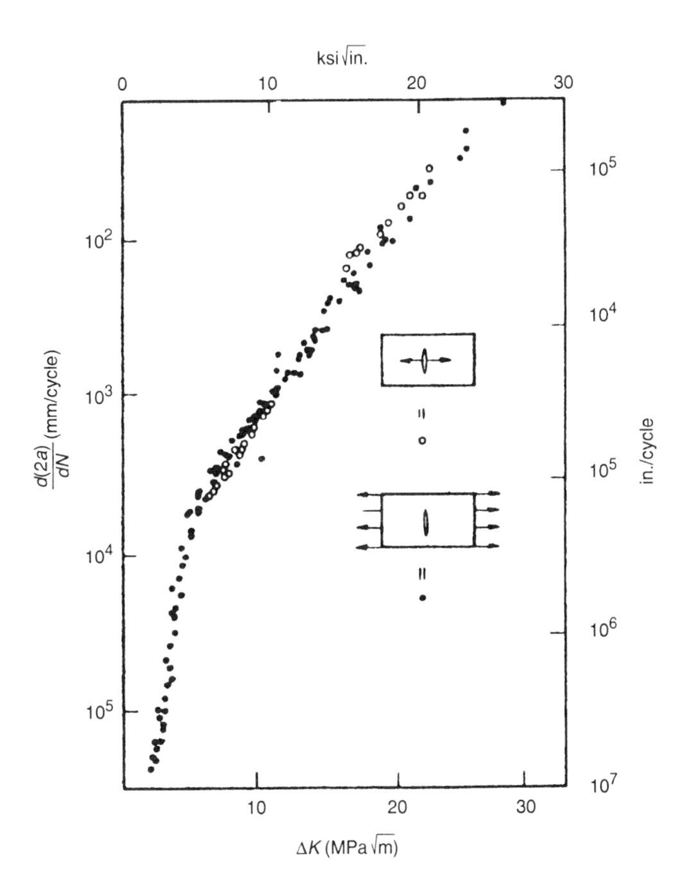

Fig. 3-7 Fatigue crack growth behavior of 7075-T6 aluminum under remote and crack line loading conditions. Source: Ref 3-3

words, the stress intensity at the crack tip controls the rate of fatigue crack growth.

To further clarify this concept, consider two center-cracked panels of the same size, subjected to different maximum sinusoidal stress levels S_1 and S_2 with $S_1 > S_2$ and $R_1 = R_2$. The fatigue crack length versus cycles histories are schematically shown in Fig. 3-5(a). Reduction of these data to *da*/*dN* versus ΔK will result in a graph such as that shown schematically in Fig. 3-5(b). The reason for the shift in these curves shown in Fig. 3-5(b) is self-explanatory in Fig. 3-5(a). Because both tests have the same initial and final crack lengths, a lower *K*-level is expected at the beginning and the end of a test that has a lower applied stress. In the center portion of these *da*/*dN* curves, crack growth rates generated at S_1 and S_2 will coincide, because the condi-

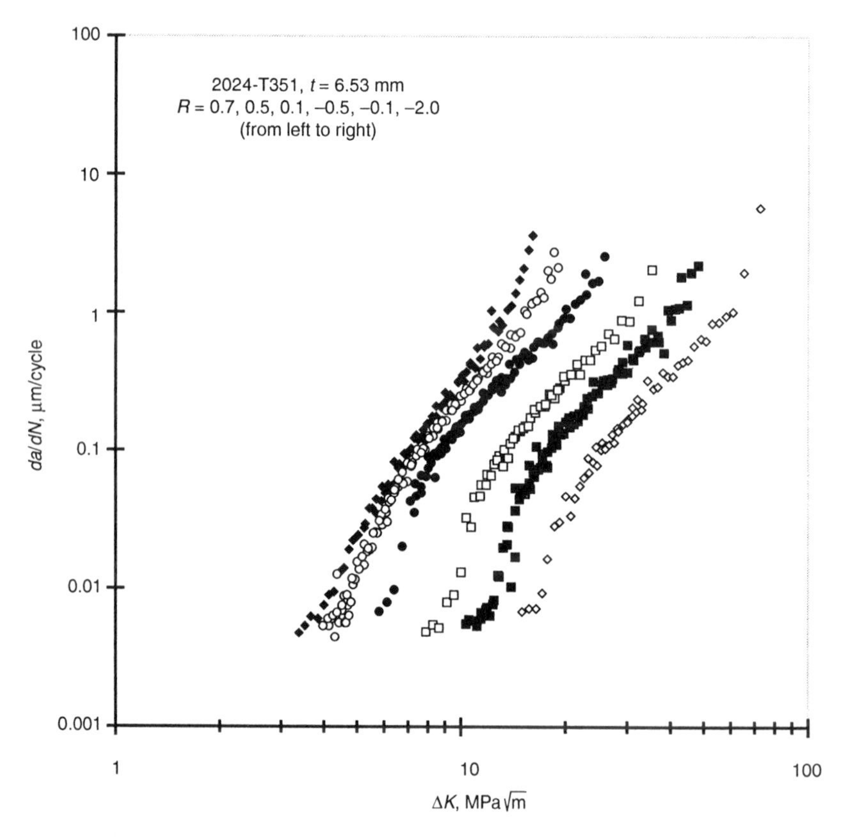

Fig. 3-8 Fatigue crack growth rates data of 2024-T351 aluminum with six *R*-ratios. Source: Ref 3-10

tions at both crack tips are identical. That is, the same maximum stress intensity and fluctuation will produce identical crack growth rates.

A generalization of this concept is further illustrated in Fig. 3-6. The applicability of K is demonstrated by comparing three crack geometry and loading combinations, i.e., a center crack or an edge crack subjected to farfield applied stress, and a center crack loaded by a pair of concentrated forces. Despite the fact that K increases with crack length in farfield loading and decreases with crack length in crack-face loading, there is full agreement among the crack growth rates for all three cases when da/dN is plotted as a function of ΔK (or K_{max}). A material will exhibit the same rate of crack growth under the same K-level, regardless of the loading condition and/or geometry that produced the K-value. In other words, the result of one con-

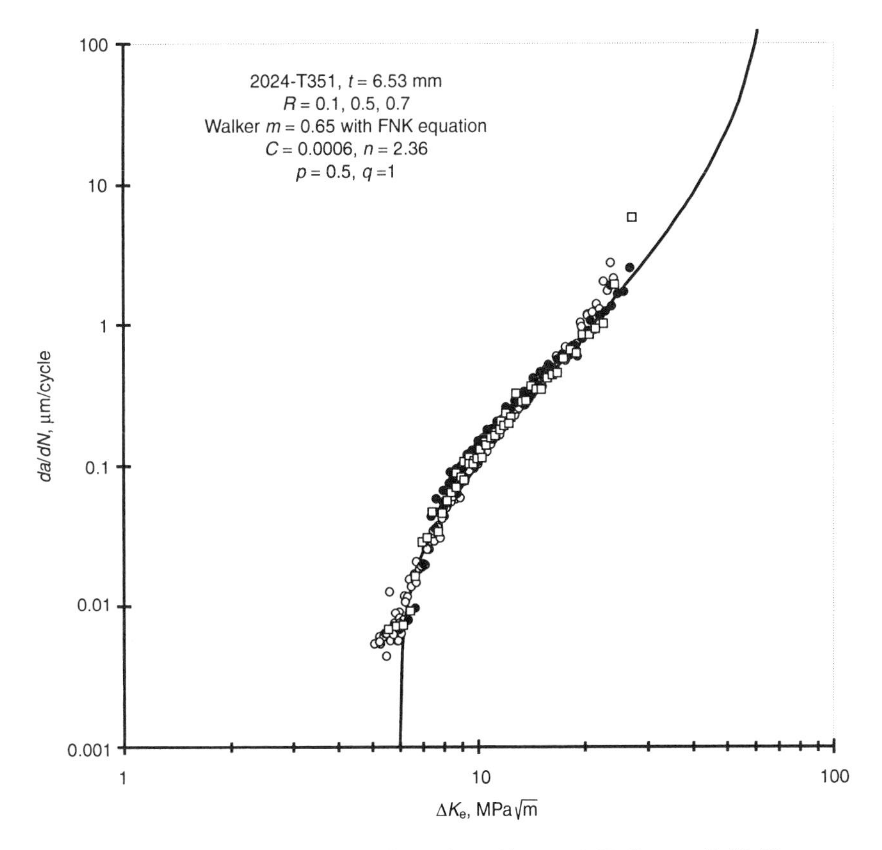

Fig. 3-9 The compressed 2024-T351 data points with $m^{+} = 0.65$. Source: Ref 3-10

figuration can be predicted if the result of another configuration is known, provided that correct K-expressions for both configurations are available. Actual experimental data for validating this concept are presented in Fig. 3-7. Therefore, using a stress intensity factor that correctly represents the geometry and loading under consideration is extremely important in performing crack growth analysis. Stress intensity solutions for some common structural configurations are compiled in Chapter 5.

Effective ΔK

It appears that the actual ΔK acting on a stress cycle does not have the same magnitude as the applied ΔK. The differences between the actual and

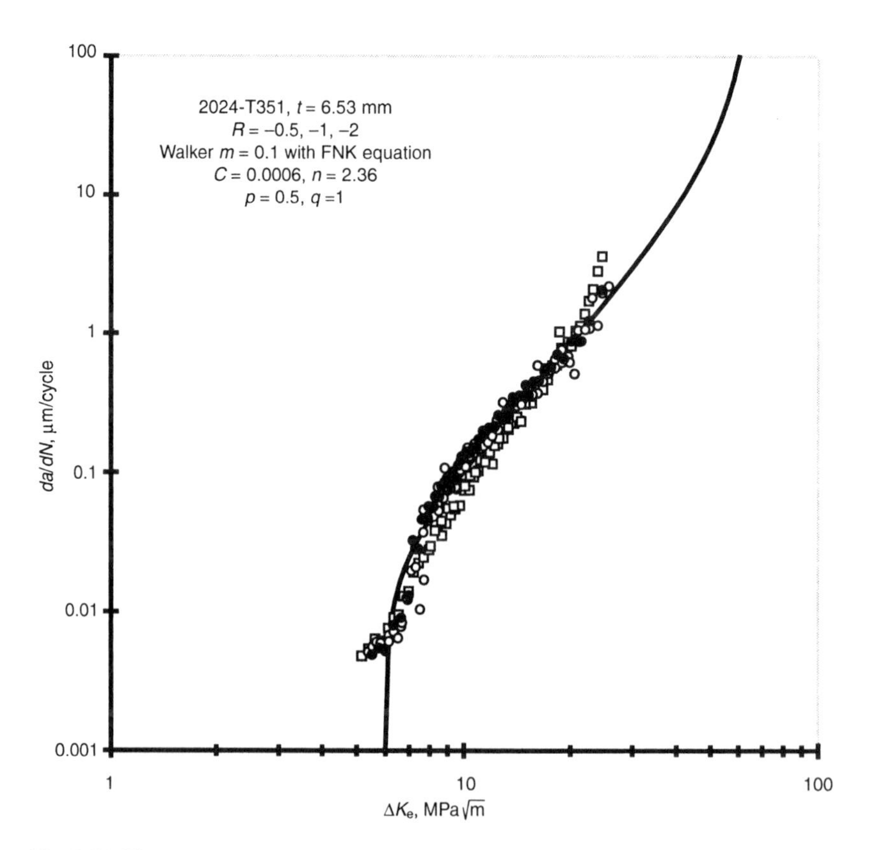

Fig. 3-10 The compressed 2024-T351 data points with m^- = 0.1. Source: Ref 3-10

the effective ΔK vary among different R-ratios. In order to account for the mean stress effect on da/dN, we need to develop a method to estimate the effective level of an applied K for a given R-ratio. Following an extensive survey of the existing methods published in the literature, it seems that the sound approach would be to develop a relationship between *effective* ΔK and R based on the concept of *crack closure.* But first, we will discuss a simple collapse function proposed by Walker (Ref 3-9), because it produces accurate results and is easy to use. Then we will discuss two crack-closure-based effective ΔK functions, those of Liu (Ref 3-10) and Newman (Ref 3-11).

The Walker *m*

The Walker effective ΔK has been defined as:

$$\Delta K_e = S_{max}^{1-m} (\Delta S)^m \sqrt{\pi a} \qquad \text{(Eq 3-4)}$$

where m is an empirical constant. This relation eventually reduces to:

$$\Delta K_e = \Delta K/(1 - R)^{1-m} \qquad \text{(Eq 3-4a)}$$

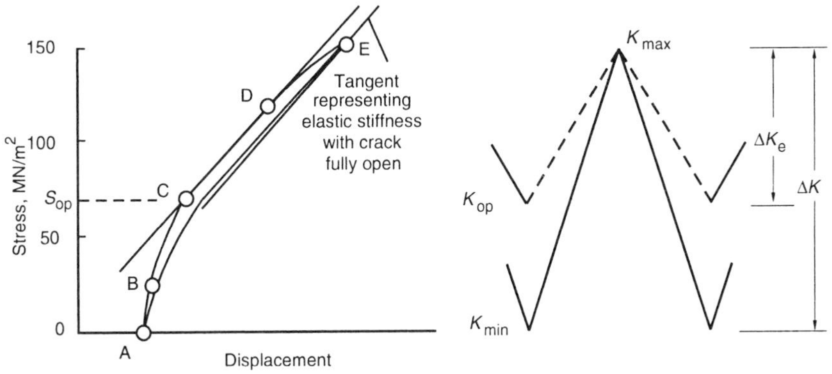

150
100
50
0
S_{op}
Stress, MN/m²
Displacement
A
B
C
D
E
Tangent representing elastic stiffness with crack fully open
K_{max}
K_{op}
K_{min}
ΔK_e
ΔK

$$U = \frac{\Delta K_e}{\Delta K} = \frac{K_{max} - K_{op}}{K_{max} - K_{min}}$$

$$R = K_{min}/K_{max}$$

$$R_e = K_{op}/K_{max}$$

$$U = \frac{1 - R_e}{1 - R}$$

Fig. 3-11 Schematic illustration of crack closure

or

$$\Delta K_e = K_{max} \cdot (1 - R)^m \qquad \text{(Eq 3-4b)}$$

An actual example demonstrating the effectiveness of Eq 3-4(a) is shown in Fig. 3-8 to 3-10. Figure 3-8 is a set of fatigue crack growth rate data for 2024-T351 aluminum. There are two good reasons for choosing this data set as a medium for demonstration. This data set covers a wide range of R-ratios (from –2 to 0.7), and it contains data points of extremely low crack growth rates, below 0.005 μm/cycle (2×10^{-7} in./cycle), needed for establishing a ΔK threshold. In the remainder of this chapter, we will repeatedly use this data set to make correlation with other R-functions and da/dN equations.

Using Eq 3-4(a), the 2024-T351 data set has been compressed into two groups (Fig. 3-9 and 3-10). As shown, the Walker method requires two m values, one for the positive R and another for the negative R. These are $m^+ = 0.65$ and $m^- = 0.10$. In all the equations and figures in this book the applied ΔK is defined as the full ΔK (from valley to peak) of a stress cycle. Therefore, for negative R-ratios the compression part of a stress cycle is not deleted (i.e., R is not set to zero) as is customarily done by some fracture mechanics analysts. It also should be noted that the data points in Fig. 3-9 and 3-10 have been compressed to a common scale, as indicated by the empirical constants of the fitted curves.

The Walker ΔK_e is usually used along with the Paris equation, i.e.,

$$\frac{da}{dN} = C(\Delta K_e)^n \qquad \text{(Eq 3-5)}$$

However, the line that fits through the compressed data points in Fig. 3-9 and 3-10 has been obtained by using a sigmoidal function, which will be discussed below in the section on curve plotting. The empirical constants used for curve fitting are labeled in the figures. Excepting the Walker m, all the other constants are common to both groups of data.

Effective ΔK as a Function of Crack Closure

The concept of crack closure (Ref 3-12) considers that material at the crack tip is plastically deformed during fatigue crack propagation. During unloading of a stress cycle, some contact between the crack surfaces will occur due to the constraint of surrounding elastic material. The crack surfaces will not open immediately at the start of the following cycle; instead, they remain closed for some time during the cycle. The hypothesis goes further and states that crack propagation takes place only at the raising portion of a stress cycle, and that the crack is unable to propagate while it remains closed. This has led to the assumption that the net effect of closure is to reduce the apparent ΔK to some effective level ΔK_e ($K_{max} - K_{op}$). In other words, K_{op} is the minimum stress intensity level of that cycle required to reopen the crack, and its magnitude is usually higher than the applied K_{min} (Fig. 3-11). In Ref 3-12, Elber defines the effective stress intensity range as:

$$\Delta K_e = U \cdot \Delta K \tag{Eq 3-6}$$

where

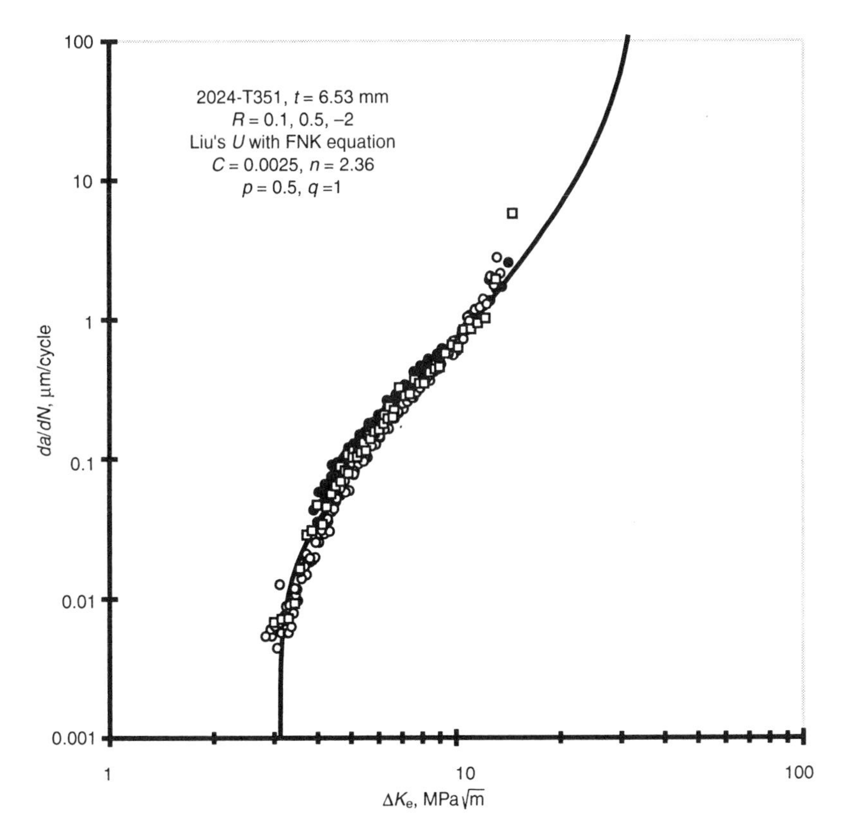

Fig. 3-12 The compressed 2024-T351 data points with $U_0 = 0.12$ and $\beta = 0.45$, part 1. Source: Ref 3-10

$$U = \frac{K_{max} - K_{op}}{K_{max} - K_{min}} \qquad \text{(Eq 3-7)}$$

Using a set of crack growth rate data for the 2024-T3 aluminum, which contains R-values between –0.1 and 0.7, Elber showed that U is a linear function of R. Numerous investigators have since sought to develop experimental techniques to measure crack opening loads for a variety of materials or to investigate the physical mechanism of crack closure.

The experimental aspects of K_{op} measurements (or S_{op}, the stress counterpart of K_{op}) are quite complicated. Investigators often disagree among themselves on issues regarding fundamental techniques as well as the definitions

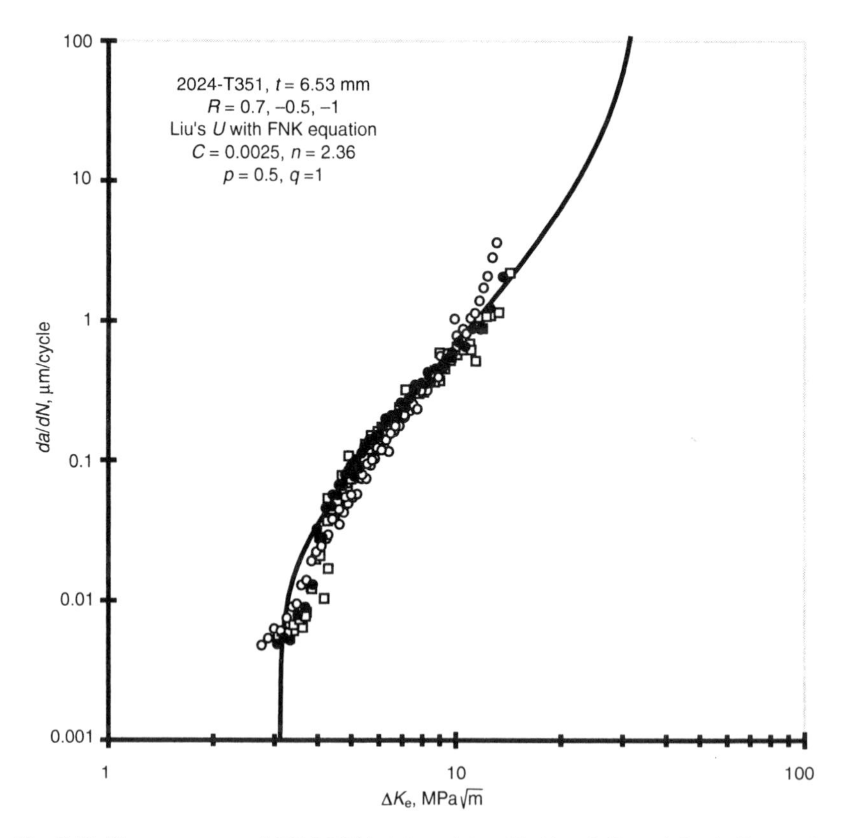

Fig. 3-13 The compressed 2024-T351 data points with $U_0 = 0.12$ and $\beta = 0.45$, part 2. Source: Ref 3-10

for *crack opening load* and *closure load.* Discussion of these is beyond the scope of this book. For information about experimental aspects of fatigue crack closure, a review paper (Ref 3-13) may be helpful. Inconsistencies in measured K_{op} values have frequently been reported in the literature. References 3-14 through 3-16 point out that K_{op} is dependent on material thickness, R-ratio, K_{max} level, crack length, and environment (including frequency and temperature). Test procedure and the method for interpreting crack opening load from the load trace record of a test also are important variables. It has been reported that for a given crack length, K_{op} changes from the surface toward the mid-thickness of the test coupon. The value of K_{op} is also dependent on the smoothness or flatness of the crack faces, which is often associated with small crack sizes or low stress intensity levels, as well as the formation of shear lips at longer crack

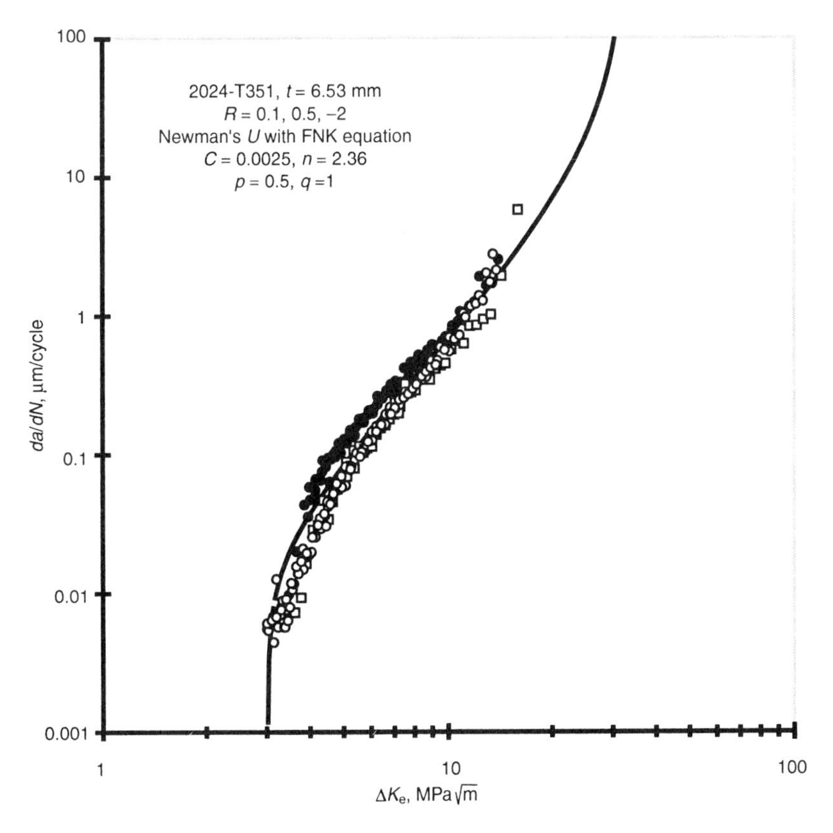

Fig. 3-14 The compressed 2024-T351 data points with $\alpha = 1.0$ and $S_{max}/S_0 = 0.23$, part 1

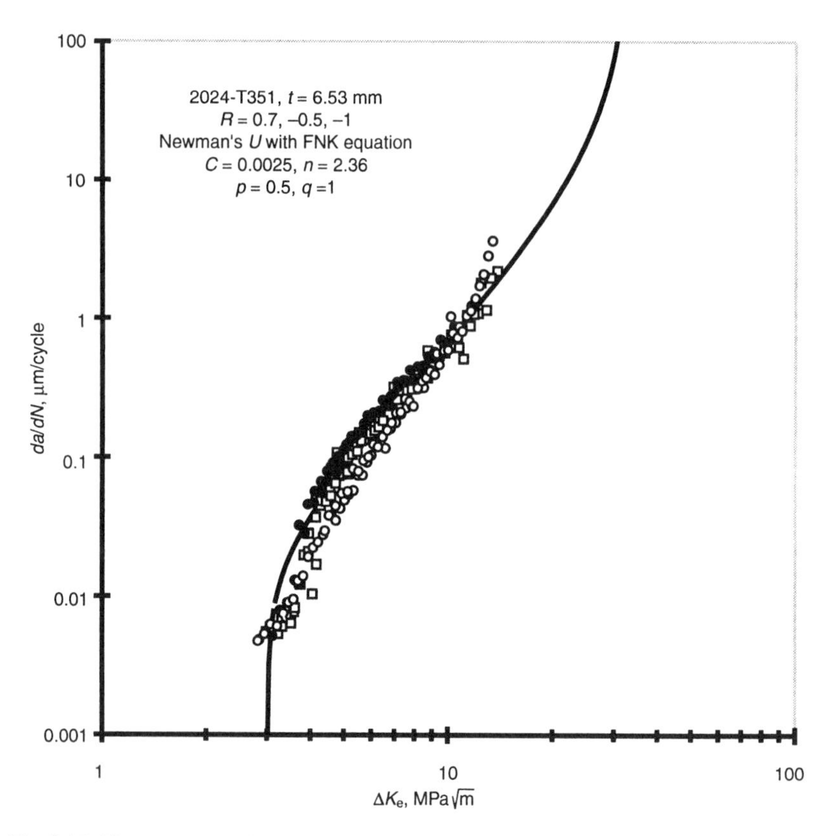

Fig. 3-15 The compressed 2024-T351 data points with α = 1.0 and S_{max}/S_0 = 0.23, part 2

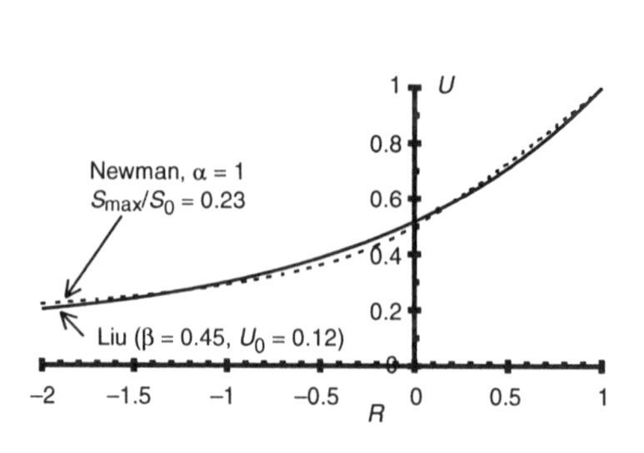

Fig. 3-16 Effective stress intensity range ratio as a function of R

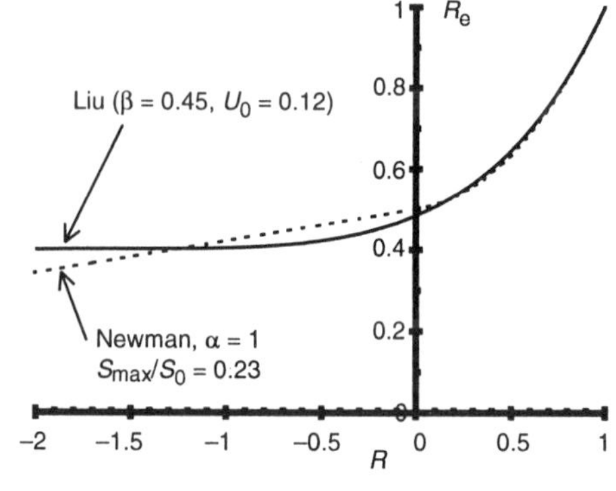

Fig. 3-17 Effective R-ratio as a function of R

lengths. Therefore, it would be impossible to develop a normalizing function for ΔK_e that correlates engineering crack growth rate data from experimental crack closure data. The challenge to a fracture mechanics analyst is to select an analytical function for U as a function of R and then pair it with a crack growth rate equation to obtain correlations with reasonable degrees of accuracy and consistency. It is desirable to have a U-function that can fit to data sets that cover a wide range of R-ratios, i.e., the R-ratios extending from an approximate unity to an extreme negative value. The following discusses the criteria for setting up such a U-function.

Examining Eq 3-6 and 3-7, it is clear that for a given R-ratio the value of U should be between 0 and 1, i.e., $0 < U \leq 1$. Physically speaking, $U = 1$ implies that the crack is fully opened (i.e., $K_{op} = K_{min}$), whereas $U = 0$ would mean that the crack is fully closed (i.e., $K_{op} = K_{max}$). For a given stress cycle, one may define the effective R-ratio as:

$$R_e = K_{op}/K_{max} \qquad \text{(Eq 3-8)}$$

where $K_{op} \geq K_{min}$. Because $R = K_{min}/K_{max}$, therefore $R_e \geq R$. In addition, the finite element analysis results of Newman (Ref 3-17) have shown that R_e should decrease with R. It is also considered that R_e can be a negative value, because K_{op} can be a negative value as long as it is above or equal to K_{min}. Therefore K_{op} (or R_e) should not be set to zero for compression-tension stress cycles.

Because both Eq 3-7 and 3-8 contain a K_{op} term, there is a unique relationship between U and R_e. Dividing both the denominator and the numerator of Eq 3-7 by K_{max}, one may obtain:

$$U = (1 - R_e)/(1 - R) \qquad \text{(Eq 3-9)}$$

Therefore, $R \leq R_e < 1$ because $0 < U \leq 1$.

Comparison of Two *U*-Functions and Correlation with Test Data

To obtain a U-function that covers a full range of R-ratios, Liu (Ref 3-10) postulated that U might reach an asymptotic limit U_0 when R approaches $-\infty$. His expression for U is denoted here as:

$$U = U_0 + (1 - U_0) \cdot \beta^{(1-R)} \qquad \text{(Eq 3-10)}$$

where β and U_0 are empirical constants whose values depend on material, thickness, and environment. This equation satisfies the above-stated criteria and leads to $U \to U_0$ for $R \to -\infty$, and $U \to 1$ for $R \to 1$. By definition, U_0 is greater than zero. By trial and error, a pair of β and U_0 values can be determined for a given data set.

As a tip for implementing Eq 3-10 to compress a set of *da/dN* data, it is recommended that U_0 be taken as a small value somewhere in the vicinity of 0.1. It is conceivable that U is in the neighborhood of 0.5 for $R = 0$. Therefore an initial estimate of β can be easily obtained by using Eq 3-10 with $U = 0.5$ and $R = 0$. Visual assessment to judge the fairness of fit can be made by displaying the normalized data points on the monitor of a personal computer. A final pair of U_0 and β values that provide the best fit can be easily reached by making a few iterations.

To demonstrate the usefulness of this equation, the 2024-T351 data set (previously shown in Fig. 3-8) has been normalized by using $\beta = 0.45$ and $U_0 = 0.12$. For purposes of clarity, the normalized data points have been arbitrarily divided into two groups, randomly mixed with positive and negative R, and are presented in Fig. 3-12 and 3-13. It can be seen that Eq 3-10 has effectively compressed the entire data set of six R-ratios into a common scale. All data points displayed in both figures are fitted by a common *da/dN* curve (having the same *da/dN* constants). The method used to plot the curve will be discussed below in the curve plotting section. For now, it is enough to demonstrate that the entire data set (covering R-ratios ranging from –2 to 0.7) has been normalized by a single pair of empirical constants ($\beta = 0.45$ and $U_0 = 0.12$).

Newman's approach to this problem was to derive an equation for R_e and then use Eq 3-9 to obtain U. The Newman equation for R_e is defined as:

$$R_e = A_0 + A_1 R + A_2 R^2 + A_3 R^3 \qquad \text{(Eq 3-11a)}$$

for $R \geq 0$, and

$$R_e = A_0 + A_1 R \qquad \text{(Eq 3-11b)}$$

for $-1 \leq R < 0$. The polynomial coefficients are defined as:

$$A_0 = (0.825 - 0.34\alpha + 0.05\alpha^2) \cdot \{\cos[(\pi/2) \cdot (S_{max}/S_0)]\}^{1/\alpha} \qquad \text{(Eq 3-12a)}$$

$$A_1 = (0.415 - 0.071\alpha) \cdot S_{max}/S_0 \qquad \text{(Eq 3-12b)}$$

$$A_2 = 1 - A_0 - A_1 - A_3 \qquad \text{(Eq 3-12c)}$$

$$A_3 = 2A_0 + A_1 - 1 \qquad \text{(Eq 3-12d)}$$

where α is a material constant, its value varying from 1 for plane stress to 3 for plane strain; S_{max} is the peak stress of a stress cycle, and S_0 is supposedly the material flow stress (i.e., the average of the tensile yield and ultimate strengths). In practice, Forman treated both α and S_{max}/S_0 as fitting constants (Ref 3-18). By letting $\alpha = 1.0$ and $S_{max}/S_0 = 0.23$, the 2024-T351 data can be compressed, with the results presented in Fig. 3-14 and 3-15. Again, curve fitting is discussed below.

To validate a given set of empirical constants against the criteria for U and R_e, simply compute U and R_e using the empirical constants in question and display the results on a graph. To illustrate this with the 2024-T351 data set, the graphical solution of Eq 3-10 (with $\beta = 0.45$ and $U_0 = 0.12$) is plotted as the solid line in Fig. 3-16. The dotted line is the solution for the Newman U with $\alpha = 1.0$ and $S_{max}/S_0 = 0.23$. The corresponding R_e values are plotted in Fig. 3-17. The abilities of these empirical constants to meet the criteria are validated by the trends displayed in both figures. Despite the differences in these two approaches, there is a very close agreement between their U and R_e values. Although Newman has set a limit of application for his R_e at $R \geq -1$, good correlation with test data in a range outside this limit (down to $R = -2$) has been obtained. In fact, in the NASA/FLAGRO computer program, Forman did lower the limit for Eq 3-11(b) to –2 (Ref 3-18).

Crack Growth Rate Equations

The second element in a crack growth rate equation is the mathematical function that describes the characteristics of the da/dN data over all three regions. The plot of the equation should display a close fit to the already normalized data points. As mentioned earlier, the da/dN versus ΔK (or ΔK_e) data usually form a sigmoidal curve. Some material may have a longer linear segment than the other, and some material may not have an identifiable threshold value. In the following paragraphs, both sigmoidal and nonsigmoidal descriptions will be discussed. However, only one equation will be chosen from each category for discussion.

The FNK Equation

To plot a sigmoidal curve through a normalized (da/dN versus ΔK_e) data set, simply use an equation that possesses the same format as Eq 3-3. Among the many sigmoidal equations available in the literature, the FNK equation is considered the most suitable for the present purpose. Here,

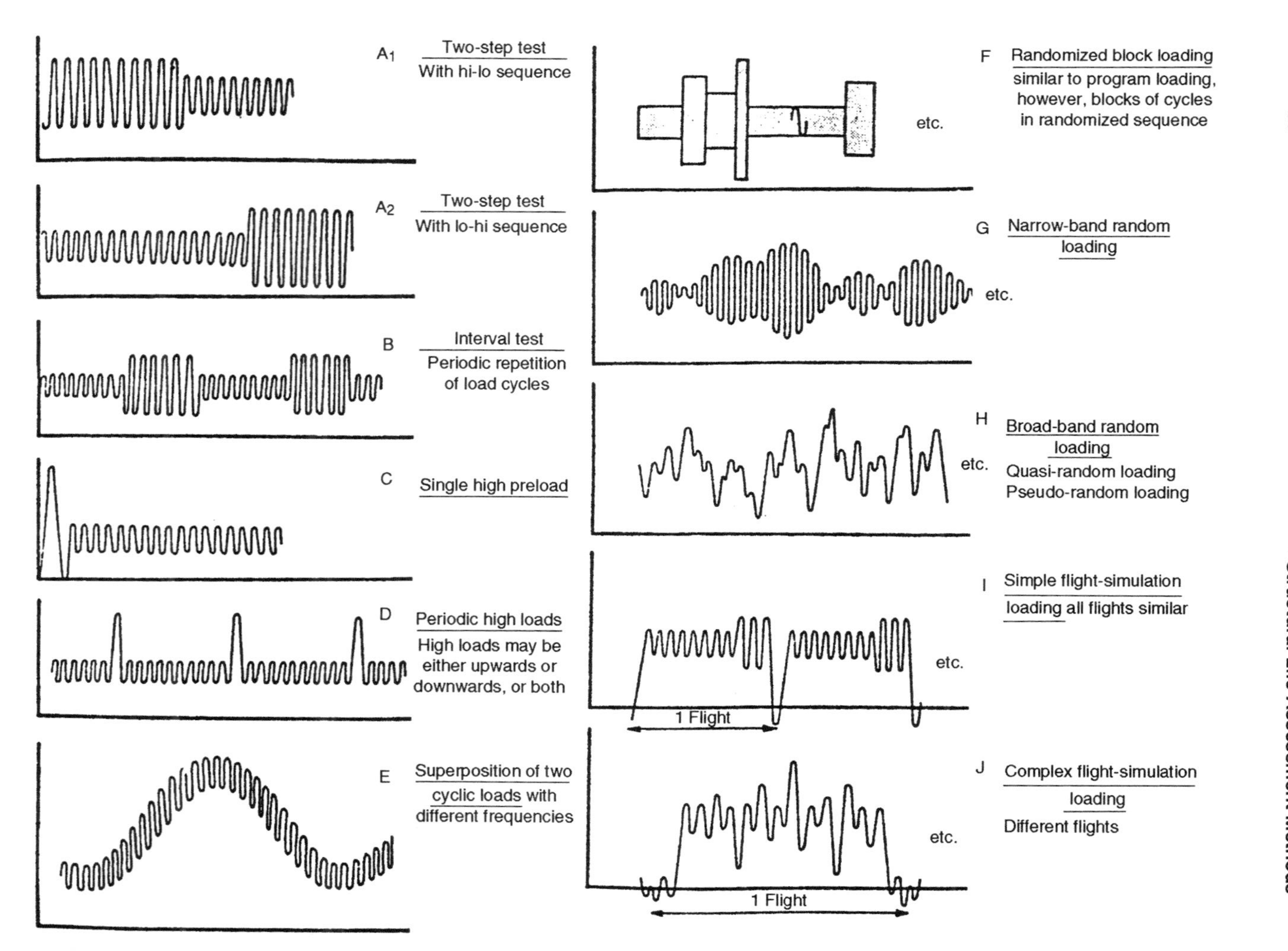
A1
Two-step test
With hi-lo sequence
A2
Two-step test
With lo-hi sequence
B
Interval test
Periodic repetition of load cycles
C
Single high preload
D
Periodic high loads
High loads may be either upwards or downwards, or both
E
Superposition of two cyclic loads with different frequencies
F
Randomized block loading
similar to program loading, however, blocks of cycles in randomized sequence
etc.
G
Narrow-band random loading
etc.
H
Broad-band random loading
Quasi-random loading
Pseudo-random loading
etc.
I
Simple flight-simulation loading all flights similar
etc.
1 Flight
J
Complex flight-simulation loading
Different flights
etc.
1 Flight

"FNK" stands for Forman, Newman, and de Konig. The equation is used in the most recent release of the crack growth prediction program, NASA/FLAGRO (Ref 3-18). Similar to Eq 3-3, the equation is written as:

$$\frac{da}{dN} = C\left(\frac{1-f}{1-R}\Delta K\right)^n \cdot \frac{[1-(\Delta K_{th}/\Delta K)]^p}{[1-(\Delta K/(1-R)K_c)]^q} \quad \text{(Eq 3-13)}$$

where f is Newman's R_e given in Eq 3-11(a) and 3-11(b), and ΔK_{th} is the threshold value of ΔK for a given R. Because $U = (1-f)/(1-R)$, and $\Delta K_e = U \cdot \Delta K$, Eq 3-13 can be rewritten as:

$$\frac{da}{dN} = C(\Delta K_e)^n \cdot \frac{[1-(U \cdot \Delta K_{th}/\Delta K_e)]^p}{[1-(\Delta K_e/U \cdot (1-R)K_c)]^q} \quad \text{(Eq 3-13a)}$$

Thus, this equation consists of two major elements, the ΔK_e term for handling the R-ratio effect and the p and q exponents for shaping the *da*/*dN* curve. In the following paragraphs, we will demonstrate that any ΔK_e function can be placed in this equation to achieve the same results.

We have shown three types of effective ΔK functions: the Walker m and the U-functions of Liu and Newman. In the paragraphs that follow, we will demonstrate how each of these functions can work together with Eq 3-13(a) to be used in plotting a *da*/*dN* curve.

To make use of Eq 3-13(a), one should start with using an effective ΔK function to normalize all the *da*/*dN* data points in a given data set. The chosen effective ΔK function can be the Walker m or any U as appropriate. Once this process is completed, i.e., after all the constants and coefficients associated with ΔK_e are determined and the normalized data points are

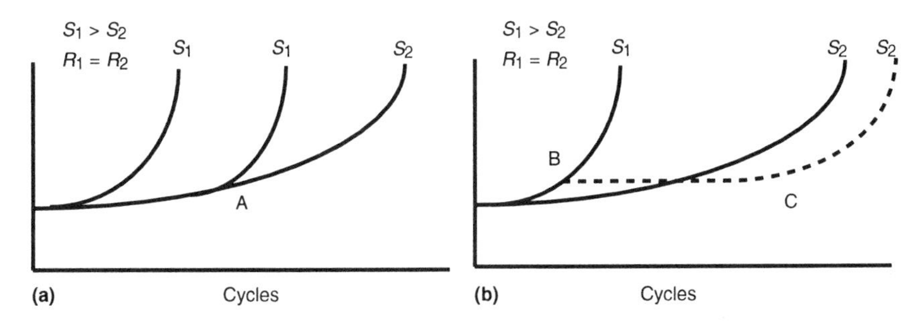

Fig. 3-19 Effect of applied stress on fatigue crack propagation

satisfactorily displayed in a log-log plot, ΔK_e can be regarded as an independent variable in Eq 3-13(a).

To make use of Eq 3-13(a) and treat ΔK_e as an independent variable, the easiest way is to let $R = 0$, so that Eq 3-13(a) becomes:

$$\frac{da}{dN} = C(\Delta K_e)^n \cdot \frac{[1 - (\bar{U} \cdot \Delta K_0/\Delta K_e)]^p}{[1 - (\Delta K_e/\bar{U} \cdot K_c)]^q} \qquad \text{(Eq 3-13b)}$$

where ΔK_0 is ΔK_{th} at $R = 0$ and $\bar{U}$ is U at $R = 0$. The quantities $\bar{U} \cdot \Delta K_0$ and $\bar{U} \cdot K_c$ can be regarded as the effective values of ΔK_0 and K_c, respectively. When ΔK_e is a function of the Walker m, we can treat $1/(1 - R)^{1-m}$ as U, which results in $\bar{U} = 1$. When ΔK_e is a function of Liu's U, $\bar{U} = [U_0 + (1 - U_0) \cdot \beta]$. When ΔK_e is a function of Newman's U, $\bar{U} = 1 - A_0$, and A_0 is given by Eq 3-12(a).

Now, we can use Eq 3-13(b) to plot those curves shown in Fig. 3-9, 3-10, and 3-12 to 3-15. First, we adopt the recommended values for p and q, i.e., $p = 0.5$ and $q = 1.0$, for the 2024-T351 aluminum (Ref 3-18). Next, we select a pair of values for ΔK_0 and K_c, that is, $\Delta K_0 = 6.04$ MPa$\sqrt{\text{m}}$ and $K_c = 65.93$ MPa$\sqrt{\text{m}}$ (Ref 3-10). The curves in all six figures are plotted by using these coefficients. In a case when Liu or Newman's U is used, the curve is plotted with $C = 0.0025$ and $n = 2.36$. Comparing the graphs in these figures, one may find that the test data points in Fig. 3-9 and 3-10, which were normalized by the Walker m, are off to the right as compared with the data points in the other four figures. This is just because the Walker m does not operate the same way as U. The Walker m normalizes the data from both ends of R toward $R = 0$. The U-function normalizes all R-ratios toward $R = 1$. Therefore, a different C value is needed to plot through the data points properly. The data points in Fig. 3-9 and 3-10 are fitted with $C = 0.0006$ instead of 0.0025.

At this point, it is important to note that not all of the effective ΔK and crack growth rate coefficients used in this exercise will provide a perfect fit for this batch of material, and certainly they are not intended to be used as material allowables. They are used here to demonstrate how to work with these crack growth rate equations.

The Three-Component Model

For crack growth rate data that do not exhibit a sigmoidal shape or do not have a clearly defined threshold, Sexena and Hudak (Ref 3-19) implemented an alternate method to plot a curve through the entire ΔK range of the da/dN data. The equation:

$$\frac{1}{(da/dN)} = \frac{A_1}{\Delta K^{n_1}} + A_2 \left\{ \frac{1}{\Delta K^{n_2}} - \frac{1}{[(1-R)K_c]^{n_2}} \right\} \quad \text{(Eq 3-14)}$$

serves to connect all three segments of a da/dN curve into one. The first term on the right-hand side of the equation corresponds to the slow-growing region (without a threshold). The second and third terms correspond to the linear and terminal regions, respectively. This method also can simultaneously take care of the R-ratio effect if the intercept term (instead of the ΔK term) is adjusted in the da/dN equation. It was stated above that the slopes n_1 and n_2 are independent of R-ratios, but the intercepts (denoted here as A_1 and A_2) are dependent on R. The method starts with determining the coefficients of the intercepts for each individual R, that is, a different pair of A_1 and A_2 is determined for each R. By using a linear equation to connect these coefficients, a da/dN data set containing several R-ratios can then be fitted collectively for all R-ratios. The drawback of this approach is the lack of any fixed functional form that can relate A_1 and A_2 to R.

Variable-Amplitude Loading

Structural components are often subjected to variable-amplitude loading. The load history during the lifetime of a given structural component consists of various load patterns as a result of the in-service applications. The sequence of the high, low, and compression ($S_{min} < 0$) loads in a load history is

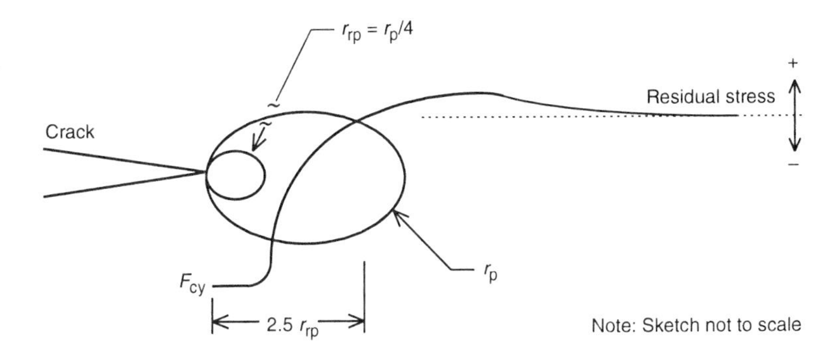

Fig. 3-20 Schematic comparison of the monotonic and cyclic plastic zones

one of the key elements that controls crack growth behavior. Figure 3-18 shows several typical loading patterns that are often used for developing experimental crack growth data as a means to study crack growth behavior.

A simple illustration of the load sequence effect on crack growth is schematically presented in Fig. 3-19. The crack growth histories shown here are supposedly developed from identical panels. Both tests start at the same arbitrary crack length, but one of the panels is subjected to a series of higher constant-amplitude cyclic stress levels than the other panel. The crack propagation curves for these two panels are shown as solid lines. If a third panel starts at the lower of the two stress levels (point A) and then the stress is increased, the rate of fatigue crack propagation will increase to the expected rate for the higher stress at that crack length, as indicated by the dashed line in Fig. 3-19(a).

However, if the sequence of loading is reversed, that is, if the stress level is changed from the higher to the lower stress at point B in Fig. 3-19(b), the new rate of crack propagation is not what would be expected for the lower stress for that crack length. Actually, the crack remains dormant for a period of time. However, after a number of load cycles (point C) the crack will

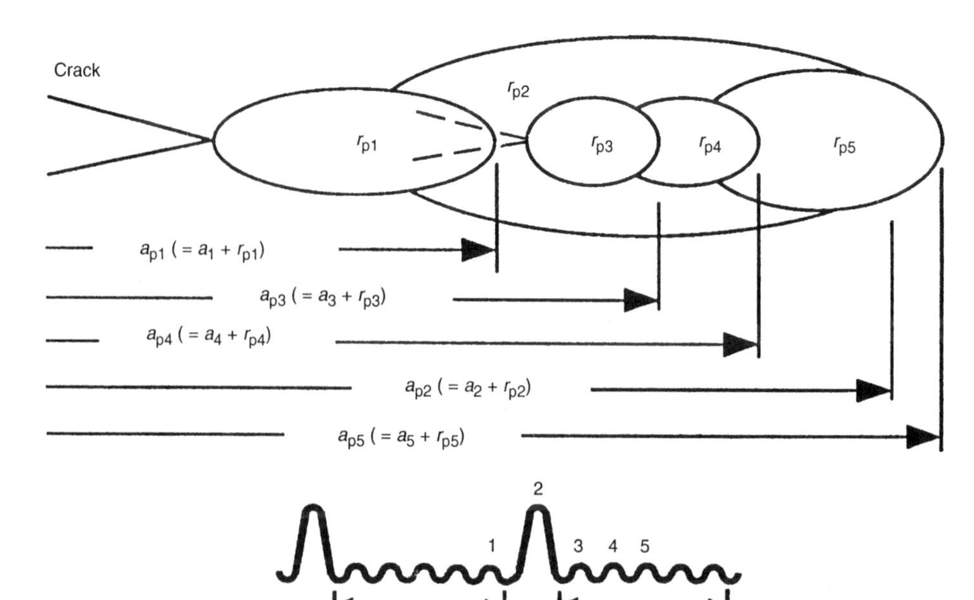

Fig. 3-21 Schematic definition of load interaction zones

resume its normal growth behavior as if the change of load levels had never occurred during the test. For a given material, the amount of time that crack propagation remains inactive depends on the level of the higher stress and the difference between the two stress levels. It can be said that the action of normal crack propagation behavior for the lower loads is being deferred by the presence of the high loads. The number of load cycles spent on this deferring action is called the number of *delay cycles*. The overall phenomenon of this event is called *delay,* better known today as *crack growth retardation.* As will be shown later in this chapter, retardation is not the only matter of concern in prediction of spectrum crack growth lives. Under certain circumstances the magnitude of retardation might be minimized, and accelerated crack growth rates have also been observed.

Approaches for Predicting Spectrum Crack Growth

Three general methods are used for predicting fatigue crack growth histories of spectrum loaded structures. One method utilizes cycle-by-cycle accounting procedures to determine the advance of a crack for each applied load cycle (Ref 3-21 to 3-24). This approach takes into account the load interaction between each pair of consecutive load cycles and its effect on the crack growth behavior of the subsequent load cycles.

The second method characterizes the spectrum with statistically-based load variables. Analysts normally replace service loading histories with equivalent constant-amplitude stress histories prior to performing life prediction (Ref 3-25 to 3-27). The applicability of this method is limited to load spectra that impose insignificant effects on load sequence interactions.

The third method groups the entire spectrum into several mini-blocks. This approach can be considered the combination of the first and second procedures. The approach hinges on identifying the statistically repeating stress group that approximates the loading and sequence effects for the complete spectrum. Service load-spectra, once derived (or measured), can be described as an ordered sequence of stress events. They normally have a few identifiable repetitions within a certain period. However, most of the stress events can only be regarded as random-variable stress events having a defined probability of occurrence between the repeatable events. The repeatable stress events in a service spectrum can be used to separate the spectrum into a series of consecutive stress mini-blocks. This method is limited to short-spectra loading histories. It is not intended for long-spectra loading in which a high overload cycle is followed by a huge number of lower load cycles. The core element of this method is to compute the crack growth per loading block instead of calculating crack growth by summing growth for each stress cycle. Therefore, very significant saving in computer time is realized. Demonstration of the step-by-step details of this method is out of the scope of this chapter and is left to the open literature (Ref 3-28 to 3-30).

In the following sections, several of the most often used crack growth prediction models are presented. The discussion is limited to the so-called load interaction models (Ref 3-21, 3-22), and one statistically based damage accumulation model (Ref 3-25).

The Basic Concept of Load Interaction

The above simple explanation of load sequence effect on crack growth has led researchers to believe that some interaction takes place between each pair of consecutive loading steps. The plastic zone (at the crack tip) associated with each loading step is responsible for such interaction and thereby for the crack growth behavior. Attempts have been made to develop crack growth modeling techniques to elucidate how the stress-strain field in the plastic zone around the tip of a crack responds to variable-amplitude loads and thereby influences subsequent crack growth. An objective that is common to all the existing load interaction models (or, the so-called retardation models) is to predict variable-amplitude crack growth rate behavior by using the constant-amplitude crack growth rate data. Prior to discussing some of these models, some physical background on crack tip plastic zones and load interactions and their roles on crack growth retardation is reviewed in the following paragraphs.

Crack Tip Plastic Zones

When a crack is subjected to a cyclic stress, due to the high stress concentration at the tip of a sharp crack, even though the applied stress level might have been very low, the bulk of material in front of the crack will have undergone plastic yielding. As shown in Chapter 1, the size of a crack tip plastic zone can be expressed as:

$$r_p = \frac{\beta}{\pi}(K_{max}/F_{ty})^2 \qquad \text{(Eq 3-15)}$$

where F_{ty} is the material tensile yield strength and β accounts for the degree of plastic constraint at the crack tip. For example, $\beta = 1$ for a through-the-thickness crack in a thin sheet because the stress state at the crack tip is plane stress. The value of β decreases to 1/3 as the stress state changes from plane stress to plane strain.

Equation 3-15 represents the crack tip plastic zone that is formed during the upward excursion of a load cycle. This plastic zone is usually referred to as the monotonic plastic zone. The rate of crack growth at each load cycle depends on the crack tip constraint and the magnitude of the stress intensity. After removal of the applied load, or at the valley of each load cycle, there is a residual plastic zone remaining at the crack tip. This is called the reverse plastic zone. A schematic diagram showing the monotonic and reverse plastic zones is given in Fig. 3-20. The symbols r_p and r_{rp} in this figure refer to monotonic and reverse plastic zone, respectively. The residual stress distribution ahead of the crack (associated with one completed cycle) is also shown in the figure. The crossover point, from negative residual stress to positive residual stress, is theoretically 2.5 times the reverse plastic zone. In Fig. 3-20, the term F_{cy} is the material compression yield strength. Its absolute magnitude is approximately the same as the tensile yield strength of the same material. During unloading, the material immediately in front of the crack tip experiences an excursion load. Its absolute magnitude is equivalent to twice that of F_{ty} (because the load is dropping from F_{ty} to F_{cy}). Following Eq 3-15, the reverse plastic zone can be written as:

$$r_{rp} = \frac{\beta}{\pi}(K_{max}/2F_{ty})^2 = r_p/4 \qquad \text{(Eq 3-16)}$$

Therefore, the size of the reverse plastic zone is one-quarter of the monotonic plastic zone. The reverse plastic zone is also called the cyclic plastic zone. However, a cyclic plastic zone is sometimes defined as:

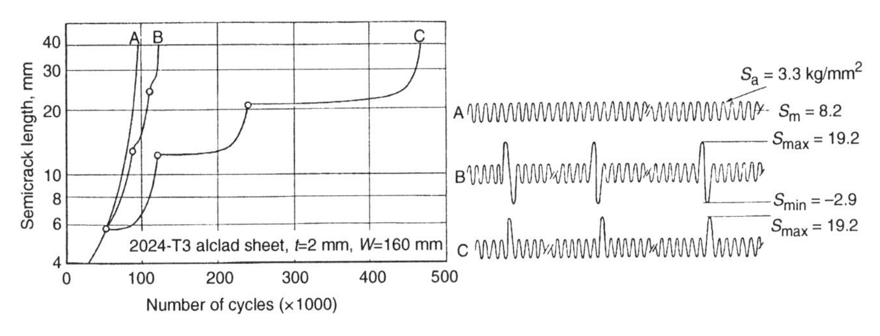

Fig. 3-22 Comparison of the delaying effects between two spectrum profiles: overload only vs. overload followed by underload. Source: Ref 3-20

$$r_{cp} = \frac{\beta}{\pi}(\Delta K/F_{ty})^2 \quad \text{(Eq 3-17)}$$

because the crack tip experiences an excursion of stresses in between S_{min} and S_{max} (not always in between S_{max} and zero load).

Now we are dealing with three types of plastic zones: the monotonic zone based on K_{max}, the reversed zone based on K_{max}, and the cyclic zone based on ΔK. Most of the experimental data available today indicate that the crack growth rate behavior is better correlated with the monotonic plastic zone induced by the overload. While some investigators have used the full zone (r_p) to make their correlations, others have used the plastic zone radius r_y ($= r_p/2$). In any event, it is customarily assumed that load interaction takes place inside an area that is related to the monotonic plastic zone, although its exact size has to be determined. This area is frequently referred to as the overload affected zone, or the load interaction zone. For the purpose of making life assessment, either approach would make little difference in the predicted life as long as the method is consistent.

Current crack growth interaction models always incorporate ways to compute crack growth rates inside the load interaction zone. The approaches taken by investigators generally have one of the following purposes: (a) to modify the stress intensity factors (Ref 3-21), (b) to directly compute the instantaneous crack opening load and residual stress field at the crack tip (Ref 3-22, 3-23), or (c) to empirically adjust the crack growth rates (Ref 3-24). Prior to discussing some of these models, we will review some physical background on crack growth retardation behavior.

Load Interaction

During crack propagation under variable-amplitude loading, or spectrum loading, there is a crack tip plastic zone associated with each stress cycle. Upon unloading, residual stresses remain on the crack plane and interact with the singular stress field of the subsequent stress cycle. The actual driving force for crack growth will be an effective stress field that accounts for the contributions of the applied stress and the residual stresses.

Prior to discussing the methods for determining the instantaneous effective stress field that results from mechanical load interaction, it is necessary to illustrate the load interaction phenomena by going through a scenario that involves a hypothetical stress profile, depicted in Fig. 3-21.

In a given loading step, the plastic zone is calculated using Eq 3-15 (or using one-half of this magnitude, as one might prefer). The calculation is based on the current crack length and the load applied to it, and the computed plastic zone is placed in front of that crack length, as shown in Fig. 3-21. In this manner, there will be no load interaction if the elastic-plastic boundary of the current plastic zone exceeds the largest elastic-plastic boundary of the preceding plastic zones. In other words, for a crack to propagate at its baseline constant-amplitude rate, the advancing boundary of the crack tip plastic zone due to the current tensile loading (the lower loads) must exceed the greatest prior elastic-plastic interface created by a prior higher tensile load. Load interaction exists as long as the plastic zone for the current stress cycle is inside the elastic-plastic boundary remaining from any of the preceding stress cycles.

Consider a stress level S_1, which is being applied to a crack length a_1. This action will create a plastic zone r_{p1} in front of the crack, as shown in Fig. 3-21. The elastic-plastic boundary for this stage in the crack growth history will be a_{p1} ($= a_1 + r_{p1}$). Meanwhile, the crack will grow from a_1 to a_2. The stress event that follows, S_2, will create another plastic zone r_{p2} in

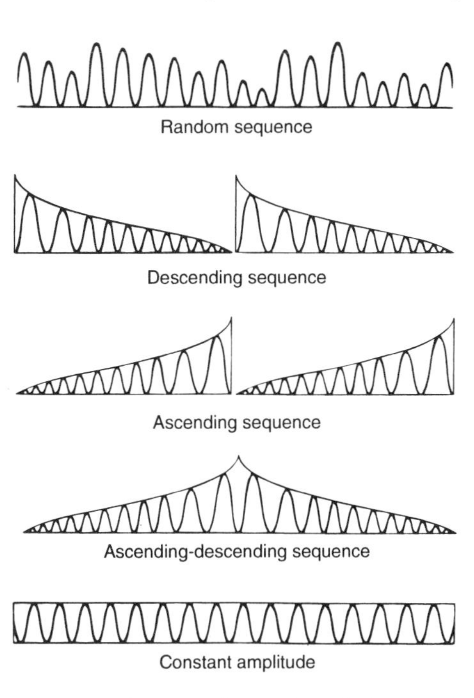

Fig. 3-23 Schematic representation of loading profiles of fatigue crack growth rate testing. Source: Ref 3-25

front of a_2, so that the elastic-plastic boundary will become a_{p2} $(= a_2 + r_{p2})$. Because $a_{p2} > a_{p1}$, no load interaction will occur while the crack is extend-

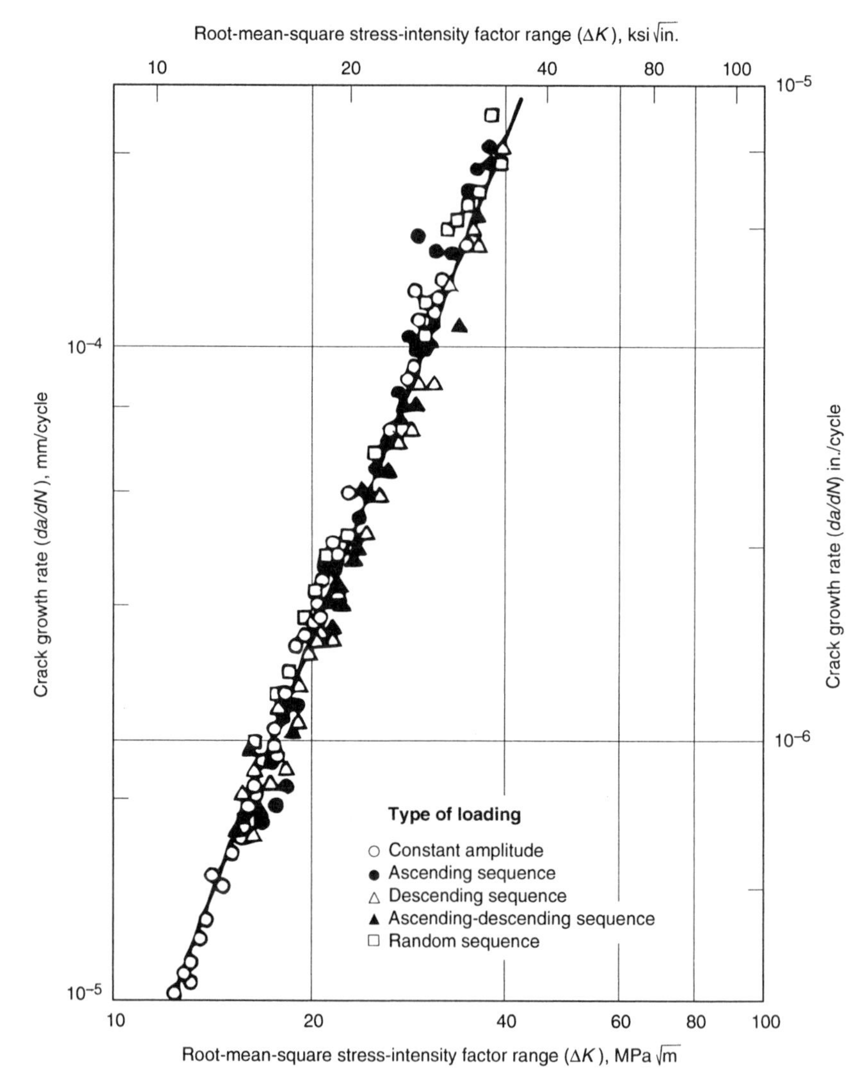

Fig. 3-24 Effect of loading sequence on fatigue crack growth rate. The loading patterns shown in Fig. 3-23 were imposed on specimens of A514B steel; rationalizing the stress intensity factor range by a root-mean-square method caused the data to fall within a single band, independent of loading pattern. Source: Ref 3-25

ing from a_2 to a_3. In the following three steps, both a_{p3} and a_{p4} are smaller than a_{p2}, but a_{p5} finally travels through the previous elastic-plastic boundary, i.e., $a_{p5} > a_{p2}$. Therefore, load interactions exist in the two loading steps that are associated with a_{p3} and a_{p4}. Crack growth rates will be affected by the residual stress field associated with a_{p2} while the crack is propagating from a_3 to a_4, and from a_4 to a_5. By definition, there is no load interaction between a_{p5} and a_{p2} because $a_{p5} > a_{p2}$. Thus, the growth of a_5 to a_6 (not shown) will be solely due to S_5.

Load Interaction Models

There are many ways to determine the instantaneous effective stress field that results from mechanical load interaction and to compute the cumulative damage (i.e., crack growth history) under variable-amplitude fatigue loading. The analytic models of Ref 3-21, 3-22, and 3-24 are representative of those being used in the aircraft/aerospace industry today. These models consider either the residual crack tip stress or crack closure. Among the analytic models mentioned earlier, the Willenborg model (Ref 3-21) and the Wheeler model (Ref 3-24) are the two models most often used by fracture

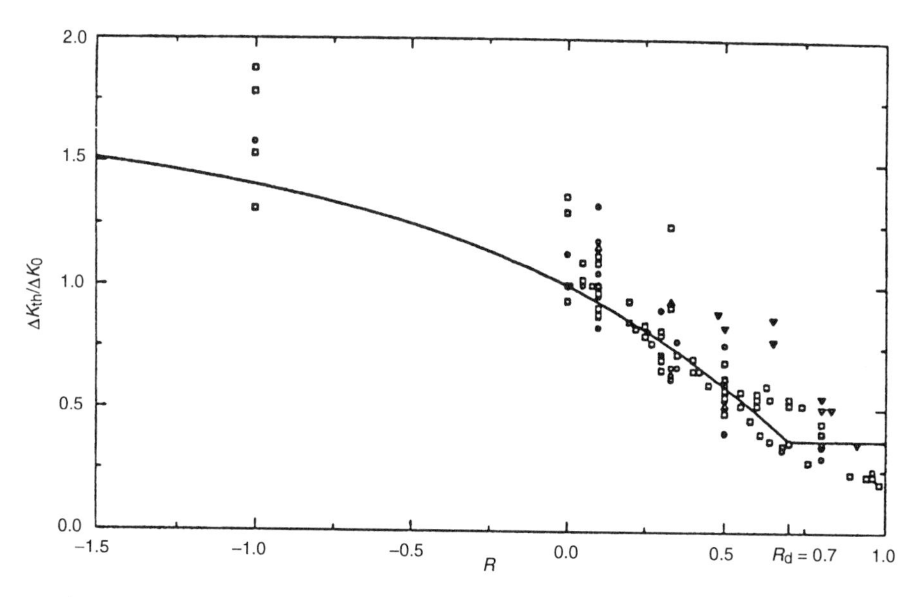

Fig. 3-25 Effect of stress ratio on fatigue crack growth rate threshold for several aluminum alloys. Source: Ref 3-18

mechanics analysts. Both models operate on the same platform, i.e., treating the crack tip plastic zone as the source and sink for load interaction.

The Willenborg Model

The Willenborg model handles crack growth retardation by computing an effective stress to reduce the applied stress and hence the crack tip elastic stress intensity factor. The model does not rely on empirically derived parameters. If the plastic zone of a given loading step is surrounded by a prior plastic zone, crack propagation inside the load interaction zone will be a function of the effective stresses. The effective stress would have a lower magnitude as compared with the applied stress (i.e., σ_{max}, or σ_{min}). The level of reduction is defined as:

$$\sigma_{red} = \sigma_{ap} - \sigma_{max} \quad \text{(Eq 3-18)}$$

where σ_{ap} is an imaginary stress that creates a plastic zone whose boundary coincides with the prior largest plastic zone boundary, a_p, and is defined as:

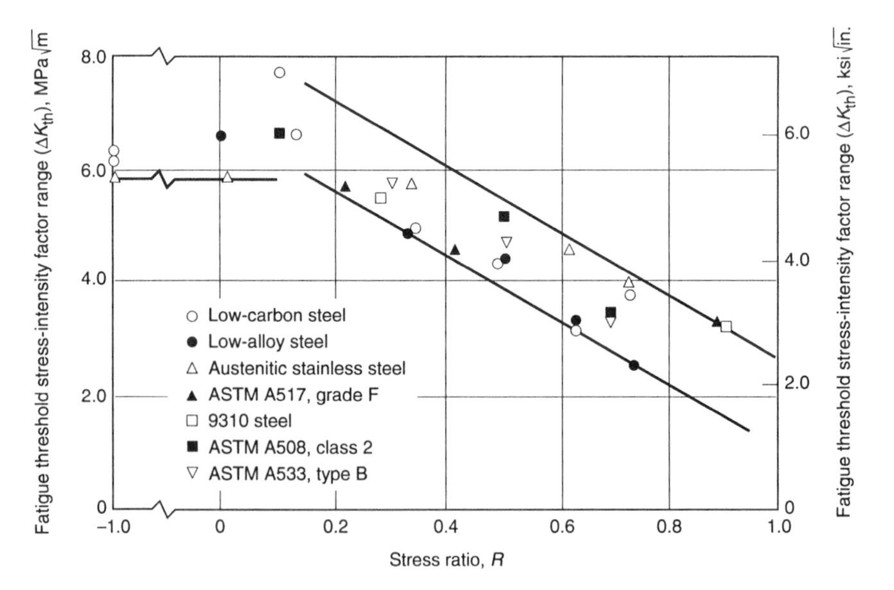

Fig. 3-26 Effect of stress ratio on fatigue crack growth rate threshold for several steels. Source: Ref 3-25

$$\sigma = \frac{F_{ty}}{\phi_i}\sqrt{\frac{2(a_p - a)}{a}} \qquad \text{(Eq 3-19)}$$

where ϕ_i is the stress intensity geometry factor. The effective stresses are then obtained as:

$$\sigma_{max,eff} = \sigma_{max} - \sigma_{red} \qquad \text{(Eq 3-20a)}$$

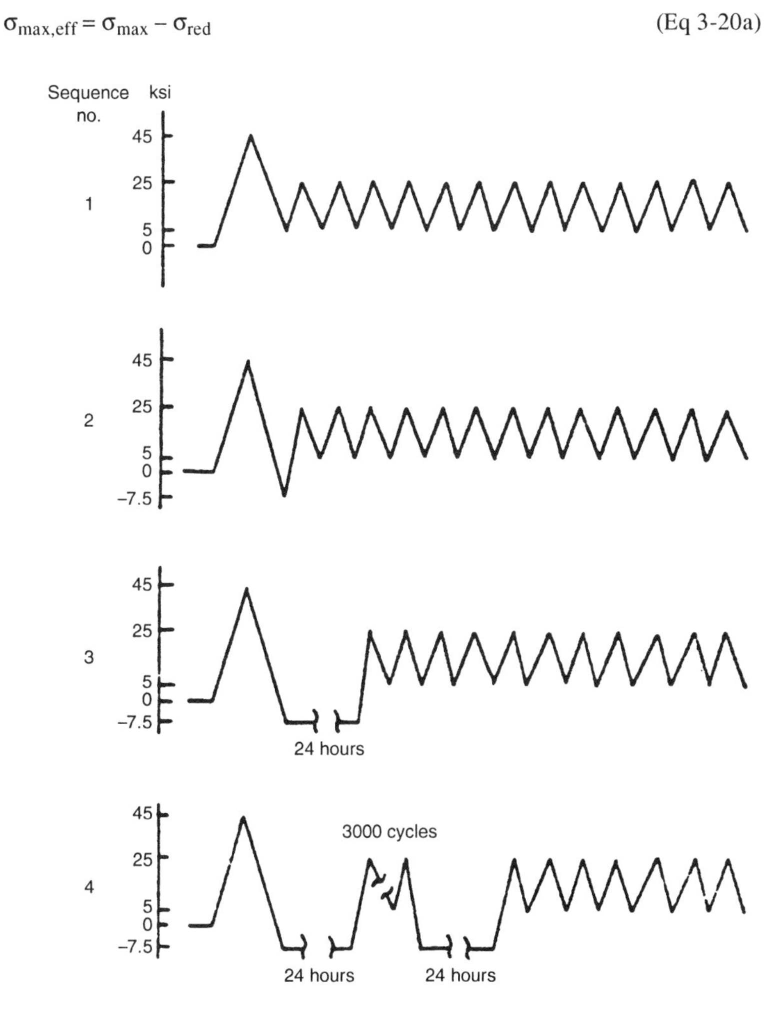

Fig. 3-27 Schematic stress profiles for fatigue crack propagation testing showing the effects of overload/underload with hold time. Source: Ref 3-38

Table 3-1 Fatigue crack propagation test results for overload/underload with hold time

Sequence No.	Specimen life, 1000 cycles	Mean life, 1000 cycles
1	1322.0	910.33
	1009.2	
	399.2	
2	212.2	214.4
	171.0	
	260.0	
3	158.4	126.05
	112.2	
	97.8	
	135.8	
4	118.7	130.5
	124.4	
	148.4	

Source: Ref 3-38

$$\sigma_{min,eff} = \sigma_{min} - \sigma_{red} \qquad \text{(Eq 3-20b)}$$

After the effective stresses are obtained, the crack growth rate is determined from the effective stress intensity range, the effective stress intensity ratio, and the material baseline constant-amplitude crack growth rate data.

By definition, $\Delta K_{eff} = K_{max,eff} - K_{min,eff}$, and $R_{eff} = K_{min,eff}/K_{max,eff}$ (or $\sigma_{min,eff}/\sigma_{max,eff}$). Equations 3-20(a) and (b) lead us to $\Delta K_{eff} = \Delta K$ and $R_{eff} \neq R$. Because the effective ΔK is the same as the applied ΔK, changing R to R_{eff} is the only source for crack growth retardation. Thus, when using constant-amplitude crack growth rate data to predict spectrum crack growth behavior, the crack growth rate equation representing the baseline material must include a term to account for the R-ratio effect.

The Willenborg model also specifies that both $\sigma_{max,eff}$ and $\sigma_{min,eff}$ must be non-negative quantities. In other words, any negative effective stress ratio will be set to zero, i.e., $R_{eff} = 0$ if $\sigma_{min} < \sigma_{red}$ and $\sigma_{max} \geq \sigma_{red}$. This situation leads to $\Delta K_{eff} = K_{max,eff}$. Finally, when $\sigma_{max} \leq \sigma_{red}$, complete crack arrest occurs (i.e., $\Delta K_{eff} = 0$). This situation arises when the overload is two times or more higher than that of the following applied load. This is called the "shut-off." However, test data have shown that complete retardation is unrealistic. Therefore modification to the model is necessary.

Generalization of the Willenborg Model

The first modification of the Willenborg model is to go around the "shut-off" by imposing an empirical term onto σ_{red}, i.e., to replace Eq 3-18 by letting:

$$\sigma_{red} = \Phi \cdot (\sigma_{ap} - \sigma_{max}) \qquad \text{(Eq 3-21)}$$

where

$$\Phi = \frac{1 - (\Delta K_0 / K_{max})}{S - 1} \qquad \text{(Eq 3-22)}$$

Where ΔK_0 is the threshold value for $R = 0$ and S is the empirically determined overload shut-off ratio (Ref 3-30, 3-31).This version of the Willenborg model is known as the generalized Willenborg model.

In addition to adding a shut-off ratio to the Willenborg model, the following treatments have been implemented by some fracture mechanics analysts. All or parts of these approaches can be used together with the generalized Willenborg model. Recognizing the effect of negative R on da/dN, the second modification to the Willenborg model is to remove the assumption of $R_{eff} = 0$ for $\sigma_{min} < \sigma_{red}$. The result of this treatment leads us back to $\Delta K_{eff} = \Delta K$; and $R_{eff} = K_{min,eff}/K_{max,eff}$.

The third modification is made to provide flexibility in choosing the interaction zone size. The factor "2" inside the square root of Eq 3-19 comes from the plane-stress plastic zone radius, r_y. Thus, one can rewrite Eq 3-19 as:

$$\sigma ap = \frac{F_{ty}}{\phi_i} \sqrt{\frac{n(a_p - a)}{a}} \qquad \text{(Eq 3-23)}$$

The value for n can be as high as 6.0 for the plane-strain plastic zone radius and as low as 1.0 for the plane-stress full zone.

If there is a compressive stress cycle (sometimes called the *underload cycle*) between the overload and the normal load cycles, the extent of retardation will be reduced. This is attributed to the fact that the overload plastic zone has been counteracted by the plastic zone of the underload. Figure 3-22 compares the effect on crack growth behavior of tests conducted with and without an underload.

The fourth modification intends to quantitatively account for underload effect. The proposed method assumes that the crack tip plastic zone associated with the underload is smaller than the monotonic plastic zone associated with the overload. Therefore the r_p (or r_y) of the overload is replaced by an effective overload retardation zone (Ref 3-32):

$$(r_{OL})_{eff} = (1 + R_{eff}) \cdot r_{OL} \qquad \text{(Eq 3-24)}$$

where r_{OL} can be r_p or r_y, depending on the preference of the fracture mechanics analyst. The applicability of Eq 3-24 is limited to $-1.0 < R_{eff} < 0$.

The Wheeler Model

In the Wheeler model (Ref 3-24), the load interaction zone size is assumed to be the plane-strain plastic zone radius ahead of the crack tip. Without changing the basic formulation of the stress intensity and its range, crack growth retardation is characterized by a growth rate coefficient, C_p, so that:

$$\frac{da}{dN} = C_p \cdot f(\Delta K, R) \qquad \text{(Eq 3-25)}$$

If $(a + r_y) < a_p$:

$$C_p = \left(\frac{r_y}{a_p - a} \right)^m \qquad \text{(Eq 3-26)}$$

where m is an empirical shaping constant, its value falling between 1.3 to 3.4 depending on the material and applied stress spectrum. If $(a + r_y) \geq a_p$, C_p is set equal to 1. Consequently, the value of C_p ranges from 0 to 1, corresponding to full retardation and no retardation, respectively. This approach totally relies on determining the Wheeler m by correlation with spectrum crack growth test data.

The Method of Root Mean Square

Depending on the application, or perhaps the combination of loading profile and material, fatigue crack growth under block loading does not always exhibit crack growth retardation. If stress overloads or load history (sequence effects) do not significantly affect fatigue crack growth, then crack growth increment (Δa) in each individual cycle of variable-amplitude loading can be estimated from the baseline da/dN versus ΔK curve. Using a complicated cycle-by-cycle load interaction model would be unnecessary.

Summing the values of Δa, while keeping track of the number of cycles, is a straightforward way of estimating fatigue crack propagation life under variable-amplitude loading. To correlate variable-amplitude data with constant-amplitude data, some method of normalizing the varying stress intensity factor ranges in the block spectrum is necessary. One such normalization scheme is to use the root mean square (rms) value of the stress intensity factor. Here ΔK_{rms} replaces ΔK in the constant-amplitude crack growth rate equation where ΔK_{rms} is given by:

$$\Delta K_{rms} = \frac{\sqrt{\sum_{i=1}^{N} (\Delta K)^2}}{N} \quad \text{(Eq 3-27)}$$

Barsom (Ref 3-25) used the loading sequences shown in Fig. 3-23 to establish a database for evaluation of the ΔK_{rms} normalization procedure. The results, shown in Fig. 3-24, indicate excellent correlation between the constant-amplitude data, ΔK, and the variable-amplitude data, ΔK_{rms}.

Other Factors Affecting Cyclic Crack Growth

Unlike tensile yield strength and ultimate strength, fatigue crack growth behavior does not have consistent material characteristics. Fatigue crack growth is influenced by many uncontrollable factors, and as a result, a certain amount of scatter exists. Therefore, the fairness of the predicted crack growth lives should be judged based on circumstantial factors that are relevant to the service conditions. Among the many factors that affect crack propagation, the following should be taken into consideration for crack growth predictions:

Metallurgical and processing variables

- Alloy composition
- Microstructure
- Batch (i.e., heat-to-heat variation)
- Distribution of alloy elements and impurities
- Grain size
- Preferred orientation (texture)
- Product form
- Orientation with respect to grain direction
- Heat treatment
- Mechanical or thermal-mechanical treatment
- Residual stress
- Manufacturer

Geometrical Variables

- Thickness
- Crack geometry
- Component geometry
- Stress concentrations (e.g., the presence of a notch)

Mechanical variables

- Cyclic stress amplitude (R-ratio)
- Loading condition (biaxial load, pin load, load transfer, etc.)
- Cyclic load frequency, wave form, and hold time
- Load interactions in spectrum loading
- Pre-existing residual stress

Environmental Variables

- Type of aggressive environment (gas, liquid, liquid metal, etc.)
- Concentration of aggressive species (pH)
- Partial pressure of gaseous species
- Electrochemical potential
- Temperature
- Cyclic load frequency, wave form, and hold time

No attempt is made to illustrate the effects of all these factors with test data, because some factors have largely different effects on different materials. Rather, some general trends are highlighted below.

Metallurgical and processing variables pertain to the material under consideration. The crack propagation characteristics for a particular alloy differ in plate, extrusion, and forging. The latter may exhibit a rather large anisotropy, which may have to be considered in the growth of surface flaws and corner cracks, which grow simultaneously in two perpendicular directions. Stretched plates normally exhibit a faster crack growth rate, primarily due to increased tensile yield strength, which causes a smaller plastic zone under load. Consequently, there will be less crack closure (i.e., crack opening starts at a lower applied load level), which results in a larger effective ΔK. The higher tensile yield strength resulting from prestraining also causes a smaller shear lip (flatter fracture surface), which further enhances crack propagation (Ref 3-33, 3-34). For a given product form, the grain orientation effects on fatigue crack growth rates usually are ranked in the following order with respect to applied loads: short transverse has the fastest rate; long transverse is ranked medium; longitudinal will exhibit the slowest rate. Closely related to this are the other processing variables, heat treatment in particular. Alloys of nominally the same composition but produced by different manufacturers may have largely different crack growth rate properties. Variations in crack growth rates may occur between different batches of the same alloy produced by the same manufacturer. Residual stresses resulting from metallurgical or mechanical processing variables may significantly affect the crack growth rate of a given material (Ref 3-35). Extensive coverage of the effects of metallurgical and processing variables on fatigue crack growth is found in *ASM Handbook*, Volumes 19 and 20, as well as in many other technical publications in the open literature.

Geometric Variables. Thickness has a small but systematic effect on crack growth rate behavior. Crack growth rates are higher in thicker plates. Excepting the thickness, which is primarily related to material processing variations and the stress state at the crack tip (i.e., plane stress or plane strain), the other geometric variables can be handled by using appropriate stress intensity expressions. However, residual stress is often a result of load excursions at stress concentration. Its effect on crack growth behavior is discussed in Chapter 4.

Stress ratio is an important variable in baseline constant-amplitude crack growth. The resultant stress ratios attributed to overload, underload, or residual stress should not be ignored in crack growth analysis. In the case of constant-amplitude crack growth, both ΔK threshold and the termination point of the da/dN curve are functions of the R-ratio. Empirically determined ΔK_{th} versus R relationships for a number of engineering alloys are available. Figures 3-25 and 3-26 are compilations of the data for aluminum alloys and steels, respectively.

Biaxial Stress Effect. A detailed discussion of the biaxial stress effect on fatigue crack propagation is given in Chapter 7.

Load Transfer. Fatigue crack growth analysis of splices, joints, lugs, etc., falls into a category of problems involving load transfer through fasteners from one part to another. This load transfer will tend to change the basic fatigue spectrum in the local area of the load transfer. For example, the lug hole does not recognize the compressive load contained in the fatigue spectrum. Therefore, all the compression loads in the spectrum may be truncated (i.e., set the minimum load level to zero). In a spliced joint where the fastener holes are loaded holes, various degrees of load transfer take place at these holes while the structural member is subjected to uniform farfield loading. Therefore, a spectrum modification logic must be incorporated into one's life prediction methodology to realistically account for the structural response to the applied loads.

Hold Time. Applying hold time on a given fatigue load cycle may alter the crack growth retardation characteristics of the subsequent load cycles. When delay tests are carried out on titanium (Ti-6Al-4V, mill annealed) in room-temperature air, and hold time is applied at the peak of an overload cycle, the number of delay cycles N_D will be increased. However, if the hold time is held at K_{min} (either at zero load or at the valley of an underload), it would result in a reduced N_D as compared to no hold time (Ref 3-36). It is believed that the reduction of delay cycles is attributable to relaxation of the crack tip residual stresses.

Validation of this phenomenon is supported by two other experimental works reported in the literature (Ref 3-37, 3-38). Aluminum alloy 7075-T651 was used on both occasions. Reference 3-37 reports that residual stress relaxation occurred when the tests were conducted in vacuum and the load was held at zero (below K_{min}). The tests of Ref 3-38 were conducted in room-temperature air. The test sequences (loading profiles) and the results of the tests are presented in Fig. 3-27 and Table 3-1, respectively. The stress levels labeled in Fig. 3-27 were net section stresses and the specimens used contained a circular hole. Again, it is quite clear that both underloading and holding the load at K_{min} did cause reduction of delay cycles. In light of these observations, i.e., the possibility of erasing the beneficial effect of retardation due to residual stress relaxation, it is conceivable that using a retardation model for crack growth life prediction might result in an overestimated life. More research in this area is warranted.

Environment. The effect of environment on crack growth (including temperature, cyclic load frequency, wave form, and hold time) is discussed in Chapter 6.

REFERENCES

3-1. J.C. McMillan and R.M.N. Pelloux, "Fatigue Crack Propagation under Programmed and Random Loads," Report D1-82-0553, Boeing Scientific Research Laboratory, July 1966

3-2. P.C. Paris, M.P. Gomez, and W.E. Anderson, A Rational Analytic Theory of Fatigue, *The Trend in Engineering,* Vol 13, University of Washington, 1961, p 9–14

3-3. P.C. Paris, The Fracture Mechanics Approach to Fatigue, *Fatigue—An Interdisciplinary Approach,* Syracuse University Press, 1964, p 107–132

3-4 Metals Test Methods and Analytical Procedures, *Annual Book of ASTM Standards,* Vol 03.01, *Metals—Mechanical Testing; Elevated and Low Temperature Tests; Metallography,* American Society for Testing and Materials, 1995

3-5. A. Saxena and C.L. Muhlstein, Fatigue Crack Growth Testing, *ASM Handbook,* Vol 19, *Fatigue and Fracture,* ASM International, 1996, p 168–181

3-6. "Metallic Materials and Elements for Aerospace Vehicle Structures," MIL-HDBK-5E, U.S. Department of Defense, June 1987
3-7. M.S. Miller and J.P. Gallagher, An Analysis of Several Fatigue Crack Growth Rate (FCGR) Descriptions, *Fatigue Crack Growth Measurement and Data Analysis,* STP 738, American Society for Testing and Materials, 1981, p 205–251
3-8. J. Schijve, "Four Lectures on Fatigue Crack Growth," Delft University of Technology, Delft, The Netherlands, Oct 1977
3-9. K. Walker, The Effect of Stress Ratio during Crack Propagation and Fatigue for 2024-T3 and 7075-T6 Aluminum, *Effects of Environment and Complex Load History on Fatigue Life,* STP 462, American Society for Testing and Materials, 1970, p 1–14
3-10. A.F. Liu, Application of Effective Stress Intensity Factors to Crack Growth Rate Description, *J. Aircr.,* Vol 23, 1986, p 333–339
3-11. J.C. Newman, Jr., A Crack Opening Stress Equation for Fatigue Crack Growth, *Int. J. Fract.,* Vol 24, 1984, p R-131 to R-135
3-12. W. Elber, The Significance of Fatigue Crack Closure, *Damage Tolerance in Aircraft Structures,* STP 486, American Society for Testing and Materials, 1970, p 37–45
3-13. S. Banerjee, "A Review of Crack Closure," Report AFWAL-TR-84-4031, Air Force Materials Laboratory, Wright-Patterson Air Force Base, Dayton, OH, April 1984
3-14. K.D. Unangst, T.T. Shih, and R.P. Wei, Crack Closure in 2219-T851 Aluminum Alloy, *Eng. Fract. Mech.,* Vol 9, 1977, p 725–734
3-15. P.C. Paris and L. Hermann, Twenty Years of Reflection on Questions Involving Fatigue Crack Growth, Part II: Some Observation of Crack Closure, *Fatigue Thresholds,* Vol I, Engineering Materials Advisory Services Ltd., U.K., 1982, p 11–32
3-16. S. Sunesh and R.O. Ritchie, On the Influence of Environment on the Load Ratio Dependence of Fatigue Thresholds in Pressure Vessel Steel, *Eng. Fract. Mech.,* Vol 18, 1983, p 785–800
3-17. J.C. Newman, Jr., A Finite Element Analysis of Fatigue Crack Growth, *Mechanics of Crack Growth,* STP 590, American Society for Testing and Materials, 1976, p 281–301
3-18. R.G. Forman, V. Shivakumar, and J.C. Newman, Jr., "Fatigue Crack Growth Computer Program NASA/FLAGO Version 2.0," Report JSC-22267A, National Aeronautics and Space Administration, May 1994
3-19. A. Sexena and S.J. Hudak, Evaluation of the Three Component Model for Representing Wide-Range Fatigue Crack Growth Rate Data, *J. Test. Eval.,* Vol 8, 1980, p 113–118
3-20. J. Schijve, "The Accumulation of Fatigue Damage in Aircraft Materials and Structures," AGARDDograph AGARD-AG-157, North Atlantic Treaty Organization, London, Jan 1972
3-21. J.D. Willenborg, R.M. Engle, Jr., and H.A. Wood, "A Crack Growth Retardation Model Using Effective Stress Concept," Report AFFDL-TM-71-1-FBR, Air Force Flight Dynamics Laboratory, Wright-Patterson Air Force Base, Dayton, OH, Jan 1971
3-22. H.D. Dill and C.R. Saff, *Fatigue Crack Growth under Spectrum Loads,* STP 595, American Society for Testing and Materials, 1976, p 306–319

3-23. J.C. Newman, Jr., "FASTRAN II—A Fatigue Crack Growth Structural Analysis Program," TM 104159, National Aeronautics and Space Administration, Washington, D.C., Feb 1992
3-24. O.E. Wheeler, *J. Basic Eng., Trans. ASME,* Series D, Vol 94, 1972, p 181-186
3-25. S.T. Rolfe and J.M. Barsom, *Fracture and Fatigue Control in Structures—Application of Fracture Mechanics,* Prentice-Hall, 1977
3-26. S.H. Smith, Fatigue Crack Growth under Axial Narrow and Broad Band Random Loading, *Acoustical Fatigue in Aerospace Structures,* Syracuse University Press, 1965, p 331–360
3-27. S.H. Smith, Random-Loading Fatigue Crack Growth Behavior of Some Aluminum and Titanium Alloys, *Structural Fatigue in Aircraft,* STP 404, American Society for Testing and Materials, 1966, p 74–100
3-28. T.R. Brussat, Rapid Calculation of Fatigue Crack Growth by Integration, *Fracture Toughness and Slow-Stable Cracking,* STP 559, American Society for Testing and Materials, 1974, p 298–331
3-29. J.P. Gallagher, Estimating Fatigue-Crack Lives for Aircraft: Techniques, *Exp. Mech.,* Vol 16, 1976, p 425–433
3-30. J.P. Gallagher and H.D. Stalnaker, Predicting Flight by Flight Fatigue Crack Growth Rates, *J. Aircr.,* Vol 12, 1975, p 699–705
3-31. J.P. Gallagher and T.F. Hughes, "Influence of Yield Strength on Overload Affected Fatigue Crack Growth Behavior in 4340 Steel," Report AFFDL-TR-74-1-27, Air Force Flight Dynamics Laboratory, Wright-Patterson Air Force Base, Dayton, OH, July 1974
3-32. J.B. Chang and R.M. Engle, Improved Damage-Tolerance Analysis Methodology, *J. Aircr.,* Vol 21, 1984, p 722–730
3-33. J. Schijve, The Effect of Pre-Strain on Fatigue Crack Growth and Crack Closure, *Eng. Fract. Mech.,* Vol 8, 1976, p 575–581
3-34. T.S. Kang and H.W. Liu, The Effect of Pre-Stress Cycles on Fatigue Crack Growth—An Analysis of Crack Growth Mechanism, *Eng. Fract. Mech.,* Vol 6, 1974, p 631–638
3-35. R.J. Bucci, Effect of Residual Stress on Fatigue Crack Growth Rate Measurement, *Fracture Mechanics: Thirteenth Conf.,* STP 743, American Society for Testing and Materials, 1981, p 28–47
3-36. O. Jonas and R.P. Wei, An Exploratory Study of Delay in Fatigue-Crack Growth, *Int. J. Fract. Mech.,* Vol 7, 1971, p 116–118
3-37. W.P. Slagle, D. Mahulikar, and H.L. Marcus, Effect of Hold Times on Crack Retardation in Aluminum Alloys, *Eng. Fract. Mech.,* Vol 13, 1980, p 889–895
3-38. D. Simpkins, R.L. Neulieb, and D.J. Golden, Load-Time Dependent Relaxation of Residual Stresses, *J. Aircr.,* Vol 9, 1972, p 867–868

Chapter 4

Life Assessment and Improvement Methods

To obtain a valid crack growth history prediction, proper consideration must be given to all major aspects of the problem. The initial flaw size and shape must be described, and the anticipated load history must be defined. Additionally, crack growth must be considered as a step-by-step process of accumulated damage in which crack extension progresses as a stochastic process to some point of instability. Therefore, the following tools and data must be available:

- The anticipated load history and the technique of spectrum representation
- The stress intensity factors that associate with the local stress field, crack morphology and size, and the local geometry of the structure
- The fatigue crack growth rate data for the material under consideration
- An analytical model that keeps track of damage accumulation while accounting for the interaction effects of the applied loads
- The material residual strength (fracture toughness) data and a failure criterion for onset of catastrophic failure

Each industry (aircraft, nuclear, automobile, offshore drilling, etc.) has its own service applications and therefore its own way of handling the problem. Even within one industry, the loads spectra for different products have different characteristics (e.g., fighter plane versus passenger plane). Spectrum editing should be performed in order to sort out the irregular peaks and valleys in the raw spectrum. While making a life assessment for a given product, there is always a strong desire to simplify the edited spectrum further in order to minimize computation time. The key issue in spectrum simplification is the equivalency between the simplified spectrum and the original spectrum. That is, simplified and original spectra should yield the same amount of damage whether they are being used for structural testing or life assessment. Once the spectrum is properly simplified, accelerated structural verification tests can be performed to obtain meaningful data in a timely manner.

Spectrum development and representation techniques are out of the scope of this chapter. Readers are referred to the references listed at the end of this chapter. A compilation of stress intensity solutions for many common structural geometries is given in Chapter 5. The physical and analytical aspects of the other three areas listed above are covered in other chapters of this book. In the remainder of this chapter, we will present some engineering methods that can be used for determining stress intensity factors for nonstandard structural geometries under generalized loading conditions. Other techniques are discussed, such as reducing the stress intensity factor in a loaded structural member, and the methods for implementing the elastic-plastic and fully plastic fracture indices. Lastly, discussion of fracture behavior of vessels and shells is also included. Although the methods presented here are oriented toward aircraft, the same principles and techniques are applicable to other types of structures.

Damage Tolerance Analysis Methodology

Safe crack growth and damage tolerance are the key considerations in today's structural design practices. The structural requirement is specified in terms of longevity. It assumes that the structural element originally contains a certain flaw at a critical location (e.g., crack(s) emanating from a fastener hole). Assessing the damage tolerance of a given design against its anticipated usage has become an integral step during design/development and structural sizing (Ref 4-1).

Sometimes a probable initial crack size can be determined by gathering nondestructive inspection data from periodic in-service inspections or inspection during full-scale testing. If none of these are applicable, a reasonable initial crack size has to be selected. The actual criticality of a potentially fracture critical part is determined by checking the fracture mechanic analysis results against the design criteria. Trade studies can be conducted at the same time in order to attain an optimized design.

For vehicles already in service, periodic reassessment during the service life is essential. The initial crack size to be used will be determined from periodic ground inspections. These exercises are necessary for establishing and revising inspection intervals, confirming the original design life, and determining the feasibility of extending the service life.

Therefore, a fracture-mechanics-oriented damage tolerance design/analysis procedure consists of the following steps:

1 Establish design criteria suitable to the expected use of the vehicle.
2 Identify structural elements and the expected loading conditions, magnitude, and in-service environments.
3 Develop a systematic means of identifying the criticality of these structural elements.
4 Check the actual criticality of the questionable structural elements by using the best available fracture mechanics methodology, and perform trade studies regarding safety, weight, and cost. The fracture mechanics analysis also helps to establish reliable in-service inspection intervals and to periodically update the fracture critical parts list.

In actual application, it is necessary to specify the extent and type of damage and the load level to be achieved with the damage present. These basic criteria are assigned on the basis of engineering judgment. The goal is to ensure that damage is readily detectable before the strength is impaired beyond the point of safety. Once the basic criteria are defined, the structures must be designed to meet the criteria. This chapter discusses the methodologies needed to achieve the design goal.

Classification of Airframe Structures

Prior to assessing the damage tolerance capability of an airframe structure, it is worthwhile to examine the inherent damage tolerance characteristics. Generally, a structure can be classified under one or two categories:

- *Safe Crack Growth:* From the standpoint of damage tolerance, this classification of safety is also called safe-life. It has the same meaning as "safe-life" for fatigue. The difference is that damage tolerance life is the life for a structure in which a crack is initially present, whereas fatigue life implies that the structure initially does not contain any crack. In this category, structures are designed such that initial damage will grow at a stable, slow rate under service conditions and not reach a size large enough to cause rapid unstable propagation. Damage tolerance (and thus safety) is ensured only by the maintenance of a slow rate of damage growth, a residual strength capacity. It is also necessary to ensure that subcritical damage will be detected at the depot or will not reach unstable dimensions within a specified design lifetime.

- *Fail-Safe:* In this category, structures are designed such that propagating damage is safely contained by failing a major load path or by other damage-arrestment features. Damage tolerance and thus safety is ensured by the allowance of partial structural failure, the ability to detect this failure prior to total loss of the structure, the ability to operate safely with the partial failure prior to inspection, and the maintenance of specified static residual strength throughout this period.

On the basis of geometric and functional considerations, airframe structure can be classified as:

- Monolithic structure
- Crack-arrest (crack-stopper) structure
  - Skin stringer design
  - Integral stiffener design
- Multiple-element structure
  - Multiload path-independent (i.e., multilaminate or redundant-element design)
  - Multiload path-dependent (i.e., multiplank design)

The basic differences among these three typical classes of structures arise from how the crack extension process is interrupted or altered, either by an artificial barrier in the crack arrest structure or by a geometric discontinuity in the multiple-element structure. When the crack is far away from these geometric barriers or discontinuities, the structure is essentially monolithic. Effective fracture mechanics analysis of cracked structures requires appropriate stress intensity factors representative of local structural geometries and crack morphologies. In Chapter 5 we will present and discuss mode 1 stress intensity factors for the crack geometries commonly found in structural components.

Typical fail-safe design involves multiple-element or redundant-element arrangements, with crack arrest provisions in the form of geometric boundaries or stiffening elements. Many of the stiffened or reinforced panels employed in conventional aircraft wing and fuselage structures possess an inherent fail-safe capability of significant magnitude. It is obviously desirable to make use of such characteristics and enhance them to the extent required to comply with the fail-safe design requirements to obtain maximum efficiency in applying this design concept.

As a reference for discussion, consider that the structures are inspectable in service. Also assume that a damage a_0 (an appropriate crack size parameter) is initially present. This is the size of the largest flaw that could be missed in the initial fabrication inspection or during a regularly scheduled inspection. Fatigue crack propagation characteristics (plots of damage size versus time or mission) for each structural type are schematically illustrated in Fig. 4-1 and 4-2. The mechanics of component failure, under a single monotonically increasing load (plots of load level versus crack size) for each type of these structures are schematically shown in Fig. 4-3(a) and 4-3(b). Both crack-stopper structures and multiple-element structures are inherently provided with fail-safe and safe-life (safe-crack-growth) capabilities. Unless otherwise noted, the analytical methods presented herein are intended for the monolithic, safe-crack-growth structures. More in-depth discussions about the fail-safe and safe-life design concepts can be found in Ref 4-2.

Methods for Determining Stress Intensity Factors

It has long been recognized that crack tip stress intensity, K, is the driving force for crack growth and fracture of metallic materials. The analytical techniques for obtaining stress intensity factors for cracks in isotropic materials are well established. The basic analytical techniques commonly used are the finite-element methods or various mathematical methods. Using mathematical techniques, the stress intensity factors have been obtained for

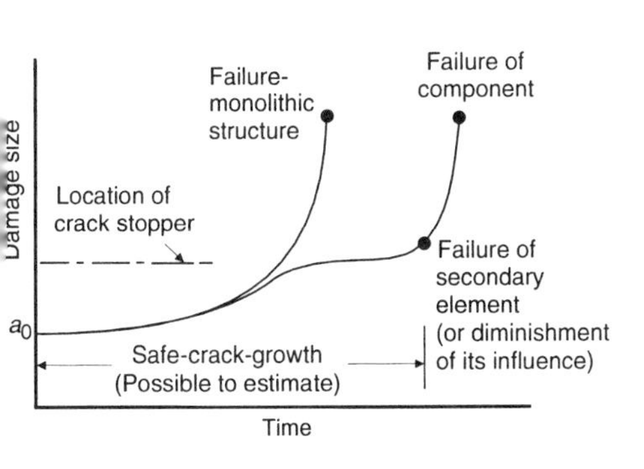

Fig. 4-1 Comparison of fatigue crack growth characteristics for monolithic structure and crack-arrest structure

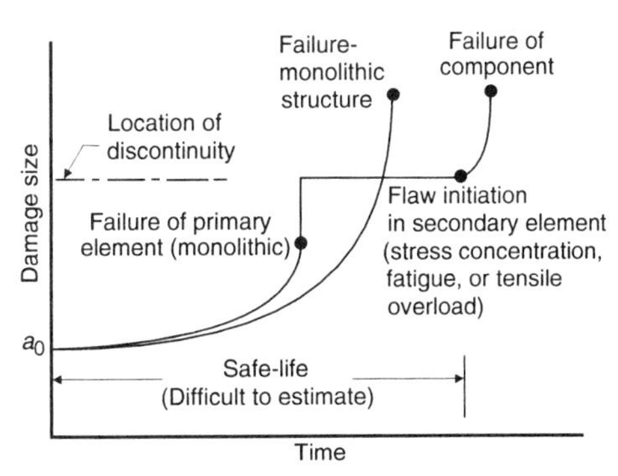

Fig. 4-2 Comparison of fatigue crack growth characteristics for monolithic structure and multiple-element structure

a variety of crack and structure geometries and are readily available in handbooks such as those listed in Chapter 5. Damage tolerance analysis (on a real structural member under a specific loading condition) usually treats the local structural area under consideration as if it were the same as a simplified model for which known stress intensity solutions exist.

For complex structural configurations where stress intensity solutions are not available, determining K becomes an integral and critical step in performing a structural life predictive analysis. Now that high-speed, workstation-based commercial structural analysis computer code has become available, the finite-element technique has become a popular tool for structural analysis, as well as for stress intensity factor determination. Following the discussion of the finite-element method, three other engineering methods will be discussed: the compounding method (which utilizes known solutions), the weight function method (for a crack under arbitrary crack face loading), and the empirical calibration method (which uses actual fatigue

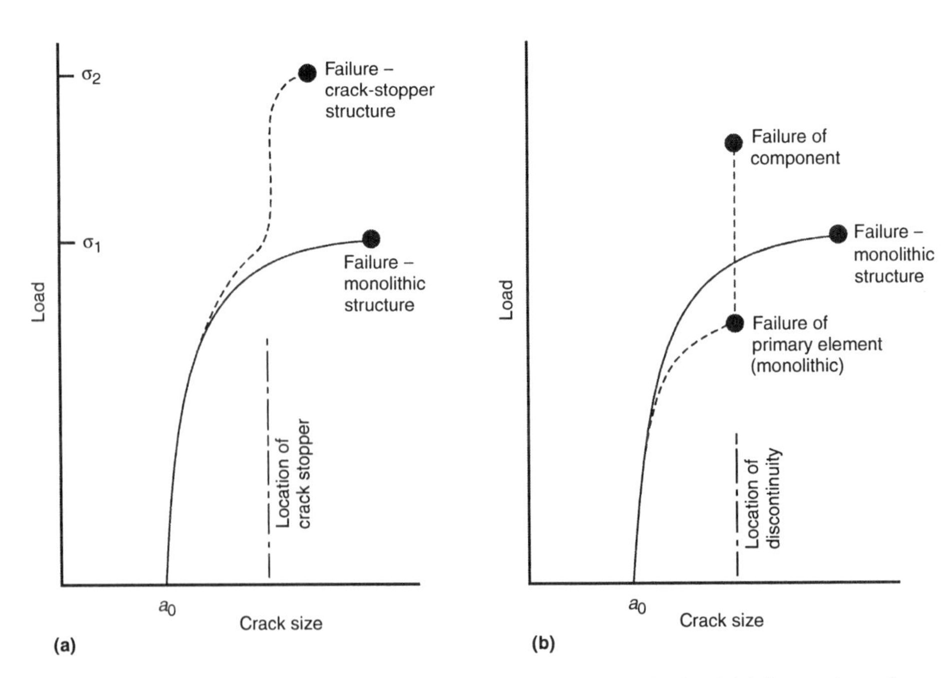

Fig. 4-3 Crack growth characteristics under monotonic increasing load. (a) Comparison of monotonic structure and crack-arrest structure. (b) Comparison of monotonic structure and multiple-element structure

crack growth test data). Each of these engineering methods has advantages and limitations.

The Finite-Element Method

Among the many analytical techniques used to determine *K*, the methods in the finite-element category are most suitable to structural analysis application. In the finite-element method, the crack is treated as an integral part of the structure, which can be modeled in as much detail as necessary to accurately reflect the structural load paths, both near to and far from the tip of the crack. This technique is particularly useful in modeling three-dimensional (solid-type) structures with flaws. Therefore, in recent years, a great deal of attention has been given to its development.

The approach to determining *K* using finite-elements is to match the nodal point displacements (or element stresses) with classical crack tip stress, or crack face displacement solutions (i.e., Eq 1-18 and 1-19). For a stationary crack of a given size, subjected to uniform remote tension load normal to the crack, the displacements along the crack face and the local stresses ahead of the crack tip (on the crack plane) are:

$$v = K \cdot \frac{4(1-\nu^2)}{E} \cdot \sqrt{\frac{r}{2\pi}} + \text{higher-order terms} \qquad \text{(Eq 4-1)}$$

for plane strain, where r is the absolute distance from the crack tip, K is the stress intensity, ν is Poisson's ratio, and E is Young's modulus. For plane strain:

$$v = K \cdot \frac{4(1-(\nu/(1+\nu))^2)}{E} \cdot \sqrt{\frac{r}{2\pi}} + \text{higher-order terms} \qquad \text{(Eq 4-2)}$$

for plane stress; and

$$\sigma_y = K/\sqrt{2\pi r} + \text{higher-order terms} \qquad \text{(Eq 4-3)}$$

regardless of plane strain or plane stress. Equations 4-1 to 4-3 are obtained from rewriting Eq 1-18, 1-19, and 1-19(a) for $\theta = 0$.

The h-Version Finite-Element Code

When performing a finite-element analysis, one method of obtaining accuracy in determining the singular stress field ahead of the crack tip is to provide a fine mesh of elements around the crack. Using a conventional (h-version) finite-element code, a model (Fig. 4-4) containing an extremely large number of elements in the vicinity of the crack tip is required, and setup and computer run time increase.

Special elements, such as the cracked element and the quarter point/collapse element, have been developed to account for crack tip singularities. The William stress function (Ref 4-3) is often built in to the element for determining the stress intensity factor. Such elements not only enhance the reliability of accurate stress intensity factor calculations but also streamline the analysis procedure. In the past thirty years, many types of special elements have been developed and published in the literature (Ref 4-3 to 4-12). These were thought to obtain a direct measure of the crack tip stress singularity, or to provide a more accurate account of the stress and displacement fields at the crack tip, or both. More than 90% of this work has dealt with plane elements, i.e., two-dimensional geometry. Published work on modeling with solid elements is very limited. Figures 4-5 and 4-6 show examples of two-dimensional cracked elements. The Lockheed element (Fig. 4-5) was developed by M. Creager and completed by S.T. Chiu. The detail about this element is documented in Ref 4-13. A detailed account of the M.I.T. element (shown in Fig. 4-6) is given in Ref 4-4.

The key features of a cracked element are that the element can have any shape, it can have many nodes on any boundary, and there is always a node directly in front of the crack tip. The crack is simulated by two opposite nodes splitting about the crack line. The computational accuracy is maximized by experimenting with the number of nodes, the positions of the nodes, and the position of the crack tip. The cracked element can be incorporated into an existing structural analysis finite-element program. Sometimes, the size of the cracked element is also important (smaller is not always better). It is the user's responsibility to make preliminary runs. When the analysis is for a mode 1 crack, a symmetric half-element can be used. An example showing the connections between the cracked element and the surrounding elements is given in Fig. 4-7. A case history of using a cracked element to determine the stress intensity factor in a cruciform specimen subjected to various biaxial loading conditions is given in Chapter 8.

The p-Version Finite-Element Code

The p-version finite-element technology, which has only recently become available, allows stress analysis to be done using coarse mesh and/or odd-shaped elements. By using the p-version code, it is conceivable that stress

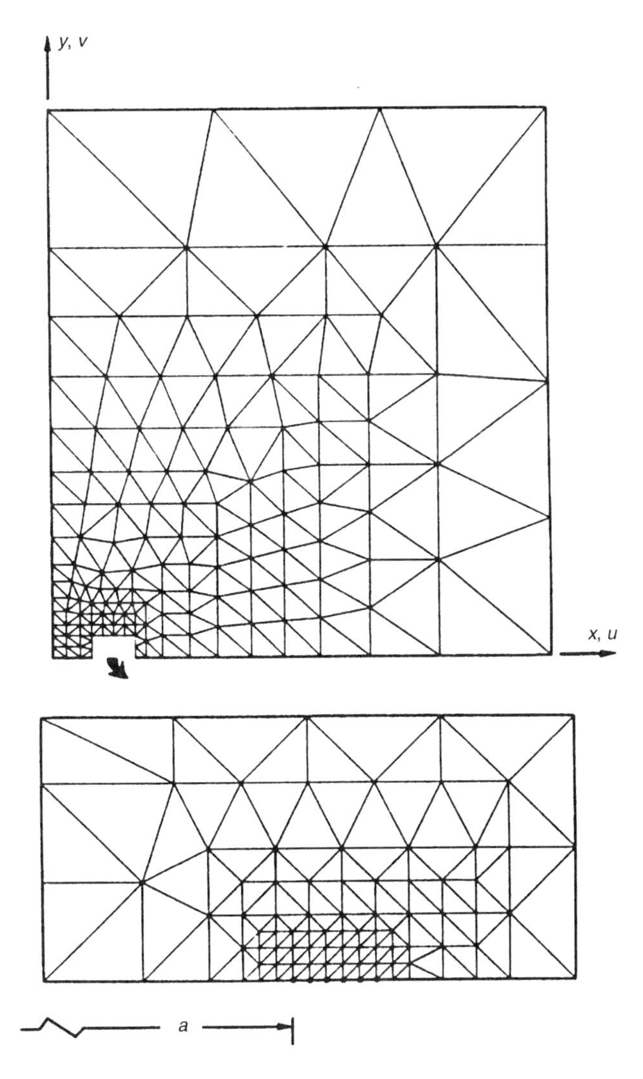

Fig. 4-4 Sample finite-element model of a center-cracked panel, using conventional h-version elements

intensity factors can be determined without incorporating any special element at the crack tip.

- In the conventional method, i.e., the h-version, the shape of the element is either rectangular, square, or triangular. In any case, the element must have a reasonable aspect ratio (the height-to-width ratio). However, the p-version technology allows the use of odd-shaped elements (e.g., a

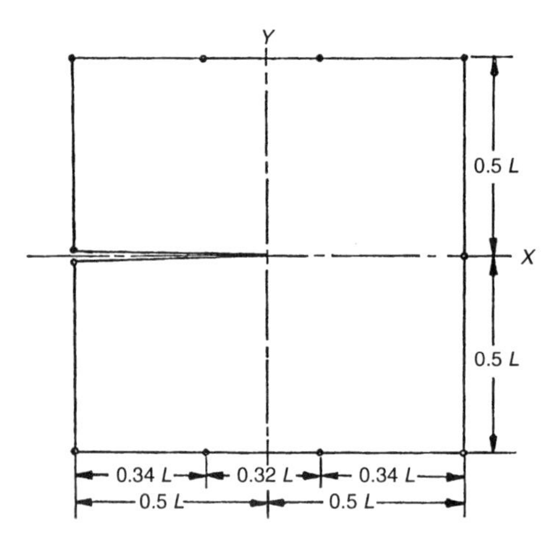

Fig. 4-5 The Lockheed element for representation of crack tip singularity

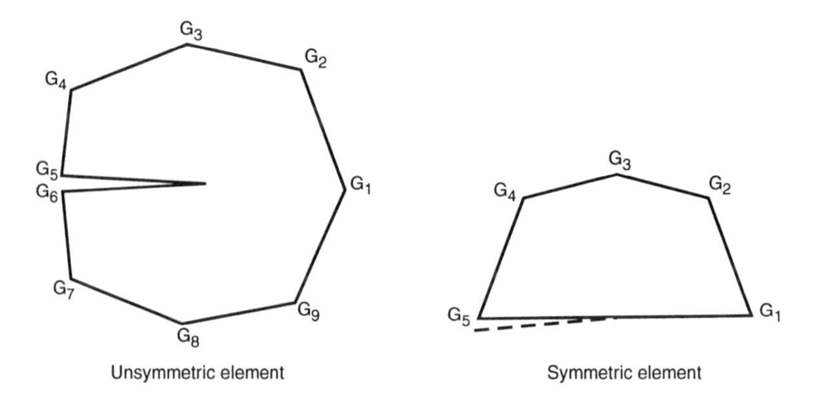

Fig. 4-6 The M.I.T. element for representation of crack tip singularity

triangle having a curved edge) and/or elements having an extreme aspect ratio.

- When using the h-version method, the error of approximation is controlled by refining the mesh. In the p-version method, the element is described by a group of shape functions of high-degree polynomials (the range can be as high as a nine-degree polynomial, as compared with linear or quadratic in the h-version), with the flexibility of varying the p-level in any designated element.
- To verify the solution, the p-version allows creation of a series of solutions without the need to refine the mesh, as is necessary with the h-version. A full sequence of polynomial levels (from 1 to 9) can be run to allow examination of strain-energy convergence and other functions of interest. The same mesh is used for all p-levels, so no additional effort by the user is required.
- The user can judge the accuracy of the solution based on convergence on stress, convergence on displacement, or both.

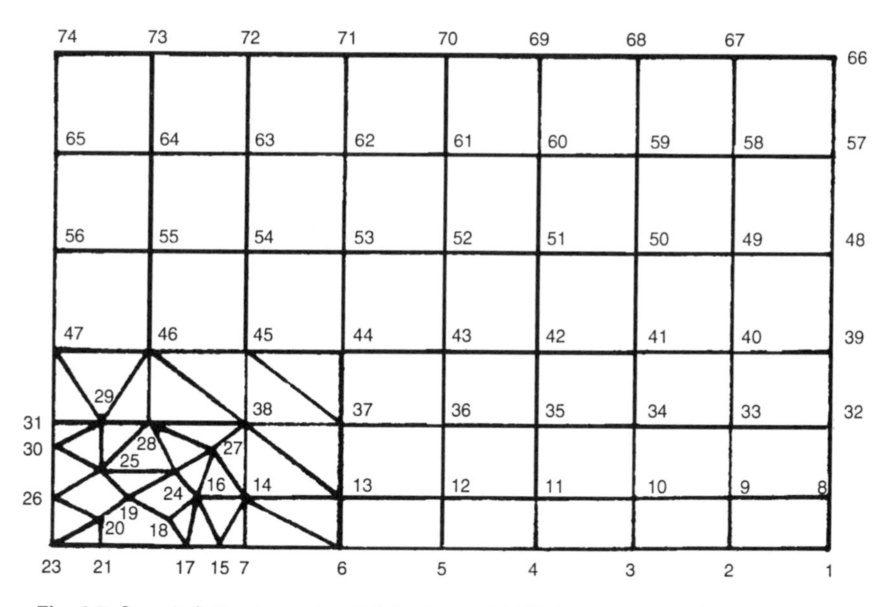

Fig. 4-7 Sample finite-element model showing an M.I.T. element surrounded by other conventional rectangular and triangular elements. The five nodes for the cracked element are 17, 18, 19, 20, and 21.

Implementation of the p-Version Code — An Example

In this section we will demonstrate how to determine stress intensity factors for a center-cracked panel by using a p-version code. The finite-element code that was used for this investigation was the Mechanica Applied Structure code, developed by the RASNA Corporation, San Jose, California. In the following we will refer to it as PFEC (p-version finite-element code) or PFEM (p-version finite-element model), as appropriate. The code we used was an older version (version 2.1) of the PFEC, which contains both plane and solid elements and runs on an IBM workstation but does not contain any fracture mechanics extraction routine. Users must rely on matching the PFEM solutions with Eq 4-1 and 4-3, or 4-2 and 4-3, whichever is appropriate. The procedures that are implemented below are equally applicable to using an h-version finite-mesh model or the crack tip stress and displacement distributions obtained by any other means. Therefore, the illustration given below is most appropriate for showing the practical meaning of classical fracture mechanics.

PFEM for the Center Cracked Panel. Figure 4-8 shows a typical two-dimensional finite-element mesh (which is made of plane elements) for a rectangular panel containing a through-the-thickness crack. The panel is geometrically symmetrical, so only one-quarter is shown. Due to the singular nature of the stresses near the crack tip, it might require a very high level of polynomial to converge. As shown in Fig. 4-8, a box of very small elements (0.254 mm in radius) is placed around the crack tip. This group of elements is designated as sacrificial elements. They are not special elements of any kind, except that verification of convergence inside these elements will not be required. Thus, computer space will not be exhausted as a result of running the program at p-levels associated with extraordinary high degrees of freedom. Parametric evaluation was not performed to obtain an optimized size for the sacrificial elements, but experience has revealed that stresses or displacements could reach reasonably good convergence at p-levels as low as 6 or 7. Beyond that, only insignificant improvements on minimizing the computational error could be achieved. Thus, quite often, sacrificial elements are unnecessary.

Six models of the center-cracked panel (Fig. 4-8) were constructed. The total width, W, was 2 in. (50.8 mm). The ratios of crack length to panel width, $2a/W$, were 0.05, 0.25, 0.4, 0.55, 0.7, and 0.9. The material properties used were $E = 10{,}300$ ksi (71,000 MPa) and $\nu = 0.33$. Each panel was subjected to 10 ksi (68.92 MPa) uniform tension stress (i.e., the applied load divided by the product of the full width times the thickness of the panel).

Data Reduction. For each panel, convergence was separately checked for convergence on stress and convergence on displacement. Based on the reduced data, the impact of convergence criteria on the solutions for stresses, or displacements, was insignificant. Therefore, convergence criteria were not an issue here. For data reduction, the PFEC allows the user to extract 10 pairs of data points (of stress and displacement) along any preselected line connecting any two nodal points of a given element (e.g., along an edge of the element). The data for all six panels are presented in Fig. 4-9 and 4-10. The K-value for a given panel was determined by using Eq 4-2 and 4-3, omitting the higher-order terms. The displacement and stress distributions were computed by assigning a K-value to these equations. Iterations were

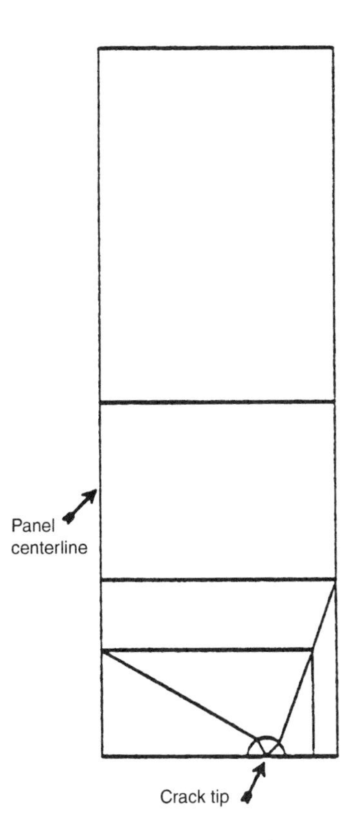

Fig. 4-8 Finite-element mesh for one quarter of a center-cracked panel, using p-version elements

made, assigning a different *K*-value each time, until the computed *v*, or σ_y, distribution fit the PFEM data. As shown in Fig. 4-9 and 4-10, the analytical solution (for *v* or σ_y) only fits the PFEM data over a narrow range of *r*, which is not too close to, and not too far from, the crack tip. This is due, in part, to omission of the higher-order terms in the analytic equations, and the limitation of the finite-element program in producing a proper stress singularity at the crack tip.

Theoretically the *K*-value for a given crack can be directly calculated from the PFEM data (*v* or σ_y) with Eq 4-1 to 4-3. Within a range of *r* where the analytical solution matches the finite-element solution, *K* should be a single value. When *K* is plotted against *r*, *K* should be a constant in that region. The plots for this group of center-cracked panels are shown in Fig. 4-11 and 4-12. The K_v and K_s terms in these figures imply that the *K*-values are determined based on finite-element displacement and finite-element stress, respectively. Also shown in these figures are the characteristics of the K_v (or K_s) versus *r* distribution, in which the *K*-values are linear across the

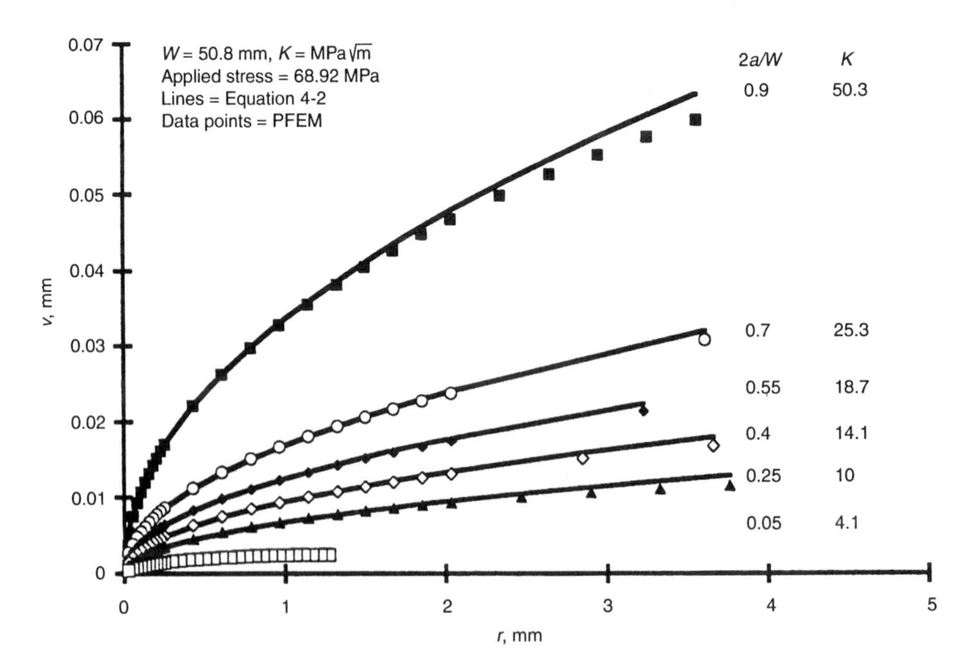

Fig. 4-9 Crack face displacements for cracks in a center-cracked panel. Source: Ref 4-16

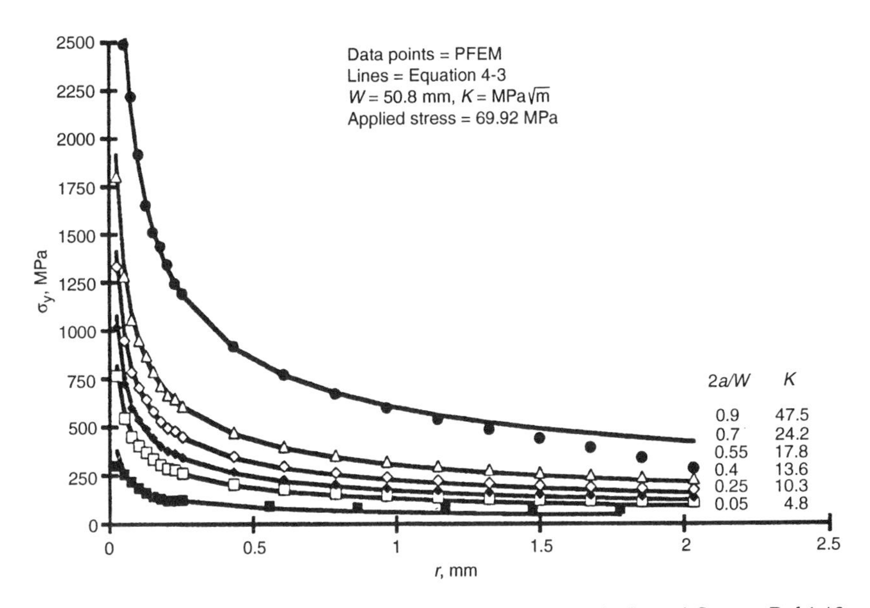

Fig. 4-10 Crack tip stress distributions for cracks in a center-cracked panel. Source: Ref 4-16

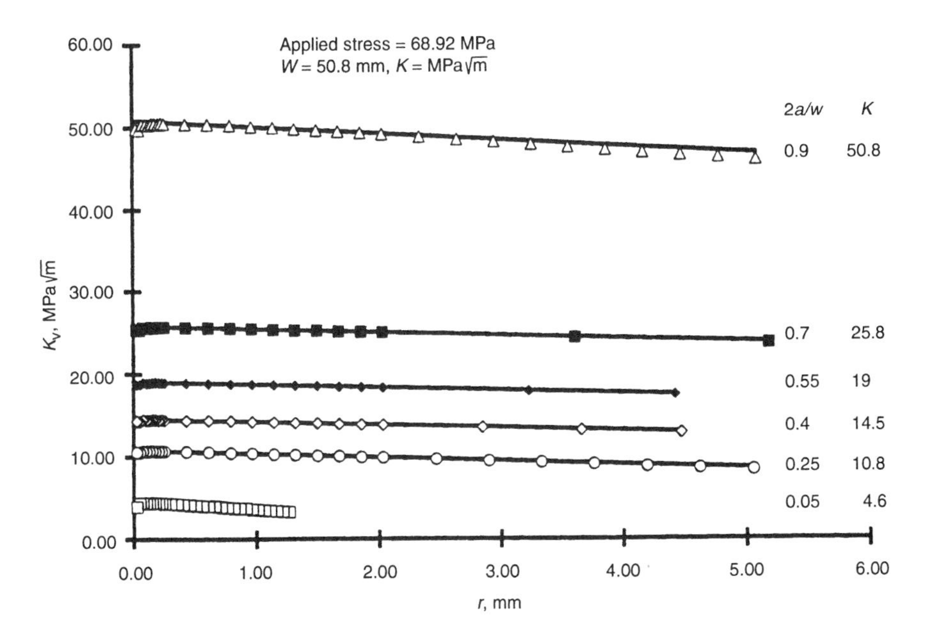

Fig. 4-11 Center-cracked panel stress intensity solutions based on the displacement extrapolation method. Source: Ref 4-16

region where the analytical and the PFEM solutions correlate. This feature is typical of any finite-element analysis results reported in the literature (e.g., Ref 4-7, 4-14, 4-15). In areas outside the linear region, the computed K-values deviate from the regression line and are considered to be invalid. The data points inside the linear region can be used as a basis for determining K.

According to Eq 4-1 to 4-3, all K-values on the regression line should have the same value (i.e., K is a constant). Although not shown here, the K_v or K_s values for many other crack configurations studied using this PFEC were indeed a single value. When K_v or K_s is not a constant, the real K-values can be obtained by rewriting Eq 4-3 as:

$$K_s = \sqrt{2\pi r} \lim_{r \to 0} \sigma_y \qquad \text{(Eq 4-4)}$$

i.e., by extrapolating the regression line to $r = 0$. The intercept point can be taken as the real K (Ref 4-14). The same approach can be applied to determine K_v.

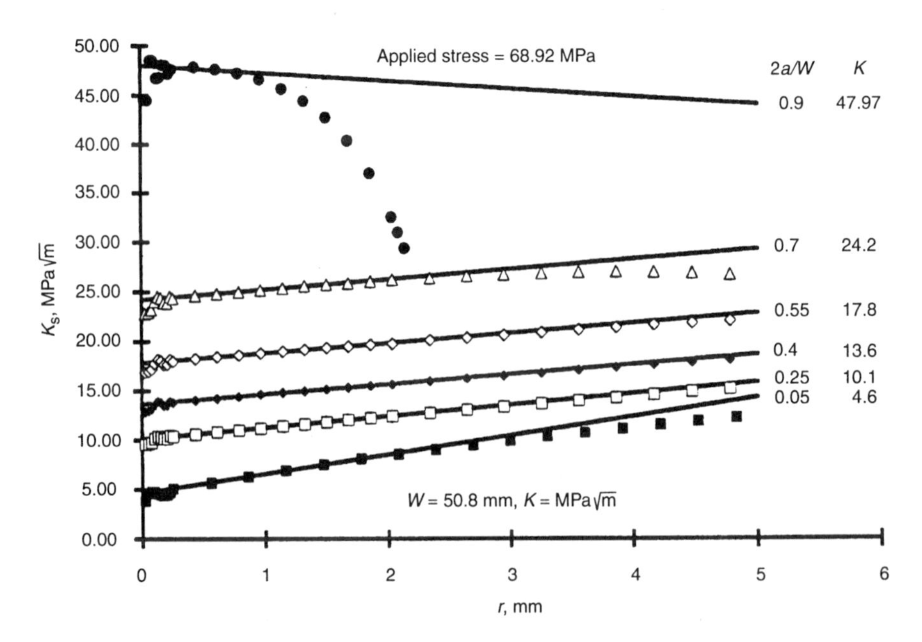

Fig. 4-12 Center-cracked panel stress intensity solutions based on the stress extrapolation method. Source: Ref 4-16

Conceptually, the four methods illustrated in Fig. 4-9 to 4-12 should give an identical K-value for a given geometry. In reality, partly due to the computational characteristics of a finite-element program, and partly dependent on the configuration under consideration for which the missing higher-order terms in the analytical solution might become significantly important, one or more of these methods may yield more dependable results than the others.

The K-values for all six panels, derived from each of the four data reduction methods, are labeled in Fig. 4-9 through 4-12 along with each fitted curve. From these figures, it appears that the displacement extrapolation method (i.e., that shown in Fig. 4-11) is consistently better than the others.

There are many stress intensity solutions for the center-cracked configuration. The Isida solution (Ref 4-17) is regarded by the fracture mechanics community as being the most accurate. For comparison purposes the Isida solution is plotted in Fig. 4-13, together with the finite-element solution, which is based on the intercept point of the K_v versus r curve. The agreement between the two solutions is excellent.

The Principles of Superposition and Compounding

Superposition and compounding are the most common and simplest techniques in use for obtaining stress intensity factors. Complex configurations and/or loading conditions are considered to be a combination of a number of separate simple configurations, with separate boundary conditions that have known stress intensity solutions. The K-factors for the simple configurations are then added or multiplied together to obtain the required solution. This is less sophisticated but perhaps more expedient than other analytical methods. The advantage is that the compounded K-factors can be conveniently expressed in a parametric closed form as functions of various dimensionless geometric variables (e.g., the ratio of crack length to panel width, the ratio of crack length to hole radius), useful to designers and analysts. This engineering equation can be easily incorporated into an operative crack growth analysis computer code. Stress intensity factors for any combinations of structural geometry, loading condition, crack length, and crack shape (even changes in crack length and crack shape while the crack growth is in progress) can be rapidly determined. Furthermore, once a set of workable engineering equations is determined, the cost of calculating a K-value is quite low.

As a rule of thumb, superposition is usually involved with combinations of loading conditions, and compounding is used for obtaining a solution for the combination of geometric variables. Many investigators use "compounding" as a general term, i.e., it might mean superposition, or compounding, or both. A compounded K-expression is usually formulated by multiplication of a number of individual K-expressions for each geometric boundary involved in a general configuration under consideration. For example, the K-expression for the center-cracked panel (Eq 1-22) can be regarded as a compounded solution that combines the basic K-solution for an infinite sheet and a so-called width correction factor. The K-expressions for the surface flaw and the corner crack(s) coming out from a hole (presented in Chapter 5) are other typical examples. In fact, a majority of the stress intensity factors presented in Chapter 5 are compounded factors.

Certain known solutions for a partially loaded crack may be superimposed, by adding or subtracting one to another, to obtain approximate solutions for cracks in arbitrary stress fields. The summation may consist of many idealized individual conditions. The only restriction is that the stress intensity factors must be associated with the same structural geometry (including crack geometry). Illustration of this principle is given in the following examples. For detailed descriptions of the principle and implementation of superposition and compounding, see the references listed at the end of this chapter. For a crack subjected to mixed mode 1 and mode 2, a failure

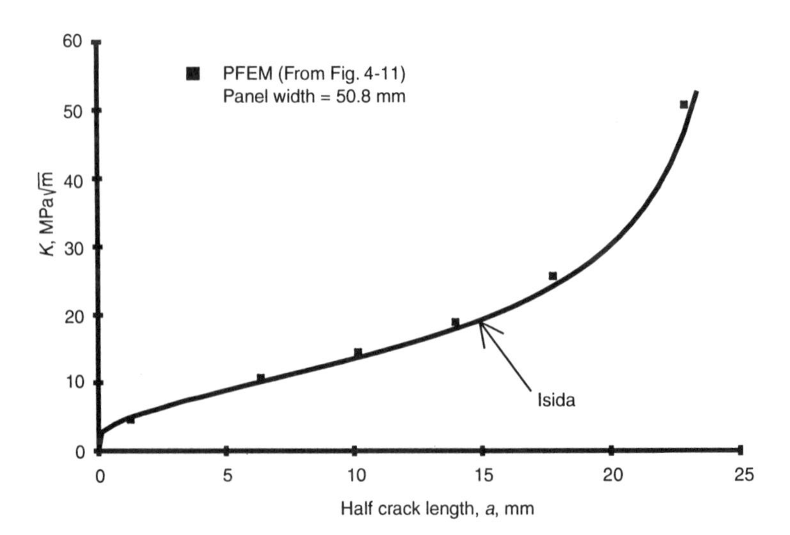

Fig. 4-13 Comparison of the PFEM and Isida solutions for the center-cracked panel. Source: Ref 4-16

criterion is required for obtaining an equivalently total mode 1 stress intensity factor. For a complicated configuration, the use of all three techniques, i.e., superposition, compounding, and equivalence, may be required.

Example 1

Consider a sheet that contains a pin-loaded hole. The pin load is balanced by the reaction stress uniformly distributed at the other end of the sheet (Fig. 4-14). To keep the problem simple, assume that the size of the pin-loaded hole is very small so that the actual contact pressure distribution between the pin and the hole is not an issue here. Secondly, ignore the small amount of mode 2 deformation at the crack tip that would have been caused by this particular loading condition.

We know that there are solutions available for a center crack subject to either uniform far-field loading or point loading applied on the crack line. We also know that a stress intensity factor for an open hole subjected to uniform far-field loading (the Bowie factor, Ref 4-18) is available. With this knowledge in hand, we can set up a superposition scheme (as depicted in Fig. 4-14) to derive an approximate stress intensity solution. Here, K^C (K for part C) is the K for a pair of concentrated loads; K^B (K for part B) is the K

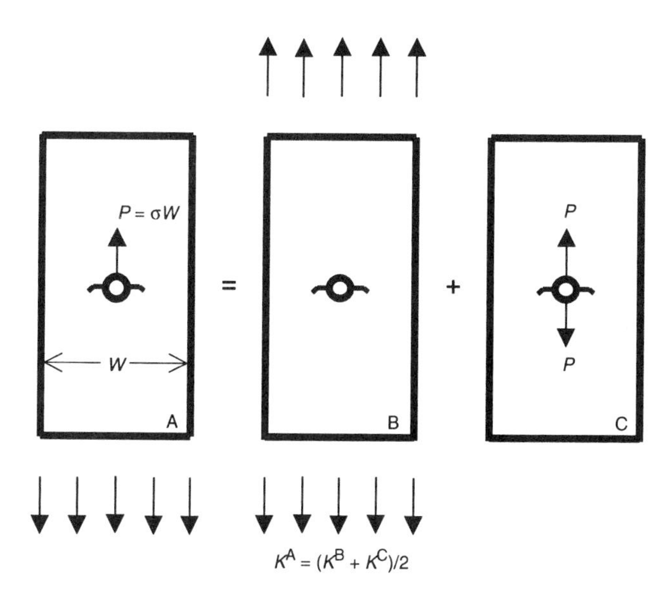

Fig. 4-14 Stress intensity factor for a pin-loaded hole obtained by superposition

for Bowie's open hole solution. These stress intensity factors are given in Chapter 5.

Example 2

The problem in Example 1 is extended to include load transfer. As shown in Fig. 4-15, the stress field can be treated as if it were in two portions; that is, a uniformly stretched strip (open hole) and a concentrated load at the fastener hole reacted by uniform stresses at the other end (loaded hole). In Fig. 4-15, α is the fraction of load transfer. Here, the open hole part is equivalent to K^B (with load = $(1 - \alpha)P$), and the loaded hole part is equivalent to K^A (with load = αP).

Example 3

Lateral tension or compression stresses would not affect the elastic stress intensity factors for a central straight crack in a plate. In other words, when a straight crack is subjected to σ_y, with or without σ_x, the K-values will be the same for either case (Ref 4-18 to 4-20). However, if a crack or cracks come out of a circular hole in a plate, compressive loading parallel to the crack can cause tensile mode 1 stress intensity factors, even without σ_y. On the other hand, tensile stresses parallel to the crack reduce the stress concen-

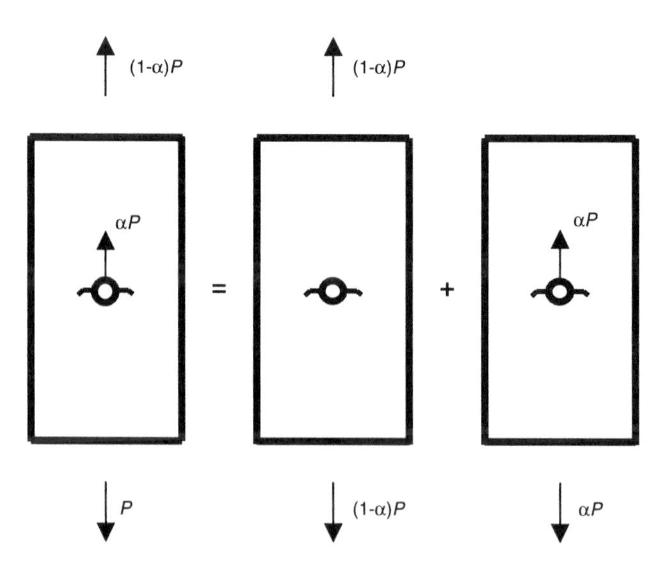

Fig. 4-15 Superposition of stress fields for a fastener hole with load transfer

tration at the hole and thus reduce the crack tip stress intensity factor. Bowie's solutions (Ref 4-18) have provided stress intensity factors for a single crack (or cracks) emanating from the edge of a circular hole, in an infinitely wide plate, under uniaxial and one-to-one biaxial loading conditions. For any other biaxial load ratios (either tension combined with tension or tension combined with compression), stress intensity factors can be developed by using superposition of the uniaxial and the equibiaxial tension solutions of Bowie, as illustrated in Fig. 4-16. Thus, stress intensity for any biaxial loading combination can be expressed as:

$$K = [(\sigma_y - \sigma_x)B_1 + \sigma_x B_2]\sqrt{\pi c} \qquad \text{(Eq 4-5)}$$

where c is the crack length measured from the edge at either side of the hole, B_1 is the Bowie factor for uniaxial loading, B_2 is the Bowie factor for one-to-one biaxial loading, σ_y is the far-field gross area stress perpendicular to the crack (always in tension), and σ_x is the far-field gross area stress parallel to the crack (either in tension or compression). Letting $\lambda = \sigma_x/\sigma_y$, Eq 4-5 can be rewritten as:

$$K = \sigma_y\sqrt{\pi c}\,[(1 - \lambda)B_1 + \lambda \cdot B_2] \qquad \text{(Eq 4-6)}$$

and the sign for λ may be either positive or negative. The values for B_1 and B_2 as functions of c/r can be found in Chapter 5. If the hole is inside a finite-width plate, the plate dimensions influence the stress concentration as well as the local stress distribution along the crack line, thereby varying the

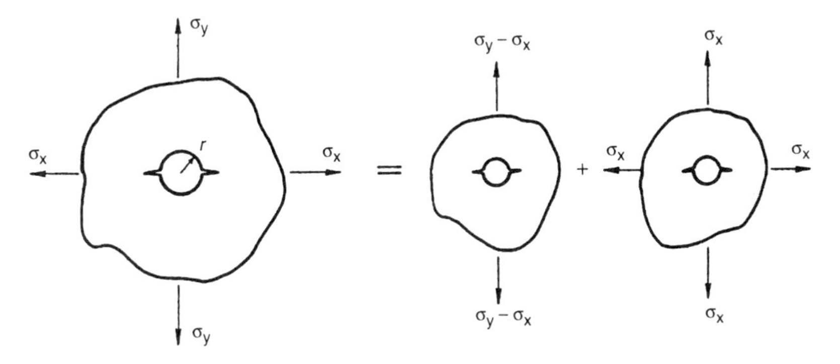

Fig. 4-16 Evaluation of Bowie's factor for biaxial loading

stress intensity factor. Furthermore, finite-width dimension also increases the crack tip stress intensity as if the hole had never been there. A compounded factor that accounts for the finite-width effect is also given in Chapter 5. Although the width correction factor given in Chapter 5 is for uniaxial loading, it is conceivable that it can also be used for biaxial load conditions.

In a previous work of the author, the validity of Eq 4-6 was examined by conducting fatigue crack growth tests on cruciform specimens with and without a hole (Ref 4-19). The overall geometry and dimensions for all the cruciform specimens (with and without a hole) were the same. Details about designing the cruciform specimen are given in Chapter 8 as a case history demonstrating how the finite-element method is used in fracture mechanics research. Eight cruciform specimens (with a hole) of the 7075-T7351 aluminum alloy were tested under biaxial cyclic stresses. The circular holes were either 6.35 mm or 19.05 mm in diameter. Each specimen was cyclically precracked from a crack starter (a very small saw cut approximately 1.0 mm in length) at both sides of the hole. The stress amplitude for the crack growth tests was 82.68 MPa and 8.268 MPa for σ_y with λ being either 0, +1, or −1. A crack growth history for each test was recorded. The K-value for each data point was computed using Eq 4-6. On the basis of cracked finite-element analyses (see Chapter 8), the K-values for a straight crack in the cruciform specimen are numerically equal to a crack in an infinite plate. Therefore Eq 4-6 can be used as is, without any finite-width correction factor. All da/dN versus ΔK data points, for all eight tests at three different biaxial ratios, are presented in Fig. 4-17. A group of da/dN versus ΔK data points (developed from uniaxial load testing of center-cracked specimens and cruciform specimens without a hole) representing the baseline crack growth rate behavior of the same material is shown in Fig. 4-18. At a given ΔK level the crack growth rates are the same regardless of the biaxial stress ratio, with or without hole. The results indicate that stress intensity factors for cracks at a hole can be accurately computed by using Eq 4-6.

The Weight Function

For one-dimensional variation of stresses acting across the potential crack plane, the concept of using a weight function to determine K can be briefly described as follows: If the crack-face displacement $u(x, a)$ and the mode 1 stress intensity factor for a reference configuration, K^{re}, are known for a symmetrical loading system on a linearly elastic body containing a crack of length a, then the stress intensity factor, K, for any symmetrical loading system on that same body may be obtained from:

$$K = \int_0^a \sigma(x) \cdot \left[\frac{H}{K^{re}} \cdot \frac{\delta u(x,a)}{\delta c} \right] \cdot dx \qquad \text{(Eq 4-7)}$$

where H is a material constant (= E for plane stress and $E/(1 - \nu^2)$ for plane strain). The term inside the bracket is called the weight function. In Eq 4-7 the integration is carried out over the length of the crack, a, and $\sigma(x)$ is the stress distribution across the plane of the crack in the uncracked body loaded by the "force system" with which K is associated. The force system can be any arbitrary applied loading (for example, the irregular crack face pressure shown in Fig 4-19). In case a crack at a circular hole is loaded by uniform far-field tension stresses, $\sigma(x)$ is the local stress distribution across

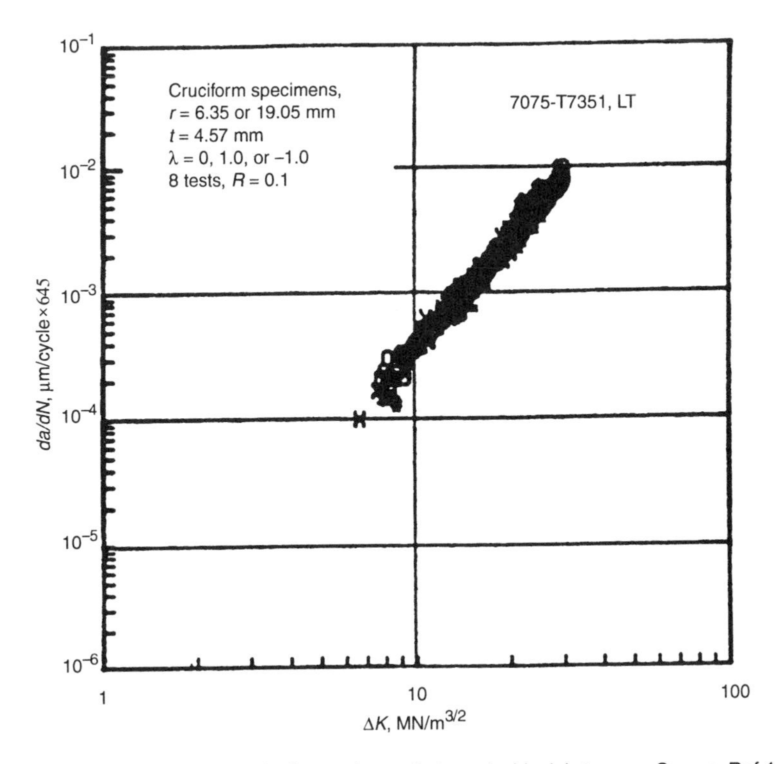

Fig. 4-17 Crack growth behavior for cracks at a hole under biaxial stresses. Source: Ref 4-19

the crack plane (without the crack present). Stress distributions are given by Kirsch (Ref 4-21) for the hole inside an uncracked, infinitely wide sheet, and by Howland (Ref 4-22) for the hole inside a finite-width strip.

For demonstration purposes we will try to obtain K-solutions for cracks coming out of a hole in an infinitely wide sheet. Both the single and double crack configurations will be considered. The sheet is loaded by far-field uniaxial tension, which causes a local stress distribution $\sigma(x)$ over the crack line. In other words, we are trying to recreate the Bowie solution by using the weight function method. We are doing this simply because a known solution is readily available for verifying the new solution.

To obtain a K-solution for one of these configurations by using Eq 4-7:

1. Obtain a mathematical representation for $\sigma(x)$.

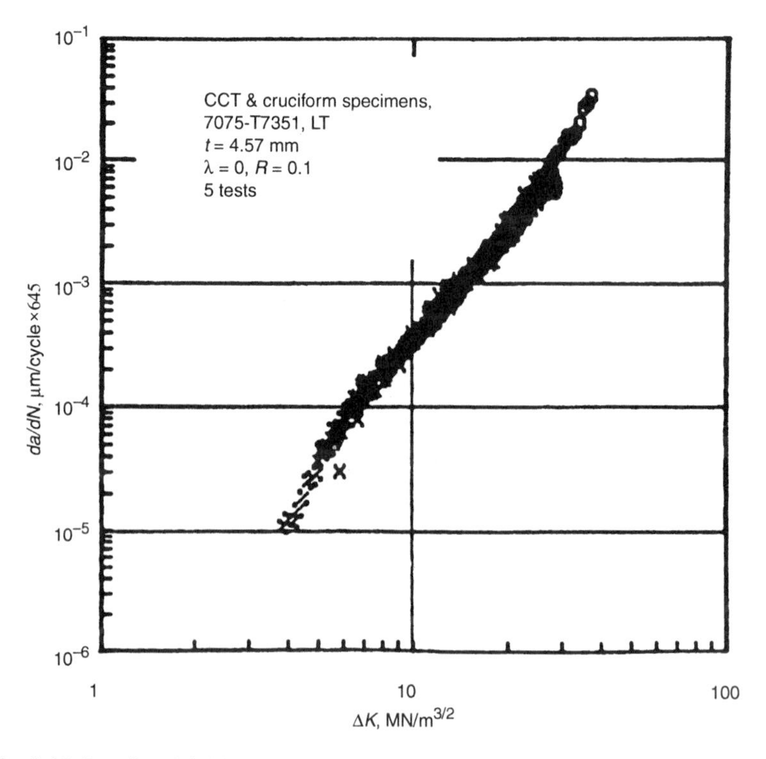

Fig. 4-18 Baseline *da/dN* behavior for 7075-T7351 aluminum. Source: Ref 4-19

2. Use the Bowie solution (for the crack in an infinite sheet subjected to far-field loading) as the reference configuration.
3. Determine the displacements along the entire crack face for the crack subjected to $\sigma(x)$.

Because this is an example of an uncracked stress distribution, $\sigma(x)$ is the same on either side of the circular hole. The Kirsch stress distribution over the crack plane can be represented by:

$$\sigma(x) = \frac{\sigma}{2}\left[2 + \left(\frac{r}{r+a}\right)^2 + 3\left(\frac{r}{r+a}\right)^4\right] \qquad \text{(Eq 4-8)}$$

where σ is the applied far-field stress, and r and a are the radius of the hole and crack length, respectively. In order to compute $\delta u/\delta a$, we need to determine the crack face displacement u by some analytical means. There are many ways to accomplish this (Ref 4-23 to 4-25). In the following we will

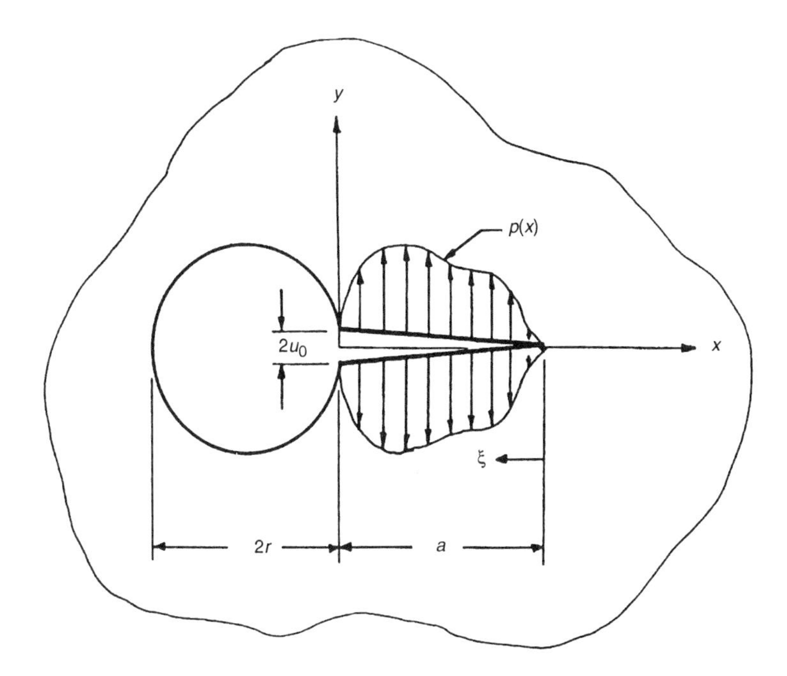

Fig. 4-19 Open hole containing a radial crack subjected to pressure $p(x)$

explain the Grandt approach in solving this problem (Ref 4-25). A finite-element model was used to determine the crack face displacements. The infinite sheet configuration was approximated by using a ratio of hole diameter to sheet width of 0.052 (Ref 4-25). It was found that the finite-element results for u can be represented by:

$$\left(\frac{u}{u_0}\right)^2 = \frac{2}{2+m}\left(\frac{\xi}{a}\right) + \frac{2}{2+m}\left(\frac{\xi}{a}\right)^2 \tag{Eq 4-9}$$

where u_0 is the displacement at the crack mouth ($\xi = a$) and m is the conic section coefficient found from:

$$m = \pi(Hu_0/2\sigma aY)^2 - 2 \tag{Eq 4-10}$$

where σ is associated with the reference configuration, i.e., it is the far-field stress that goes with the K of Bowie, and Y is obtained from the Bowie factor:

$$Y = K/\sigma\sqrt{a} = \sqrt{\pi} \cdot \left[\frac{F_1}{F_2 + a/r} + F_3\right] \tag{Eq 4-11}$$

where $F_1 = 0.8733$, $F_2 = 0.3245$, and $F_3 = 0.6762$ for a single crack, and $F_1 = 0.6865$, $F_2 = 0.2772$, and $F_3 = 0.9439$ for double symmetrical cracks.

The finite-element results for u_0 are then represented by the least square expression:

$$\frac{u_0}{r} = \sum_{i=0}^{6} D_i (a/r)^i \tag{Eq 4-12}$$

Table 4-1 Least square fit of finite-element data for crack mouth displacement

Coefficient	Single crack	Double crack
D_0	-1.567×10^{-6}	1.548×10^{-5}
D_1	6.269×10^{-4}	5.888×10^{-4}
D_2	-6.500×10^{-4}	-4.497×10^{-4}
D_3	4.466×10^{-4}	3.101×10^{-4}
D_4	-1.725×10^{-4}	-1.162×10^{-4}
D_5	3.485×10^{-5}	2.228×10^{-5}
D_6	-2.900×10^{-6}	-1.694×10^{-6}

Source: Ref 4-25

where the constants D_i are given in Table 4-1.

Using Eq 4-7 to 4-12, K-factors corresponding to crack plane stress distributions of Eq 4-8 have been computed and plotted as a function of a/r in Fig. 4-20 and 4-21. The Bowie factors corresponding to the applied far-field stresses are also plotted in these figures for comparison. It can be seen that the error is insignificant.

In summary, the weight function method, though more complicated than one would expect, is a useful engineering tool for determining the stress intensity factor for arbitrary loading conditions. It requires knowledge of the stress distribution over the crack plane and a weight function that is explicitly suitable to the configuration under consideration. The selected references listed at the end of this chapter include in-depth discussions of the theory and a vast number of examples for a variety of applications.

Empirical Method

The empirical method for determining stress intensity factor is commonly known as the James/Anderson approach (Ref 4-26). The idea is based on the principle of fatigue crack propagation, which is described in the first part of Chapter 3. That is, wherever in two different places the same magnitude of K will cause the same amount of fatigue crack growth rate. Therefore, one can simply use the actual crack growth test data obtained for the crack geometry of interest and matches them with the baseline crack growth rate

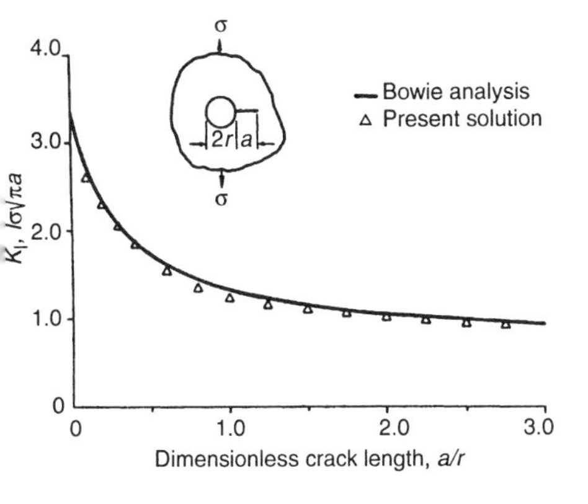

Fig. 4-20 Comparison of stress intensity factor solutions for the Bowie single crack problem. Source: Ref 4-25

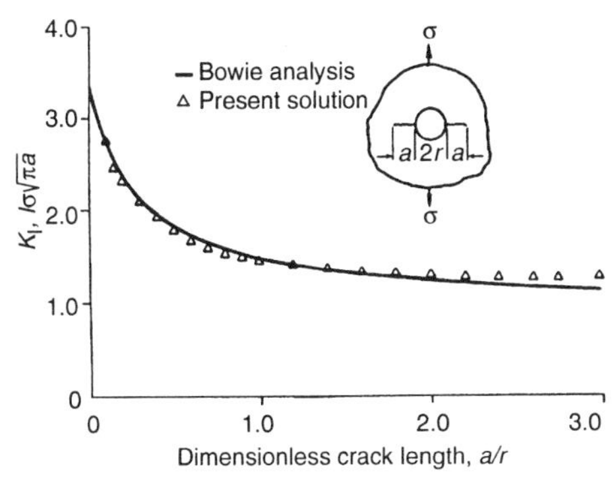

Fig. 4-21 Comparison of stress intensity factor solutions for the Bowie double cracks problem. Source: Ref 4-25

data of the same material. The K-value that pairs with a given da/dN value for the baseline material is the K-value corresponding to the new configuration. The only requirement in this method is that the specimen that is used

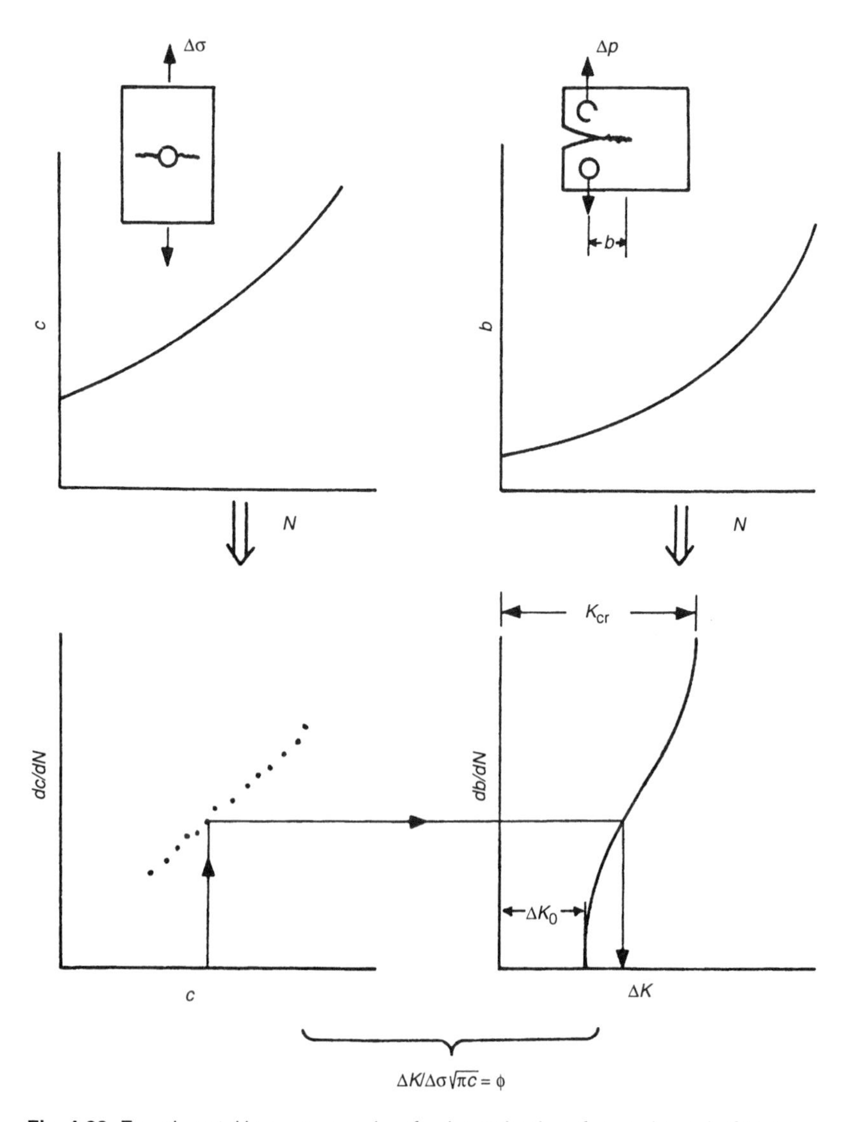

Fig. 4-22 Experimental inverse procedure for determination of stress intensity factors

for developing the baseline *da/dN* data should possess a set of known *K*-solution.

A schematic illustration of this method is given in Fig 4-22. Showing on the left side of the figure is the crack growth record (*c* versus *N*) for the configuration that involves two through-cracks coming out from a hole. This crack growth history is then reduced to *dc/dN* versus *c* format. Showing on the right of Fig. 4-22 is the baseline crack growth rate data for the same material, generated by using the compact specimen for which a known *K*-solution is available. Here we designate the crack length in a compact specimen by "*b*" just for making a distinction from those cracks coming out of a hole (which is now under consideration). As illustrated in the figure, by matching the crack growth rate of both sources, a ΔK value corresponding to each "*c*" can be found. From that we can plot another curve representing *K* versus *c*. Further, we can reduce the data (by using the relationship of $\phi = K/\sigma\sqrt{\pi c}$, or $\phi = \Delta K/\Delta\sigma\sqrt{\pi c}$) to ϕ versus *c*, or ϕ versus a dimensionless parameter (such as *c/r* as appropriate).

In case it should have been a pair of corner cracks, instead of through cracks, both *c* versus *N* and *a* versus *N* should be recorded (with *c* being the surface dimension of the corner crack, and *a* being the depth of the corner crack). After reducing these crack growth histories to *dc/dN* versus *c* and *da/dN* versus *a* format, and repeat the same procedures as described above, a pair of ϕ_C and ϕ_A can be obtained.

As a check on this approach, fatigue crack propagation tests were conducted on 15 specimens made of the 2024–T851 alloy. The specimens were 51 to 152 mm wide, containing a circular hole 6.35 or 12.7 mm in diameter, providing a wide range of ratios of plate width to hole radius. Through-the-thickness elox cuts less than 1 mm long were placed on each side of the hole, perpendicular to the loading direction. Fatigue crack propagation tests were conducted on each specimen after fatigue precracking at a relatively low stress level. The applied stress levels for fatigue crack propagation tests ranged from 51.7 to 137.8 MPa (7.5 to 20 ksi) and were kept below one-third of the material tensile yield strength. The test results were reduced as described in Fig. 4-22. Matching of *dc/dN* data points was done with baseline material crack growth rate data separately obtained from testing of standard compact specimens. The corresponding stress intensity values were then determined. A series of lumped stress intensity factors representing the combined effects due to the hole and the plate boundaries were plotted as a function of a dimensionless parameter, *c/r*, the ratio of crack length to hole radius. The empirical stress intensity factors (ϕ =

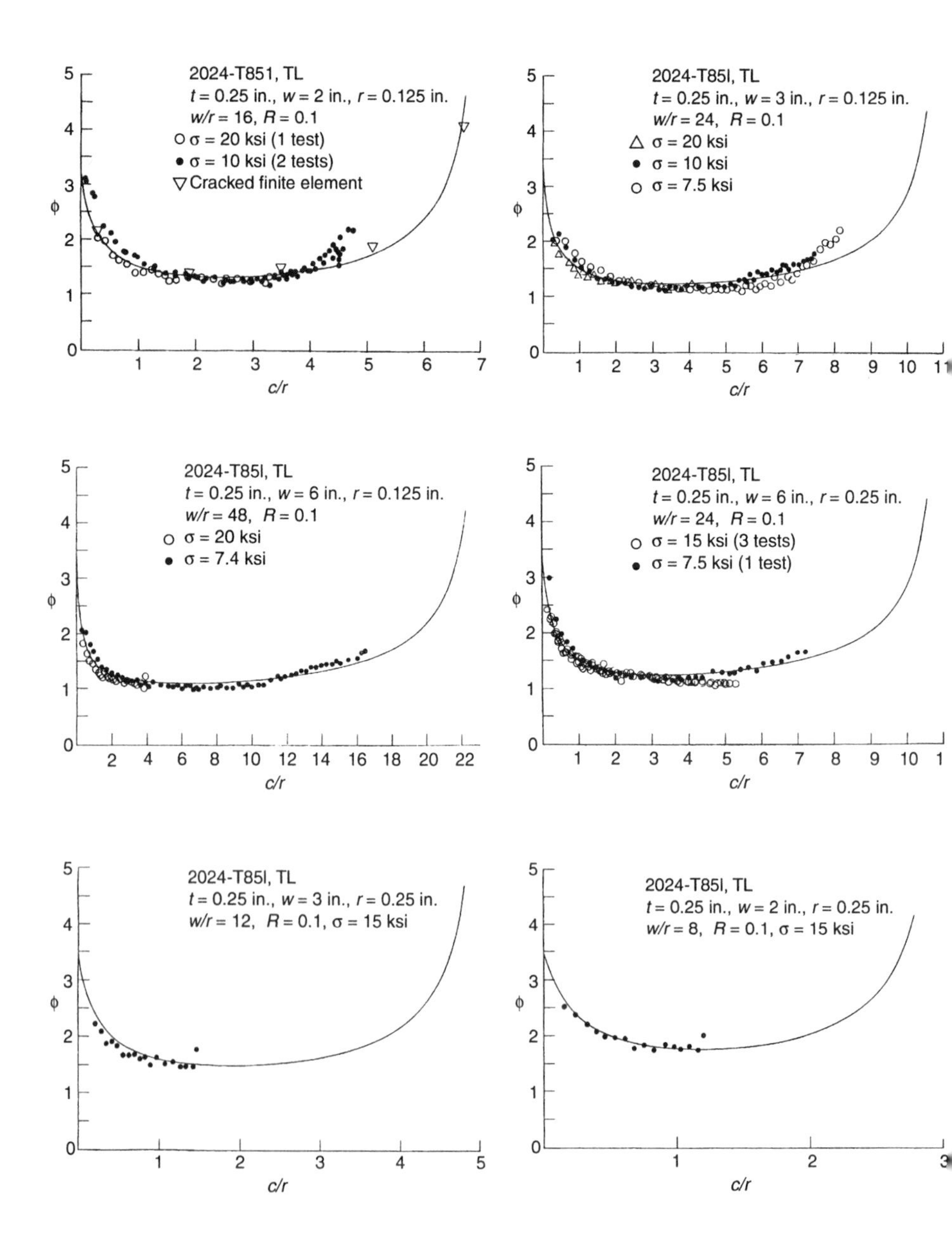

Fig. 4-23 Stress intensity factors for double cracks coming out from a circular hole; comparison of compounded solution and empirical results. Source: Ref 4-27

$K/\sigma)\sqrt{\pi c}$) are presented in Fig. 4-23 in six parts. The line that runs through the empirical data points is a compounded factor, i.e., the product of the Bowie factor and the Isida finite-width correction factor. The first part of Fig. 4-23 also contains several checkpoints computed by using the cracked finite-element method. Evidently there is agreement among all three methods. It means that the compounded factor is adequately representing the geometry and loading condition under consideration.

Though this method seems effective and efficient, one should be aware that the geometric factor extracted from the test data is a compounded factor in which all geometric variables were lumped together. Its effective range is limited. Extrapolating outside the range of the variables that were used for experimental data development may not be appropriate. Dissecting the lumped factor into parts and equating each part to a single geometric variable is desirable, but not always possible. It will become increasingly difficult for crack morphologies that involve many variables, e.g., a corner crack coming out of a hole (it involves crack shape, front and back surfaces of the plate, width of the plate, size of the hole, etc.).

Stress Intensity Reduction Techniques

Crack-Arrest Structure

The basic technique for designing crack-arrest structures is to employ an effective barrier to retard fast propagation of a crack under normal operating conditions. These barriers, or reinforcements, redistribute the stress field in the vicinity of the crack tip (i.e., they provide a region of low-stress intensity in the path of the advancing crack front). The barriers can be attached stiffeners or the riser in an integrally stiffened plank. The attached stiffeners are normally stringers or flat straps riveted onto the sheet skin. Flat straps adhesively bonded onto sheet skin are also considered to be attached stiffeners, provided that the bond has sufficient flexibility. In either case, the idea is to determine the pattern of crack tip stress intensity at various locations respective to the stiffener position. Fatigue crack growth analysis can be performed by using this pattern to modify the crack tip stress intensity so that the safe-crack-growth period can be estimated from the material fatigue crack propagation rate curve (da/dN versus ΔK curve). The residual strength (which can be interpreted in terms of fail-safe load levels) for the stiffened panel can also be estimated by using the same type of crack tip stress intensity analysis. The methods of analysis for either type of crack-arrest structures are discussed in the following paragraphs.

Attached Stiffener

Figure 4-24 shows a typical configuration for the attached stiffener structure. For this type of structural arrangement, the stress intensity factor level decreases as the crack approaches the reinforcement and significantly decreases when the crack tip is right at the vicinity of the reinforcement. The K-value will increase again as the crack propagates past the region wherein the reinforcement is effective. This temporary reduction of crack tip stress intensity is due to the reaction of the rivet forces (i.e., a portion of the skin load is transmitted through the fastener and will be carried by the stiffener). Consequently, the general stress intensity factor for this case will consist of two terms: the term involved with the overall stress acting on the skin (based on uniform stress and crack length only) and the term involved with

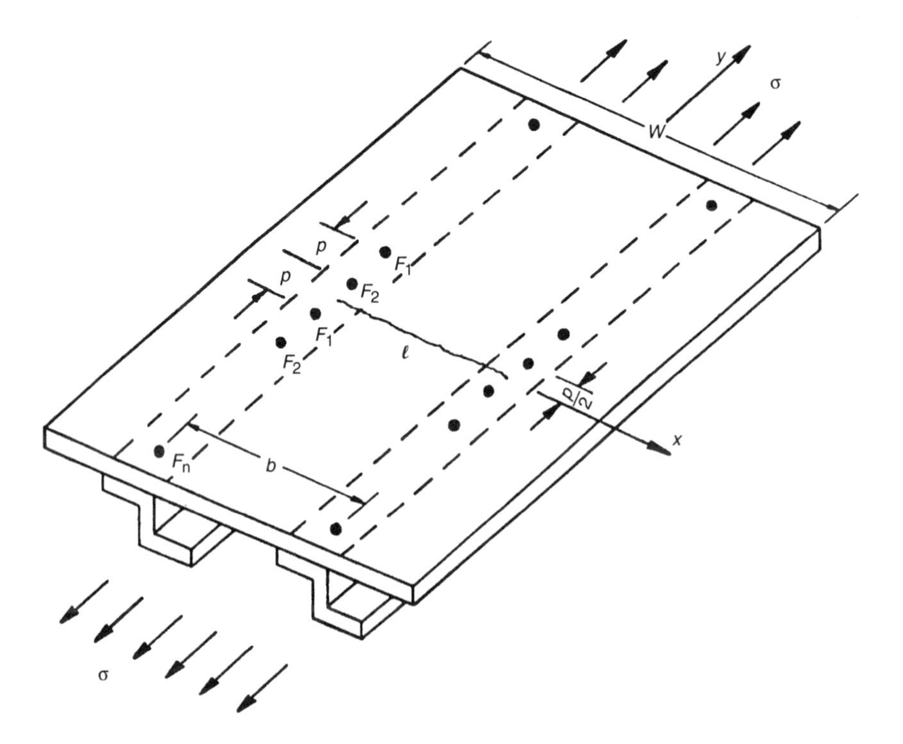

Fig. 4-24 Through-the-thickness crack in a skin-stiffened panel

the transmitted load in the reinforcement. For an infinitely wide panel, the K-expression can be written as:

$$K = \left[\sigma - \sum_{j=1}^{n} f(F_j) \right] \cdot \sqrt{\pi \cdot l/2} \quad \text{(Eq 4-13)}$$

where F_j is the fastener load of the jth fastener and l is the total crack length (i.e., same as the conventional notation, $2a$). The minus sign refers to the reduced crack tip stress intensity due to the effect of the reinforcement (the rivet forces are acting in the opposite direction respective to the applied load). A typical example of this type of analysis is schematically presented in Fig. 4-25. Here, the variations in K due to the effect of reinforcement (dotted lines) are compared with the normal values of K for the plain sheet alone (solid lines) in two arbitrarily chosen stress levels.

There have been many investigations, both analytical and experimental, involving the damage tolerance of reinforced flat panels. Only a few of these investigations have attempted to systematically study the effect of configuration and material variables. References 4-28 to 4-44 are pertinent to the present study. Three analytical investigations (Ref 4-31, 4-39, and

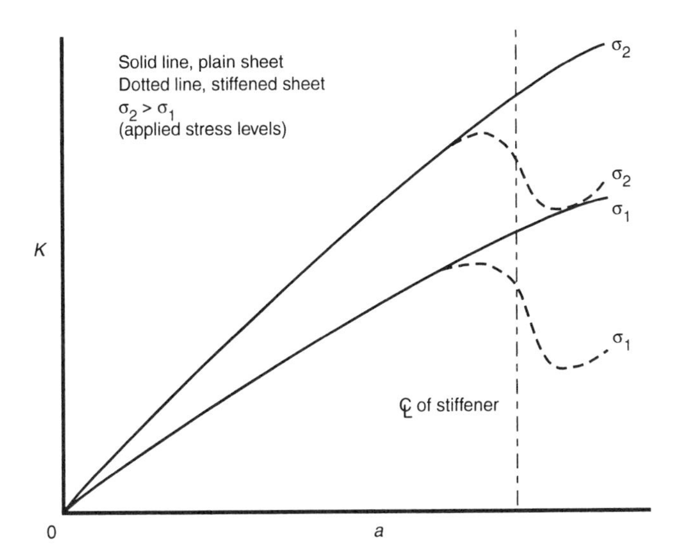

Fig. 4-25 Stress intensity analysis of unstiffened and stiffened panels

4-40), modeling an infinite linearly elastic cracked plate with linearly elastic reinforcements and rigid attachments, have established the effect of elastic material properties and geometric variables. Some highlights of Poe's results (Ref 4-31) are given in Fig. 4-26 and 4-27. In these illustrations, the stress intensity modification factor, *C*, is essentially the ratio of the stress intensity factor for the stiffened panel (Eq 4-13) to the stress intensity factor for the plain sheet (without stiffener). Two configurations are shown in these figures: the crack is either in the middle of a bay, between two stiffeners, or occupying two adjacent bays, starting at a rivet hole beneath the

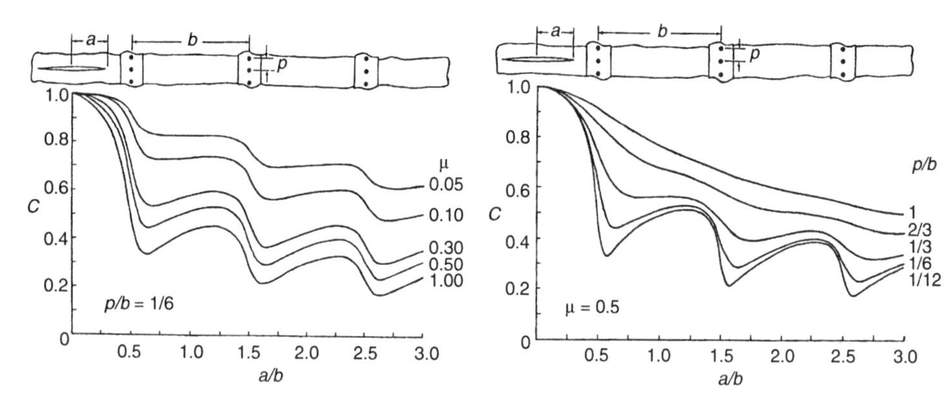

Fig. 4-26 Stress intensity factors for a one-bay crack centered between two stiffeners. Source: Ref 4-31

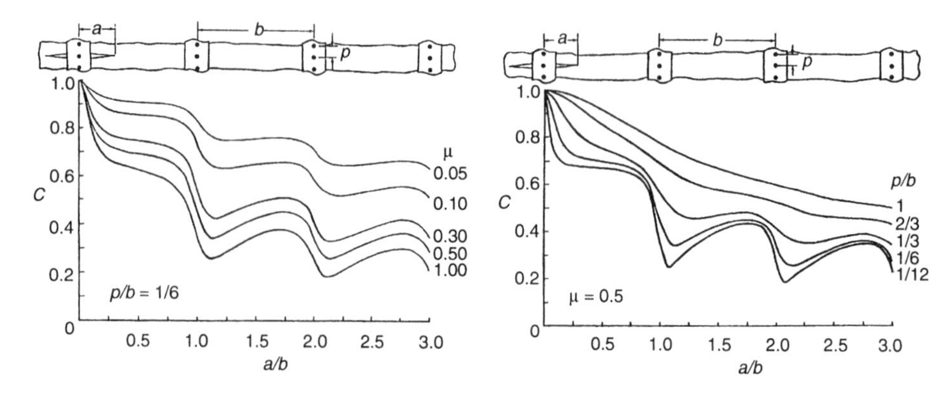

Fig. 4-27 Stress intensity factors for a two-bay crack extended equally on both sides of a stiffener. Source: Ref 4-31

center stiffener. The C-factor is interpreted to be the function of the following variables:

- Attachment spacing (rivet pitch), p
- Reinforcement spacing, b
- A relative stiffness parameter, μ, which is defined as:

$$\mu = \left[1 + \frac{b \cdot t}{A_e} \cdot \frac{E}{E_s} \right]^{-1} \quad \text{(Eq 4-14)}$$

where t is the sheet thickness, A_e is the effective cross-sectional area of the stiffener, and E and E_s are the Young's modulus for the sheet and the stiffener, respectively. For a flat strap, A_e is approximately equal to the actual size (gross area) of the strap. For the case of a stringer, an estimate for A_e is given by Ref 4-45 as:

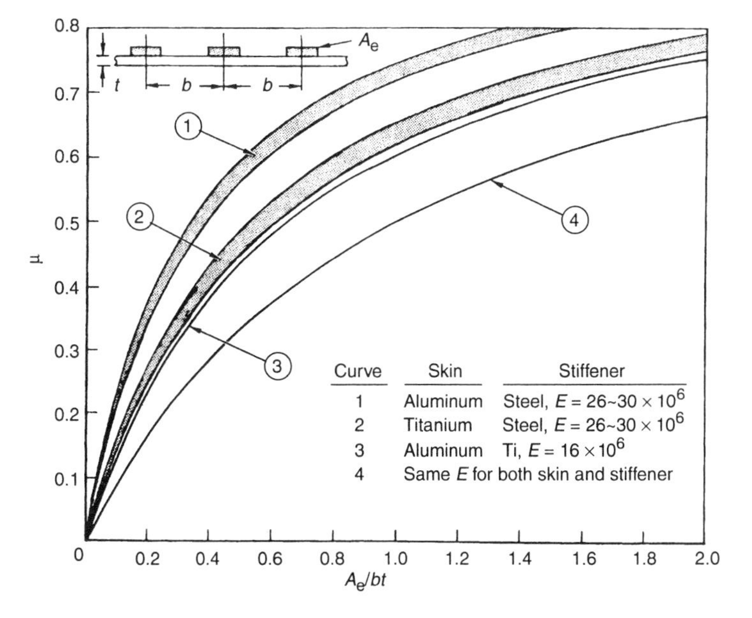

Fig. 4-28 Stiffener stiffness parameter as a function of effective stiffener area. Source: Ref 4-41

$$A_e = A\left[1 + \left(\frac{y}{\rho}\right)^2\right]^{-1} \qquad \text{(Eq 4-15)}$$

where A is the cross-sectional area of the stringer, y is the distance from the inner surface of the sheet to the centroid of the stringer, and ρ is the radius of gyration of the stringer. A plot of μ as a function of A_e to $(b \cdot t)$ ratio for several typical skin and stiffener material combinations is presented in Fig. 4-28. This set of graphs demonstrates that the ratio of stiffener modulus to sheet modulus is the key element for improving reinforcement efficiency. All other dimensions being equal, a higher μ value will result in a greater reduction in K.

In actual applications, the safe-crack-growth period can be computed by modifying the crack tip stress intensity factor with a series of appropriate C-factors, given in Fig. 4-26 and 4-27 (interpolate or extrapolate as required). The general stress intensity factor thus has the form:

$$K = \sigma\sqrt{\pi a} \cdot C \cdot \Pi\phi_i \qquad \text{(Eq 4-16)}$$

where ϕ_i is the appropriate factor accounting for the structural geometry and crack morphology. Equation 4-16 is the same as Eq 1-20, except that an extra factor C is present. In Chapter 5, more of Poe's C-factors will be given.

The mechanism of crack growth and fracture of a reinforced panel under monotonically increasing load must be understood in order for the residual strength of a reinforced structure to be estimated. In addition, other variables that might be significantly affecting the reinforcement efficiency have to be considered and properly accounted for in the stress intensity calculations. Referring to Fig. 4-3(a) and 4-25, for an unreinforced center-cracked panel (monolithic structure) the stress intensity K at the crack tip increases linearly with the normal stress component acting on the panel. As the K level increases, some point will be reached at which the crack will start to increase in length. As illustrated in Fig. 4-3(a), a crack in sheet material with sufficiently high fracture toughness will extend gradually as the load continues to increase, until it reaches the critical size at which rapid fracture occurs. As for the reinforced structure (the dotted lines in both figures), it is shown in Fig. 4-25 that the stress intensity level at the vicinity of the reinforcement is drastically reduced. The crack will not extend until a higher load level is reached. That is why a reinforcement is often called the crack stopper. If the crack in a sheet will normally grow to failure at a final stress level σ_1, a crack of the same initial size in a reinforced panel will do the following: It will grow under monotonically increasing load and slow down (K reduced), and will be capable of taking up more load until the reinforcement has reached its limit of effectiveness (e.g., may be at a stress level of σ_2). Then the crack will extend again, and probably rapid fracture will follow immediately. Therefore, the effect of the reinforcement on the residual strength is to increase the critical failure curve (σ_c versus a for a constant K_c) for the stiffened panel over the sheet alone. The fail-safe capability (the residual strength) for the stiffened panel can be estimated simply by applying a reinforcement efficiency factor onto the basic allowable stress σ_c for the plain sheet. Thus, the residual strength for the reinforced panel is:

$$\sigma_s = \gamma \cdot \sigma_c \quad \text{(Eq 4-17)}$$

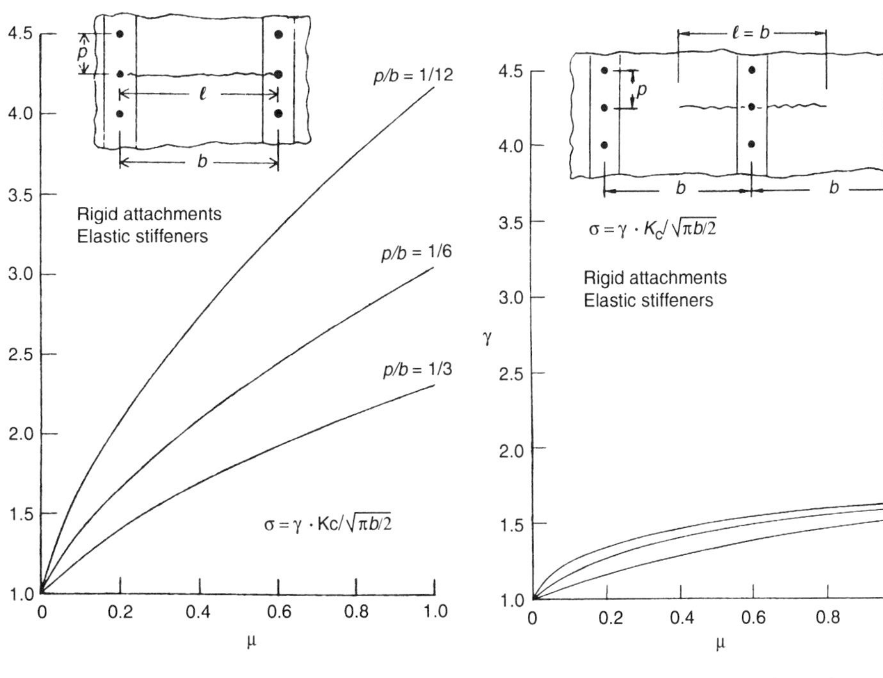

Fig. 4-29 Stiffener efficiency factor for a one-bay crack as a function of stiffener stiffness parameter

Fig. 4-30 Stiffener efficiency factor for a crack extended symmetrically on both sides of a stiffener

where $\sigma_c = K_c/\sqrt{\pi l/2}$. For all design purposes the crack length, l, can be taken as the distance between two intact stiffeners.

The value of γ for any type of reinforcement (e.g., any A_e or μ value) can be developed by experiments, by analysis, or both. For the example just discussed in Fig. 4-3(a), $\gamma = \sigma_2/\sigma_1$ and $\sigma_c = \sigma_1$.

In the analytical case (Fig. 4-3a, 4-26, 4-27), the crack is usually arrested at the vicinity of the reinforcement where the K-value is reduced to a minimum (the maximum capability for the reinforcement). In order to have failure at the same K_c value in both cases (reinforced and unreinforced, see Fig. 4-3a), assuming that the errors attributed to the slow stable tear behavior in these cases are insignificant, it is required that $\sigma_1 = \sigma_2 \cdot C_{min,}$ where C_{min} is the C value for the point at the bottom of the valley in the C versus a/b curve. Therefore:

$$\gamma = \frac{\sigma_2}{\sigma_1} = \frac{1}{C_{min}} \quad \text{(Eq 4-18)}$$

As an example, several γ versus μ curves reduced from Poe's work are presented in Fig. 4-29. The data shown here only concern the case in which the crack tips are in the vicinity of the first pair of stiffeners. In the case of a monotonically increasing load, once the crack starts to grow again the second pair of stiffeners is not likely to stop the fast-running crack.

For an airframe structure, a crack or cracks would most likely initiate from a fastener hole. Initially the crack or cracks would be developed on the skin but not on the stiffener. Under normal fatigue loading conditions, we can assume that the crack will propagate safely and be arrested at the next stiffener. The safe-crack-growth period can be estimated by using the stress intensity modification factors (as a function of crack tip location), such as those given in Fig. 4-27.

To estimate the residual strength for the same reinforced panel, there are two possibilities. At the time the monotonically increasing loads are applied onto the structure, the crack tip is either remote from the second adjacent stiffener (close to the center stiffener) or close to the second stiffener (remote from the center stiffener). Just for an exercise, we consider the first case as being a crack having a length of one bay (one-half bay length extended from each side of the center stiffener). The reinforcement efficiency factor can be developed by converting the C-factors at the midpoint of the bay given in Fig. 4-26. At this location, the efficiency for the center stiffener is maximum, and the efficiency for the second stiffener ahead of the crack is negligible. The γ factors for this case are plotted in Fig. 4-30.

For the second situation, since the crack is so long, excessive forces would have been transferred into the center stiffener, causing the center stiffener to yield or even be broken. Another possibility would be that the first pair of fasteners yielded and subsequently broke. At each crack tip, most of the load transfer activity actually took place at the first pair of fasteners, the fasteners above and below the crack line. Failure of the other fasteners will follow. This is called *fastener unzipping* (Ref 4-29).

The results of tests conducted on a wide class of stiffener configurations and material variations have indicated that for stiffeners having a small cross-sectional area, the efficiency of the reinforcement would be limited by the occurrence of reinforcement yielding (Ref 4-35). For the case of a

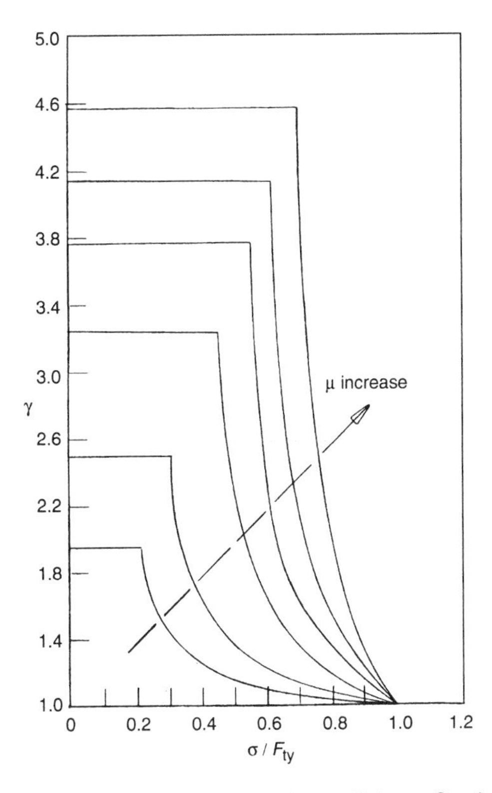

Fig. 4-31 Effects of applied stress levels on the stiffener efficiency. One-bay crack, rigid attachments. Example, not data

broken stiffener, the loads feed back into the cracked sheet, causing the K at the crack tip to increase instead of decrease. A schematic illustration of the reinforcement yielding effects is presented in Fig. 4-31. The reinforcement efficiency, γ, is plotted as a function of the ratio of the gross applied stress (σ) in the panel to the tensile yield strength of the reinforcement material (F_{ty}). The horizontal portions of the curves reflect the simple fact that until the reinforcements actually begin to yield, the yield stress does not play a role. The rapid drop in efficiency, once the reinforcement starts to yield, is clear evidence of the importance of this effect. Although this has not been proven by experiments, stiffener yielding may not be a problem for a short-crack case, which involved only with the center stiffener. Because the γ values are relatively low, very little extra load will be transferred to the center stiffener.

Therefore, at this point, it is important to consider three additional variables that significantly affect the efficiency of the reinforcement (both the C and the γ values):

- Broken stiffener
- Reinforcement yielding
- Stiffness of the attachment

The computed rivet forces, and the loads carried by the stiffener for various crack and reinforcement configurations, are also given in Ref 4-31. These data are also included in Chapter 5 of this book. For a given reinforcement configuration, the crack lengths and applied stress levels at which

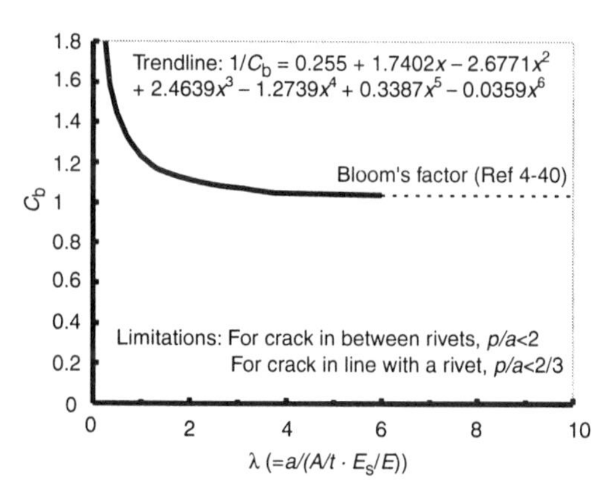

Fig. 4-32 Effect of broken stiffener on stress intensity factor. Source: Ref 4-40

the stiffener (or fastener) reaches its ultimate strength can be calculated. The effect on γ as a result of stiffener (or fastener) failure can be found accordingly. The easiest way to construct a set of parametric design curves for the case having a two-bay crack with a center stiffener broken is by superposition of two separate cases, the case without a center stiffener (Fig. 4-27) and the case involved with a broken stiffener alone. The stress intensity modification factor for the broken stiffener, C_b, taken from Ref 4-40, is presented in Fig. 4-32. An example of the result of such superposition is presented in Fig. 4-33. Because the load carried by a stiffener is proportional to its area, the full stiffener area was used to obtain the C_b values from Fig. 4-32. Each γ factor in Fig. 4-33 actually came from manipulating two separate reinforcement configurations: the C_b factor that corresponds to a single broken stiffener, and a C-factor that corresponds to a configuration adjusted to suit

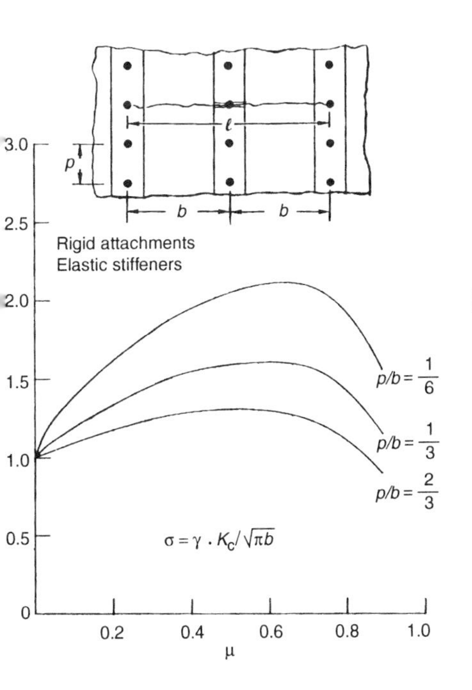

Fig. 4-33 Stiffener efficiency factor for a two-bay crack with a center broken stiffener

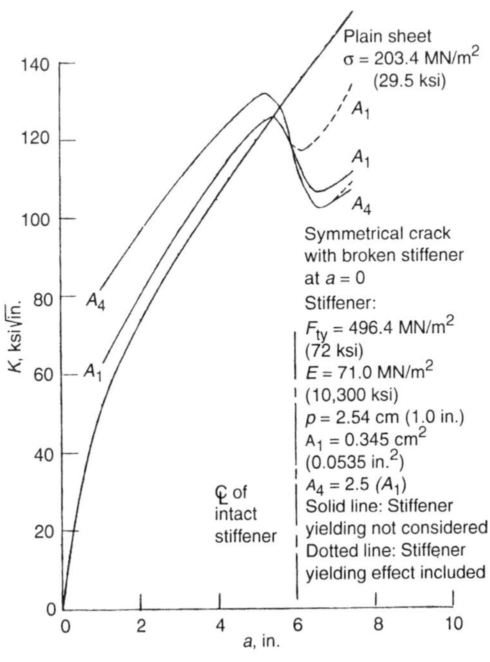

Fig. 4-34 Effects of broken stiffener and stiffener yielding on crack tip stress intensity. 1 ksi$\sqrt{\text{in.}}$ = 1.0988 MN/m$^{3/2}$. Source: Ref 4-41

the present situation. That is, the stiffener spacing for the imaginary configuration was $2b$ and therefore $p/b = 1/6$ was treated as $p/b = 1/12$.

This type of design curve provides very important information. For example, consider the case of $p/b = 1/6$ in Fig. 4-33. This curve indicates that a reinforced structure can be designed having the relative stiffness parameter μ equal 0.45 or 0.76 to obtain the same γ value (reinforcement efficiency parameter) of 2.0. If the panel is made up of aluminum skin and titanium stiffeners, the required A_e values corresponding to the μ values of 0.45 and 0.76 are 0.41 $(B \cdot t)$ and 2.05 $(B \cdot t)$, respectively. If the reinforcements are flat straps (i.e., $A_e = A$), it would mean a total difference of 500% in weight for the stiffener material.

For the curves in Fig. 4-33, C and C_b were determined assuming that both the stiffeners and the fasteners were intact and would remain elastic. The use of rigid fasteners also was assumed. There is no easy way to incorporate stiffener yielding or the effect of fastener stiffness by means of ready-made parametric factors. Insufficient data have been developed for these parameters, although limited quantities are available in the literature (Ref 4-34 and 4-35 and 4-44). To discuss these parameters, the results of a computer-aided trade study (Ref 4-34) are presented below.

The computer program essentially calculates the stress intensity as a function of crack length, as defined in Eq 4-13. The fastener load in the equation for each fastener is an unknown value. It needs to be determined as functions of all the aforementioned variables. The exercise concerns two panels, 1.2 m (48 in.) wide and 2.5 m (100 in.) long. The panels were reinforced with seven flat straps parallel to the loading direction. One of these straps was placed at the centerline of the panel. The remaining straps were placed symmetrically on either side of the centerline strap. The strap spacing was 15.24 cm (6 in.). The crack was at the midlength of the panel between two rivets with the centerline strap broken ($l = 30.48$ cm, or 12 in.). The cross-sectional area of each of the straps in one of the panels was 2½ times larger than the area of the other panel. The thickness of the skin in each case was adjusted to balance the differences in the strap area, so that the gross weight and the gross area stress of both panels would be the same. It was also assumed that flexible rivets were used in the panels (rivet stiffness was approximately equal to 700×10^6 N/m, or 4×10^6 lb/in., of defection per inch of sheet thickness).

The stress intensity factors in the vicinity of the first pair of intact stiffeners were calculated with the effect of straps yielding included (dotted lines) and ignored (solid lines). Figure 4-34 shows the comparison of the computed K-values as a function of crack length for the two reinforced panels and a plain sheet. The higher crack tip stress intensity in the reinforced sheet at the center region of the panel was primarily attributed to the damaging effect of the broken center stiffener. The resultant stress intensity modification factor, C, is plotted as dotted lines in Fig. 4-35. The K-values in Fig. 4-34 indicate that:

- In this particular example, if the effects of reinforcement yielding are ignored and the intact straps are assumed to remain elastic, the effect of

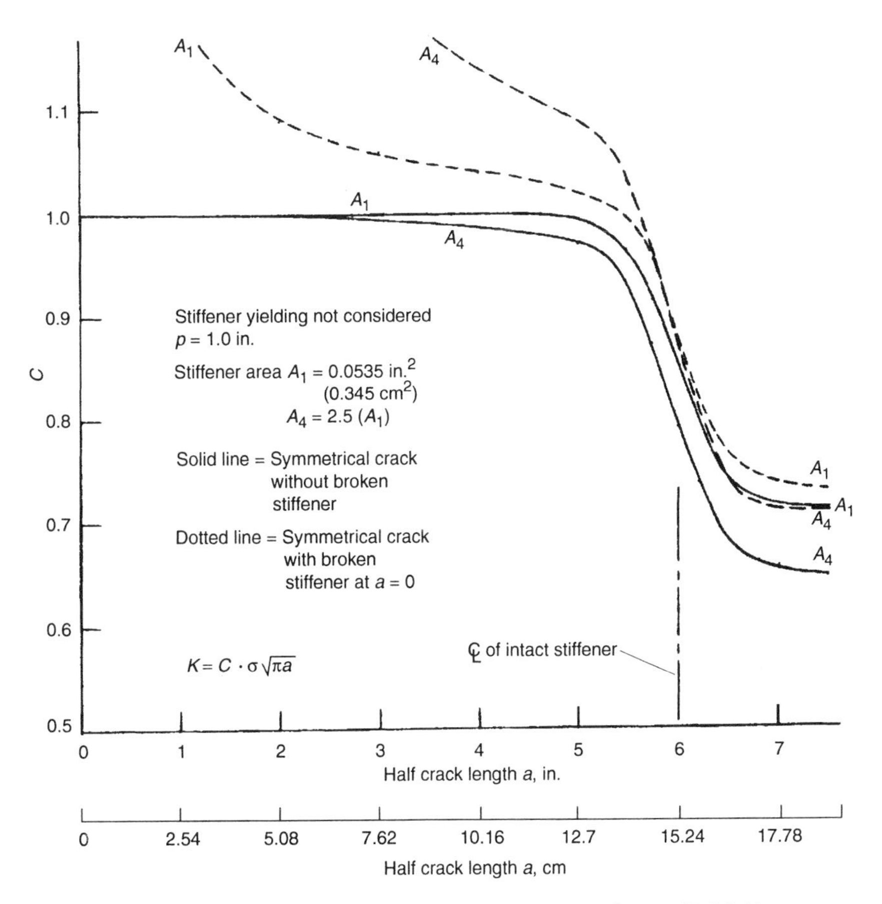

Fig. 4-35 Effect of broken stiffener on crack tip stress intensity. Source: Ref 4-41

strap cross-sectional area on the residual strength of the panel is insignificant. The gain in reduction in K (in the effective region of the intact strap) obtained by increasing the size of the strap is negligible, due in part to the broken center strap. Increasing the size of the broken strap causes higher loads to be transferred to the crack tip vicinity and, consequently, tends to increase the stress intensity factor. Had a less realistic configuration been considered, without a broken center strap, a greater effect of reinforcement area would be evident.

- When the effects of reinforcement yielding are included, the difference in reinforcement efficiency between the two sizes of straps studied is very significant.
- K-curves for the panel having a larger size of strap are almost the same for the elastic and the plastic cases. Therefore, in this particular problem, it is not necessary to design a strap area larger than that size.

In conclusion, the examples discussed above clearly demonstrate the importance of design considerations and the necessary information for imple-

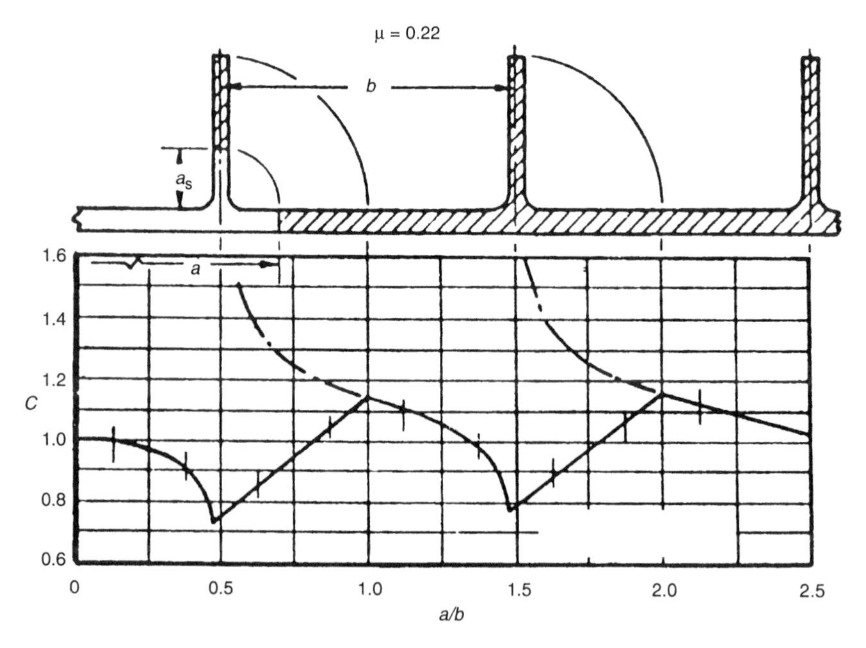

Fig. 4-36 Relationship between stress intensity factor and crack length for panel with integral stiffeners. Source: Ref 4-33

menting them in regard to structural design optimization, tradeoff, and weight savings.

Integrally Stiffened Panel

The load transfer phenomenon for the integrally stiffened panel is similar to that for the skin-stringer-type structure, but for the case of a panel with the reinforcement riveted or adhesively bonded onto the skin, the skin crack grows under the stiffener. For the case of a panel with integral risers, the crack will advance through the integral stringer as well as the sheet itself. To develop the stress intensity modification factors (the C-factors) for this type of structural configuration, the same analytical technique can be used as those used by Poe for the skin-stringer configuration. Sample C-factors (taken from Ref 4-33) are shown in Fig. 4-36 for $\mu = 0.22$. As pointed out by Poe, this C versus a/b curve was calculated by assuming a very close attachment spacing ($p/b = 1/15$) and properly accounting for the effects of the partially damaged integral stringer as the crack branches and proceeds simultaneously through the sheet and the integral riser. Fair correlation between analysis and fatigue crack growth test results was also reported (Ref 4-33). The crack arrest capabilities (under monotonically increasing load) for this type of structure are not well understood and test data of this type are not available. Therefore, no residual strength analysis procedure is presented here.

Multiple-Element Structure

Many structural designs lend themselves to partitioning in the interest of fail-safe damage tolerance, with little or no increase in cost, weight, or complications. Examples include longerons made of back-to-back channels in place of I-beams, back-to-back angles in place of T-sections, and panelization of the wing surfaces. These multi-member redundant structures, any single member of which may be completely severed, require only static strength principles to predict allowable strengths. Fatigue analysis or fatigue tests may be required to determine the safe inspection intervals after one member is broken. A special kind of multiple-element structure is a skin panel made up of several layers of thinner sheets adhesively bonded together to obtain the total desired thickness to carry the design loads.

For back-to-back members the structure must be able to support fail-safe loading conditions with one member broken. Therefore, for this type of structural arrangement, static strength analysis can be used by considering the redistribution of loads due to the broken element. For the multi-plank design case, the residual strength of the structure is determined by assuming that a crack extends completely across the width of one skin plank. The riveted splice joint provides a geometric discontinuity to serve as an effective means to interrupt crack propagation, so it is not necessary to apply fracture mechanics theory to predict residual strengths for the remaining structure. However, the failure of one skin plank will cause an elastic concentration of loading in the adjacent planks. This is basically a load redistribution problem. The fatigue crack propagation and monotonic load-carrying characteristics for these two types of structures were illustrated in Fig. 4-2 and 4-3(b). For the crack-arrest structures the propagation of the fatigue crack is a continuous process, and its safe-crack-growth period can be estimated by knowing the variations in crack tip stress intensity. However, in the case of a multiple-element structure, especially in the case of multiload path-dependent design (e.g., multiplank skins), the remaining life in the adjacent unbroken member is very difficult to estimate, because it involves such problems as reinitiation of the fatigue crack, pre-existing fatigue cracks in the adjacent elements, and so on. Therefore, only the residual strength prediction methods will be discussed here.

Laminated Sheets. A laminated panel made up of several plies of sheets or plates is classified as either:

- *A monolithic structure,* if it contains a through-the-total-thickness crack
- *A multi-element (multiload path-independent) structure,* if it contains a crack or cracks in one or some, but not all, of the plies

In either case, the advantage of lamination is that the fatigue crack growth and fracture behavior for any cracked plies in the panel are determined by the fatigue crack growth and fracture toughness properties for the individual sheet alone. An effect that is rather familiar is that both the fatigue crack growth rate and fracture toughness are functions of material thickness. Consider a wing skin plank that has a thickness of 1.27 cm (0.5 in.) in order to meet the static strength and fatigue requirements. Taking advantage of the difference in K_c values between 0.254 cm (0.1 in.) thick sheet and 1.27 cm (0.5 in.) thick plate, the wing shin plank can be made up of five plies of 0.254 cm (0.1 in.) thick sheet (in the fracture mechanics point of view only). In this design, a substantially larger crack can be tolerated because for some materials the fracture toughness of the 0.254 cm (0.1 in.) thick sheet is much higher than the fracture toughness of the 1.27 cm (0.5 in.) thick plate. The failure mode in the laminate design would be all plane-stress K_c failure, because the crack in each individual sheet would most likely be a through-the-thickness crack. Even if the crack started out as a scratch (i.e., surface flaw), it would soon grow to become a through-the-thickness crack, and so failure in plane-strain K_{Ic} mode would be avoided. Furthermore, the laminated sheet design would exhibit much longer fatigue crack growth life, because fatigue crack growth rate is usually slower in the thinner sheet. Because the critical crack length is larger, as determined by a higher K_c, it would require to take many more fatigue cycles to grow the crack. A last point to make is that statistically it is unlikely that all five sheets would contain cracks initially. Stress analysis might show that the remaining plies could take the total load after one or two plies failed.

No matter how good this approach may sound from the fracture mechanics point of view, a final decision on design changes still should be based on rigidity requirements, complexities in manufacturing, and structural functional considerations. Nonetheless, this approach to improving structural life without paying a weight penalty is applicable to many other structural members, machinery, and ground vehicle components, and worth the trouble to perform trade studies during design/sizing.

Multiple Plank Structure. In the multiload path-dependent design, the fail-safe load-carrying capability in the remaining structure after failure of one major structural element relies on the load-redistribution characteristics of the structural system. Finite-element analysis is the tool most commonly used for these purposes. Prior to the availability of the present efficient structural modeling computer software, engineers had to use semiempirical methods such as those described in Ref 4-46 and 4-47. These methods belong to the category of static strength analysis and will not be discussed here.

The Characteristics of a Yielded Hole

Residual stresses often bring about beneficial effects on structural fatigue life. This is particularly true in the case of fastener holes, or at locations of any type of stress concentration. Preyielding the material around the hole by way of cold working can set up a desired residual stress field in the vicinity of the hole. Improvement of the fatigue life of a hole usually is accomplished by implementing the technique of interference fit or cold work. Due to geometric stress concentration, residual stress also can occur at an open hole after a high load. Whatever the case may be, the mechanics of crack growth is the superposition of the applied and pre-existing stresses. Therefore, the first step in analyzing crack growth at a yielded hole is to determine the residual stress distribution of an uncracked hole. The weight function technique will then be applied to determine the stress intensity factors corresponding to a given residual stress distribution. The superimposed stress intensity factors are used for making crack growth prediction.

In the following sections some techniques for determining the residual stress distributions for these three types of yielded holes will be presented. The mechanics of crack growth in each type of these compounded stress fields will be discussed.

Open Hole

Structural details often play an important role in contributing to fatigue crack growth behavior. The local area in the vicinity of a geometric stress raiser, or a cutout, is known to experience a nonuniform distribution of stresses. Taking a circular hole, for example, the local tangential stress at the hole edge is at least three times the applied far-field stress. The local stresses gradually diminish and eventually disperse to the nominal stress

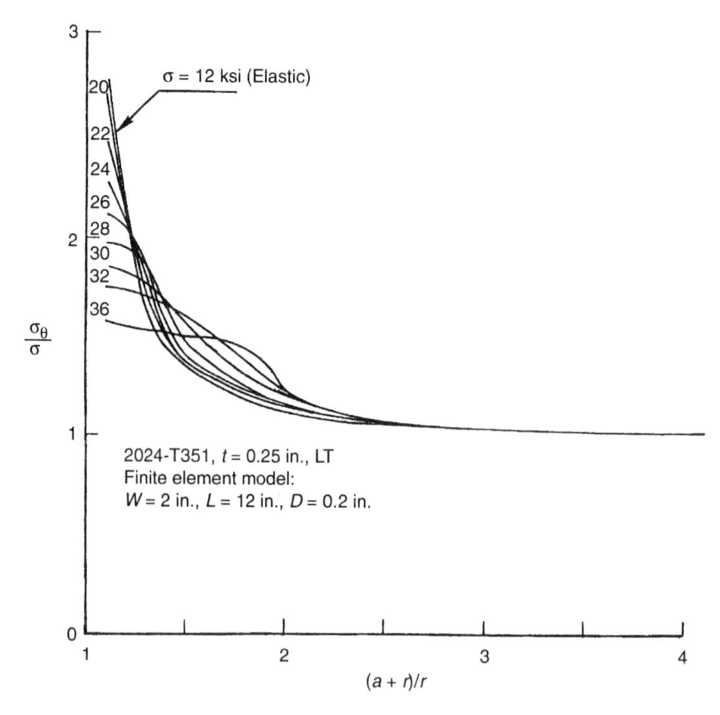

Fig. 4-37 Uncracked stress distribution in the vicinity of an open hole as a function of applied uniform far-field stress levels. Source: Ref 4-27

field at a distance away from the hole. In the theory of elasticity, this local stress distribution is represented by the commonly known Kirsch/Howland solutions (Ref 4-21, 4-22). Because the material cannot forever follow Hook's law at all levels of the applied load, in reality it follows the stress-strain relationship of the tensile stress-strain curve. Therefore, the material at the hole edge will undergo plastic deformation whenever the applied gross area stress exceeds one-third of the material tensile yield strength. Precisely speaking, nonlinear behavior starts at the proportional limit, a point on the tensile stress-strain curve. However, for all engineering applications this distinction can be regarded as unimportant.

Local Stresses around a Hole. Once the material starts to yield, the stress concentration factor will be altered, as well as the entire local stress distribution over the crack plane that is normally represented by the Kirsch/Howland solutions. Depending on the applied stress level, the stress concentration factor (the ratio of local stress to far-field stress at the hole edge) is proportionally reduced. The cross-the-width stress distribution also changes accordingly. The result of an elastic-plastic analysis has shown that the local stress distributions vary depending on the applied stress level. An example of such variation is shown in Fig 4-37. In this figure the local tangential stresses (corresponding to applied stress levels from 12 to 36 ksi) were computed by using the NASTRAN structural analysis computer code. The stepwise linear option of the NASTRAN code was used to determine the plastic stresses. The tensile stress-strain curve that was obtained for the same material used in this example is shown in Fig. 4-38. The proportional limit and the tensile yield strength for the 2024-T351 alloy were 345 MPa (50 ksi) and 372 MPa (54 ksi), respectively, so the stress distribution for a 12 ksi applied stress level is considered to be pure elastic. It is also shown in Fig. 4-37 that the local stresses deviate from the elastic stress curve as soon as the applied stress level exceeds one-third the proportional limit of the material. Should the applied stress continue to increase beyond 36 ksi, the plastic stress curve would eventually become flat, meaning that net section yielding across the entire plate width has been reached. The effect of stress concentration disappears because K_T is equal to unity.

Residual Stresses around a Hole. Upon unloading from a given point in the plastic range of the tensile stress-strain curve, the plastically deformed material in the vicinity of the hole will be subjected to restoring forces by the surrounding elastic material. In other words, the plate material follows Hook's law during unloading. The result is the creation of a residual stress field. The residual stress at a given local point in the plate (i.e., at a distance

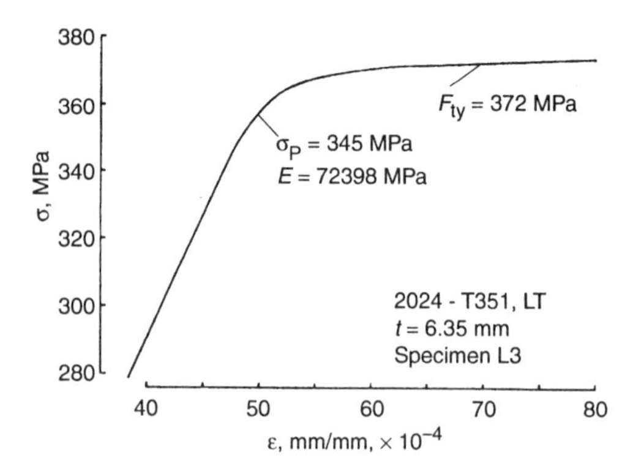

Fig. 4-38 Tensile stress-strain curve. Source: Ref 4-27

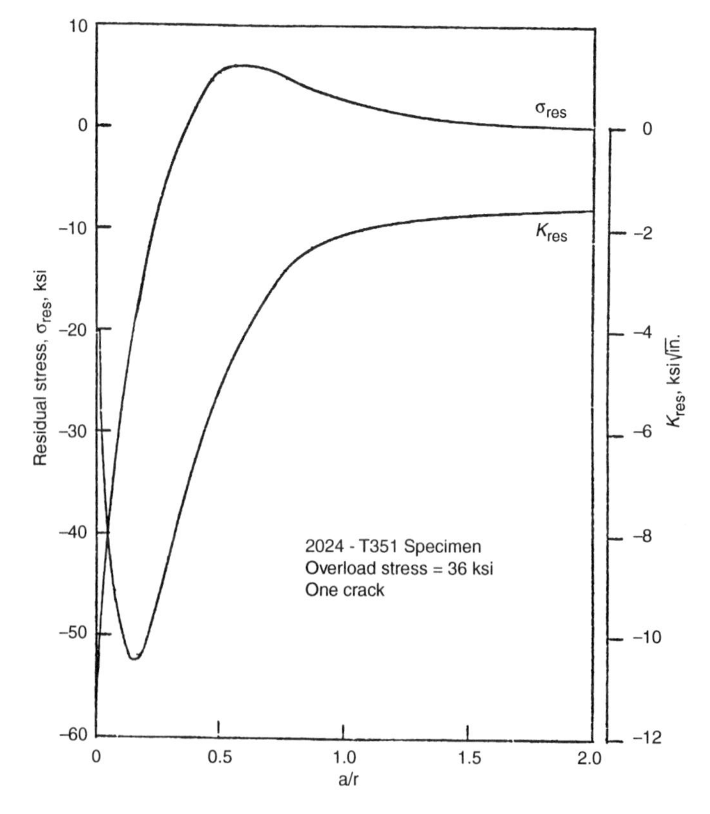

Fig. 4-39 Distribution of residual stresses and residual stress intensity factors after overload. Source: Ref 4-27

from the hole edge) is simply computed by subtracting the elastic stress (i.e., Eq 4-8 or equivalent) from the plastic stress. In this case the elastic stress distribution is the one corresponding to the 12 ksi applied stress. An example of the results, for which the residual stress corresponds to a 36 ksi overload, is depicted in Fig. 4-39. After the residual stress distribution is properly determined, its corresponding stress intensity factor K_{res} can be determined by integration of the weight function. The result (the computed K_{res} corresponding to the 36 ksi overload) is also shown in Fig. 4-39. In this case the weight function of Impellizzeri and Rich (Ref 4-24) was used, as a matter of availability and convenience.

Crack Growth in a Compounded Stress Field. Consider now the subsequent reloading of the plate. The resulting stress field in the vicinity of the hole will be the combination of the residual and applied stresses. If the stress field that would exist due to the new applied stress acting alone is elastic, i.e., when the subsequently applied stress level is low enough, then the resulting stress field may be determined by superposition.

In the case of constant-amplitude loading, the applied maximum and minimum stress levels are held constant so that the applied R-ratio is a constant while crack propagation is in progress. However, the pre-existing residual stress is not a constant across the specimen width. Therefore, the effective R-ratio (R_{eff}) corresponding to the superimposed stresses is a func-

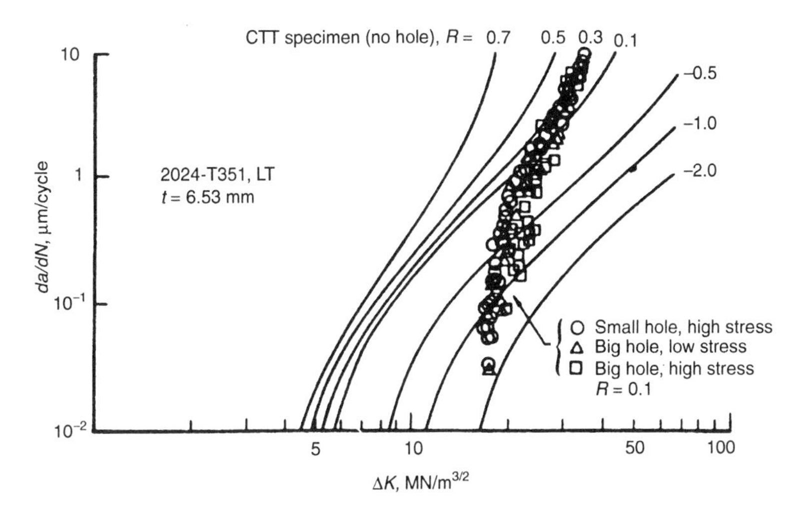

Fig. 4-40 Crack growth rate behavior of specimens containing a preyielded hole. Source: Ref 4-48

tion of crack length. With the assumption that the applied stress does not significantly alter the state of residual stress during crack propagation, the effective stress intensity range at a crack length along the entire crack propagation path can be formulated as:

$$\begin{aligned}(\Delta K)_{eff} &= (K_{max})_{eff} - (K_{min})_{eff} \\ &= (K_{max,\sigma} + K_{res}) - (K_{min,\sigma} + K_{res}) \\ &= K_{max,\sigma} - K_{min,\sigma}\end{aligned} \quad \text{(Eq 4-19)}$$

In other words, the operational stress intensity range had never changed; that is:

$$(\Delta K)_{eff} = (\Delta K)_{\sigma} \quad \text{(Eq 4-20)}$$

Therefore, the effective stress intensity ratio is:

$$R_{eff} = \frac{K_{min,\sigma} + K_{res}}{K_{max,\sigma} + K_{res}} \neq R_{\sigma} \quad \text{(Eq 4-21)}$$

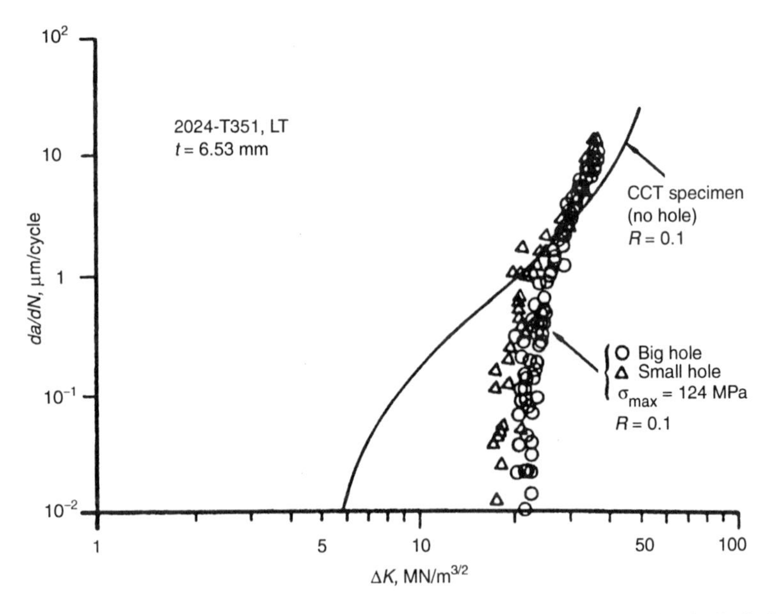

Fig. 4-41 Crack growth rate behavior of specimens containing an overloaded hole. Source: Ref 4-48

Here, the subscripts "σ" and "res" denote the applied stress and the residual stress, respectively. It is clear from Eq 4-20 and 4-21 that the change in crack growth rate is due to the change from R_{σ} to R_{eff}. Therefore, crack growth in a compounded stress field can be considered as if it were propagating under a sequence of applied elastic stresses, i.e., $(\Delta K)_{\sigma}$, but having its corresponding R-ratios modified by the pre-existing residual stresses.

To demonstrate this type of crack growth behavior, the results of an experimental test program, which was designed to generate cyclic crack growth rate data to test the hypothesis (Ref 4-27, 4-48), are presented below.

The test matrix consisted of specimens with and without a hole. The specimens were made of 2024-T351 aluminum, LT, $t = 6.33$ mm, $W = 152.4$ mm. The material baseline data were developed by using center-cracked specimens. The raw data, which were shown in Fig. 3-9, are presented here as Fig. 4-40. For clarity, only the fitted *da/dN* curves (for seven R-ratios) are shown in the figure; the raw test data points are not included. Here again, the independent variable, ΔK, is defined to be the full stress intensity range, i.e., $(\Delta K)_{\sigma} \neq K_{max}$ for $R < 0$.

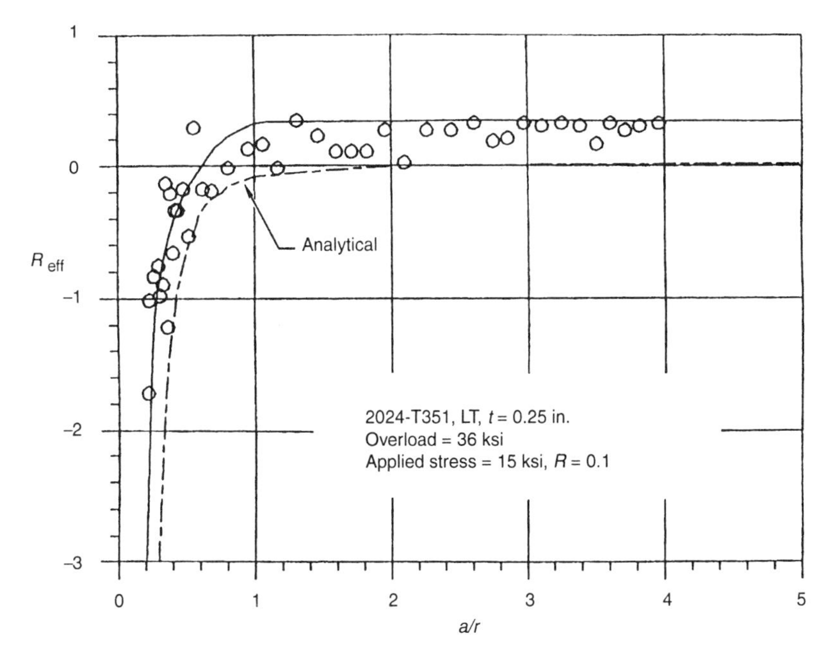

Fig. 4-42 Comparison of experimental and analytical effective-R distribution adjacent to a yielded hole. Source: Ref 4-27

In addition to the center-cracked specimens, eight specimens of the same material were fabricated with a circular hole at the center. The hole size was either 12.7 or 19.05 mm in diameter. Five of these specimens were preloaded to a gross area stress level of 248 MPa (2/3 F_{ty}). Following unloading, after the preload was removed, an elox cut was made at one side of the hole, and the specimen was then precracked by applying fatigue cycling. The other three specimens were also subjected to a 248 MPa preload; however, the high load was applied after precracking, simulating the classic crack growth retardation testing. In the remainder of this chapter, the former will be referred to as the preyielded hole and the latter will be referred to as the overloaded hole. The terms *cracked hole* and *yielded hole* are applicable to either type of these specimens. All eight specimens were then subjected to crack propagation testing at constant-amplitude stress cycles (σ_{max} = 103.35 or 124.03 MPa with $R_\sigma = 0.1$).

All the recorded "crack length versus cycles" data points were reduced to the *da*/*dN* versus ΔK format. The *K*-values were computed based on the applied far-field stress, with applicable stress intensity solutions given in Chapter 5. The data points for crack growth rate versus $(\Delta K)_\sigma$ are divided into two groups, one for the preyielded hole and the other for the overloaded hole. The cracked hole data, superimposed with the material baseline data, are presented in Fig. 4-40 and 4-41. It is apparent that the crack growth rate behavior for the cracked hole specimens did not resemble the crack growth

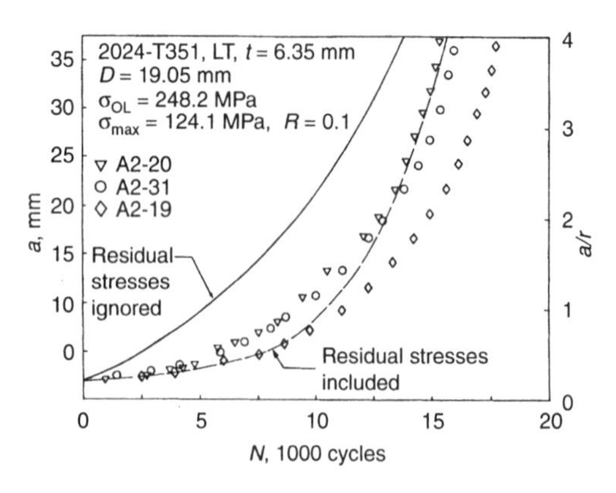

Fig. 4-43 Comparison between test data and prediction for crack growth from a preyielded hole. Source: Ref 4-27

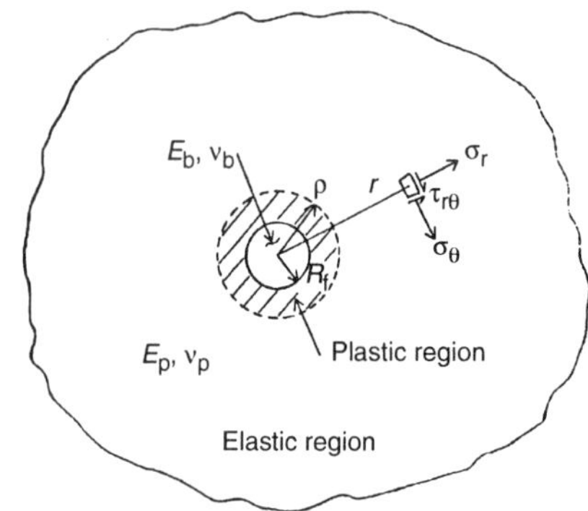

Fig. 4-44 Coordinate system for hole with interference-fit bolt

rates for the center-cracked specimens. The *da/dN* data points for the cracked hole specimens at $R_\sigma = 0.1$ behaved as if the crack were propagating under a series of *R*-ratios (the effective *R*-ratio changed from –2.0 to +0.3), as had been hypothesized. Although there is experimental crack growth rate scatter corresponding to each ΔK level (i.e., for a given crack length), the variation of test data points on the ΔK scale was basically due to the effects of specimen geometry (hole size, crack size) and cyclic stress level. By converting the *da/dN* data points to R_{eff}, and plotting them as a function of *a/r*, we can compare the test results with those derived from Eq 4-21. The results for one of the preyielded hole specimens, which is typical of all the cracked hole specimens tested are presented in Fig. 4-42. The analytically determined R_{eff} values fall on the lower bound of the experimental scatter.

In making correlations with test data, we can simply compute the crack growth history of a test using the stress intensity factors and effective *R*-ratios per Eq 4-20 and 4-21. Just for the purpose of illustration, test data for three specimens (out of the eight tested) are plotted in Fig. 4-43. These specimens had the same geometry, overload stress level, and applied constant-amplitude stress levels. The analytical crack growth histories, computed with and without residual stresses, are also shown in this figure for comparison. Because the initial crack sizes (after precracking) for these specimens were not identical, the data were plotted using a common initial crack size having stress cycles adjusted. Evidently significant improvement on the predicted life was obtained by including the residual stresses. However, the effective *R*-ratios used in this life prediction are the hand-fitted average of the experimental data, not Eq 4-21. The predicted life would have been slightly longer if Eq 4-21 had been used here.

Interference-Fit Hole

Residual stresses around an interference-fit hole are the result of inserting an oversized pin (e.g., oversized fastener) into the hole. The interference between the bolt and the plate is defined by:

$$\delta = R_b - R_p \qquad \text{(Eq 4-22)}$$

where R_b and R_p are the initial radius of the bolt and the hole, respectively. Both the strength and modulus of the bolt are greater than (or at least equal to) those for the plate, so that the bolt will deform elastically. For sufficiently large interference, the plate material around the bolt hole will undergo plastic yielding due to buildup of excessive pressure between the plate and the bolt. The size of the yielded zone and the stress distribution in the plate depend on the magnitude of δ. Figure 4-44 shows a sketch of a bolt inserted into a circular hole in a plate. The area bounded by a radius ρ and the final radius of the hole R_f depicts the plastic region in the plate. The dimension of R_f is equal to the sum of the initial radius of the hole and the radial displacement in the plate.

Note there are two limiting cases on the contact pressure. The first one occurs when the stress in the bolt reaches the tensile yield strength of the bolt. The bolt cannot impose any more pressure on the hole, so that P (the pressure induced by the interference fit) reaches a maximum value. The second case is that when the contact pressure approaches a magnitude equivalent to 115.5% of the tensile yield strength of the plate, excessive yielding occurs in the plate. The process of building up higher contact pressure by further increase in interference will be retarded (Ref 4-49).

While the bolt remains in the hole, the internal stresses around the hole that were created by the contact pressure also remain in the plate. For convenience, we call these internal stresses the interference-fit stresses (i.e., the stresses induced by the interference of the bolt and the hole). During fatigue loading, crack growth is a function of a compounded stress field (i.e., the combination of the applied stresses and the pre-existing interference-fit stresses). Therefore, the objective of the present analysis is to determine the interference-fit stress distribution in terms of δ.

There are many solutions of interference-fit-stresses available in the literature. Some of them are listed in the "Selected References" section at the end of this chapter. Most of the literature presents a huge set of fairly involved equations, making it almost impossible to present them here in a systematic manner. Worse, many of the solutions present the interference-fit stresses as a function of contact pressure, not δ. This type of solution is not directly applicable to the present problem. To put this problem in a clear perspective, a simple, straightforward derivation of a solution for the interference-fit stress in terms of δ is provided in the following.

Derivation of the Interference-Fit Stress Distribution. Consider a plate equivalent to a flat ring, having its outside circumference at infinity. Assuming an elastic/ideally plastic material for the plate and a plane-stress condition, the equilibrium equation governing the stress distribution in the plate is (Ref 4-49):

$$\frac{d\sigma_r}{dr} + \frac{\sigma_r - \sigma_\theta}{r} = 0 \quad \text{(Eq 4-23)}$$

The elastic stresses at a point (r, θ) in the elastic region (i.e., $r > \rho$) are given as:

$$\sigma_\theta = -\sigma_r = \frac{F_{ty}}{\sqrt{3}}\left(\frac{\rho}{r}\right)^2 \quad \text{(Eq 4-24)}$$

where F_{ty} is the tensile yield strength for the plate. In the plastically deformed region, i.e., $r < \rho$, the solution for Eq 4-23 is:

$$\ln\left(\frac{r}{\rho}\right) = \frac{\sqrt{3}}{2}\sin^{-1}\left(\frac{\sqrt{3}}{2}\cdot\frac{\sigma_r}{F_{ty}}\right) - \frac{1}{2}\ln\left[\sqrt{1-\frac{3}{4}\left(\frac{\sigma_r}{F_{ty}}\right) - \frac{1}{2}\left(\frac{\sigma r}{F_{ty}}\right)}\right] - \frac{\sqrt{3}}{2}\sin^{-1}\left(\frac{-1}{2}\right) + \frac{1}{2}\ln\left(\frac{2}{\sqrt{3}}\right) \quad \text{(Eq 4-25)}$$

This solution satisfies the boundary condition, $\sigma_r = -F_{ty}/\sqrt{3}$ at $r = \rho$, and the Von Mises-Henkey criterion of plastic yielding:

$$\sigma_r^2 - \sigma_\theta\sigma_r + \sigma_\theta^2 = F_{ty}^2 \quad \text{(Eq 4-26)}$$

Both σ_r and ρ in Eq 4-25 are dependent variables. Their values, in terms of δ, can be determined as follows.

The plate yields at $P = F_{ty}/\sqrt{3}$ and the bolt yields at $P = F_{tyb}$, where P is the pressure induced by the interference fit and F_{tyb} is the tensile yield strength of the bolt. It is assumed that the bolt deforms elastically and that the plate material near the hole may deform plastically, depending on the severity of interference, so the displacement on the bolt will be (Ref 4-50):

$$U_b = \frac{-P}{E_b}\cdot R_b(1-\nu_b) \quad \text{(Eq 4-27)}$$

where E_b and ν_b are the Young's modulus and Poisson's ratio of the bolt, respectively. The plastic displacement of the plate, at the bolt and plate interface, will be:

$$U_p = \frac{1}{2\sqrt{3}}\cdot\frac{F_{ty}}{G_p}\cdot\frac{\rho^2}{R_p} \quad \text{(Eq 4-28)}$$

G_p is the shear modulus of the plate. Because:

$$|U_b| + U_p = \delta \quad \text{(Eq 4-29)}$$

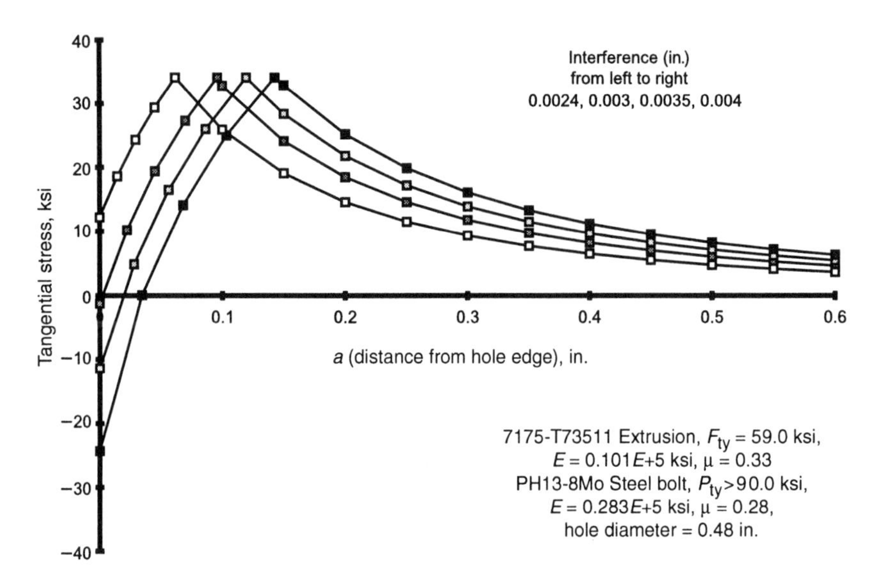

Fig. 4-45 Tangential stress distributions in the vicinity of an interference-fit hole, for aluminum plate and steel bolt

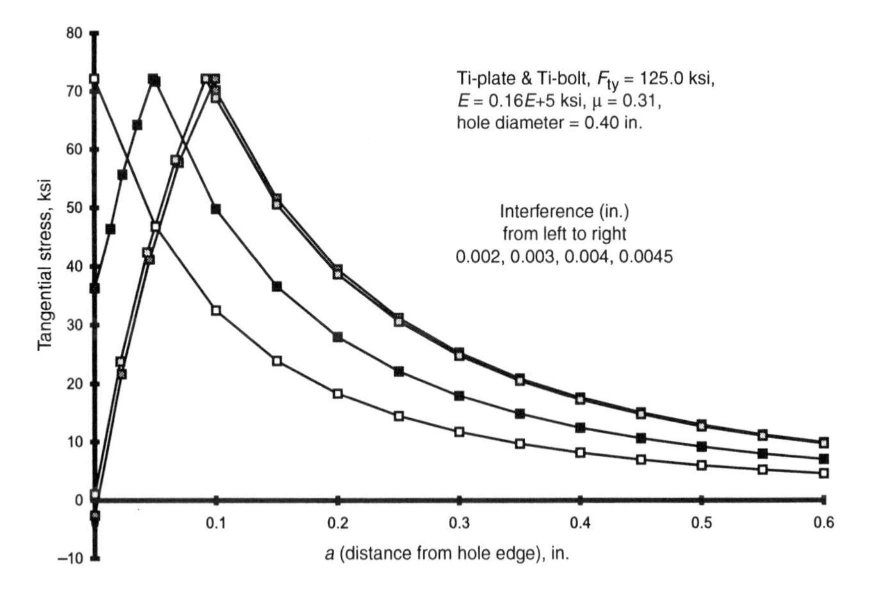

Fig. 4-46 Tangential stress distributions in the vicinity of an interference-fit hole, for titanium plate and bolt

therefore, after adding Eq 4-27 and 4-28, and solving for ρ, we have:

$$\rho^2 = \frac{2\sqrt{3}G_p R_p}{F_{ty}}\left[\delta - P\frac{R_b}{E_b}(1-\nu_b)\right] \quad \text{(Eq 4-30)}$$

Substituting ρ into Eq 4-25, and relating P and δ through the boundary condition, i.e., $-\sigma_r = P$ at $r = R_p$, Eq 4-25 becomes:

$$\frac{1}{2}\ln\left\{\frac{2\sqrt{3}\,G_p}{F_{ty}R_p}\left[\delta - P\frac{R_b}{E_b}(1-\nu_b)\right]\right\}$$

$$= \frac{1}{2}\ln\left[\frac{1}{2}\frac{P}{F_{ty}} + \sqrt{1-\frac{3}{4}\left(\frac{P}{F_{ty}}\right)^2}\right] - \frac{\sqrt{3}}{2}\sin^{-1}\left(-\frac{\sqrt{3}}{2}\frac{P}{F_{ty}}\right) - 0.52537 \quad \text{(Eq 4-31)}$$

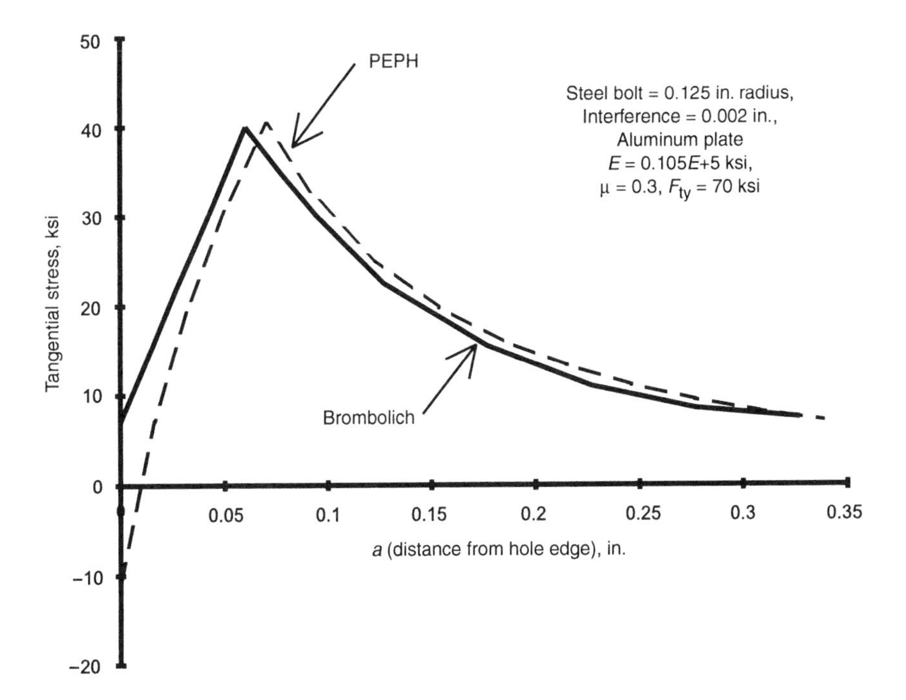

Fig. 4-47 Comparison between plasticity solution and elastic-plastic finite-element result, for aluminum plate and steel bolt

For a given δ and R_p, P can be obtained by iteration. The value of P is then substituted back into Eq 4-30 to solve for ρ. Finally, the radial stress σ_r inside the plastic region can be determined by using Eq 4-25. The plastic tangential stress σ_θ is obtained through the Von Mises-Henkey relationship, i.e., Eq 4-26.

A computer code, PEPH, which contains numerical procedures for computing the radial and tangential stress distributions based on Eq 4-22 to 4-31, is available in the public domain (Ref 4-51). By using this code, the results of two sample problems have been obtained and are presented in Fig. 4-45 and 4-46. These figures present the tangential stress distributions for various levels of interference between an aluminum plate and a steel bolt, and the case in which both plate and bolt are made of titanium. These figures also show that the tangential stress at the edge of the hole decreases as the level of interference increases. The peak tangential stress remains

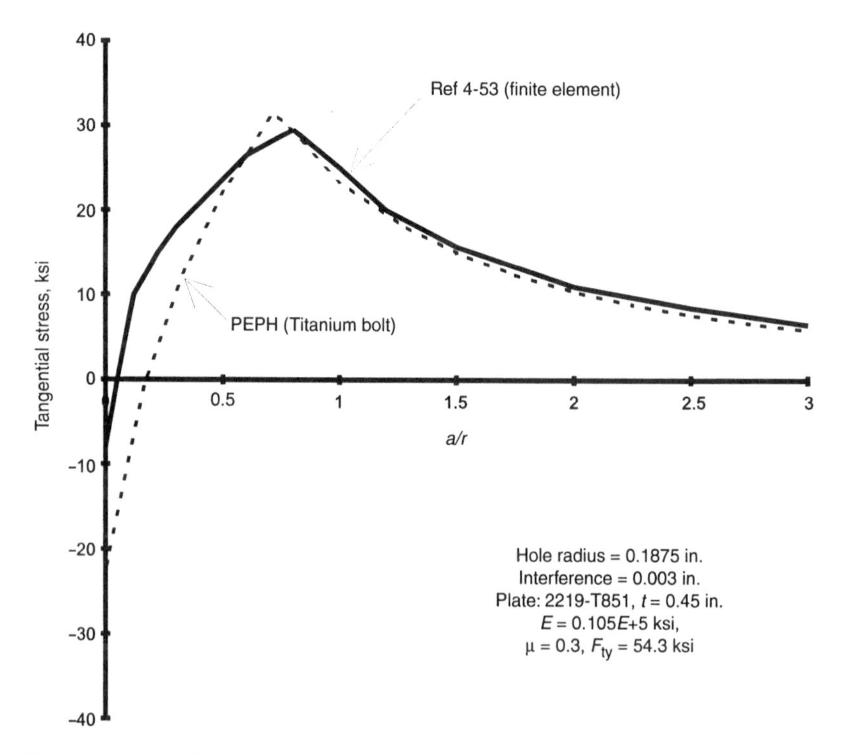

Fig. 4-48 Comparison between plasticity solution and elastic-plastic finite-element result, for aluminum plate and titanium bolt

practically constant, but it is located farther away from the hole as the level of interference increases. The stresses on the left of the peak on each stress distribution curve are plastic stresses; the stresses on the right of the peak are elastic. For the titanium case, it is shown that plastic yielding does not occur if the radial interference is equal to or less than 0.002 in., because the resulting contact pressure was lower than 57.7% of the tensile yield strength of the plate ($<F_{ty}/\sqrt{3}$). The limiting case is reached at an interference level of 0.0045 in., due to yielding of the bolt.

Comparison of PEPH and Finite-Element Solutions. Stresses in the vicinity of an interference-fit hole can be obtained by using finite-element techniques. The stress distributions shown in Fig. 4-47 and 4-48 are a couple of examples extracted from the literature (Ref 4-52, 4-53).

The example shown in Fig. 4-47 is for the combination of a steel bolt and an aluminum plate (Ref 4-52). The tensile yield strength of the plate was 70 ksi. An elastic ideally plastic behavior of the plate material was assumed.

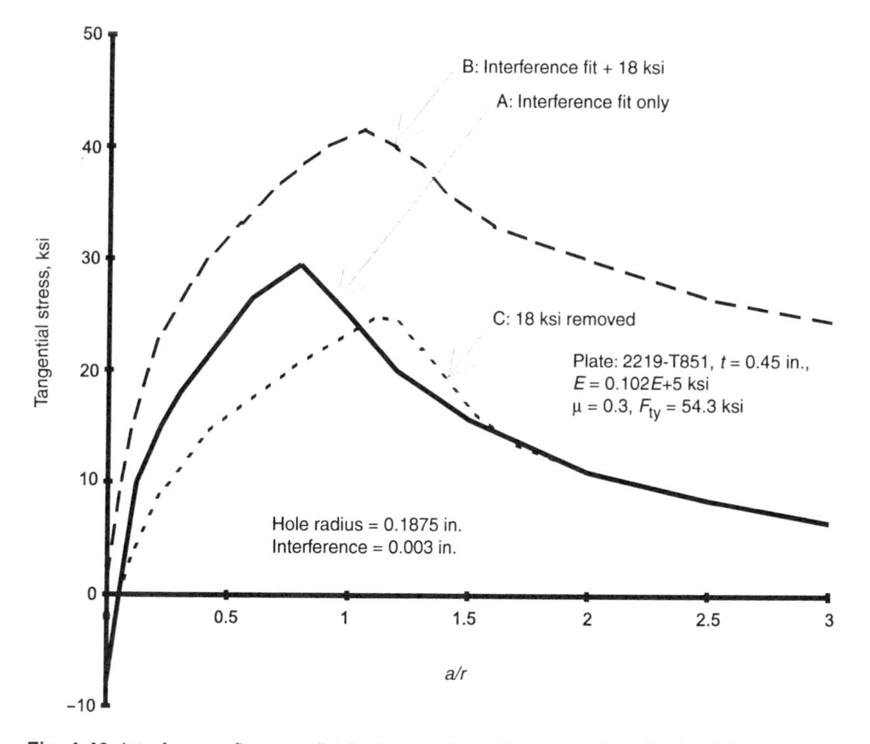

Fig. 4-49 Interference-fit stress distribution as affected by external applied load. Source: 4-53

The bolt was 0.25 in. in diameter. The radial interference (the difference between the radius of the plate and the radius of the bolt) was 0.002 in. To compare with the finite-element solution, PEPH was used to obtain an interference-fit stress distribution for this configuration. Because Ref 4-52 did not report the other material properties for the plate and the bolt, typical values were inputted into PEPH: $E_p = 10{,}500$ ksi, $\nu_p = 0.33$, $E_b = 28{,}300$ ksi, $\nu_b = 0.28$. Close agreement between the two solutions has been obtained (see Fig. 4-47). The slight discrepancy between the two solutions shown in Fig. 4-47 is probably due to the assumed material properties used in PEPH. If the material properties used in PEPH had been identical to those used by Brombolich, the results would have been identical.

The example shown in Fig. 4-48 is for 2219-T851 aluminum plate (F_{ty} = 54.3 ksi, $E_p = 10{,}500$ ksi, $\nu_p = 0.3$). The finite-element model represents one-quarter of a plate (Ref 4-53). The full size of the plate was 4 by 8 in., containing a circular hole ($r = 3/16$ in.) at the center. The radial interference (the difference between the radius of the hole and the radius of the bolt) was 0.003 in. A bilinear elastic-plastic behavior of the plate material was assumed. The modulus for the plastic portion of the tensile stress-strain curve

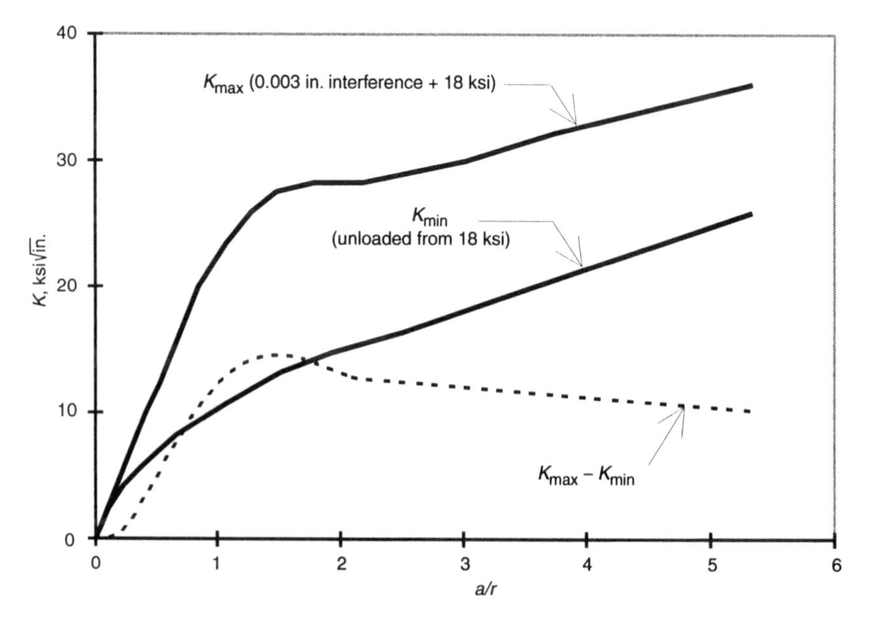

Fig. 4-50 Stress intensity factors corresponding to the stress distributions of Fig. 4-49. Source: Ref 4-53

was 192.8 ksi and the modulus for unloading was 7280 ksi. To compare with the finite-element solution, PEPH was used to obtain interference-fit stress distributions for this configuration. Because Ref 4-53 did not report any material properties for the bolt, it was assumed that the bolt was made of titanium and that its elastic modulus and Poisson's ratio were 16,000 ksi and 0.33, respectively. Close agreement on the elastic stress distribution (for $r > \rho$) between the two solutions has been obtained (see Fig. 4-48). Because the computed plastic zone size matches the finite-element result, the assumption of a titanium bolt appears to have been appropriate. The discrepancy between the two solutions in the plastic region is due to the fact that an assumed elastic ideally plastic material was used in PEPH. The bilinear elastic-plastic material used in the finite-element analysis in Ref 4-53 was capable of taking more load in the plastic region, thereby resulting in a higher stress.

The Mechanics of Crack Growth from an Interference-Fit Hole. The behavior of fatigue crack propagation in a plate containing an interference-fit hole is quite different than that in a plate with an open hole. The main

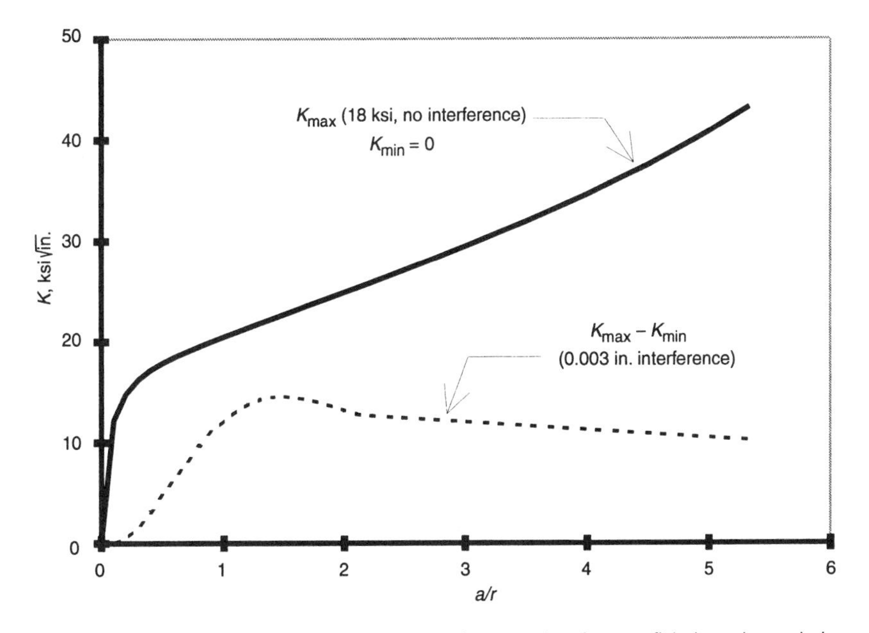

Fig. 4-51 Comparison of stress intensity ranges between interference-fit hole and open hole

reason is that the interference-fit stresses that are remaining in the plate have never been unloaded. The entire stress distribution is partially plastic (i.e., in the region near the hole, where $r < \rho$), and partially elastic (i.e., in the area outside the plastic zone). The local stress distribution corresponding to the applied far-field fatigue loads is not the same as those obtained from Eq 4-8 or Fig. 4-19. The current version of PEPH is incapable of determining the effective stresses for fatigue crack propagation of a loaded plate containing pre-existing interference-fit stresses. The finite-element results for loading the plate that is shown in Fig. 4-48 to a remote stress level of 124 MPa (18 ksi) and then having the load removed are shown in Fig. 4-49. It is seen that the peak of the on-loading resultant tangential stress has been pushed farther away from the hole because the new load causes more yielding in the plate. In the plastic region, the magnitude of the resultant stress is only a fraction higher than the pre-existing stress, because the already-yielded material has limited ability to respond to the applied load. In the elastic region, the magnitude of the resultant stress is approximately equal to the addition of the applied stress and the pre-existing stress. When the far-field loading is removed, the residual tangential stress is approximately the same as the one caused by the interference alone. However, the peak decreases slightly and is located even farther away from the hole.

Therefore, the effective stress amplitude for fatigue crack propagation is equal to the finite-element results of curves B and C in Fig. 4-49. The corresponding K_{max} and K_{min}, determined by integration of the weight function given by Ref 4-23, are shown in Fig. 4-50. The effective ΔK along the crack plane are also shown in the same figure. The ΔK determined by finite-element analysis is presented again in Fig. 4-51 , for comparison to the conventional open hole solution obtained for the applied stress only. Significant reduction in effective ΔK as a result of pin/hole interference is very evident.

Expanded Hole

An expanded cold-worked hole usually is considered to be a plastically deformed hole. In many cases, the hole is expanded by a slightly oversized mandrel. Radial pressure is applied onto the hole while the tapered mandrel is drawn through the hole. Upon release of the pressure, i.e., after the mandrel is removed, residual stresses remain at the vicinity of the hole.

This type of cold working is much the same as removing an interference-fit bolt from a hole. Therefore, one can assume that the plate material follows Hook's law during unloading. The residual stresses around the hole would be the difference between the stresses calculated from the plasticity solution and the elasticity solution.

The solution for elastic deformation of a hole under internal pressure P is given by Ref 4-50 as:

$$\sigma_\theta = -\sigma_r = P\left(\frac{R_p}{r}\right)^2 \quad \text{(Eq 4-32)}$$

The sum of the elastic displacements on the bolt (or mandrel) and the plate is:

$$\delta = P\left[\frac{R_b}{E_b}(1-\nu_b) + \frac{R_p}{2G_p}\right] \quad \text{(Eq 4-33)}$$

Therefore:

$$\sigma_\theta = \delta\left(\frac{R_p}{r}\right)^2\left[\frac{R_b}{E_b}(1-\rho_b) + \frac{R_p}{2G_p}\right]^{-1} \quad \text{(Eq 4-34)}$$

Equations 4-32 and 4-34 represent stresses in an elastically strained plate under a pressure that would have caused partial yielding of the hole. The

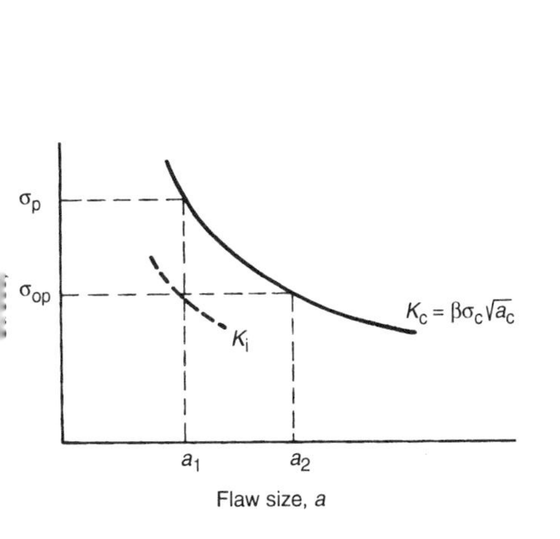

Fig. 4-52 Simplified proof test logic

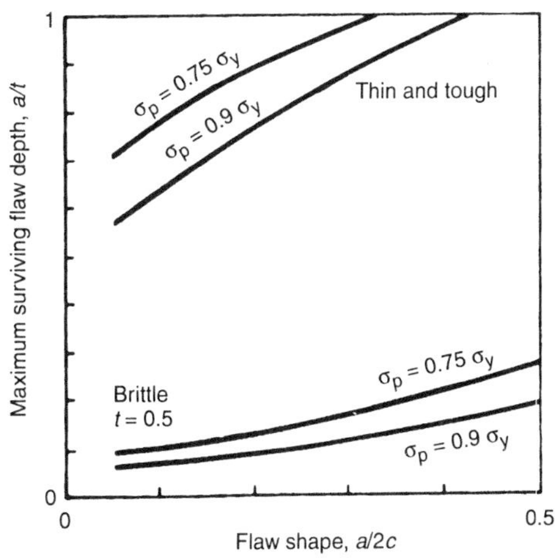

Fig. 4-53 Matrix of flaws to survive a given proof stress for brittle and ductile materials

residual stresses for $r < \rho$ are obtained by subtracting the results calculated from Eq 4-34 from the results of Eq 4-25 and 4-26. Likewise, Eq 4-34 and 4-24 are used to compute residual stresses for $r > \rho$.

Once the mandrel is removed from the hole, the plate that contains a cold-worked hole is the same as an open hole with residual stresses in it. The mechanics of fatigue crack growth in a cold-worked hole is the same as for a yielded open hole, described earlier. The only difference is the method of obtaining the residual stress field. The same weight function can be used to obtain stress intensity factors for the residual stress distribution. The Bowie factor is applicable to the applied fatigue stresses.

Pressurized Cylinders and Vessels

Pressure Vessel

There are excellent articles discussing the mechanism of pressure vessel failures and the current state of the art in handling this problem (Ref 4-54 to 4-58). Successful implementation of linear elastic fracture mechanics to predict pressure vessel failure has been demonstrated. In these documentations, only one problem has been considered: the case of a part-through crack emanating from one side (surface) of the pressure vessel wall. In this case, if the crack length on the surface, $2c$, is small compared to the mean radius of curvature of the vessel, and the crack depth is also small compared to the wall thickness of the vessel, plastic yielding would be restricted to the close neighborhood of the crack periphery. The state of stresses around the leading edge of the crack is approximately plane strain. The residual strength and the fatigue crack propagation rate behavior can be estimated by applying linear elastic fracture mechanics technology. This is particularly true in cryogenic temperature, in which the material becomes more brittle and the small-scale yielding condition is met. Although some uncertainties arise concerning problems such as black surface yielding (or breakthrough) exhibited in relatively high-toughness materials, correlation between experimental data and analysis has been successful.

A number of rational paths may be followed to meet a safe-crack-growth life requirement for a pressure vessel. The predicted life is a function of the initial flaw size. However, initial flaw size varies, depending on the method that is used to determine it. It is also dependent on the definition of failure, i.e., whether the critical crack size is based on complete catastrophic failure or leakage. Pressure vessel design often requires that pre-existing flaws grow to through-the-thickness cracks so that leakage will occur prior to catastrophic failure. Leakage or wall breakthrough can occur as a result of crack propagation under cyclic load or sustained load, or during slow stable tear caused by monotonically increasing load. Therefore a variety of combinations is possible. In instances where hazardous or indispensable contents are involved, leakage cannot be tolerated but will often be a preferred or required failure mode when the alternative is rupture.

Often, a proof test is considered a reliable method for screening pre-existing flaws in pressure vessels. Reference 4-56 is a comprehensive report documenting the pros and cons of the proof test logic. Several key points extracted from that document are presented below.

- Given an ideally brittle material of sufficient thickness with an absolute and invariant measure of its fracture toughness, the stress and flaw size relationship is dictated by Eq 1-20. As shown in Fig. 4-52, a_1 is identified as the maximum flaw size that could survive an applied proof stress σ_p, and a_2 is the crack size that would precipitate final failure at an actual operating stress, σ_{op}. Because the proof stress is chosen to be higher than the operating stress, the initial stress intensity level corresponding to σ_{op} and a_1 would be K_i, which is lower than K_{Ic}. Following a successful proof test, the life of the structure (i.e., time or number of

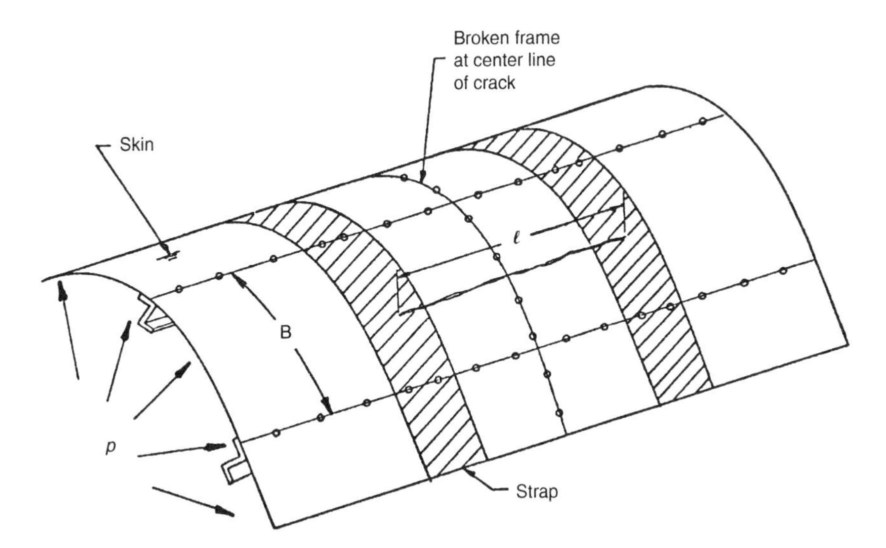

Fig. 4-54 Sketch of a typical segment of an aircraft fuselage

stress cycles required to grow a crack from a_1 to a_2) at the operating stress level can be calculated.

- Actually, the required proof stress level is determined by working the above steps backward. Given a design operating stress level and a known fracture toughness of the material, crack growth analysis is conducted with an assumed initial flaw size. After several iterations (with different assumed initial flaw sizes), the tolerable initial flaw size is determined by matching the predicted life with the required structural life. The proof stress that corresponds to a_1 can then be computed by equating the known fracture toughness value to a known stress intensity solution. Thus, if the vessel survives this proof stress it is likely that the pre-existing initial flaw size is smaller than a_1.
- In the case of a part-through flaw (i.e., surface flaw), many combinations in initial flaw shape (a/c ratio) and flaw size can result in the same predicted life. Screening of initial flaws in pressure vessels basically

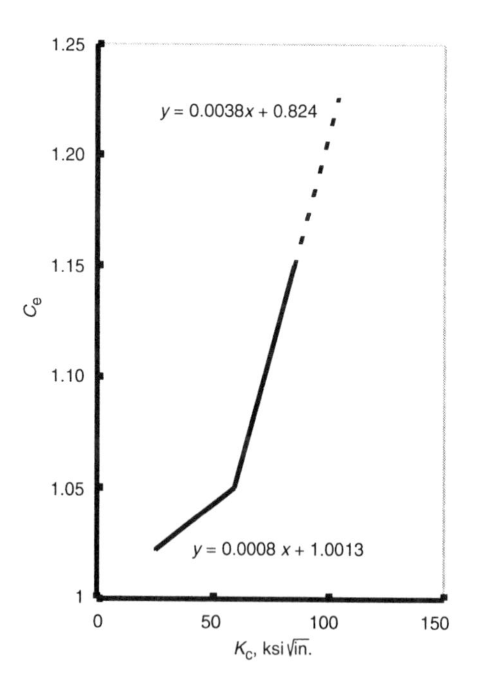

Fig. 4-55 Longitudinal crack extension parameter for pressurized fuselage panel. Source: Ref 4-47

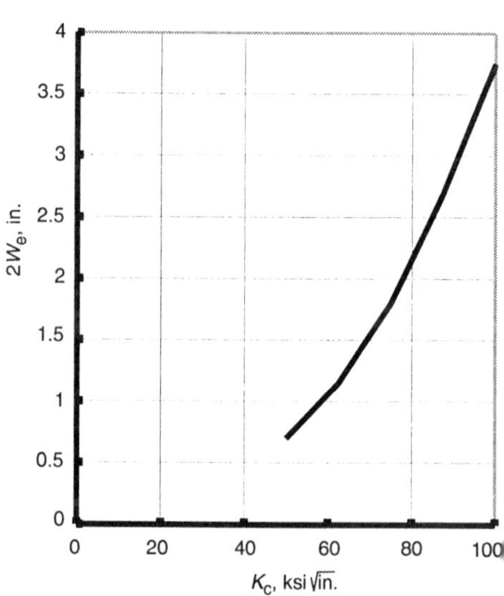

Fig. 4-56 Effective width for stringerless sidewall fuselage panel. Source: Ref 4-47

deals with part-through flaws, so a_1 in Fig. 4-52 is not a single value. Figure 4-53 shows that, for a more brittle material, a relatively moderate proof test would provide information regarding the entire range of flaw shapes (from aspect ratios near zero to semicircular flaws). For thin-gage and/or ductile material, it is illustrated that some flaws of high aspect ratio may not result in fracture, even as near-through cracks.

- In the case of structural sizing, a lower-bound fracture toughness value is often used to obtain a conservative estimate of the predicted life. However, this practice is prohibited for flaw screening, because using a lower fracture toughness value will lead to totally unconservative results when applied to proof test logic. Crack growth analysis using a low fracture toughness value will indicate a smaller initial crack size after proof, in addition to a smaller critical crack size.
- As mentioned in Chapter 2, the fracture toughness value that we are using for routine engineering tasks is K_{IE}. For a ductile material, subcritical crack extension may occur during application of the proof load. A crack of an unknown size that is actually larger than the anticipated initial crack size might have been unknowingly left behind. This would result in an actual life that is shorter than expected.

Failure of Thin-Walled Cylinder or Shell

For a through-the-thickness crack in a relatively thin-walled shell, the crack tip stress intensity is drastically increased as compared to that in a flat plate containing a crack of the same size. The cracked shell might fail at a substantially lower load or exhibit a considerably higher fatigue crack growth rate. Such increases in stress intensity or decreases in burst strength are partly attributed to interaction of the structural geometry (crack size, thickness, curvature, etc.) with the inherent properties of the material, and the effect of bulging at the crack tip that is caused by internal pressure. In typical applications, the shells or cylinders might be an airplane fuselage, a submarine hull, a pressure vessel, a pressure tube in a nuclear reactor, or natural gas pipelines. These structures cover a wide range of combinations in structural geometries, size, and materials. Consequently, a wide variety of failure modes is possible, and the failure mode might be quite different for each class of geometry and material combinations.

In the late 1960s several theories were proposed to classify the possible failure modes in cylindrical shells containing an axial crack and to develop failure criteria suitable to each class of failure mode (Ref 4-59 to 4-61). It was hypothesized that the pressurized tube or vessel would undergo either a fracture-mechanics-based fracture mechanism or a "plastic zone instability" mechanism, depending on the geometry and applied stress level. Correspondingly, the applicable fracture indices would be either the linear elastic stress intensity factor, K_c, or the flow stress $\bar{\sigma}$ of the material. These failure criteria were needed then, when application of J-integrals to fracture under large-scale yielding was not fashionable, and the fracture indices C^* and C_t (for crack growth and fracture at high temperature) were nonexistent. The current state of the art of fracture mechanics analysis has advanced so far that a full range of crack tip deformation modes can be handled by fracture-mechanics-based fracture indices (i.e., K_c, J, or C_t). Therefore, the aforementioned failure criteria and using $\bar{\sigma}$ as a fracture index have become unnecessary. Stress intensity solutions that are applicable to tubes and shells are given in Chapter 5. The solutions for J and C^* (and C_t) are given in this chapter and Chapter 6, respectively.

Fracture Strength of a Reinforced Shell

The reinforcements in a shell are frames and longitudinal stringers either parallel or perpendicular to the crack, depending on the crack orientation. For a circumferential crack, the horizontal stringers might act as crack stoppers. For a longitudinal crack, the frames and the other attached or bonded stiffeners act as crack stoppers. The applicability of the analysis method for skin-stringer structures presented earlier is limited to flat sheets and slightly curved panels. In case of a pressurized aircraft fuselage, which usually has frames and longitudinal stringers, implementation of an alternative method is necessary. Theoretically the hoop stress in a pressurized cylinder is twice that of the axial stress. This ratio can change, depending on the presence of stringers and the added axial stress due to the flight load. In any event it is fair to assume that an analysis based on a longitudinal crack is always conservative. An empirical method is available for the longitudinal crack (Ref 4-47) and is presented in the following.

When uniform far-field stresses are applied normal to the crack, the crack faces are bent to open. For a longitudinal crack, the stringer that parallels the crack is normally not considered a crack stopper. However, it helps the crack to resist bending, thereby reducing crack opening. In turn, the crack tip stress intensity is reduced to a magnitude that is lower than that without a parallel stiffener. For a typical section such as those shown in Fig. 4-54, having a crack parallel to stringers, and arrested at stiffeners between frames, the allowable hoop stress can be computed by:

$$F_{pg} = 1.2F_{tus}\left[\frac{2\overline{W}_e + \Sigma A_e/t}{C_e \cdot l + 2\overline{W}_e}\right] \quad \text{(Eq 4-35)}$$

where F_{tus} is the tensile ultimate strength of the sheet and C_e is given in Fig. 4-55 as a function of K_c. The stiffener effective area, A_e, can be computed by using Eq 5-17 to 5-17(c). If the stringer spacing B is smaller than 108 cm (42.5 in.), the effective width, $2\overline{W}_e$, which is a function of the longitudinal stringer spacing and fracture toughness of the sheet material, can be determined by:

$$2\overline{W}_e = 2W_e \cdot (1.84 - 0.01975 \cdot \text{B}) \quad \text{(Eq 4-36)}$$

where $2W_e$ is the effective width of a curved panel whose stringer spacing is 108 cm (42.5 in.) or larger. Values for $2W_e$ are given in Fig. 4-56.

If the reinforcements are absent (i.e., $A_e = 0$), Eq 4-35 simply reduces to:

$$F_{po} = 1.2F_{tus}\left[\frac{2\overline{W}_e}{C_e \cdot l + 2\overline{W}_e}\right] \quad \text{(Eq 4-37)}$$

The above formulas were derived from test data. The geometries of the pressurized cylinders were within the range of $0.15 \le 1/\sqrt{Rt} \le 0.5$. In addition, the effect of a broken frame has also been taken into account. In structural sizing, it is always a good practice to check on both the short and long crack configurations. The long crack configuration consists of a crack arrested at the flat straps with the center frame broken. Thus, Eq 4-35 is applicable. The short crack configuration is the one having a crack symmetrically extended half bay on each side of the center frame. Because the crack tip is far away from the straps, it is equivalent to $A_e = 0$. Therefore Eq 4-37 is applicable.

Elastic-Plastic Fracture

As mentioned in Chapter 1, the total J is a superposition of two parts, an elastic part and a fully plastic part:

$$J = J_e + J_p \quad \text{(Eq 4-38)}$$

For mode 1, the term J_e is directly related to K_1 by way of:

$$J_e = K_1^2/E'$$ (Eq 4-39)

where $E' = E$ for plane stress and $E' = E/(1 - \nu^2)$ for plane strain. The existing handbook solution for K can be used for a structural configuration under consideration. Chapter 5 provides many K-solutions that are being used by fracture mechanics analysts today.

Close-form solutions for J_p are available. Solutions for many laboratory specimen geometries and a number of common structural configurations are given in Ref 4-62 and 4-63. Three of these solutions are presented in the following section. The specimen configurations are the center-cracked plate, the single-edge-cracked plate, and the compact specimen. Integration of Eq

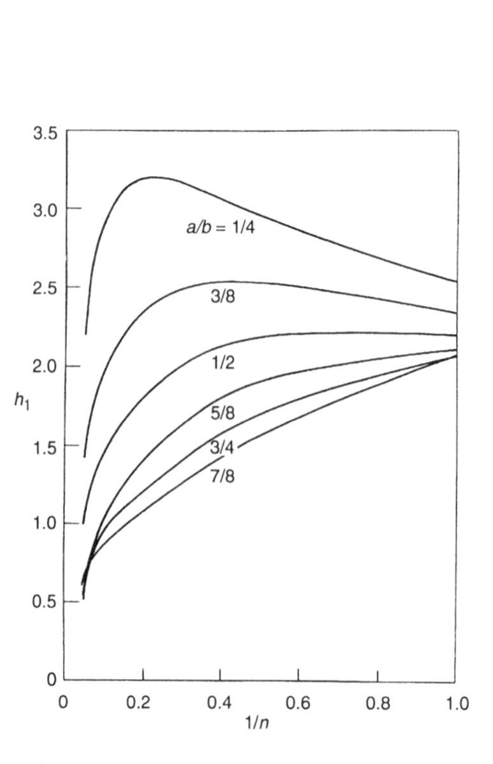

Fig. 4-57 h_1 vs. $1/n$ for a center-cracked panel in tension, plane stress. Source: Ref 4-62

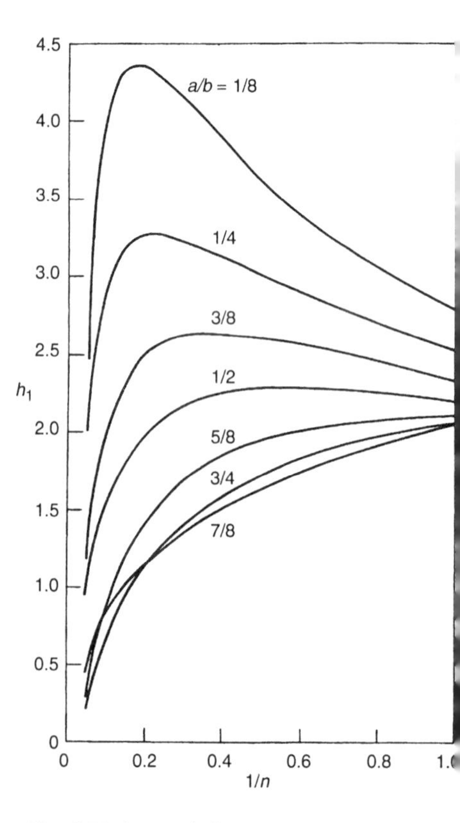

Fig. 4-58 h_1 vs. $1/n$ for a center-cracked panel in tension, plane strain. Source: Ref 4-62

1-40 is required for those cases for which a solution for K (for J_e) or J_p is not available.

The Fully Plastic Solutions

In the following paragraphs the fully plastic solutions for the center-cracked and single-edge-cracked plates and the compact specimen are presented. Following that, a procedure for doing the numerical integration will be presented along with a sample problem.

The Center-Cracked Panel

For a center through-the-thickness crack subjected to far-field uniform tension, e.g., the M(T) specimen configuration, having a total crack length $2a$ and width W, the fully plastic solution for J is:

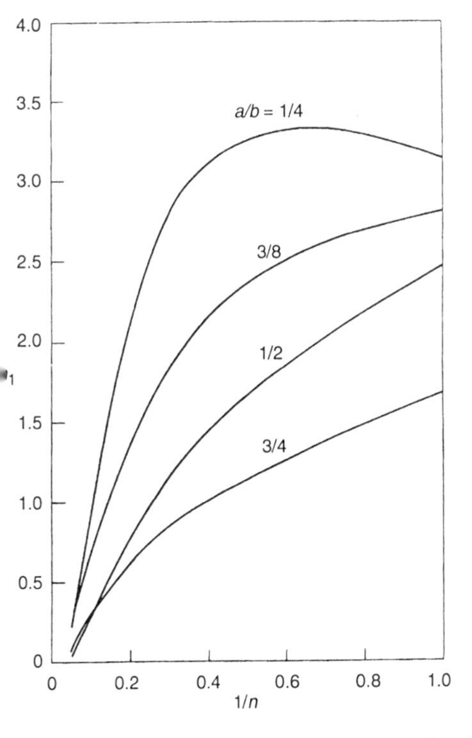

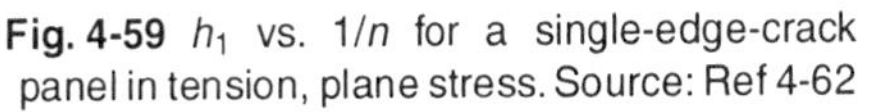
Fig. 4-59 h_1 vs. $1/n$ for a single-edge-crack panel in tension, plane stress. Source: Ref 4-62

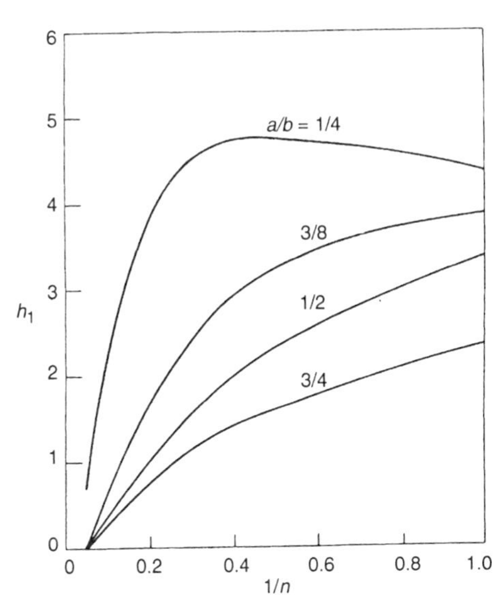

Fig. 4-60 h_1 vs. $1/n$ for a single-edge-crack panel in tension, plane strain. Source: Ref 4-62

$$J_p = \alpha\sigma_0\varepsilon_0 a \cdot (1 - 2a/W) \cdot h_1 \cdot \left(\frac{P}{P_0}\right)^{n+1} \quad \text{(Eq 4-40)}$$

where n and α are material constants defined by the pure stress-strain law of Eq 1-38, σ_0 is the yield stress, ε_0 is the yield strain, and h_1 is a function of $2a/W$ and n; its value is given in Fig. 4-57 (for plane stress) and Fig. 4-58 (for plane strain). In these figures, the dimension b is equal to $W/2$. P is the applied load per unit thickness, and:

$$P_0 = \psi \cdot (W - 2a) \cdot \sigma_0 \quad \text{(Eq 4-41)}$$

where ψ is 1 for plane stress and $2\sqrt{3}$ for plane strain.

Single Edge Crack in Uniform Tension

For a single-edge-cracked plate subjected to far-field uniform tension, having a crack length a and width W, the fully plastic solution for J is:

$$J_p = \alpha\sigma_0\varepsilon_0 \cdot (W - a) \cdot (a/W) \cdot h_1 \cdot \left(\frac{P}{P_0}\right)^{n+1} \quad \text{(Eq 4-42)}$$

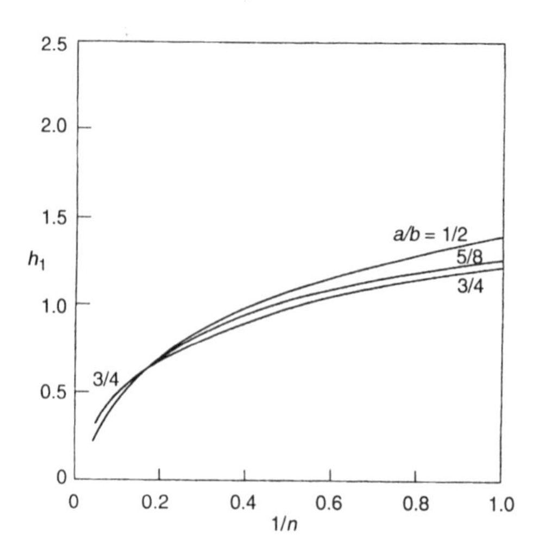

Fig. 4-61 h_1 vs. $1/n$ for a compact specimen, plane stress. Source: Ref 4-62

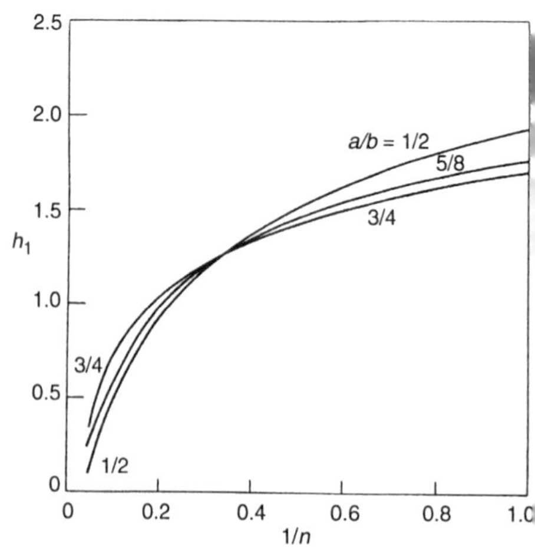

Fig. 4-62 h_1 vs. $1/n$ for a compact specimen, plane strain. Source: Ref 4-62

where h_1 is a function of a/W and n; its value is given in Fig. 4-59 (for plane stress) and Fig. 4-60 (for plane strain). In these figures, the dimension b is the same as W of Fig. 5-2. P is the applied load per unit thickness, and:

$$P_0 = \psi \cdot \eta \cdot (W - a) \cdot \sigma_0 \qquad \text{(Eq 4-43)}$$

where ψ is 1.072 for plane stress and 1.455 for plane strain, and:

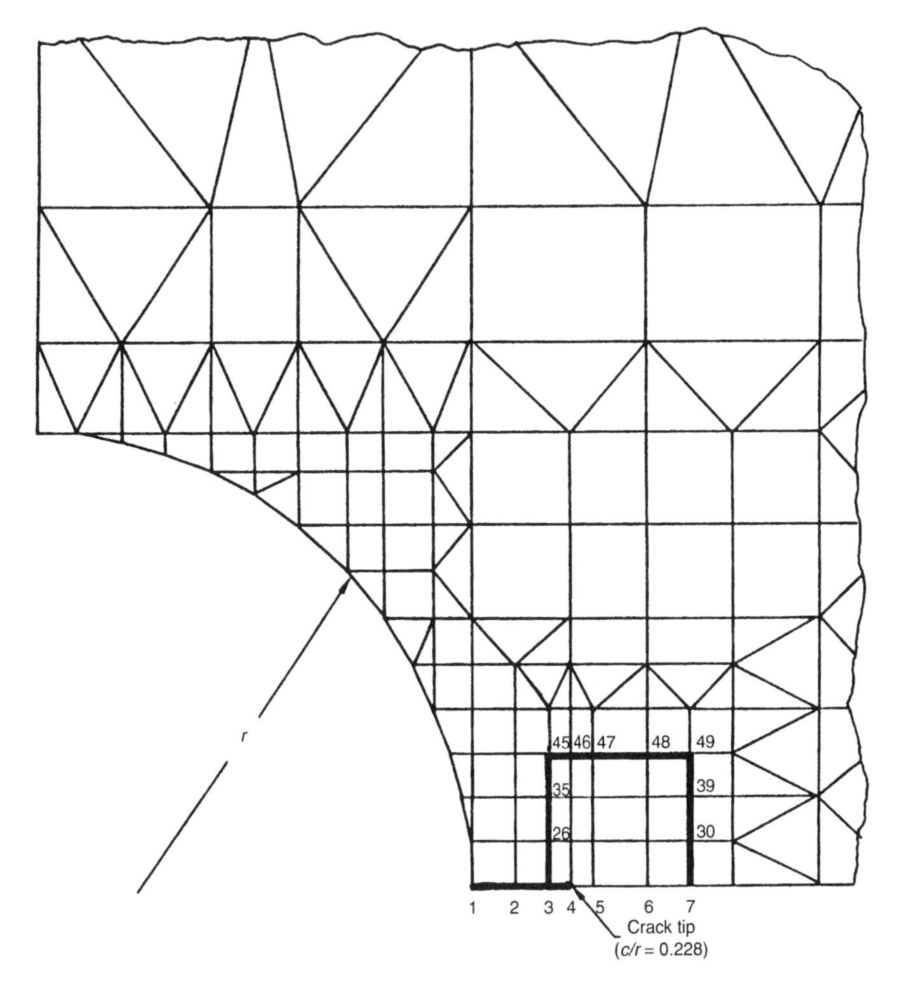

Fig. 4-63 Finite-element mesh near a crack coming out from a hole. Source: Ref 4-65

$$\eta = \left[1 + \left(\frac{a}{W-a}\right)^2\right]^{1/2} - \frac{a}{W-a} \qquad \text{(Eq 4-44)}$$

The Compact Specimen

The fully plastic J solution for the ASTM standard compact specimen is:

$$J_p = \alpha\sigma_0\varepsilon_0 \cdot c \cdot h_1 \cdot \left(\frac{P}{P_0}\right)^{n+1} \qquad \text{(Eq 4-45)}$$

where h_1 is a function of a/W and n; its value is given in Fig. 4-61 (for plane stress) and Fig. 4-62 (for plane strain). In these figures, the dimension b is the same as W of Fig. 5-8. P is the applied load per unit thickness, and :

$$P_0 = \psi \cdot \eta \cdot c \cdot \sigma_0 \qquad \text{(Eq 4-46)}$$

where ψ is 1.072 for plane stress and 1.455 for plane strain, and :

$$\eta = \left[\left(\frac{2a}{c}\right)^2 + 2\left(\frac{2a}{c}\right) + 2\right]^{1/2} - \left(\frac{2a}{c} + 1\right) \qquad \text{(Eq 4-47)}$$

A General Procedure for Determining J

The finite-element method can be used to determine the stress and strain for points on the path of the selected contour Γ. The results are then substituted into Eq 1-40 and 1-41 for integration. In order to carry out the integration indicated in Eq 1-40 and 1-41, a relationship between stress and strain that realistically describes the plastically deforming material is needed. It has been shown in Ref 4-64 that Eq 1-41(b) can be reduced to:

$$W = \frac{1}{2E}\left(\sigma_x + \sigma_y\right)^2 + \left(\frac{1+\nu}{E}\right)\left(\tau_{xy}^2 - \sigma_x \cdot \sigma_y\right) + \int_0^{\bar{\varepsilon}_p} \bar{\sigma} \cdot (d\bar{\varepsilon}_p) \qquad \text{(Eq 4-48)}$$

where $\bar{\sigma}$ and $\bar{\varepsilon}_p$ are equivalent stress and equivalent plastic strain, respectively, and:

$$\bar{\sigma} = \left(\sigma_x^2 - \sigma_x\sigma_y + \sigma_y^2 + 3\tau_{xy}^2\right)^{1/2} \qquad \text{(Eq 4-49)}$$

For segments of Γ on which $\bar{\sigma}$ is greater than the material proportional limit and dy is nonzero, a value of $\bar{\varepsilon}_p$ is obtained from the stress-strain curve and substituted into Eq 4-48 so that the plastic component of J can be evaluated. Physically, the integral in Eq 4-48 may be interpreted as the area under the plastic stress-strain curve. Therefore, the plastic part of J is the integral in Eq 4-48, and the elastic part of J includes all the terms in Eq 1-40 excluding the integral term in Eq 4-48.

For very ductile materials, considerable care must be taken in evaluating the stress $\bar{\sigma}$. A small error in $\bar{\sigma}$ will lead to large errors in estimating $\bar{\varepsilon}_p$, and thus may contribute large errors to the J-integral. In studying Eq 4-48 it is clear that W will have a unique value only if unloading is prohibited at every point in the structure. Monotonic loading conditions prevail throughout a cracked body under steadily increasing applied loads, provided that crack extension does not occur. Because in any calculation of J the crack length is held constant, W will be unique and a valid J-integral is obtained.

In the following we will demonstrate the integration procedure by running through a numerical exercise of a sample problem. The problem dealt with a rectangular plate specimen made of 2024-T351 aluminum, 6.35 mm thick, 152.4 mm wide, and 457.2 mm long. A 12.7 mm diameter hole was located in the middle of the specimen. A through-the-thickness radio crack, 1.45 mm long, was at the bore of the hole.

In this analysis, the NASTRAN structural analysis computer code was used. A finite-element model of the specimen was constructed, using constant-strain triangles and quadrilaterals, with a very fine mesh in the neigh-

Table 4-2 Calculation of J for a 1.45 mm crack at a 12.7 mm diameter hole under 245 MPa applied stress

One-half of the J path			NASTRAN results					Calculated results			
Grid points	dx, mm	dy, mm	σ_x, MPa	σ_y, MPa	τ_{xy}, MPa	$\frac{\delta u}{\delta x} \times 10^3$	$\frac{\delta v}{\delta x} \times 10^3$	$\bar{\sigma}$, MPa	$\bar{\varepsilon}_p$	$\frac{1}{2}J_e$, MN/m	$\frac{1}{2}J_p$, MN/m
7–30	0	0.635	200.2	439.2	12.5	−0.470	−1.34	381.5	0.01200	4.943	11.230
30–39	0	0.635	151.6	427.3	5.0	−0.450	−3.26	375.3	0.00870	5.277	5.796
34–49	0	0.635	91.3	408.0	−11.7	−0.990	−4.16	371.4	0.00630	4.716	2.053
49–48	−0.635	0	65.8	396.3	−25.8	−1.360	−5.02	370.6	0.00608	8.558	0
48–47	−0.762	0	42.0	378.9	−51.2	−1.740	−6.44	370.6	0.00605	12.355	0
47–46	−0.330	0	18.8	342.7	−93.3	−1.950	−8.32	370.7	0.00610	6.077	0
46–45	−0.305	0	15.9	288.3	−134.7	−1.420	−8.67	365.0	0.00528	4.850	0
45–35	0	−0.635	32.8	276.6	−140.9	−0.997	−9.60	357.9	0.00500	2.032	0.121
35–26	0	−0.635	85.4	270.2	−160.6	−0.734	−15.00	366.9	0.00540	6.246	−0.483
26–3	0	−0.635	139.7	268.3	−172.4	+0.625	−27.00	378.4	0.01060	16.369	−8.573
										$\Sigma = 71.423$	9.902
										Total $J = 162.650$	

Source: Ref 4-65

borhood of the crack. Figure 4-63 shows the portion of the finite-element mesh adjacent to the hole and the crack. The actual material stress-strain curve shown in Fig. 4-38 was input to each element in the model using a piecewise linear format.

As shown in Fig. 4-63, one-half of a selected contour Γ starts from grid point No. 7 and ends at grid point No. 3. Therefore, one-half of the J-value is equal to:

$$\frac{J}{2} = \int_7^{49} \left[\frac{1}{2E}(\sigma_x + \sigma_y)^2 + \frac{1+\nu}{E}\left(\tau_{xy}^2 - \sigma_x \cdot \sigma_y\right) - \left\{ \sigma_x \left(\frac{\delta u}{\delta x}\right) + \tau_{xy}\left(\frac{\delta v}{\delta x}\right) \right\} + \int_0^{\bar{\varepsilon}_p} \bar{\sigma} \cdot (d\bar{\varepsilon}_p) \right] dy$$

$$- \int_{49}^{45} \left[\tau_{xy}\left(\frac{\delta u}{\delta x}\right) + \sigma_y \left(\frac{\delta v}{\delta x}\right) \right] dx$$

$$+ \int_{45}^{3} \left[\frac{1}{2E}(\sigma_x + \sigma_y)^2 + \frac{1+\nu}{E}\left(\tau_{xy}^2 - \sigma_x \cdot \sigma_y\right) - \left\{ \sigma_x \left(\frac{\delta u}{\delta x}\right) + \tau_{xy}\left(\frac{\delta v}{\delta x}\right) \right\} + \int_0^{\bar{\varepsilon}_p} \bar{\sigma} \cdot (d\bar{\varepsilon}_p) \right] dy$$

(Eq 4-50)

The terms associate with $\delta u/\delta x$ or $\delta v/\delta x$ are those coming from the second term of Eq 1-40. The rest belong to Eq 4-48. The double integral

$$\iint_0^{\bar{\varepsilon}_p} \bar{\sigma} \cdot (d\bar{\varepsilon}_p)\, dy \qquad \text{(Eq 4-51)}$$

is the "fully plastic J." The NASTRAN output and the calculated J-value for this configuration (under an applied far-field stress of 245 MPa) are presented in Table 4-2.

To confirm the path independence of J, a contour that was farther away from the crack tip (i.e., far from the plastic zone and primarily elastic), was also evaluated. Although the analysis result for the second path is not shown here, the values of J calculated from the two paths were found to agree within 1% and path independence was deemed to be verified.

REFERENCES

4-1. J.C. Ekvall, T.R. Brussat, A.F. Liu, and M. Creager, Preliminary Design of Aircraft Structures to Meet Structural Integrity Requirements, *J. Aircr.*, Vol 11, 1974, p 136–143
4-2. J.P. Gallagher, F.J. Giessler, A.P. Berens, and R.M. Engle, Jr., "USAF Damage Tolerant Design Handbook: Guidelines for the Analysis and Design of Damage Tolerant Aircraft Structures," Report AFWAL-TR-82-3073, Flight Dynamics Laboratory, Air Force Wright Aeronautical Laboratories, Wright-Patterson Air Force Base, Dayton, OH, May 1984
4-3. M.L. Williams, On the Stress Distribution at the Base of a Stationary Crack, *J. Appl. Mech.*, (*Trans. ASME*), Series E, Vol 24, 1957, p 109–114
4-4. P. Tong, T.H.H. Pian, and S. Lasry, A Hybrid-Element Approach to Crack Problems in Plane Elasticity, *Int. J. Numerical Methods in Eng.*, Vol 7, 1973, p 297–308
4-5. A.R. Ingraffea and C. Manu, Stress Intensity Factor Computation in Three Dimensions with Quarter-Point Elements, *Int. J. Numerical Methods in Eng.*, Vol 15, 1980, p 1427–1445
4-6. C.F. Shih, H.G. deLorenzi, and M.D. German, Crack Extension Modeling with Singular Quadratic Isoparametric Elements, *Int. J. Fract.*, Vol 12, 1976, p 647–651
4-7. L. Banks-Sills and D. Sherman, Comparison of Methods for Calculating Stress Intensity Factors with Quarter Point Elements, *Int. J. Fract.*, Vol 32, 1986, p 127–140
4-8. L. Banks-Sills and O. Einav, On Singular, Nine-Noded, Distorted, Isoparametric Elements in Linear Elastic Fracture Mechanics, *Comput. Struct.*, Vol 25, 1987, p 445–449
4-9. R.D. Henshell and K.G. Shaw, Crack Tip Finite Elements Are Unnecessary, *Int. J. Numerical Methods in Eng.*, Vol 9, 1975, p 495–507
4-10. D.M. Tracey, Finite-elements for Determination of Crack Tip Elastic Stress Intensity Factors, *Eng. Fract. Mech.*, Vol 3, 1971, p 255–265
4-11. I.S. Raju and J.C. Newman, Jr., "Three-Dimensional Finite-Element Analysis of Finite-Thickness Fracture Specimens," TN D-8414, National Aeronautics and Space Administration, Langley Research Center, Hampton, VA, May 1977
4-12. J.C. Newman, Jr. and I.S. Raju, "Analysis of Surfaces Cracks in Finite Plates under Tension or Bending Loads," TP-1578, National Aeronautics and Space Administration, Washington, D.C., Dec 1979
4-13. J. Robinson, *Integrated Theory of Finite Element Methods*, Wiley, 1973
4-14. S.K. Chan, I.S. Tuba, and W.K. Wilson, On the Finite Element Method in Linear Fracture Mechanics, *Eng. Fract. Mech.*, Vol 2, 1970, p 1–17
4-15. C.B. Buchalet and W.H. Bamfort, Stress Intensity Factor Solutions for Continuous Surface Flaws in Reactor Pressure Vessels, *Mechanics of Crack Growth*, STP 590, American Society for Testing and Materials, 1976, p 385–402
4-16. A.F. Liu and J.J. Gurbach, Application of a p-Version Finite Element Code to Analysis of Cracks, *AIAA Journal*, Vol 32, 1994, p 828–835
4-17. M. Isida, Stress Intensity Factor of the Tension of an Eccentrically Cracked Strip, *J. Appl. Mech.*, (*Trans. ASME*), Series E, Vol 33, 1966, p 674–675

4-18. P.C. Paris and G.C. Sih, Stress Analysis of Cracks, *Symposium on Fracture Toughness Testing and its Applications*, STP 381, American Society for Testing and Materials, 1965, p 30–83
4-19. A.F. Liu and D.F. Dittmer, "Effect of Multiaxial Loading on Crack Growth," Report AFFDL-TR-78-175 (three volumes), Air Force Flight Dynamics Laboratory, Wright-Patterson Air Force Base, Dayton, OH, Dec 1978
4-20. K.J. Miller and A.P. Kfouri, An Elastic - Plastic Finite Element Analysis of Crack Tip Fields under Biaxial Loading Conditions, *Int. J. Fract.*, Vol 10, 1974, p 393–404
4-21. S. Timoshenko and J.N. Goodier, *Theory of Elasticity*, McGraw-Hill, 1951, p 80
4-22. R.C. Howland, On the Stresses in the Neighborhood of a Circular Hole in a Strip under Tension, *Philos. Trans. R. Soc. (London) A,* Vol 119, 1930, p 49–86
4-23. T.M. Hsu and J.R. Rudd, Green's Function for Thru-Crack Emanating from Fastener Holes, *Fracture 1977*, Vol 3, 1977, p 139–148
4-24. L.F. Impellizzeri and D.L. Rich, *Fatigue Crack Growth under Spectrum Loads*, STP 595, American Society for Testing and Materials, 1976, p 320–336
4-25. A.F. Grandt, Jr., Stress Intensity Factors for Some Thru-Cracked Fastener Holes, *Int. J. Fract.*, Vol 11, 1975, p 283–294
4-26. L.A. James and W.E. Anderson, *Eng. Fract. Mech.,* Vol 3, 1969, p 601–605
4-27. A.F. Liu, "The Effect of Residual Stresses on Crack Growth from a Hole," Report NOR 79-74, Northrop Corporation, Aircraft Division, Hawthorne, CA, Aug 1979
4-28. J.I. Bluhm, Fracture Arrest, *Fracture*, H. Liebowitz, Ed., Academia Press, 1971, p 1–62
4-29. T. Swift, Fracture Analysis of Stiffened Structure, *Damage Tolerance of Metallic Structures*, STP 842, American Society for Testing and Materials, 1984, p 69–107
4-30. T. Swift, Widespread Fatigue Monitoring—Issues and Concerns, Report CP-3274, Part 2, *FAA/NASA Int. Symposium on Advanced Structural Integrity Methods for Airframe Durability and Damage Tolerance,* National Aeronautics and Space Administration, Langley Research Center, Hampton, VA, Sept 1994, p 829–870
4-31. C.C. Poe, Jr., "Stress-Intensity Factor for a Cracked Sheet with Riveted and Uniformly Spaced Stringers," Report TR-R-358, National Aeronautics and Space Administration, Langley Research Center, Hampton, VA, May 1971
4-32. C.C. Poe, Jr., "The Effect of Broken Stringers on the Stress Intensity Factor of a Uniformly Stiffened Sheet Containing a Crack," Report TM X-71947, National Aeronautics and Space Administration, Langley Research Center, Hampton, VA, 1973
4-33. C.C. Poe, Jr., Fatigue Crack Propagation in Stiffened Panels, *Damage Tolerance in Aircraft Structures*, STP 486, American Society for Testing and Materials, 1971, p 79–97
4-34. M. Creager and A.F. Liu, The Effect of Reinforcement on the Slow Stable Tear and Catastrophic Failure of Thin Metal Sheet, *J. Eng. Mater. Technol.*, (*Trans. ASME*), Series H, Vol 96, 1974, p 49–55
4-35. A.F. Liu and J.C. Ekvall, Material Toughness and Residual Strength of Damage Tolerant Aircraft Structures, *Damage Tolerance in Aircraft Structures*, STP 486, American Society for Testing and Materials, 1971, p 98–121

4-36. H. Vlieger, The Residual Strength Characteristics of Stiffened Panels Containing Fatigue Cracks, *Eng. Fract. Mech.*, Vol 5, 1973, p 447–477

4-37. C.K. Gunther and J.T. Wozumi, Critical Failure Modes in Cracked Mechanically Fastened Stiffened Panels, *Design of Fatigue and Fracture Resistant Structures*, STP 761, American Society for Testing and Materials, 1982, p 310–327

4-38. R.C. Shah and F.T. Lin, Stress Intensity Factors of Stiffened Panels with Partially Cracked Stiffeners, *Fracture Mechanics: Fourteenth Symposium,* Vol I, *Theory and Analysis,* STP 791, American Society for Testing and Materials, 1983, p I-157 to I-171

4-39. R. Grief and J.L. Sanders, Jr., The Effect of a Stringer on the Stress in a Cracked Sheet, *J. Appl. Mech.*, (*Trans. ASME*), Series E, Vol 32, 1965, p 59–66

4-40. J.M. Bloom and J.L. Sanders, Jr., The Effect of Riveted Stringer on the Stress in a Cracked Sheet, *J. Appl. Mech.*, (*Trans. ASME*), Series E, Vol 33, 1966, p 561–570

4-41. A.F. Liu, "Fracture Control Methods for Space Vehicles, Vol I, Fracture Control Design Methods," Report CR-134596, National Aeronautics and Space Administration, Washington, D.C., Aug 1974

4-42. T. Nishimura, Stress Intensity Factors of Multiple Cracked Sheet with Riveted Stiffeners, *J. Eng. Mater. Technol.*, (*Trans. ASME*), Series H, Vol 113, 1991, p 280–284

4-43. K.-J. Wang and D.J. Cartwright, A Crack Near Doubly Riveted Stiffeners, *Fracture 1977*, 1977, p 647–656

4-44. S.V. Shkarayev and E.T. Mayer, Jr., Edge Cracks in Stiffened Plates, *Eng. Fract. Mech.*, Vol 27, 1987, p 127–134

4-45. W.J. Crichlow, "The Optimum Design of Shell Structure for Static Strength, Stiffness, Fatigue and Damage Tolerance Strength," paper presented at AGARD Symposium on Structural Optimization (Istanbul, Turkey), 6–8 Oct 1969

4-46. P. Kuhn, *Stresses in Aircraft and Shell Structures*, McGraw-Hill, 1956

4-47. W.J. Crichlow, The Ultimate Strength of Damaged Structure—Analysis Methods with Correlating Test Data, *Full Scale Fatigue Testing of Aircraft Structures*, Pergamon Press, 1960, p 149–209

4-48. A.F. Liu, Effect of Residual Stresses on Crack Growth from a Hole, *AIAA J.*, Vol 22, 1984, p 1784–1785

4-49. A. Nadai, *Theory of Flow and Fracture of Solids*, 2nd ed., Vol 1, McGraw-Hill, 1950, p 472–481

4-50. S. Timoshenko, *Strength of Materials,* Part II, *Advanced Theory and Problems*, 2nd ed., D. Van Nostrand Co., 1941, p 236–241

4-51. K.K. Chan and A.F. Liu, "Residual Stress at Hole-Fortran Computer Program PEPH," New Technology Report MSC-19735, National Aeronautics and Space Administration, Nov 1975

4-52. L.J. Brombolich, "Elastic-Plastic Analysis of Stresses Near Fastener Holes," Paper 73–252, presented at the AIAA 11th Aerospace Sciences Meeting (Washington, D.C.), 10–12 Jan 1973

4-53. J.L. Rudd, T.M. Hsu, and J.A. Aberson, "Analysis and Correlation of Crack Growth from Interference-Fit Fastener Holes," paper presented at Numerical Methods in Fracture Mechanics (West Glamorgan, U.K.), 9–13 Jan 1978

4-54. C.F. Tiffeny and J.N. Masters, Applied Fracture Mechanics, *Symposium on Fracture Toughness Testing and its Applications*, STP 381, American Society for Testing and Materials, 1965, p 249–278
4-55. C.F. Tiffeny, "Fracture Control of Metallic Pressure Vessels," Report SP-8040, National Aeronautics and Space Administration, Washington, D.C., 1970
4-56. R.M. Ehret, "Fracture Control Design Methods for Space Vehicles, Vol II, Assessment of Fracture Mechanics Technology for Space Shuttle Applications," Report CR-134597, National Aeronautics and Space Administration, NASA Lewis Research Center, 1974
4-57. R.L. Jones, T.U. Marston, S.W. Tagart, D.M. Norris, and R.E. Nickell, Applications of Fatigue and Fracture Damage Tolerant Design Concepts in the Nuclear Power Industry, *Design of Fatigue and Fracture Resistant Structures*, STP 761, American Society for Testing and Materials, 1982, p 424–444
4-58. R.A. Ainsworth, I. Milne, A.R. Dowling, and A.T. Stewart, Assessing the Integrity of Structures Containing Defects by the Failure Assessment Diagram Approach of the CEGB, *Fatigue and Fracture Assessment by Analysis and Testing*, PVP 103, American Society of Mechanical Engineers, 1986, p 123–129
4-59. G.T. Hahn and M. Sarrate, Failure Criteria for Through-Cracked Vessels, *Practical Fracture Mechanics for Structural Steel*, R.W. Nichols, Ed., 1969, p P1 to P15
4-60. G.T. Hahn, M. Sarrate, and A.R. Rosenfield, Criteria for Crack Extension in Cylindrical Pressure Vessels, *Int. J. Fract. Mech.*, Vol 5, 1969, p 187–210
4-61. J.A. Vazquez and P.C. Paris, "The Application of the Plastic Zone Instability Criterion to Pressure Vessel Failure," paper presented at The Fourth National Symposium on Fracture Mechanics (Pittsburgh, PA), Aug 1970
4-62. V. Kumar, M.D. German, and C. F. Shih, "An Engineering Approach for Elastic-Plastic Fracture Analysis," Report NP-1931, Electric Power Research Institute, Palo Alto, CA, July 1981
4-63. A. Zahoor, "Ductile Fracture Handbook, Vol 1, Circumferential Throughwall Cracks," Report NP-6301-D, Electric Power Research Institute, Palo Alto, CA, 1989
4-64. M.M. Ratwani and D.P. Wilhem, Development and Evaluation of Methods of Plane Stress Fracture Analysis, Part II, Vol I, A Technique for Predicting Residual Strength of Structure, Report AFFDL-TR-73-42, Air Force Flight Dynamics Laboratory, Wright-Patterson Air Force Base, Dayton, OH, Aug 1977
4-65. D. Shows, A.F. Liu, and J.H. FitzGerald, Application of Resistance Curves to Crack at a Hole, *Fracture Mechanics: Fourteenth Symposium,* Vol II, *Testing and Applications*, STP 791, American Society for Testing and Materials, 1983, p II-87 to II-100

SELECTED REFERENCES

Load Spectrum Development, Editing, and Simplification

- J.A. Bannantine, J.J. Comer, and J.L. Handrock, *Fundamentals of Metal Fatigue Analysis,* Prentice-Hall, 1990, p 184–196

- H.D. Dill and H.T. Young, "Stress History Simulation," AFFDL-TR-76-113, Air Force Flight Dynamics Laboratory, Wright-Patterson Air Force Base, Dayton, OH, Nov, 1976
- W. Elber, Equivalent Constant-Amplitude Concept for Crack Growth under Spectrum Loading, *Fatigue Crack Growth Under Spectrum Loads*, STP 595, American Society for Testing and Materials, 1976, p 236–250
- H.O. Fuchs and R.I. Stephens, *Metal Fatigue in Engineering*, John Wiley & Sons, 1980
- J.P. Gallagher, F.J. Giessler, A.P. Berens, and R.M. Engle, Jr., "USAF Damage Tolerant Design Handbook: Guidelines for the Analysis and Design of Damage Tolerant Aircraft Structures," Report AFWAL-TR-82-3073, Flight Dynamics Laboratory, Air Force Wright Aeronautical Laboratories, Wright-Patterson Air Force Base, Dayton, OH, May, 1984
- D.V. Nelson and H.O. Fuchs, Prediction of Fatigue Crack Growth under Irregular Loading, *Fatigue Crack Growth under Spectrum Loads*, STP 595, American Society for Testing and Materials, 1976, p 267–291
- P.G. Porter and A.F. Liu, "A Rapid Method to Predict Fatigue Crack Initiation, Vol I, Technical Summary," Report NADC-81010-60, Naval Air Development Center, 1983
- J.M. Potter and R.T. Watanabe, Ed., *Development of Fatigue Loading Spectra,* STP 1006, American Society for Testing and Materials, 1989
- R.M. Wetzel, Ed., *Fatigue under Complex Loading: Analyses and Experiments,* Vol AE-6, Society of Automotive Engineers, 1977

Principle of Compounding with Sample Problems

- D.J. Cartwright and D.P. Rooke, Approximate Stress Intensity Factors Compounded from Known Solutions, *Eng. Fract. Mech.*, Vol 6, 1974, p 563–571
- J.P. Gallagher, F.J. Giessler, A.P. Berens, and R.M. Engle, Jr., "USAF Damage Tolerant Design Handbook: Guidelines for the Analysis and Design of Damage Tolerant Aircraft Structures," Report AFWAL-TR-82-3073, Flight Dynamics Laboratory, Air Force Wright Aeronautical Laboratories, Wright-Patterson Air Force Base, Dayton, OH, May, 1984
- K. Katherisan, T.M. Hsu, and T.R. Brussat, "Advanced Life Analysis Methods—Crack Growth Analysis Methods for Attachment Lugs," Report AFWAL-TR-84–3080, Vol II, Flight Dynamics Laboratory, Wright-Patterson Air Force Base, Dayton, OH, Sept 1984
- D.P. Rooke, Compounded Stress Intensity Factors for Cracks at Fastener Holes, *Eng. Fract. Mech.*, Vol 19, 1984, p 359–374
- D.P. Rooke and D.J. Cartwright, The Compounding Method Applied to Cracks in Stiffened Sheets, *Eng. Fract. Mech.*, Vol 8, 1976, p 567–573

Theory and Application of Weight Functions

- H.F. Bueckner, Field Singularities and Related Integral Representations, *Mechanics of Fracture,* Vol 1, *Methods of Analysis and Solutions of Crack Problems*, G.C. Sih, Ed., Noordhoff International Publishing, Leyden, The Netherlands, 1973

- H.F. Bueckner, The Weight Functions of Mode I of the Penny-Shaped and of the Elliptic Crack, *Fracture Mechanics and Technology*, Vol II, G.C. Sih and C.L. Chow, Ed., Sijthoff and Noordhoff International Publishers, 1977, p 1069–1089
- P.M. Besuner, Residual Life Estimates for Structures with Partial Thickness Cracks, *Mechanics of Crack Growth*, STP 590, American Society for Testing and Materials, 1976, p 403–419
- T.A. Cruse and P.M. Besuner, *J. Aircr.*, Vol 12, 1975, p 369–375
- R.G. Forman, S.R. Mettu, and V. Shivakumar, Fracture Mechanics Evaluation of Pressure Vessels and Pipes in Aerospace Applications, *Fatigue, Fracture, and Risk*, PVP Vol 241, Book G00676, American Society of Mechanical Engineers, 1992, p 25–36
- A.F. Grandt, Jr., Stress Intensity Factors for Cracked Holes and Rings Loaded with Polynomial Crack Face Pressure Distributions, *Int. J. Fract.*, Vol 14, 1978, p R221 to R229
- A.F. Grandt, Jr. and T.E. Kullgren, Tabulated Stress Intensity Factor Solutions for Flawed Fastener Holes, *Eng. Fract. Mech.*, Vol 18, 1983, p 435–451
- K. Katherisan, T.M. Hsu, and T.R. Brussat, "Advanced Life Analysis Methods—Crack Growth Analysis Methods for Attachment Lugs," Report AFWAL-TR-84–3080, Vol II, Flight Dynamics Laboratory, Wright-Patterson Air Force Base, Dayton, OH, Sept 1984
- S.R. Mettu and R.G. Forman, Analysis of Circumferential Cracks in Circular Cylinders Using the Weight-Function Method, *Fracture Mechanics: Twenty-Third Symposium*, STP 1189, American Society for Testing and Materials, 1993, p 417–440
- P.C. Paris, R.M. McMeeking, and H. Tada, The Weight Function Method for Determining Stress Intensity Factors, *Cracks and Fracture*, STP 601, American Society for Testing and Materials, 1976, p 471–489
- J.R Rice, Some Remarks on Elastic Crack Tip Stress Fields, *Int. J. Solids Struct.*, Vol 8, 1972, 751–758
- G.T. Sha, Stiffness Derivative Finite Element Technique to Determine Nodal Weight Functions with Singularity Elements, *Eng. Fract. Mech.*, Vol 19, 1984, p 685–699
- G.T. Sha and C.T. Yang, Weight Functions of Radial Cracks Emanating from a Circular Hole in a Plate, *Fracture Mechanics: Seventeenth Volume*, STP 905, American Society for Testing and Materials, 1986, p 573–600
- V. Shivakumar and G. Forman, Green's Function for a Crack Emanating from a Circular Hole in an Infinite Sheet, *Int.J. Fract.* Vol 16, 1980, p 305–316
- S.T. Xiao, M.W. Brown, and K.J. Miller, Stress Intensity Factors for Cracks in Notched Finite Plates Subjected to Biaxial Loading, *Fatigue and Fracture of Engineering Materials and Structures*, Vol 8, 1985, p 349–372

Stresses around a Hole and Their Effect on Crack Growth

- W.H. Cathey and A.F. Grandt, Jr., Fracture Mechanics Consideration of Residual Stresses Introduced by Coldworking Fastener Holes, *J. Eng. Mater. Technol.*, (*Trans. ASME*), Series H, Vol 102, 1980, p 85–91
- J.H. Crews, Jr., "An Elastic Analysis of Stresses in a Uniaxially Loaded Sheet Containing an Interference-Fit Bolt," TN D-6955, National Aeronautics and Space Administration, Washington, D.C., Sept 1972

- J.H. Crews, Jr., "An Elastoplastic Analysis of a Uniaxially Loaded Sheet with an Interference-Fit Bolt," TN D-7748, National Aeronautics and Space Administration, Washington, D.C., Oct 1974
- J.H. Crews, Jr. and N.H. White, "Fatigue Crack Growth from a Circular Hole with and without High Prior Loading," TN D-6899, National Aeronautics and Space Administration, Washington, D.C., Sept 1972
- A.J. Dunelli and C.A. Sciammarella, Elastoplastic Stress and Strain Distribution in a Finite Plate with a Circular Hole Subjected to Unidimensional Load, *J. Appl. Mech.*, (*Trans. ASME*), Series E, 1963, p 115–121
- A.F. Grandt, Jr. and J.P. Gallagher, Proposed Fracture Mechanics Criteria to Select Mechanical Fasteners for Long Service Lives, *Fracture Toughness and Slow-Stable Cracking*, STP 559, American Society for Testing and Materials, 1974, p 283–297
- Y.C. Hsu and R.G. Forman, Elastic-Plastic Analysis of an Infinite Sheet Having a Circular Hole under Pressure, *J. Appl. Mech.*, (*Trans. ASME*), Series E, 1975, p 347–352
- W.C. Huang, Theoretical Study of Stress Concentrations at Circular Holes and Inclusions in Strain Hardening Materials, *Int. J. Solids Struct.*, Vol 8, 1972, p 149–192
- D.V. Nelson, Effects of Residual Stress on Fatigue Crack Propagation, *Residual Stress Effects in Fatigue,* STP 776, American Society for Testing and Materials, 1982, p 172–194
- J.A. Regalbuto and O.E. Wheeler, Stress Distributions from Interference Fits and Uniaxial Tension, *Exp. Mech.*, July 1970, p 274–280
- D.L. Rich and L.F. Impellizzeri, Fatigue Analysis of Cold-Worked and Interference Fit Fastener Holes, *Cyclic Stress-Strain and Plastic Deformation Aspects of Fatigue Crack Growth,* STP 637, American Society for Testing and Materials, 1977, p 153–175
- R.C. Shah, On Through Cracks at Interference Fit Fasteners, *J. Pressure Vessel Technol.*, (*Trans. ASME*), Feb 1977, p 75–82
- R.C. Shah, "Quarter or Semi-Circular Cracks Originating at Interference Fit Fasteners," paper presented at AIAA/ASME/SAE 17th Structures, Structural Dynamics and Materials Conf. (King of Prussia, PA), 5–7 May 1976
- C.R. Smith, Interference Fasteners for Fatigue-Life Improvement, *Exp. Mech.*, Vol 5, 1965, p 19A to 23A
- E.Z. Stowell, "Stress and Strain Concentration at a Circular Hole in an Infinite Plate," NACA-TN-2073, National Advisory Committee for Aeronautics, Washington, D.C., April 1950

Chapter 5

Crack Opening Mode Stress Intensity Factor Solutions

To perform fracture mechanic analysis on a cracked structure (or machine part), a set of stress intensity factors that appropriately represents the global and local geometries of the part under consideration (including crack morphology and loading condition) is required. Stress intensity factors for many generic geometries in combination with different types of loading conditions have been calculated and compiled in various handbooks. This chapter presents and discusses mode 1 (crack opening mode) stress intensity factors for those crack geometries commonly found in structural components. Some of these stress intensity factor solutions have only become available recently. They are either an updated solution for an old problem or a solution for a configuration that has never been included in any existing handbook, including those listed at the end of this chapter.

LEFM Geometry Factors

For a given structural configuration and applied stress, the definition of K can be determined by a number of analytical methods. For purposes of illustration and a simple introduction, however, first consider a plate of width W containing a through-the-thickness crack of length $2a$ (Fig. 5-1). As long as the plate response to the stress is elastic, the local crack tip stress at any point is proportional to the applied stress σ (i.e., $\sigma_y \propto \sigma$). It must be expected that the crack tip stress will also depend on the crack size. Because σ_y depends on $1/\sqrt{r}$ according to Eq 1-18, it is inevitable that it depends on $\sqrt{a}$; otherwise, the dimensions would be wrong. Hence, a simple argument shows that for $\theta = 0$:

$$\sigma_y \propto \sqrt{a}\,/\sqrt{2\pi r} \qquad \text{(Eq 5-1)}$$

This proportional relation can be used to define an explicit relation between the applied stress σ and the crack tip stress σ_y, such that:

$$\sigma_y = \beta\sigma\sqrt{\pi a}/\sqrt{2\pi r} \qquad \text{(Eq 5-2)}$$

where β is a dimensionless constant. The crack tip stresses will be higher when W is narrower. Thus, β must depend on W. It is known that β must be dimensionless, yet β cannot be dimensionless and depend on W at the same time, unless β depends on W/a or a/W, that is, $\beta = f(a/W)$. Comparison of Eq 5-2, then, shows that:

$$K = \sigma \cdot \sqrt{\pi a} \cdot \beta\left(\frac{a}{W}\right) \qquad \text{(Eq 5-3)}$$

where the third term β is expressed as a function of a/W. If the crack tip stress is affected by other geometric parameters—for example, if a crack emanates from a hole, the crack tip stress will depend on the size of the

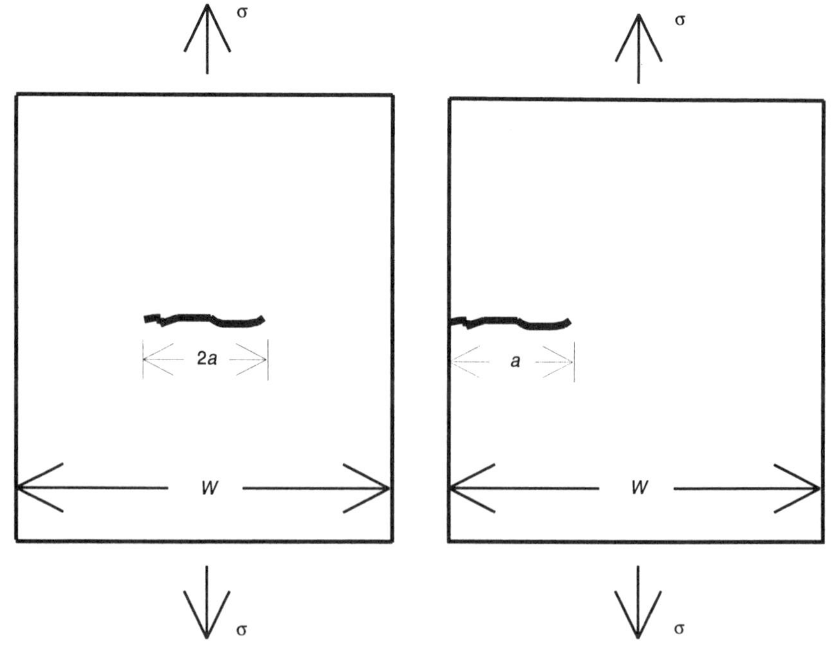

Fig. 5-1 Center-cracked panel subjected to uniform far-field tension

Fig. 5-2 Single edge crack subjected to uniform far-field tension

hole—the only effect on stress intensity factor (and the crack tip stress) will be in β. Consequently, β will be a function of all geometric factors affecting the crack tip stresses: β = f(*a/W, a/L, a/D*, etc.), where *W, L,* and *D* are plate width, plate length, and diameter of a circular hole, respectively. Crack tip stresses are always given by Eq 1-18; the value of *K* in Eq 1-18 is always given by Eq 5-3 (or its alternate form, Eq 5-4 or 5-5, shown below). All effects of geometry are reflected in a geometric parameter β. In this chapter we will write a general expression for stress intensity as:

$$K = \sigma \cdot Y \cdot \sqrt{\pi a} \qquad \text{(Eq 5-4)}$$

For a given structural configuration the definitions of σ and *a* depend on the problem under consideration. *Y* represents the complete geometric function, such as β = f(*a/W*). The value of *K* can be determined by a number of analytical methods, e.g., finite-element, boundary integral equation, weight

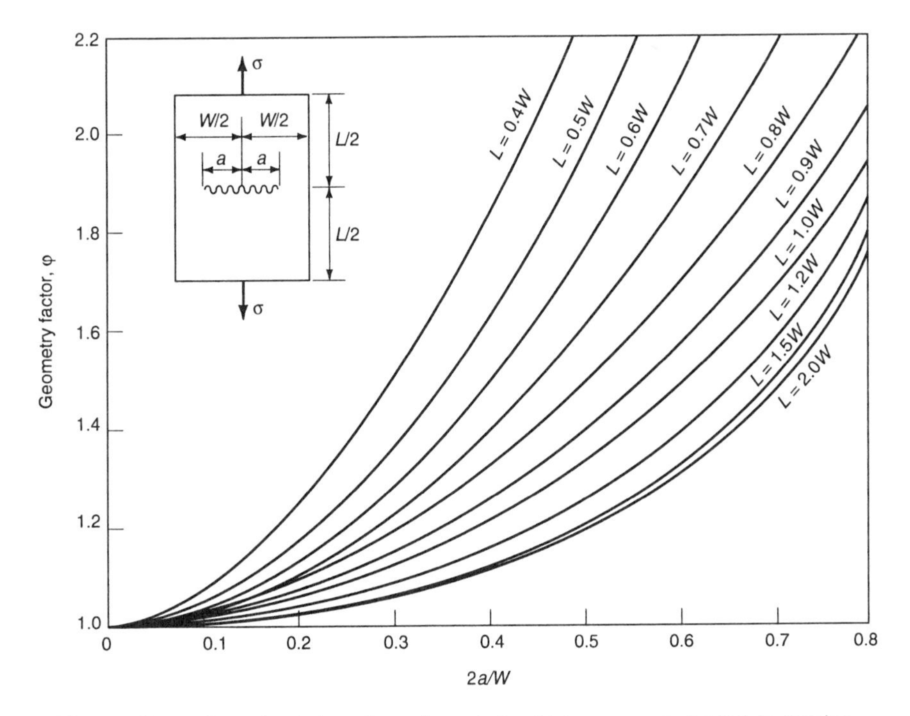

Fig. 5-3 Stress intensity factors for a through-the-thickness crack loaded in tension. Source: Ref 5-2

function, boundary collocation, etc. In this chapter we will not get into the technicality of these methods. Only the results (mainly extracted from open literature) are presented here. However, it is significant to point out that each K-value is determined for a specific crack size in an explicit structural (or test specimen) configuration. Then, a dimensionless factor Y is equal to K normalized by σ and $\sqrt{\pi a}$. After a series of K-values for various crack sizes across the entire crack plane are determined, a close-form equation can be obtained by fitting a curve through all the Y-values.

This chapter is only intended to be a summary of some mode 1 stress intensity factors for crack geometries commonly found in structural parts. In the following sections we will consider:

- A crack in a sheet (or plate)

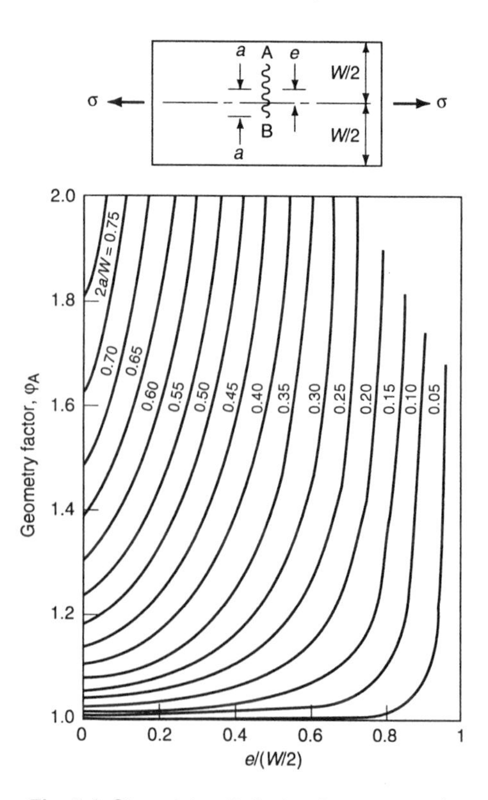

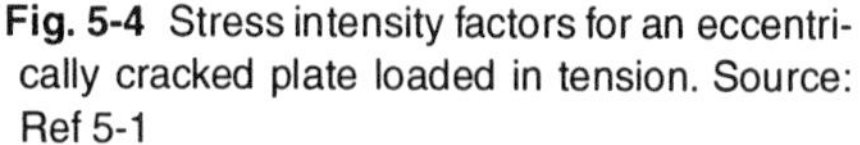

Fig. 5-4 Stress intensity factors for an eccentrically cracked plate loaded in tension. Source: Ref 5-1

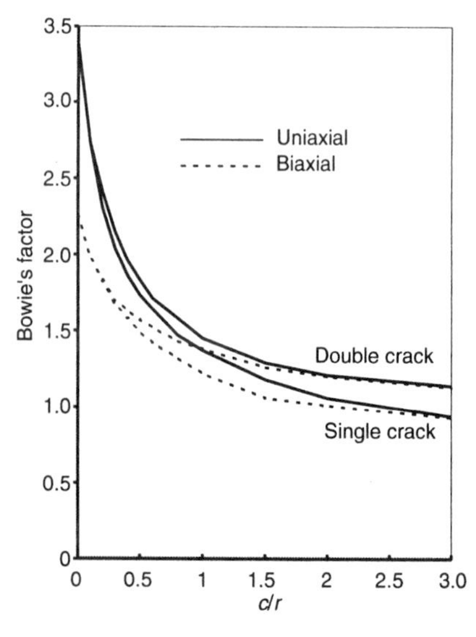

Fig. 5-5 Stress intensity factors for cracks coming out from a circular hole. Source: Ref 5-7

- A crack originated at a circular hole in a sheet (or plate)
- A cracked sheet (or plate) reinforced with attached stiffeners
- A pin-loaded lug
- A round bar (which can be a pin or a bolt)
- A hollow cylinder (i.e., a tube)

The crack may be a through-the-thickness crack or a part-through crack (commonly known as a surface crack or corner crack). The part may be subjected to uniaxial tension, or bending, or it may be loaded by a pair of concentrated forces or the combination of the pin and the reaction loads (e.g., an attachment lug). Mechanically fastened joints are important in aircraft structures as well as in many other types of applications. However, stress intensity factors for this class of structure are complex; they are not included here. See the references listed at the end of this chapter.

Compounding of Geometric Factors

When Y (or β) is a function of more than one variable, it is desirable to separate the function into a series of dimensionless parametric functions,

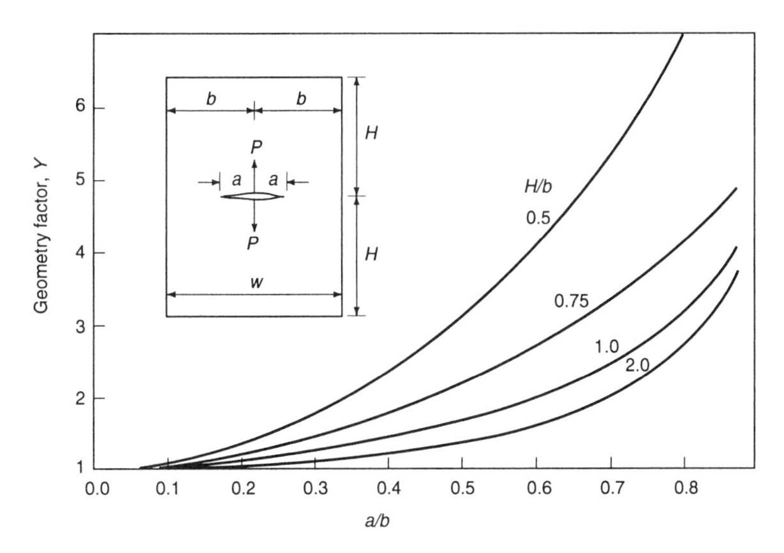

Fig. 5-6 Stress intensity factors for a central crack loaded in a rectangular plate with opposing forces at the center of the crack, where P is the force (load) per unit thickness. Source: Ref 5-9

where each segment represents an explicit boundary condition. Therefore Eq 5-4 can be rewritten as:

$$K = \sigma\sqrt{\pi a} \cdot \Pi\alpha \quad \text{(Eq 5-5)}$$

where $\Pi\alpha$ is the product of a series of dimensionless parametric functions, or factors, accounting for the influence of the part and the crack geometries and loading condition. In the absence of any geometric influence, for example, a through-the-thickness crack in an infinitely wide sheet under uniform far-field tension, $\Pi\alpha$ approaches unity ($\beta = 1$) and Eq 5-5 reduces to the basic stress intensity expression:

$$K = \sigma\sqrt{\pi a} \quad \text{(Eq 5-6)}$$

where σ is the applied stress and a is the half-length of the through-the-thickness crack. It is significant to note that K is a physical quantity (not a factor) with dimensional units of ksi$\sqrt{\text{in.}}$, or MPa$\sqrt{\text{m}}$. In this book the term *stress intensity factor* is loosely defined; it might mean K, or a geometric factor (such as α in Eq 5-5, or β or Y), as discussed earlier.

The advantage of separating a lumped geometric factor into several individual segments is obvious. The fracture mechanics analysts can build around a compounded equation to suit the crack model under consideration

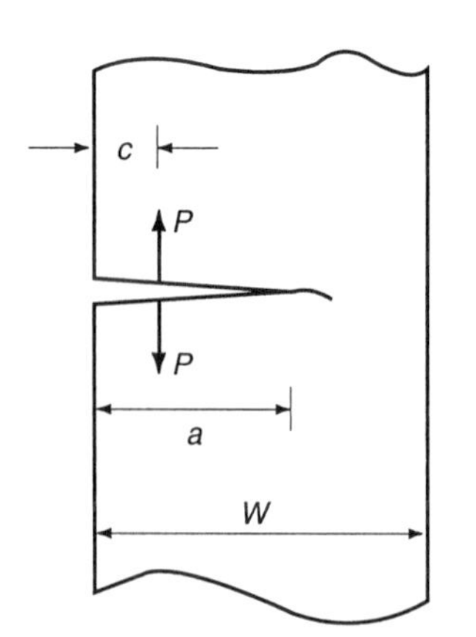

Fig. 5-7 An edge crack loaded by a pair of point forces

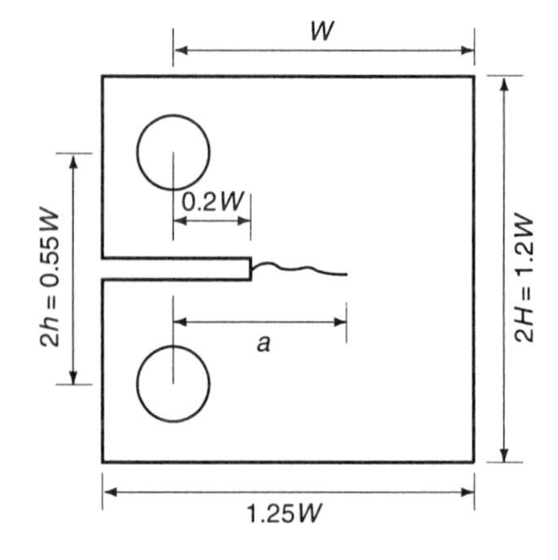

Fig. 5-8 The compact specimen configuration. Pin hole diameter is 0.25*W*.

by combining several known solutions. The method of compounding is quite simple and is discussed in Chapter 4.

Through-The-Thickness Crack in a Plate

Through-the-thickness cracks may be located in the middle of a plate (in which case they are called center cracks, Fig. 5-1), at the edge of a plate (edge cracks, Fig. 5-2), or at the edge of a hole inside a plate.

Uniform Far-Field Loading

Center Crack

There are many stress intensity factor solutions for the center crack subjected to far-field uniform tension stress. Among them, the Isida solution (Ref 5-1, 5-2) is regarded by the fracture mechanics community as being the most accurate. It accounts for the boundary effect (the free edges that define the width and length of the panel). The stress intensity expression can be written in the form of Eq 5-5 as:

$$K = \sigma\sqrt{\pi a} \cdot \varphi \qquad \text{(Eq 5-7)}$$

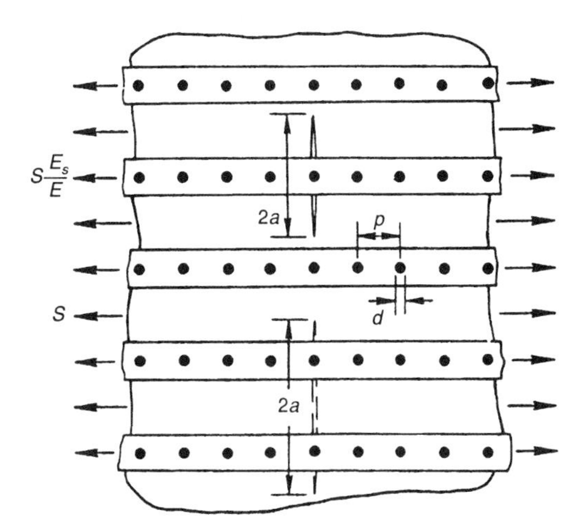

Fig. 5-9 Stiffened panel damage configurations

where σ is gross cross-sectional area stress (i.e., ignoring the crack). The geometric factor φ is a dimensionless function of the ratio of the crack length to panel width and the ratio of the panel length to panel width, as given in Fig. 5-3. For a panel having its length greater than two times the width, the effect of panel length on *K* vanishes. According to Feddersen (Ref 5-3), the finite width correction factors (i.e., the curve labeled as $L \geq 2W$ in Fig. 5-3) can be represented by a secant function. That is:

$$\varphi_w = \sqrt{\sec(\pi a/W)} \quad \text{(Eq 5-8a)}$$

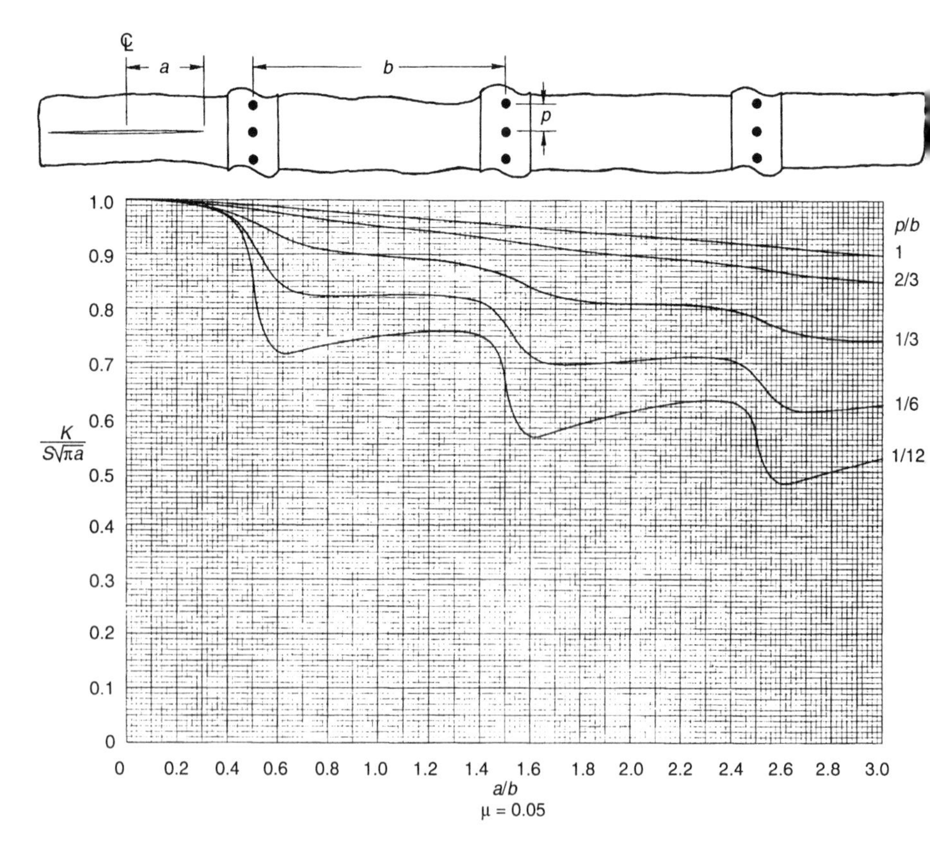

Fig. 5-10a Stiffener efficiency factors for a one-bay crack, in the middle of a bay between two stiffeners. Source: Ref 5-11

Here the secant function, in radians, is commonly known as the width correction factor for the center-cracked panel configuration. If the crack is located off the centerline of the plate, Eq 5-8(a) is replaced by the Isida eccentric crack solution (Ref 5-2). Among the two crack tips shown in Fig. 5-4, stress intensity at tip-A is higher than those at tip-B because tip-A is closer to the plate edge whereas tip-B is away from the plate edge. It is conceivable that fracture strength of the plate is dictated by the stress intensity at tip-A. in other words, the plate fails when stress intensity at tip-A reaches K_c of the material. Therefore, only the stress intensity factors for tip-A are shown in Fig. 5-4. However, stress intensity factors for both crack tips must be available for fatigue crack propagation analysis, because each

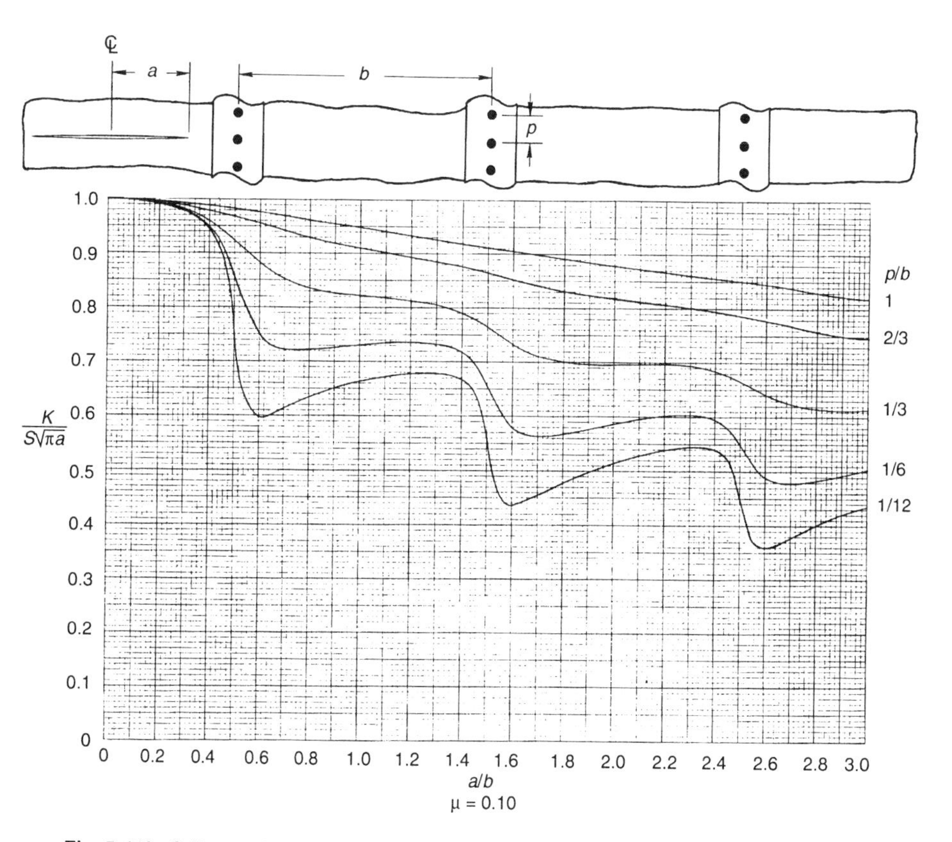

Fig. 5-10b Stiffener efficiency factors for a one-bay crack, in the middle of a bay between two stiffeners. Source: Ref 5-11

crack tip influences each other at every step of incremental growth. In a given step, each crack tip will grow a different amount of da/dN because they are subjected to different K-levels. Consequently, the eccentricity of the crack changes continuously as the crack grows. According to Newman (Ref 5-4), Isida's eccentric crack correction factors (for each crack tip) can be approximated by the following equations:

$$\varphi_A = \sqrt{\sec[(\pi a/2b_1) \cdot (1 - 0.22\,(e/b)^3]} \qquad \text{(Eq 5-8b)}$$

$$\varphi_B = \sqrt{\sec[(\pi a/2b_1) \cdot (1 - 0.37\,(e/b)^{0.5} + 0.1\,(a/b_1)^{15})]} \qquad \text{(Eq 5-8c)}$$

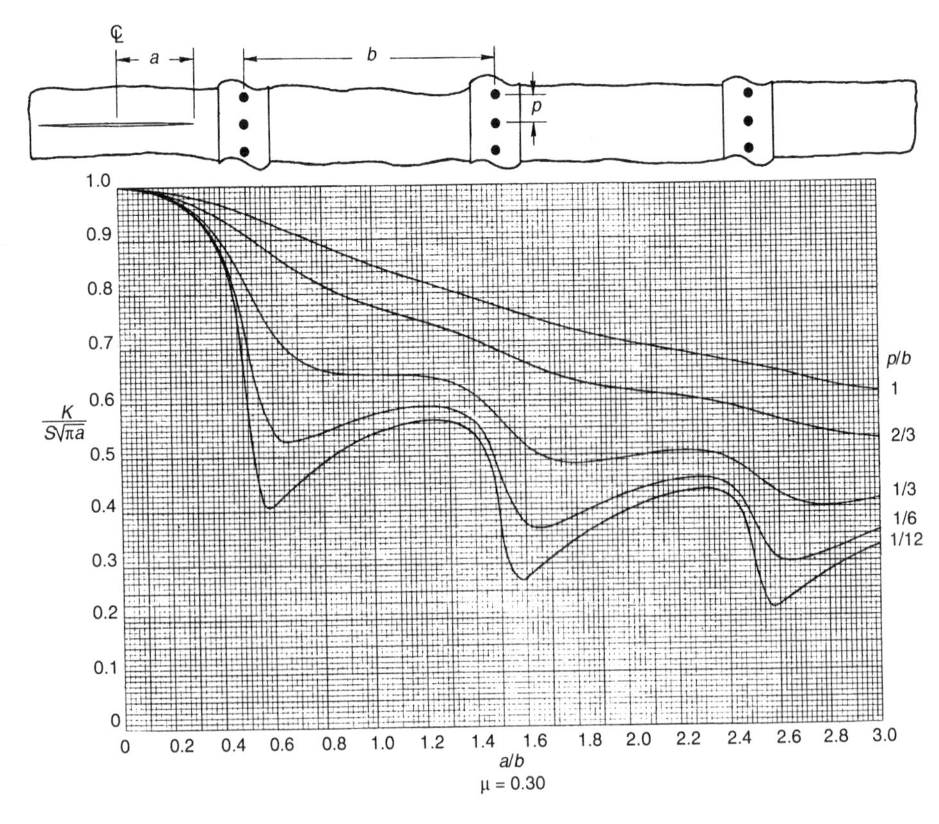

Fig. 5-10c Stiffener efficiency factors for a one-bay crack, in the middle of a bay between two stiffeners. Source: Ref 5-11

where the subscripts "A" and "B" refer to the two crack tips indicated in Fig. 5-4, e is the distance from the center of the crack to the centerline of the specimen, $b = W/2$, and $b_1 = b - e$ (the distance from the center of the crack to the nearest edge of the specimen). When $e = 0$, either Eq 5-9(b) or (c) is reduced to Eq 5-8(a).

Edge Crack

For a limiting case in which the crack is right at the edge of the plate (edge crack), the φ factor is given by Ref 5-5 as:

$$\varphi_e = \sec\beta\,[(\tan\beta)/\beta]^{1/2}\cdot[0.752 + 2.02\,(a/W) + 0.37\,(1 - \sin\beta)^3] \quad \text{(Eq 5-8d)}$$

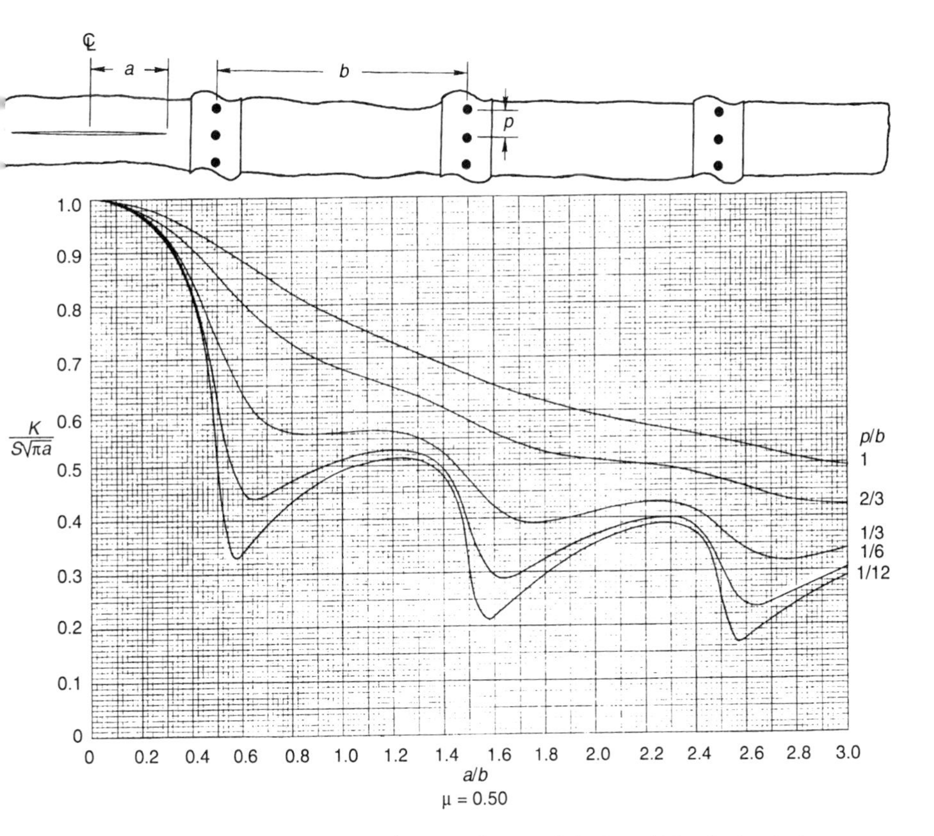

Fig. 5-10d Stiffener efficiency factors for a one-bay crack, in the middle of a bay between two stiffeners. Source: Ref 5-11

where a is the total crack length measured from the edge of the plate across the width W, and β is a specific geometric factor for this case such that $\beta = \pi a/(2W)$. The subscript "e" stands for the edge crack, $\varphi_e = 1.122$ at $a/W = 0$.

Crack Emanating from a Hole

In the case where a crack is originated at the edge of a circular hole inside an infinitely wide sheet:

$$K = \sigma\sqrt{\pi c} \cdot \varphi_n \quad \text{(Eq 5-9)}$$

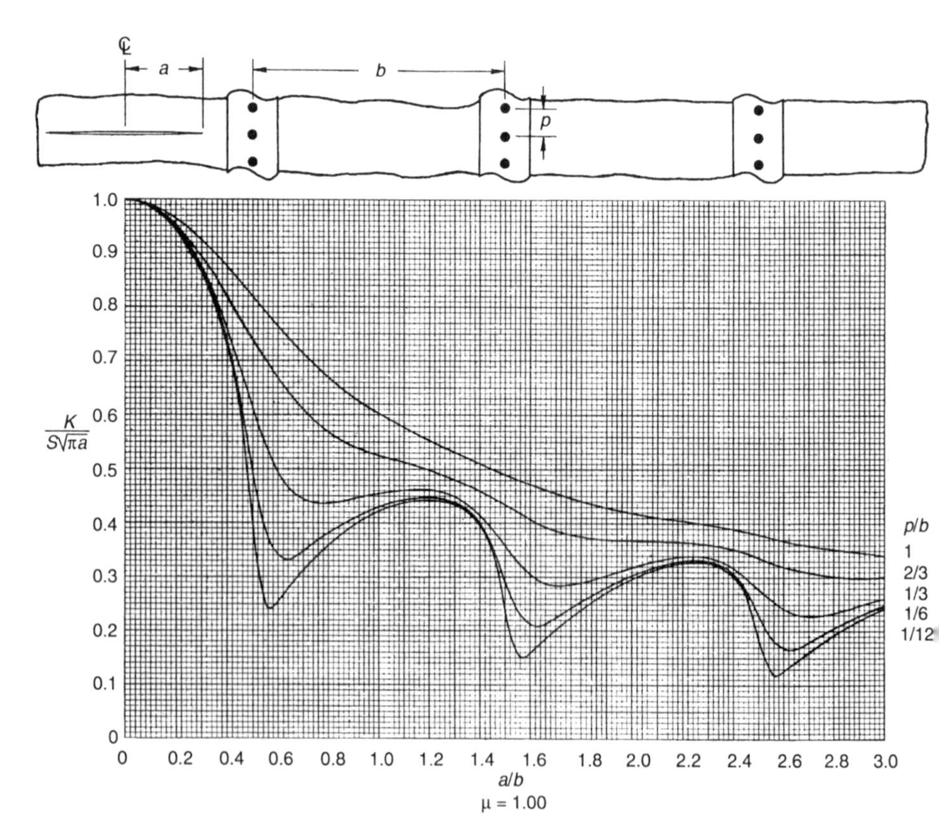

Fig. 5-10e Stiffener efficiency factors for a one-bay crack, in the middle of a bay between two stiffeners. Source: Ref 5-11

where σ is the gross cross-sectional area stress (i.e., again ignoring the hole and the crack), c is the crack length measured from the edge of the hole (not from the center of the hole), and φ_n is the Bowie solution for cracks coming out from a circular hole (Ref 5-6, 5-7). According to Ref 5-8, the Bowie solution for uniaxial tension (Fig. 5-5) can be fitted by the following equations:

$$\varphi_1 = 0.707 - 0.18 \cdot \chi + 6.55 \cdot \chi^2 - 10.54 \cdot \chi^3 + 6.85 \cdot \chi^4 \qquad \text{(Eq 5-9a)}$$

and

$$\varphi_2 = 1 - 0.15 \cdot \chi + 3.46 \cdot \chi^2 - 4.47 \cdot \chi^3 + 3.52 \cdot \chi^4 \qquad \text{(Eq 5-9b)}$$

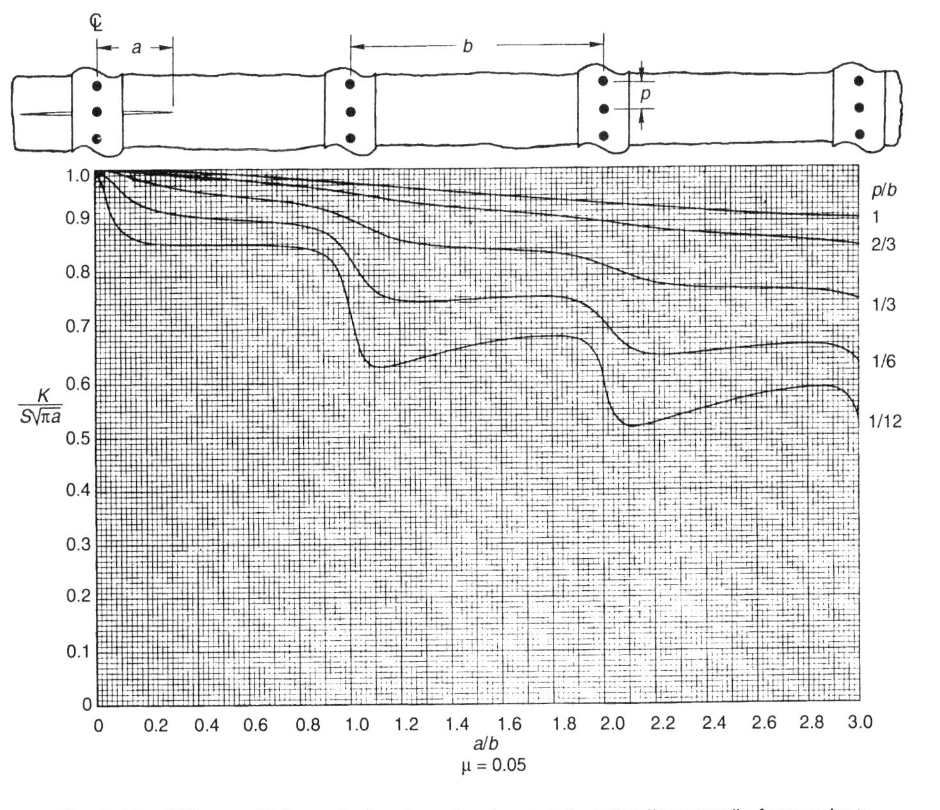

Fig. 5-11a Stiffener efficiency factors for a two-bay crack, extending equally from a rivet hole beneath the center stiffener. Source: Ref 5-11

where $\chi = (1 + c/r)^{-1}$, r is the radius of the hole. The subscripts "1" and "2" stand for a single crack and two symmetric cracks, respectively.

Now we ought to mention here an unwritten rule about making up fitting functions for stress intensity factors. For life assessment, the fracture mechanics analyst needs to use stress intensity factors across the entire cross section of the part under consideration. Accurate stress intensity solutions are difficult to obtain for cracks that are very small or are close to the free boundary of the part, but this hurdle can be overcome with sound engineering judgments. Excepting the center crack, most cracks are exposed at a free surface, i.e., they start at an edge of some sort. In the case of a crack coming out from a hole, it can be considered that the crack is an edge crack at a site of stress concentration. This notion leads to a compounding solution for a

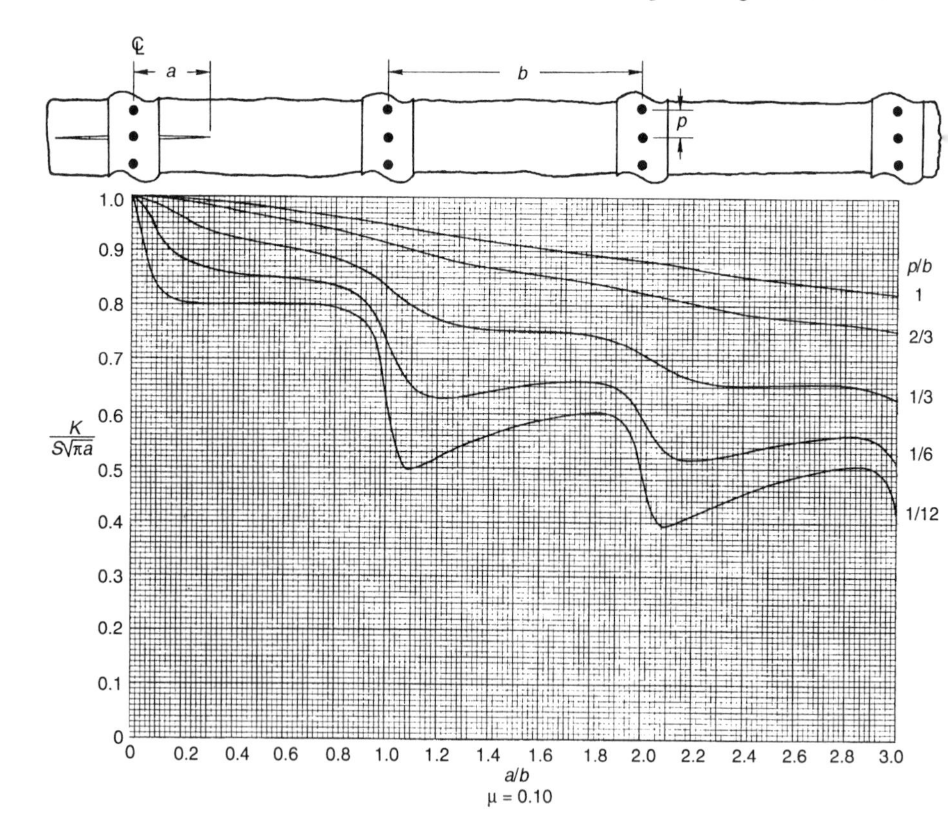

Fig. 5-11b Stiffener efficiency factors for a two-bay crack, extending equally from a rivet hole beneath the center stiffener. Source: Ref 5-11

very small crack at the edge of a circular hole. The compounding solution consists of two components: the stress concentration factor of the hole and the stress intensity factor for an edge crack with zero crack length. The stress concentration factor for a circular hole in an infinite sheet is 3.0 (the K_t based on gross area stress); the edge crack geometric factor is 1.122. Therefore the Bowie factor for $c/r = 0$ should be in the neighborhood of 3.37 (3 times 1.122). A value of 3.39 was chosen by Paris and Sih (Ref 5-7). When using Eq 5-9(a) and (b), a pair of values of 3.39 and 3.36 will be obtained.

In case the crack(s) are subjected to biaxial loading, i.e., there are forces applied parallel to the crack in addition to the forces normal to the crack, the stress concentration factor at the hole will have a value other than 3. For a

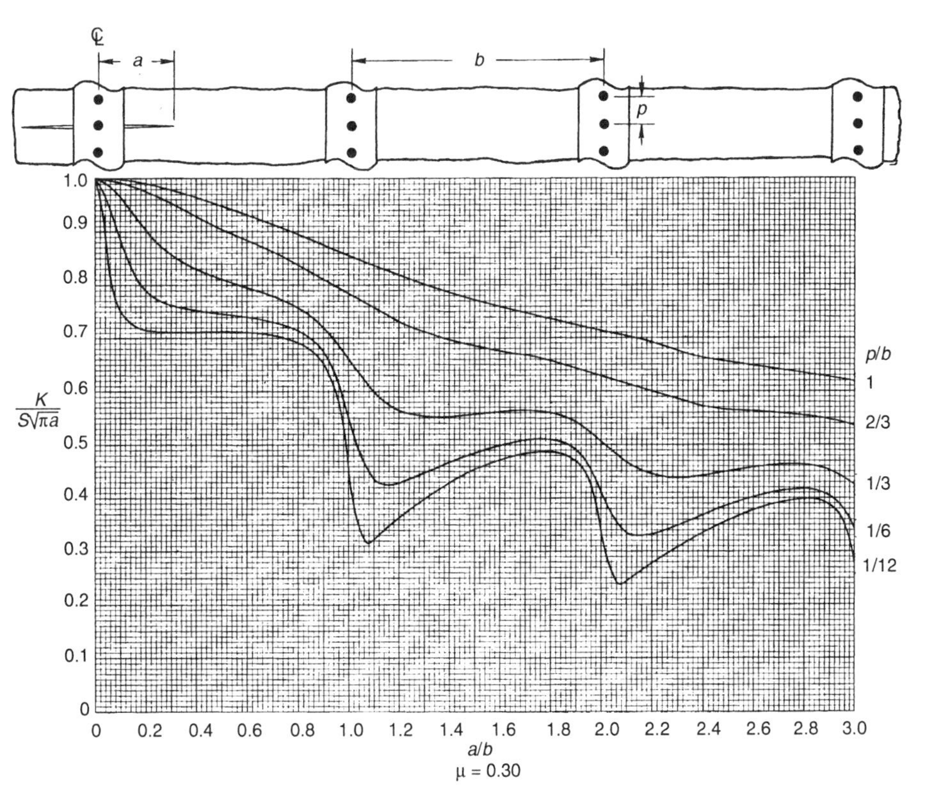

Fig. 5-11c Stiffener efficiency factors for a two-bay crack, extending equally from a rivet hole beneath the center stiffener. Source: Ref 5-11

one-to-one biaxial ratio (tension force of an equal amount in both directions), the value for K_t is 2. The Bowie factor for $c/r = 0$ would then be in the neighborhood of 2.244 (2 times 1.122). The handbook solution (Ref 5-7) starts at 2.26 and decreases to 1.0 for double cracks and 0.707 for a single crack (Fig. 5-5). Similarly, the Bowie factors for biaxial loading (Fig. 5-5) can be represented by the following polynomial equations:

$$\varphi_1 = 0.707 + 0.2744 \cdot \chi + 2.765 \cdot \chi^2 - 3.3824 \cdot \chi^3 + 1.893 \cdot \chi^4 \quad \text{(Eq 5-9c)}$$

and

$$\varphi_2 = 1 - 0.0436 \cdot \chi + 3.1256 \cdot \chi^2 - 4.46077 \cdot \chi^3 + 2.6281 \cdot \chi^4 \quad \text{(Eq 5-9d)}$$

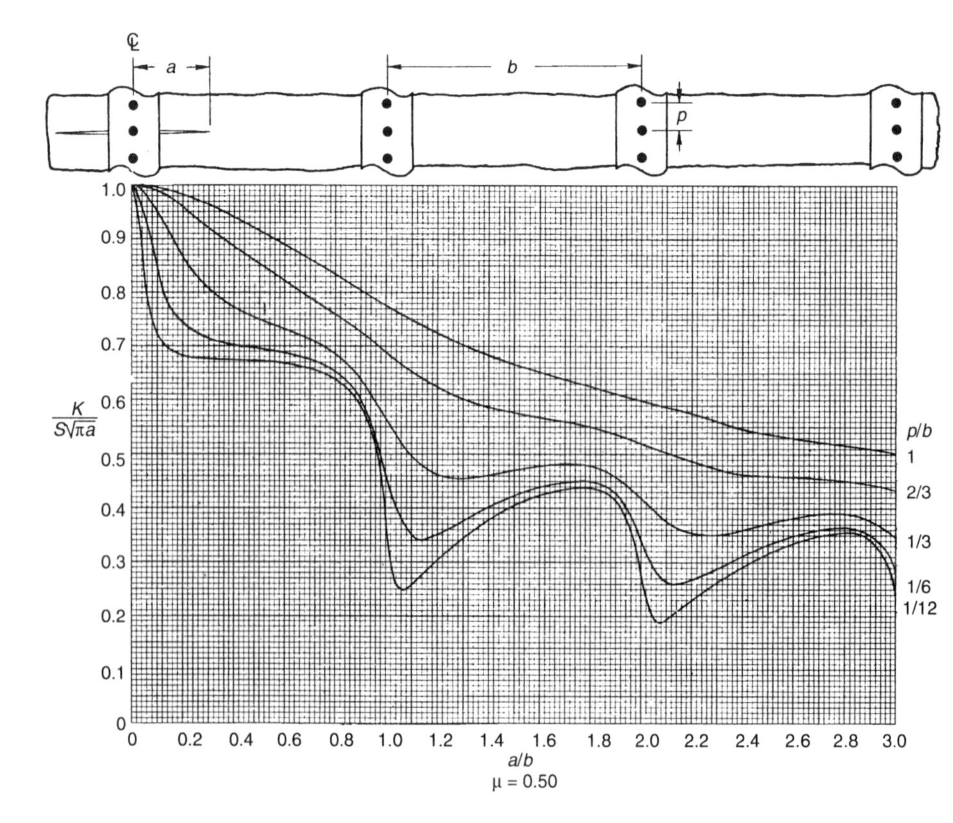

Fig. 5-11d Stiffener efficiency factors for a two-bay crack, extending equally from a rivet hole beneath the center stiffener. Source: Ref 5-11

Next, we consider the case of a long crack. The local stress distribution decreases from K_t at the hole edge to unity at locations far away from the hole, and the influence of being an edge crack vanishes, so the total crack (including the hole) can be considered as a center crack. For the double crack configuration, the total crack length can be approximated by $2c$ because the hole size becomes negligible at $c/r \to \infty$. Therefore $\varphi_2 = 1$ for $c/r \to \infty$, as given by Eq 5-9(b). For the single crack configuration the crack length c is treated as the total crack length. It means that Eq 5-9 would have become $K = \sigma\sqrt{\pi c/2}$, or $K = \sigma\sqrt{\pi c} \cdot \sqrt{1/2}$. Consequently, this leads to $\varphi_1 = 0.707$, as given by Eq 5-9(a).

To include the finite width effect (Ref 5-9):

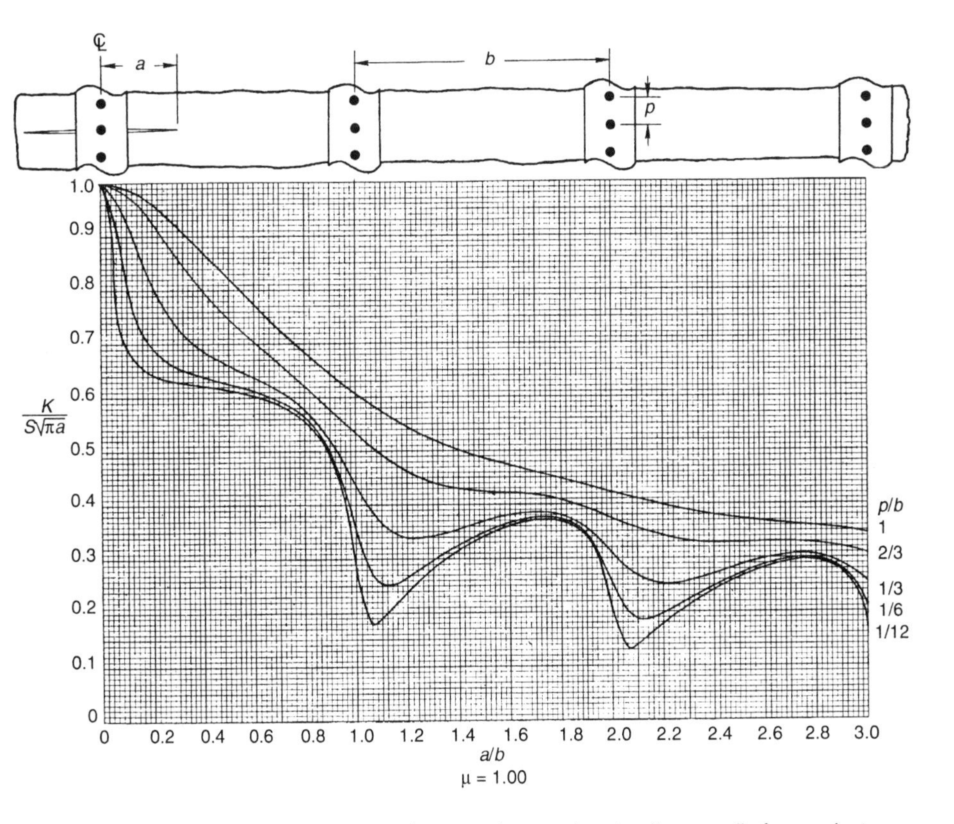

Fig. 5-11e Stiffener efficiency factors for a two-bay crack, extending equally from a rivet hole beneath the center stiffener. Source: Ref 5-11

$$K = \sigma\sqrt{\pi c} \cdot \varphi_n \cdot f_n \qquad \text{(Eq 5-9e)}$$

where

$$f_1 = \sqrt{\sec\,[(\pi\,(2r + c))/2(W - c)] \cdot \sec\,(\pi r/W)} \qquad \text{(Eq 5-9f)}$$

and

$$f_2 = \sqrt{\sec\,(\pi\,(r + c)/W) \cdot \sec\,(\pi r/W)} \qquad \text{(Eq 5-9g)}$$

where $n = 1$ for a single crack, $n = 2$ for two symmetric cracks, and the hole is located in the center of the plate.

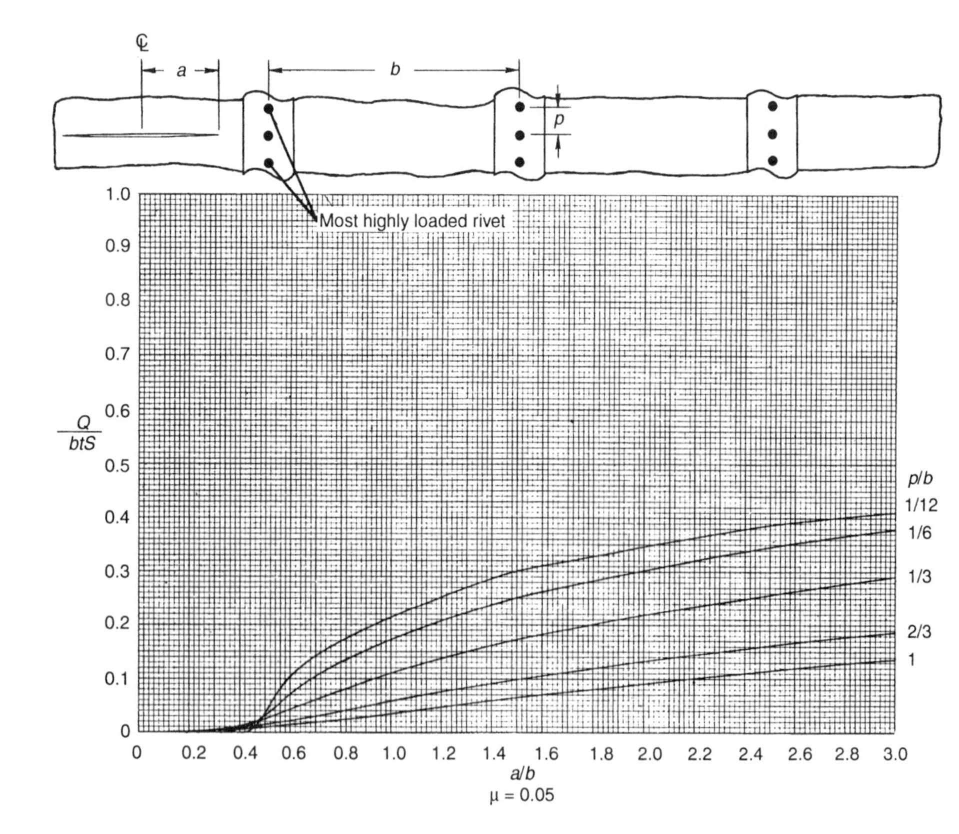

Fig. 5-12a Rivet forces in most highly loaded rivets in front of a one-bay crack. Source: Ref 5-11

Crack Line Loading

Point Loading of a Center Crack

A crack of length $2a$, subjected to forces per unit thickness P, acting at the center of the crack surfaces, is located centrally in a rectangular plate of height $2H$ and width $2b$ (Fig. 5-6). The expression for the stress intensity factor in this case is:

$$K = P \cdot Y / \sqrt{\pi a} \qquad \text{(Eq 5-10)}$$

where Y is the geometric factor, as shown in Fig. 5-6 (Ref 5-9).

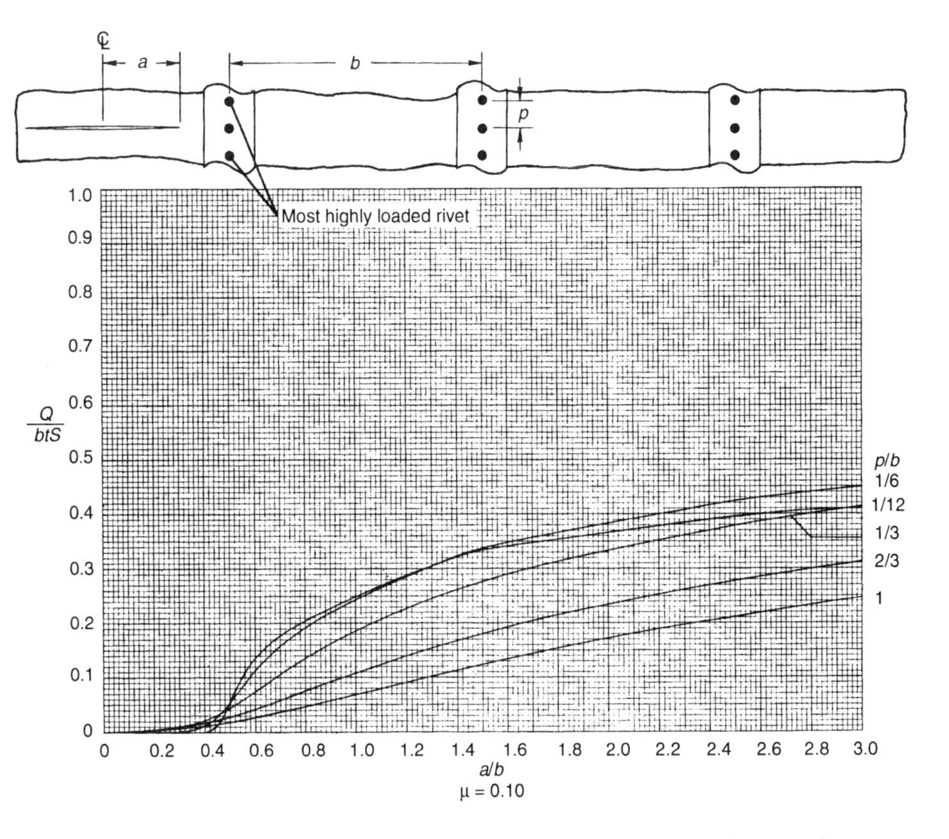

Fig. 5-12b Rivet forces in most highly loaded rivets in front of a one-bay crack. Source: Ref 5-11

Point Loading of an Edge Crack

In case the forces are applied at a point on the surface of an edge crack (Fig. 5-7), the stress intensity factor is defined as:

$$K = 2P \cdot Y/\sqrt{\pi a} \qquad \text{(Eq 5-11)}$$

where the geometric factor Y is defined by the Tada solution (Ref 5-5) as:

$$Y = \alpha_1 - \alpha_2 + \alpha_3 \cdot \alpha_4 \qquad \text{(Eq 5-11a)}$$

where

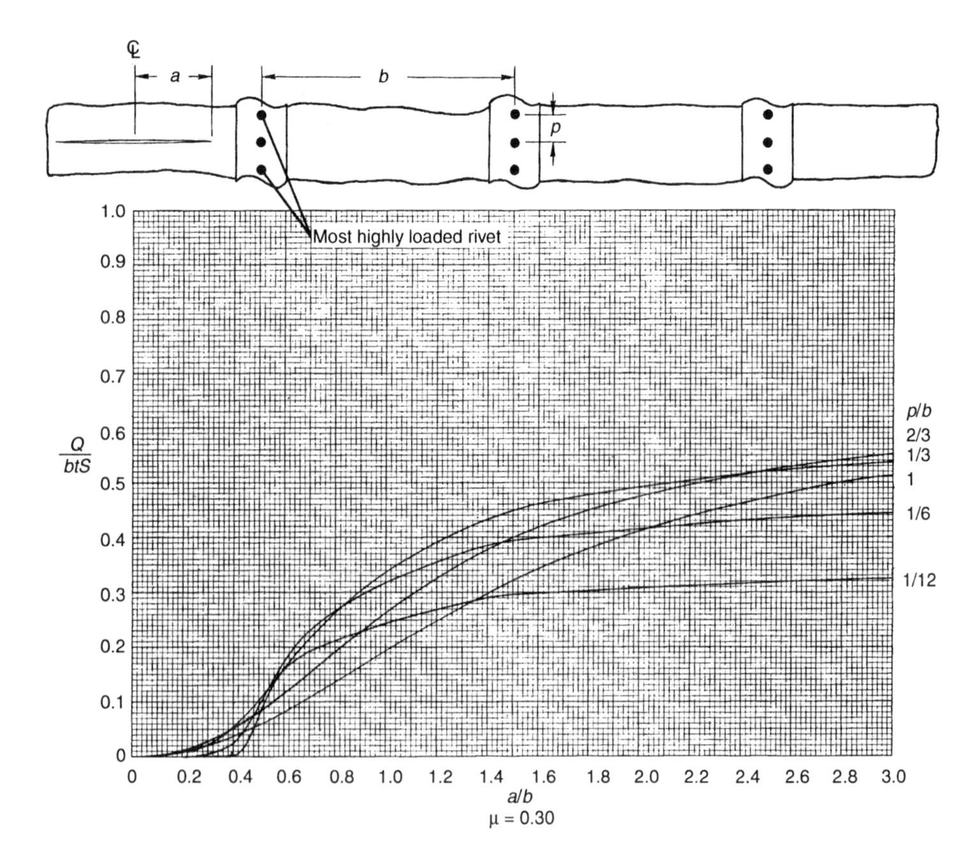

Fig. 5-12c Rivet forces in most highly loaded rivets in front of a one-bay crack. Source: Ref 5-11

$$\alpha_1 = 3.52\ (1 - \lambda)/(1 - \beta)^{3/2} \qquad \text{(Eq 5-11b)}$$

$$\alpha_2 = (4.35 - 5.28\lambda)/(1 - \beta)^{1/2} \qquad \text{(Eq 5-11c)}$$

$$\alpha_3 = [(1.3 - 0.3\lambda^{3/2})/(1 - \lambda^2)^{1/2}] + 0.83 - 1.76\lambda \qquad \text{(Eq 5-11d)}$$

$$\alpha_4 = 1 - (1 - \lambda)\beta \qquad \text{(Eq 5-11e)}$$

where $\lambda = c/a$ and $\beta = a/b$. When the load is applied right at the edge of the plate (i.e., $\lambda = 0$), Eq 5-11(a) reduces to:

$$Y = 3.52/(1 - \beta)^{3/2} - 4.35/(1 - \beta)^{1/2} + 2.13(1 - \beta) \qquad \text{(Eq 5-11f)}$$

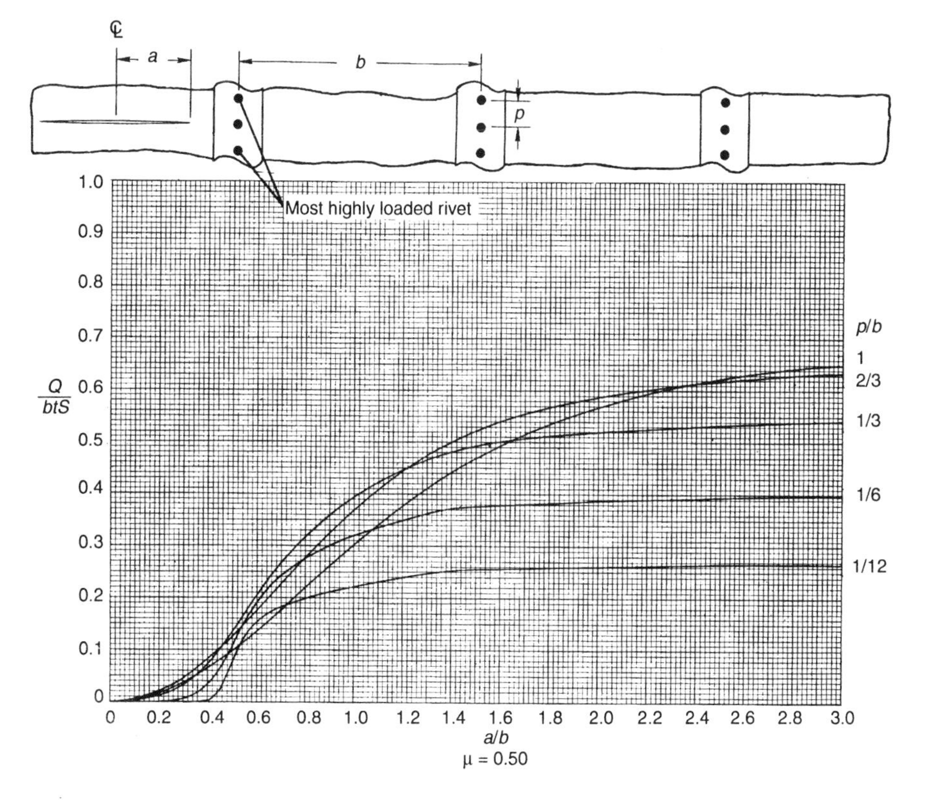

Fig. 5-12d Rivet forces in most highly loaded rivets in front of a one-bay crack. Source: Ref 5-11

The Compact Specimen

Although it is not a common configuration to be considered for structural analysis, the compact specimen (Fig. 5-8) is specified in ASTM E 399 and E 647 as the standard specimen geometry for generating material K_{Ic} and fatigue crack growth rate data. Therefore, for the convenience of readers, the ASTM-recommended stress intensity equation for the compact specimen is included here. The equation given in the *1995 Annual Book of ASTM Standards,* Volume 03.01 (Ref 5-10), is:

$$K = (P/(t\sqrt{W})) \cdot F(\alpha_1) \cdot F(\alpha_2) \quad \text{(Eq 5-12)}$$

where

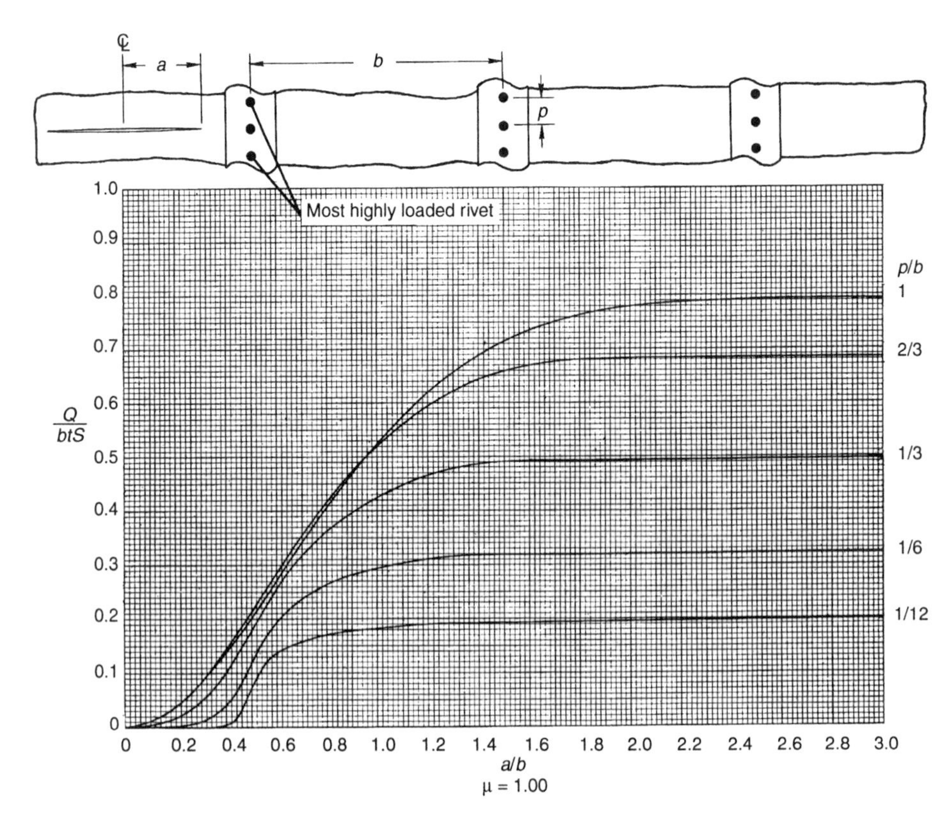

Fig. 5-12e Rivet forces in most highly loaded rivets in front of a one-bay crack. Source: Ref 5-11

$$F(\alpha_1) = 0.886 + 4.64\alpha - 13.32\alpha^2 + 14.72\alpha^3 - 5.6\alpha^4 \quad \text{(Eq 5-12a)}$$

and

$$F(\alpha_2) = (2 + \alpha)/(1 - \alpha)^{3/2} \quad \text{(Eq 5-12b)}$$

where $\alpha = a/W$ and P is the force per unit thickness. Equation (5-12) is accurate within 0.5% over the range of a/W from 0.2 to 1. As shown in Fig. 5-8, the height of the specimen (H), the length of the chevron notch, and the loading pin location (h) are functions of the specimen width (W). The ASTM-recommended dimension for the thickness B is as follows:

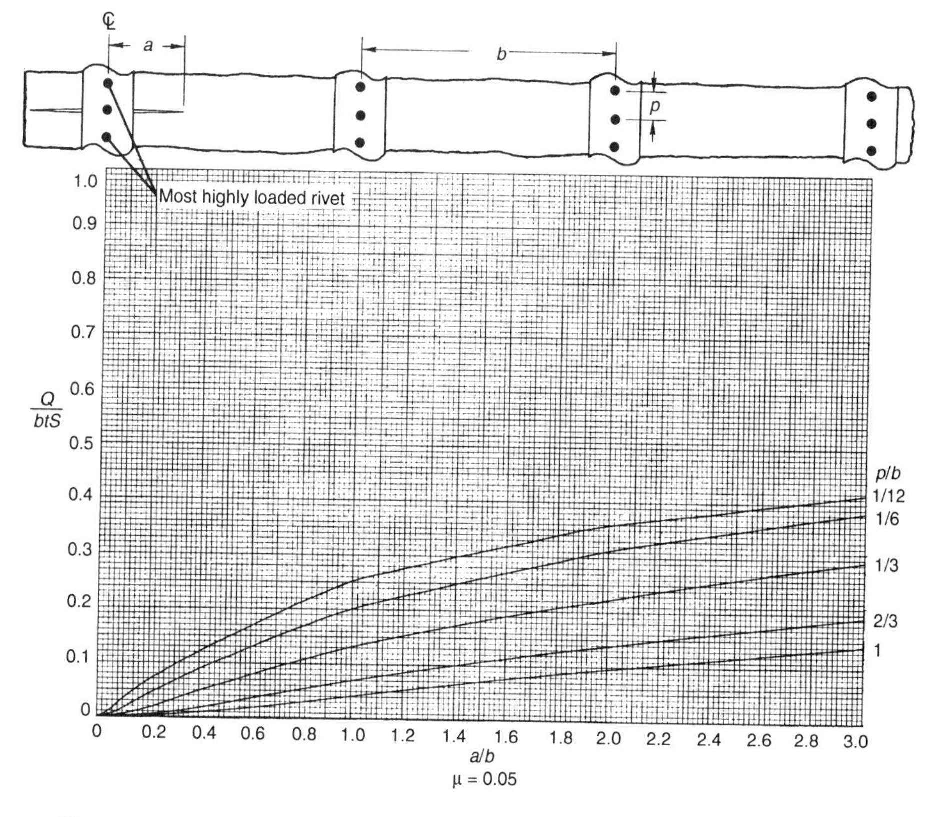

Fig. 5-13a Rivet forces in most highly loaded rivets above and below a two-bay crack. Source: Ref 5-11

- For the K_{Ic} test the preferred thickness is $W/2$. Alternatively, $2 \leq W/B \leq 4$ (with no change in other proportions) is allowed.
- For fatigue crack growth rate tests, B can be in the range of $W/20$ to $W/4$.

For test data reported in the earlier literature, an older version of the K-equation (e.g., one from the 1972 edition of the ASTM standard) might have been used. That is:

$$F(\alpha_1) = 29.6\alpha^{1/2} - 185.5\alpha^{3/2} + 655.7\alpha^{5/2} - 1017.0\alpha^{7/2} + 754.6\alpha^{9/2} \quad \text{(Eq 5-12c)}$$

for specimens where $H/W = 0.6$, or:

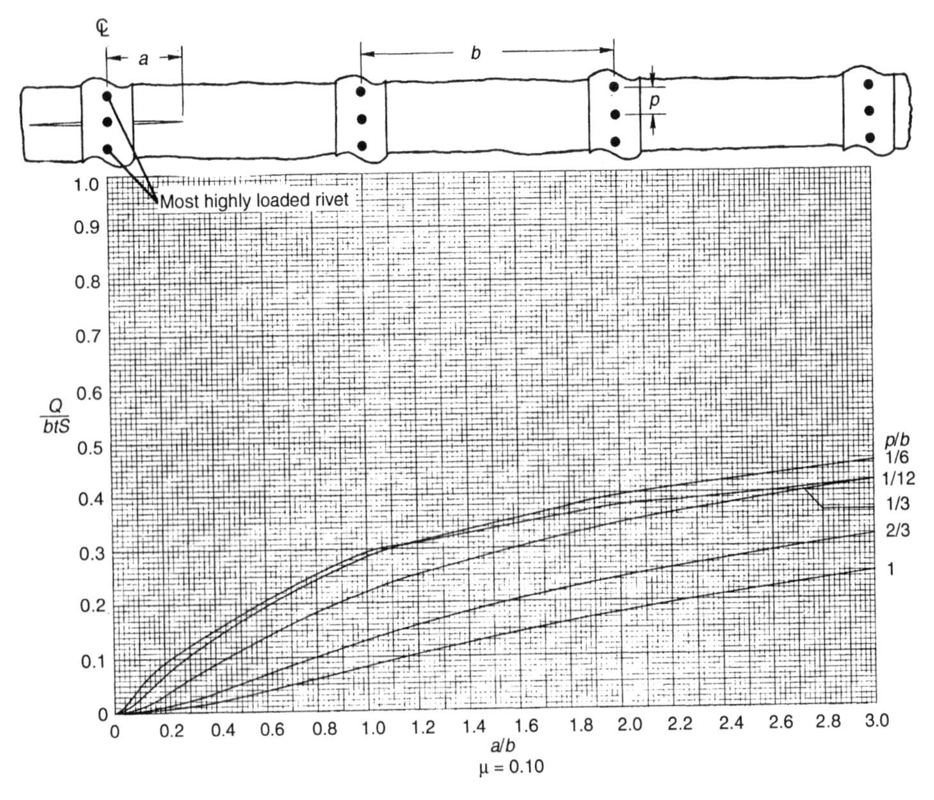

Fig. 5-13b Rivet forces in most highly loaded rivets above and below a two-bay crack. Source: Ref 5-11

$$F(\alpha_1) = 30.96\alpha^{1/2} - 195.8\alpha^{3/2} + 730.6\alpha^{5/2} - 1186.3\alpha^{7/2} + 638.9\alpha^{9/2} \quad \text{(Eq 5-12d)}$$

for specimens where $H/W = 0.486$. The second part of the geometric correction factor, $F(\alpha_2)$, did not exist in the earlier version of the equation.

A Sheet or Plate with Crack Stoppers

Figure 5-9 shows two typical arrangements for the attached stiffener structure. It shows a sheet stiffened with flat straps of equal size and spacing. The straps are attached to the sheet with equally spaced rivets. The entire skin/stiffener system is subjected to far-field uniaxial stresses. The sheet and the stiffener need not be the same material. Hence, while the stress

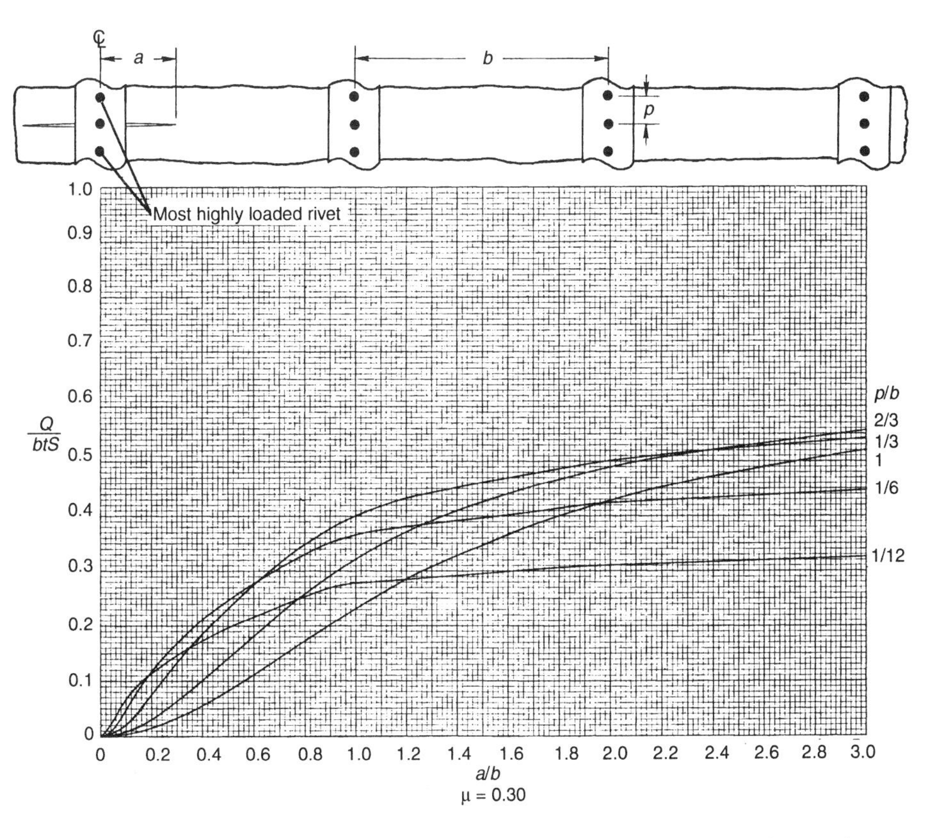

Fig. 5-13c Rivet forces in most highly loaded rivets above and below a two-bay crack. Source: Ref 5-11

on the sheet is S, the stress acting on the stiffener will be $S \cdot E_s/E$ (where E is the Young's modulus for the sheet and E_s is the Young's modulus for the stiffener) so that equal longitudinal strains at large distances from the crack will be attained.

For the two configurations shown in Fig. 5-9, i.e., for a one-bay crack or a two-bay crack with all the stiffeners intact, the analytical solutions developed by Poe (Ref 5-11) cover a wide range of geometric variables. The result of Poe's analysis can be interpreted as:

$$K = S\sqrt{\pi a} \cdot C \qquad \text{(Eq 5-13)}$$

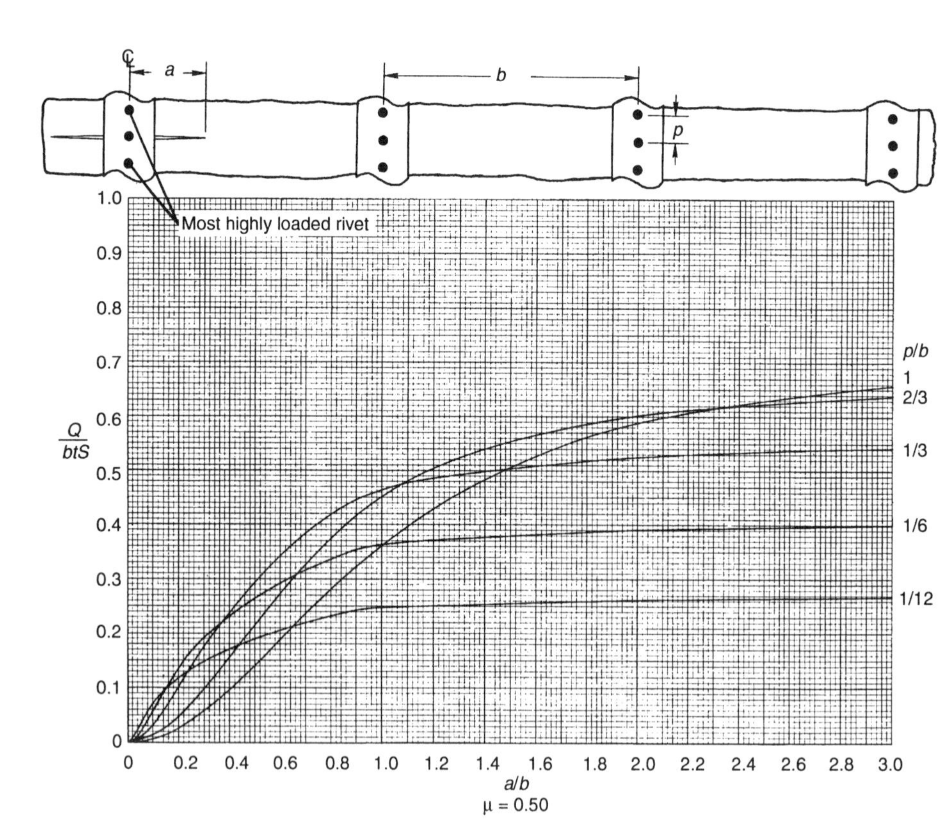

Fig. 5-13d Rivet forces in most highly loaded rivets above and below a two-bay crack. Source: Ref 5-11

where the stress intensity modification factor C is essentially the ratio of the stress intensity factor for the stiffened panel to the stress intensity factor for the plain sheet (without stiffener). The C-factor is a function of three variables: attachment spacing (rivet pitch), p, stiffener spacing, b, and a relative stiffness parameter, μ, which is defined as:

$$\mu = \left[1 + \frac{b \cdot t}{A} \cdot \frac{E}{E_s}\right]^{-1} \qquad \text{(Eq 5-14)}$$

where t is the sheet thickness and A is the cross-sectional area of the flat strap. The values for the C-factors ($=K/S\sqrt{\pi a}$ are given in Fig. 5-10 (a–e) and

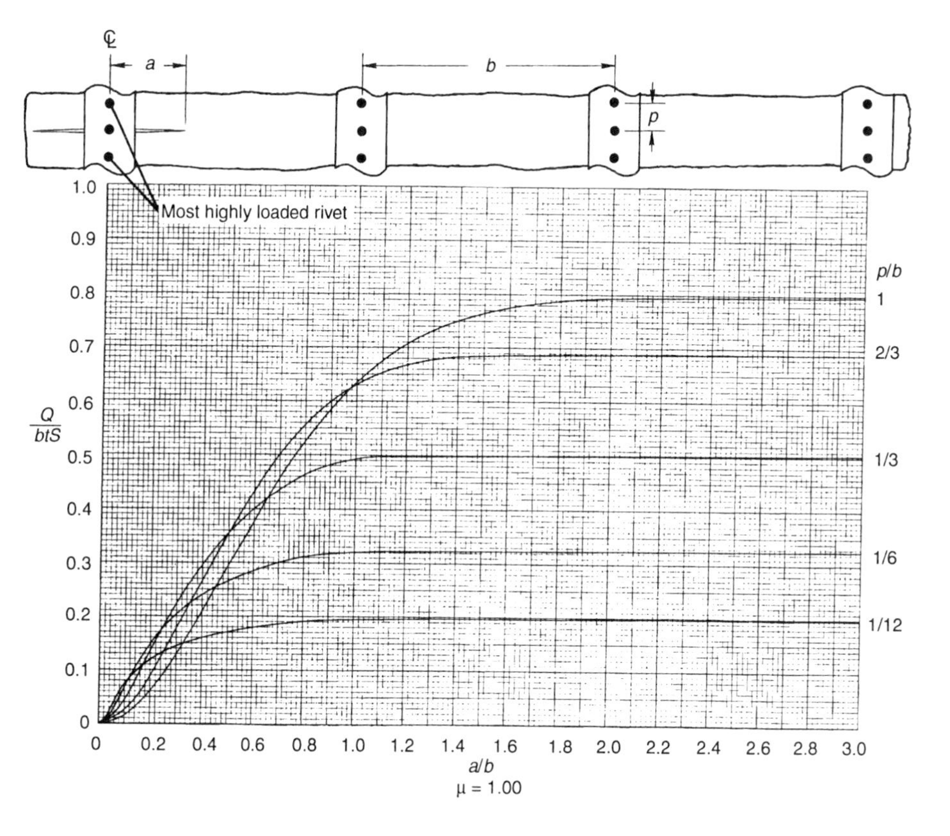

Fig. 5-13e Rivet forces in most highly loaded rivets above and below a two-bay crack. Source: Ref 5-11

5-11 (a–e). In either configuration (one-bay or two-bay), the stress intensity factor K is reduced (i.e., C decreases from 1 to less than 1) when the crack tip is in the vicinity of a stiffener. As shown in Fig. 5-10 (a–e) and 5-11 (a–e), the crack is located at one of the rivet holes, extending equally on both sides of a stiffener, or in the mid-bay between two stiffeners. In case the vertical position of the crack is between two rivets, the numerical solutions given in these figures are still useful. The first pair of rivets is most effective in contributing to K-reduction, so the important variable is the distance between the crack line and the rivet (above or below) the crack. For a given p value labeled on a C versus a/b curve, the distance can be treated as double of the p value for a configuration under consideration. For in-

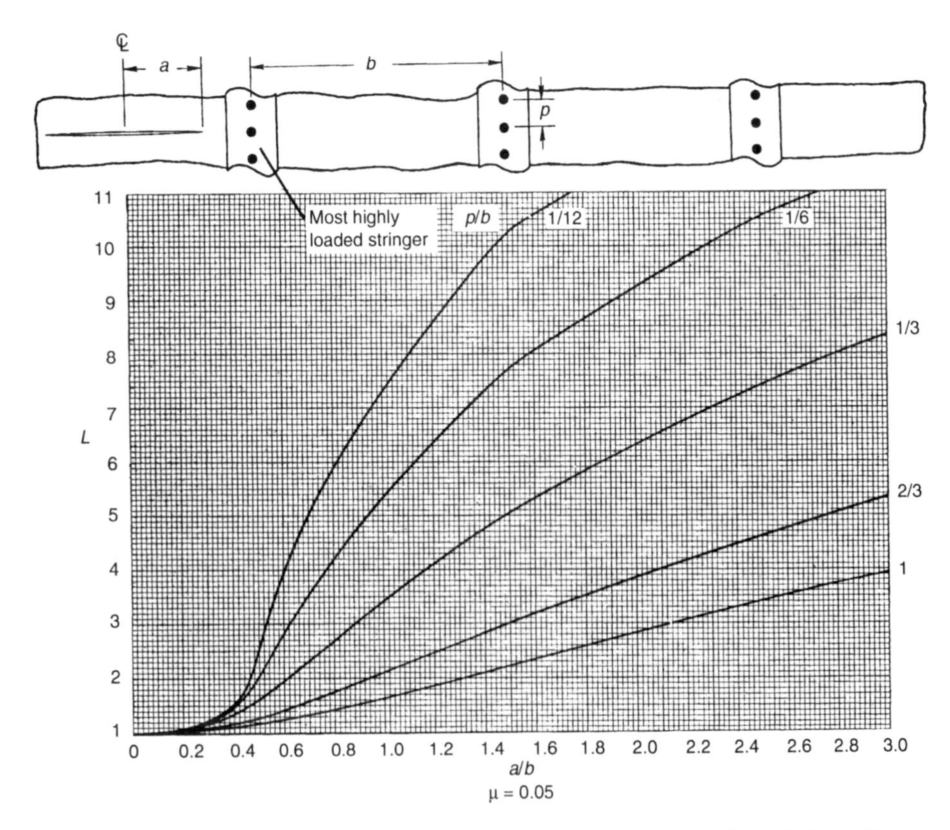

Fig. 5-14a Stiffener load concentration factor in most highly loaded stiffener in front of a one-bay crack. Source: Ref 5-11

stance, if the nominal p/b is 1/6 for a configuration under consideration, a C versus a/b curve for $p/b = 1/12$ should be used.

Poe's solution is for a crack in an infinite sheet. In actual applications, the safe-crack-growth period can be computed by compounding the C-factors with a series of other stress intensity factors. The general stress intensity factor thus has the form:

$$K = S\sqrt{\pi a} \cdot C \cdot \Pi\alpha_i \qquad \text{(Eq 5-15)}$$

where α_i is a series of other factors (or functions) accounting for the structural geometry and crack morphology. Equation 5-15 is the same as Eq 5-5, except that an extra factor C is present.

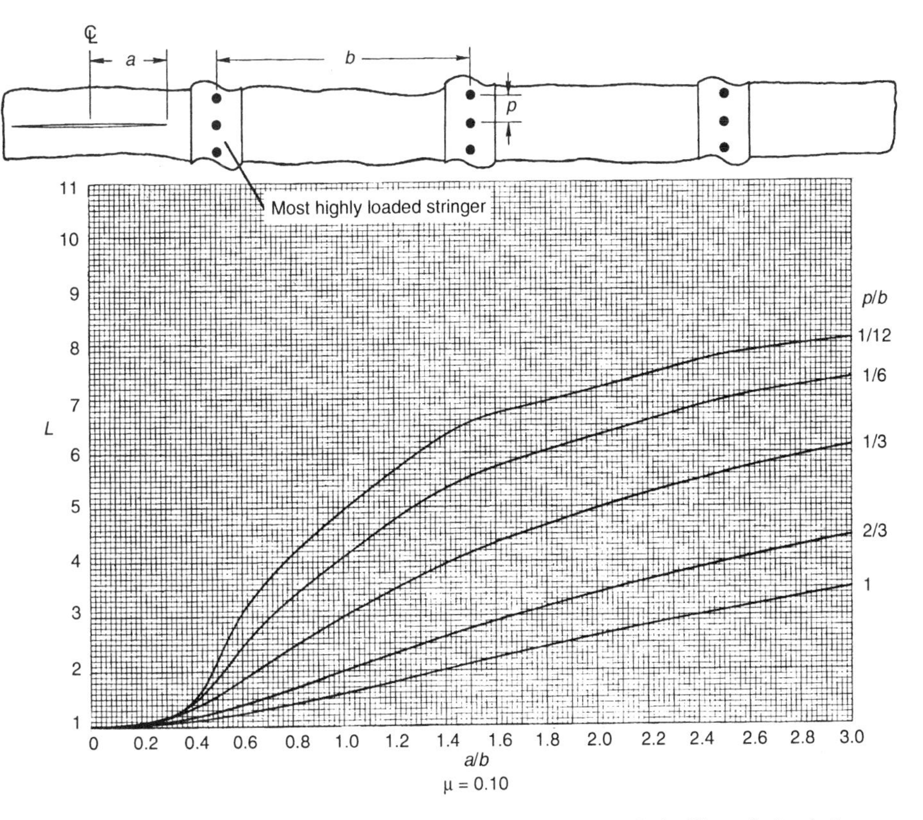

Fig. 5-14b Stiffener load concentration factor in most highly loaded stiffener in front of a one-bay crack. Source: Ref 5-11

As mentioned in Chapter 4, many factors influence the performance of the stiffener. It is impossible to address them all. However, it is certain that the stiffeners and the attachments should remain intact at all times. The following presents design graphs for calculating the rivet force acting on the most highly loaded rivet and the load-concentration factor for the most highly loaded stiffener.

Plots for the rivet force for the most highly loaded rivet Q (normalized by btS) are given in Fig. 5-12 (a–e) and 5-13 (a–e). The most highly loaded rivet is always the pair of rivets that is nearest to the crack tip as indicated in these graphs. These design graphs are only applicable to the rivet diameter that is represented by $d/p = 0.25$, $w/d = 5$, and $\nu = 0.3$.

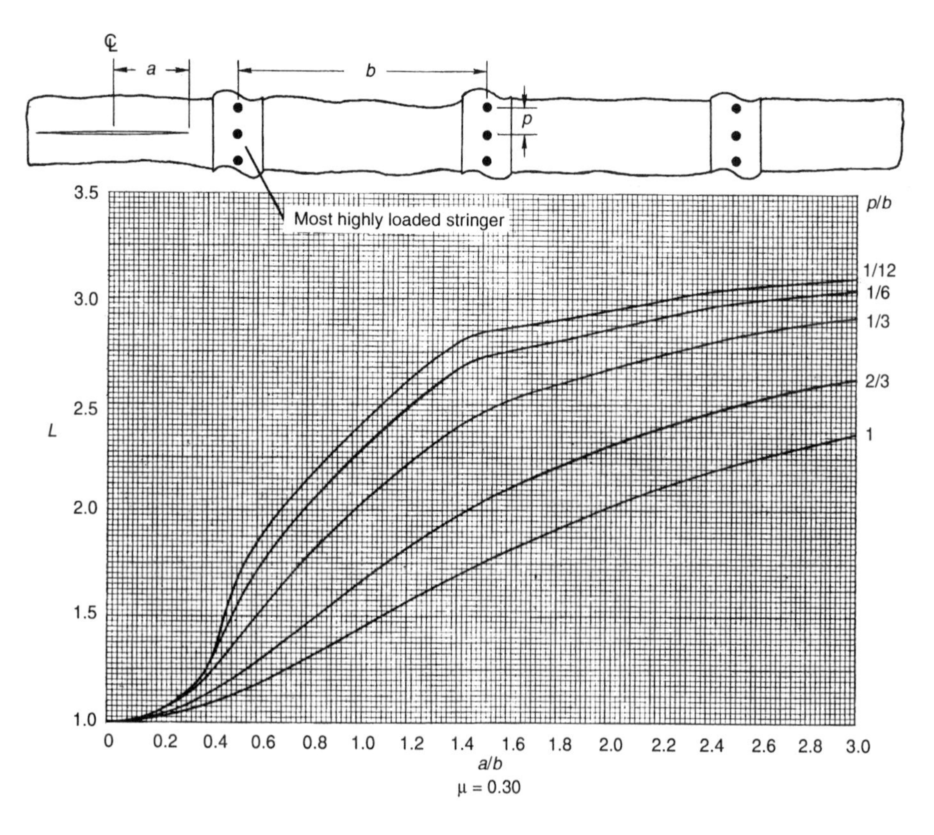

Fig. 5-14c Stiffener load concentration factor in most highly loaded stiffener in front of a one-bay crack. Source: Ref 5-11

For a stiffener that stretches over the crack, the rivet forces oppose the opening of the crack and, therefore, transfer load from the sheet to the stiffener. Thus, the maximum force in a stiffener occurs between the crack and the rivet on either side nearest the crack. The force at this location in the stiffener is:

$$F = S \cdot L \cdot A \cdot E_S/E \qquad \text{(Eq 5-16)}$$

where L is the stiffener-load-concentration factor, given in Fig. 5-14 (a–d) and 5-15 (a–d). This force can be used to compute the cross-sectional stress of the strap. By comparing the cross-sectional stress of the strap with the

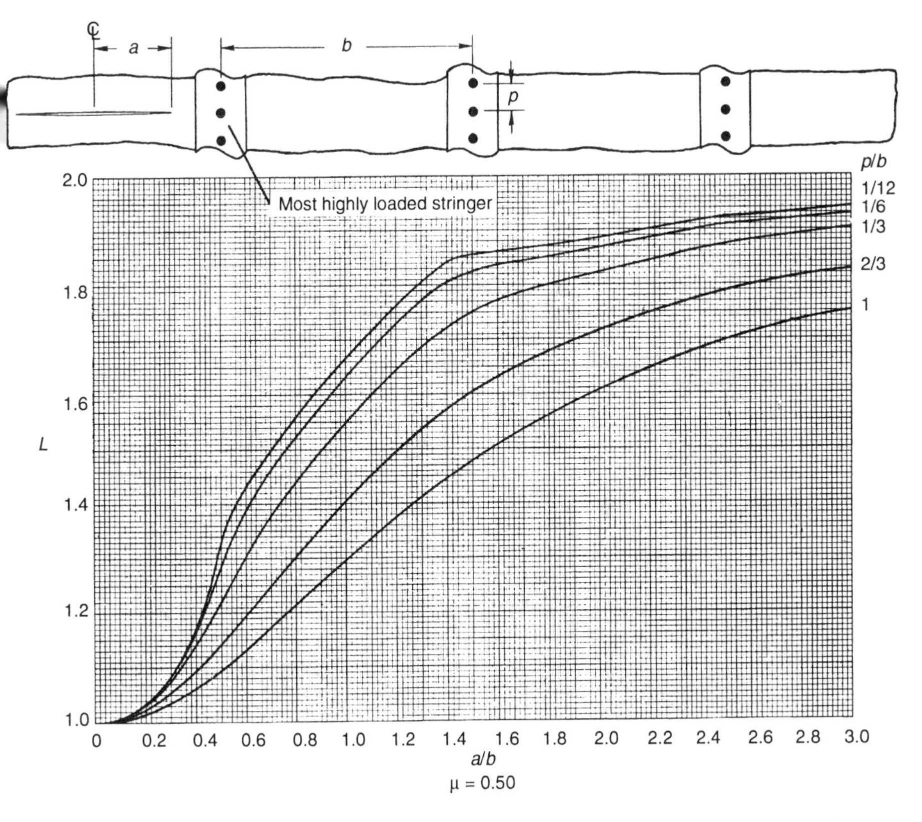

Fig. 5-14d Stiffener load concentration factor in most highly loaded stiffener in front of a one-bay crack. Source: Ref 5-11

material tensile properties, one can determine whether or not the strap is yielded or broken. Some design graphs showing the effect of a broken stiffener on reinforcement efficiency are given in Ref 5-12.

If the stiffener is a stringer, only a part of the stringer (the area that is close to the skin) makes a contribution to reducing the crack tip stress intensity. According to Ref 5-13, the effective area can be determined by the following equations.

In general:

$$A_e = A\left[1 + \left(\frac{y}{\rho}\right)^2\right]^{-1} \quad \text{(Eq 5-17)}$$

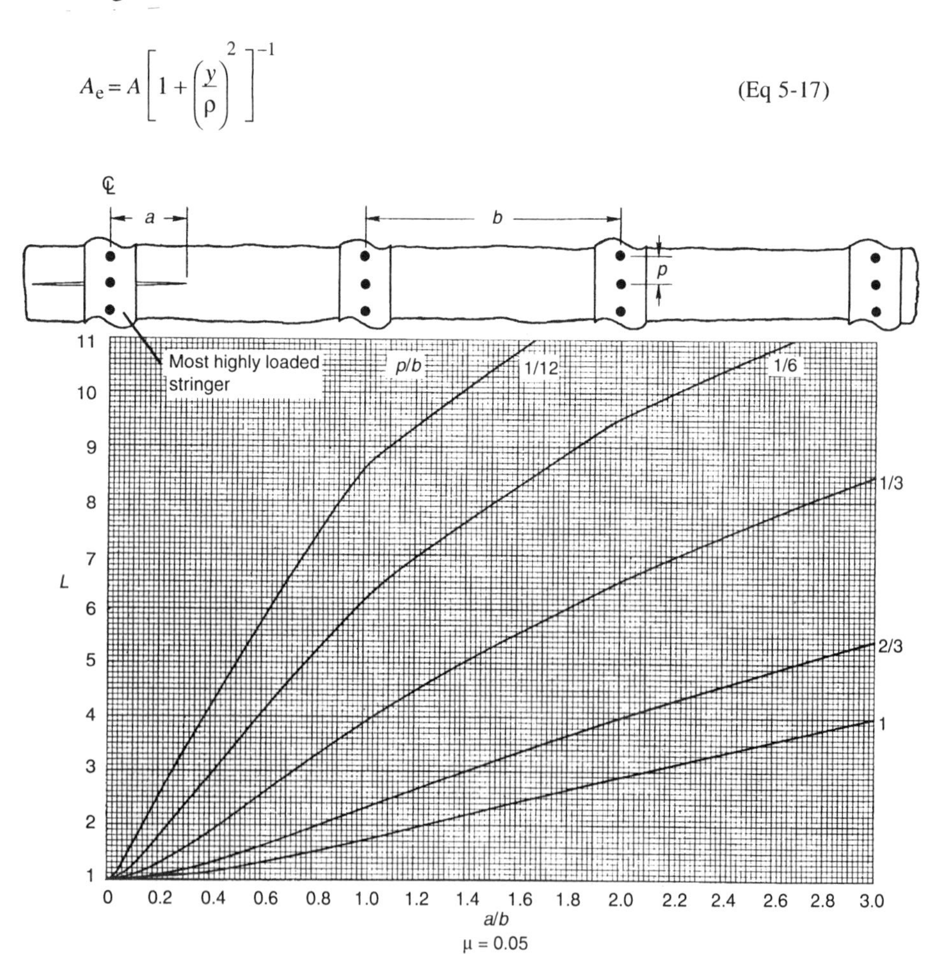

Fig. 5-15a Stiffener load concentration factor in most highly loaded stiffener at the center of a two-bay crack. Source: Ref 5-11

where y is the distance from the inner surface of the sheet to the centroid of the stringer and ρ is the radius of gyration of the stringer.

If the effective area of a channel or zee frame is very small due to the stringer cutouts, for typical cases, A_e may be estimated by:

$$A_e = A/5 \qquad \text{(Eq 5-17a)}$$

If a reinforcing angle is attached to the frame close to the stringer cutout, the effective area may be increased:

$$A_e = A/3 \qquad \text{(Eq 5-17b)}$$

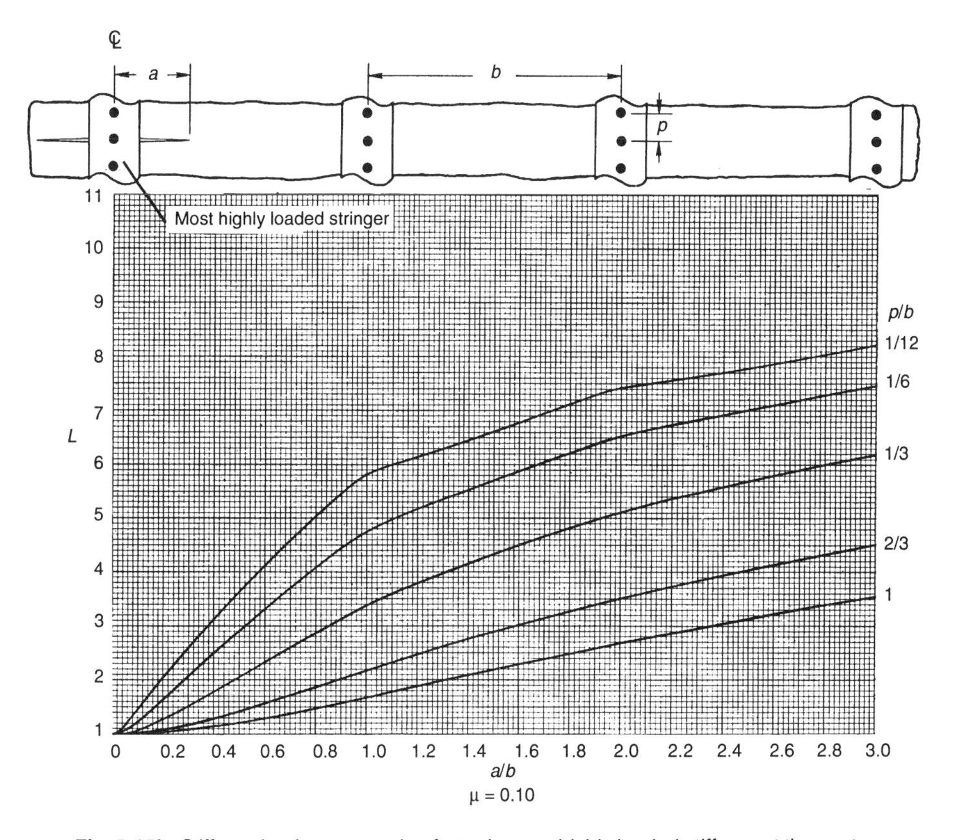

Fig. 5-15b Stiffener load concentration factor in most highly loaded stiffener at the center of a two-bay crack. Source: Ref 5-11

For flat straps attached directly to the skin by bonding or with mechanical fasteners:

$$A_e = A \cdot F_{tyf}/F_{tys} \qquad \text{(Eq 5-17c)}$$

where the subscripts "tyf" and "tys" stand for tensile yield strength for the stiffener and tensile yield strength for the skin, respectively. Therefore, for any stiffener that is not a flat strap (made of the same material as the skin), simply replace A in Eq 5-14 by A_e. However, be aware that the computed C-value may be in error if the rigidity for the stringer is not the same as for the flat strap.

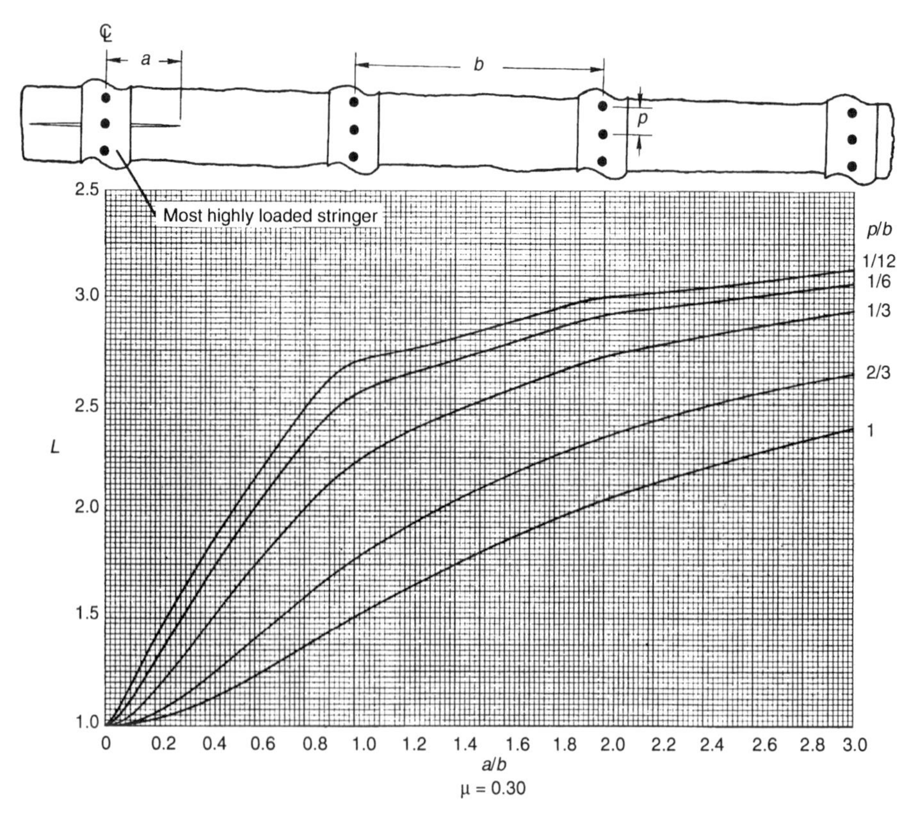

Fig. 5-15c Stiffener load concentration factor in most highly loaded stiffener at the center of a two-bay crack. Source: Ref 5-11

It is noted in Ref 5-11 that all the design curves given here (i.e., Fig. 5-10 to 5-15) are equally applicable to both riveted or spot-welded attachments. In the case of an integral stiffener, it has been suggested that this configuration is equivalent to a skin-stiffener system that has very close attachment spacing (Ref 5-14). An example is shown in Chapter 4.

Part-Through Crack in a Plate

A part-through crack that originates on the surface of a plate is usually modeled as one-half of an ellipse. As shown in Fig. 5-16, either the major or the minor axis of the ellipse may be placed on the front surface of the plate,

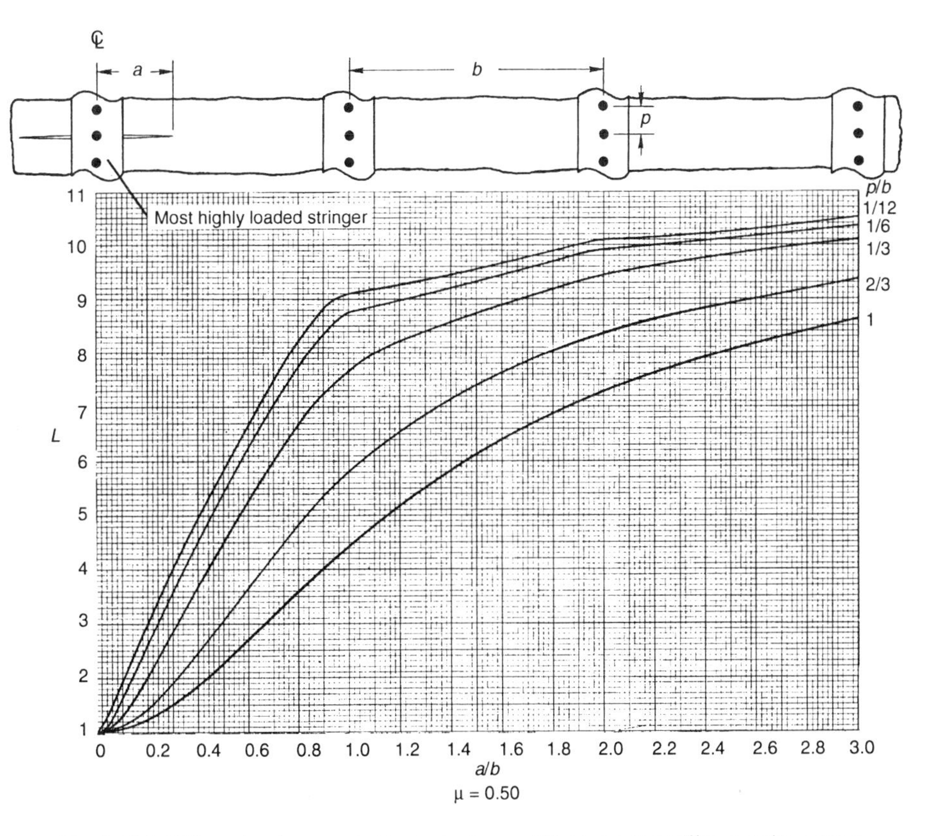

Fig. 5-15d Stiffener load concentration factor in most highly loaded stiffener at the center of a two-bay crack. Source: Ref 5-11

depending on the configuration. The length and depth of the crack are designated as $2c$ and a, respectively. The crack shape is described by an aspect ratio $a/2c$ (or a/c).

Table 5-1 Correction factors ($K/\sigma\sqrt{\pi a/Q}$) for stress intensity at shallow surface cracks under tension

		a/t				
$2c/W$	a/c	0.0	0.20	0.50	0.80	1.0
At the C-tip: tensile loading						
0.0	0.20	0.5622	0.6110	0.7802	1.1155	1.4436
0.0	0.40	0.6856	0.7817	0.9402	1.1583	1.3383
0.0	1.00	1.1365	1.1595	1.2328	1.3772	1.5145
0.1	0.20	0.5685	0.6133	0.7900	1.1477	1.5014
0.1	0.40	0.6974	0.7824	0.9456	1.2008	0.4256
0.1	1.00	1.1291	1.1544	1.2389	1.3892	1.5273
0.4	0.20	0.5849	0.6265	0.8438	1.3154	1.7999
0.4	0.40	0.7278	0.8029	1.0127	1.4012	1.7739
0.4	1.00	1.1366	1.1969	1.3475	1.5539	1.7238
0.6	0.20	0.5939	0.6415	0.9045	1.5056	2.1422
0.6	0.40	0.7385	0.8351	1.1106	1.6159	2.1036
0.6	1.00	1.1720	1.2855	1.5215	1.8229	2.0621
0.8	0.20	0.6155	0.6739	1.0240	1.8964	2.8650
0.8	0.40	0.7778	0.9036	1.3151	2.1102	2.9068
0.8	1.00	1.2630	1.4957	1.9284	2.4905	2.9440
1.0	0.20	0.6565	0.7237	1.2056	2.6060	4.2705
1.0	0.40	0.8375	1.0093	1.6395	2.9652	4.3596
1.0	1.00	1.3956	1.8446	2.6292	3.6964	4.5865
At the A-tip: tensile loading						
0.0	0.20	1.1120	1.1445	1.4504	1.7620	1.9729
0.0	0.40	1.0900	1.0945	1.2409	1.3672	1.4404
0.0	1.00	1.0400	1.0400	1.0672	1.0883	1.0800
0.1	0.20	1.1120	1.1452	1.4595	1.7744	1.9847
0.1	0.40	1.0900	1.0950	1.2442	1.3699	1.4409
0.1	1.00	1.0400	1.0260	1.0579	1.0846	1.0820
0.4	0.20	1.1120	1.1577	1.5126	1.8662	2.1012
0.4	0.40	1.0900	1.1140	1.2915	1.4254	1.4912
0.4	1.00	1.0400	1.0525	1.1046	1.1093	1.0863
0.6	0.20	1.1120	1.1764	1.5742	1.9849	2.2659
0.6	0.40	1.0900	1.1442	1.3617	1.5117	1.5761
0.6	1.00	1.0400	1.1023	1.1816	1.1623	1.0955
0.8	0.20	1.1120	1.2047	1.6720	2.2010	2.5895
0.8	0.40	1.0900	1.1885	1.4825	1.6849	1.7727
0.8	1.00	1.0400	1.1685	1.3089	1.2767	1.1638
1.0	0.20	1.1120	1.2426	1.8071	2.5259	3.0993
1.0	0.40	1.0900	1.2500	1.6564	1.9534	2.0947
1.0	1.00	1.0400	1.2613	1.4890	1.4558	1.3010

Note: These values are built into the NASA/FLAGRO program (Ref 5-20). Source: Ref 5-19

Semi-Infinite Solid

When tensile loads are applied normal to a semi-elliptical crack in a semi-infinite solid, the expression for K, for some point on the periphery of the crack, is given by Ref 5-15 as:

$$K = \sigma\sqrt{\pi a/Q} \cdot f_{\varphi} \cdot \alpha_f \quad \text{(Eq 5-18)}$$

Table 5-2 Correction factors ($K/\sigma\sqrt{\pi a/Q}$) for stress intensity at shallow surface cracks in bending

		a/t				
$2c/W$	a/c	0.0	0.20	0.50	0.80	1.0
At the C-tip: bending loading						
0.0	0.20	0.5622	0.5772	0.6464	0.7431	0.8230
0.0	0.40	0.6856	0.7301	0.7694	0.7358	0.6729
0.0	1.00	1.1365	1.0778	1.0184	0.9716	0.9474
0.1	0.20	0.5685	0.5809	0.6524	0.7646	0.8624
0.1	0.40	0.6974	0.7315	0.7856	0.8008	0.7895
0.1	1.00	1.1291	1.0740	1.0114	0.9652	0.9435
0.4	0.20	0.5849	0.5981	0.6934	0.8654	1.0249
0.4	0.40	0.7278	0.7519	0.8327	0.9312	1.0068
0.4	1.00	1.1366	1.1079	1.0634	1.0358	1.0268
0.6	0.20	0.5939	0.6158	0.7438	0.9704	1.1802
0.6	0.40	0.7385	0.7816	0.8906	1.0215	1.1211
0.6	1.00	1.1720	1.1769	1.1759	1.1820	1.1900
0.8	0.20	0.6155	0.6446	0.8320	1.1794	1.5113
0.8	0.40	0.7778	0.8386	1.0150	1.2791	1.5073
0.8	1.00	1.2630	1.3633	1.4785	1.5360	1.5431
1.0	0.20	0.6565	0.6848	0.9593	1.5053	2.0518
1.0	0.40	0.8375	0.9232	1.2285	1.7607	2.2637
1.0	1.00	1.3956	1.6821	2.0140	2.1482	2.1446
At the A-tip: bending loading						
0.0	0.20	1.1120	0.8825	0.6793	0.3063	−0.0497
0.0	0.40	1.0900	0.8292	0.5291	0.1070	−0.2489
0.0	1.00	1.0400	0.7411	0.3348	−0.1149	−0.4396
0.1	0.20	1.1120	0.8727	0.6697	0.3071	−0.0348
0.1	0.40	1.0900	0.8243	0.5170	0.1047	−0.2336
0.1	1.00	1.0400	0.7398	0.3322	−0.1172	−0.4408
0.4	0.20	1.1120	0.8683	0.6794	0.3439	0.0291
0.4	0.40	1.0900	0.8330	0.5270	0.1257	−0.1989
0.4	1.00	1.0400	0.7602	0.3572	−0.1080	−0.4543
0.6	0.20	1.1120	0.8904	0.7248	0.4033	0.0915
0.6	0.40	1.0900	0.8625	0.5803	0.1678	−0.1874
0.6	1.00	1.0400	0.7982	0.4072	−0.0856	−0.4750
0.8	0.20	1.1120	0.9191	0.7925	0.5102	0.2254
0.8	0.40	1.0900	0.8987	0.6619	0.2524	−0.1300
0.8	1.00	1.0400	0.8556	0.4981	−0.0329	−0.4960
1.0	0.20	1.1120	0.9545	0.8827	0.6666	0.4351
1.0	0.40	1.0900	0.9417	0.7723	0.3810	−0.0250
1.0	1.00	1.0400	0.9323	0.6312	0.0505	−0.5249

Note: These values are built into the NASA/FLAGRO program (Ref 5-20). Source: Ref 5-19

where the applied stress σ is equated to the applied load (force) over the full cross section of the plate (ignoring the crack). Both Q and f_φ are parameters accounting for the shape of the ellipse. Initially, Q was presented in the literature as a function of Θ and the σ/F_{ty} ratio, where Θ is a crack shape

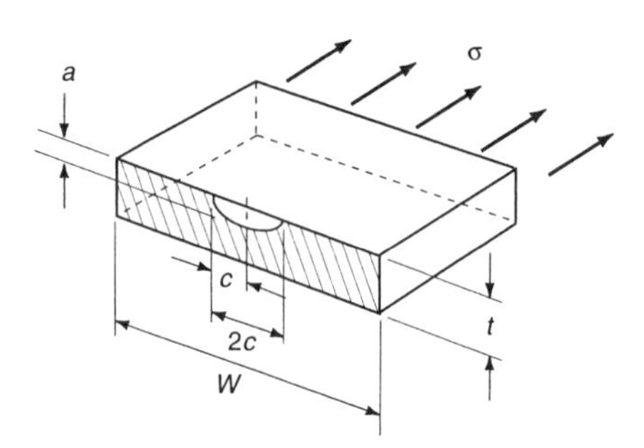

Fig. 5-16 Configuration of a semi-elliptical surface crack. This figure shows that the C-tip is at either end of the major axis (along the surface). The A-tip is at the maximum depth of the minor axis. In this book C-tips are always on the surface, and the A-tip is always at the maximum depth, whether c is greater than a, or vice versa. This definition also is applicable to the corner crack(s) at a hole configuration.

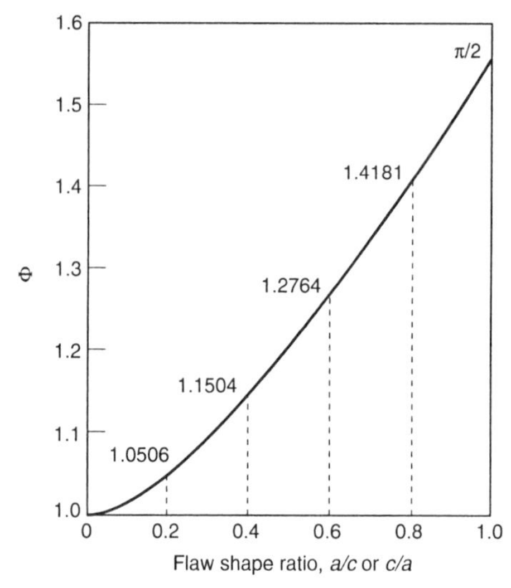

Fig. 5-17 Elliptical integral Θ

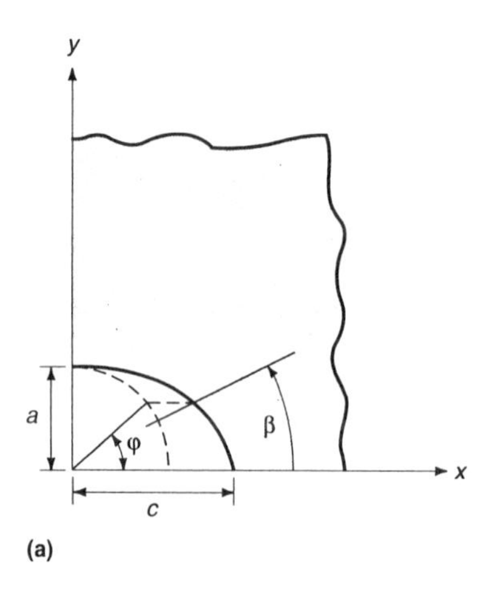

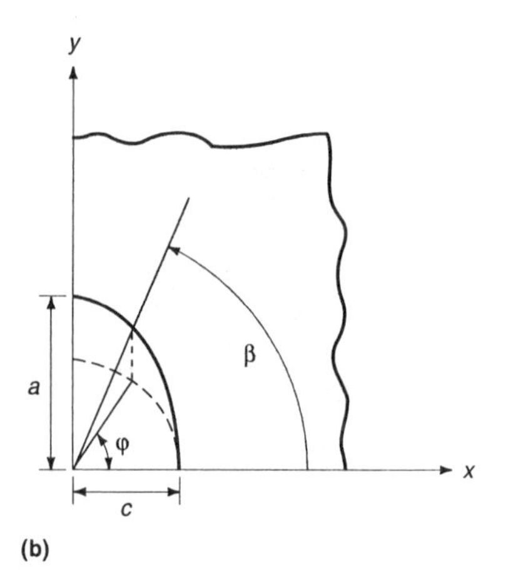

Fig. 5-18 Definition of φ and β for an elliptical crack

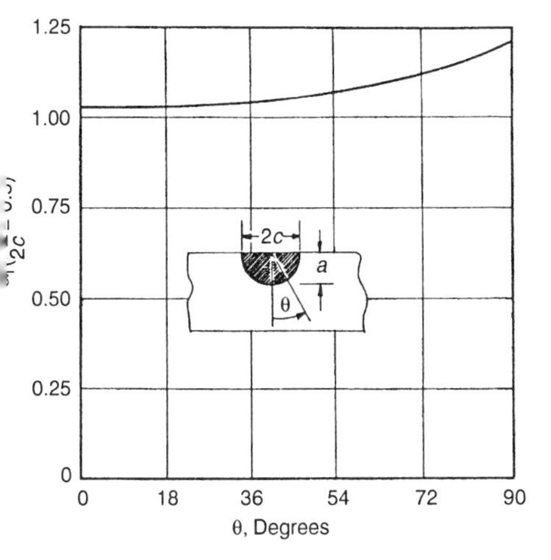

Fig. 5-19 Front face geometric coefficient for a semicircular part-through flaw. Source: Ref 5-16

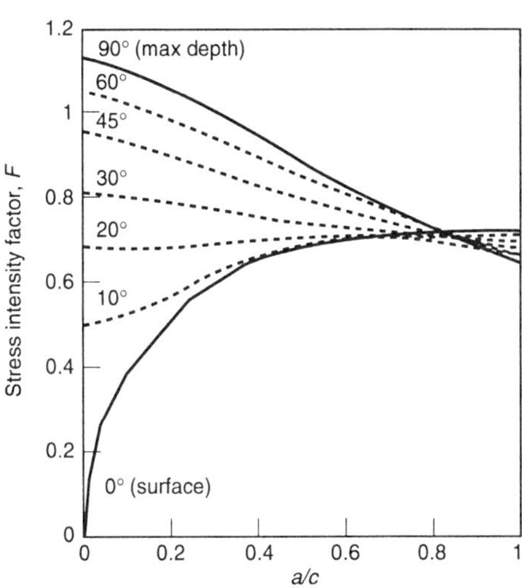

Fig. 5-20 Variation of stress intensity factors for a shallow crack in a semi-infinite solid, according to Eq 5-21 with $F = F_S\sqrt{Q}$

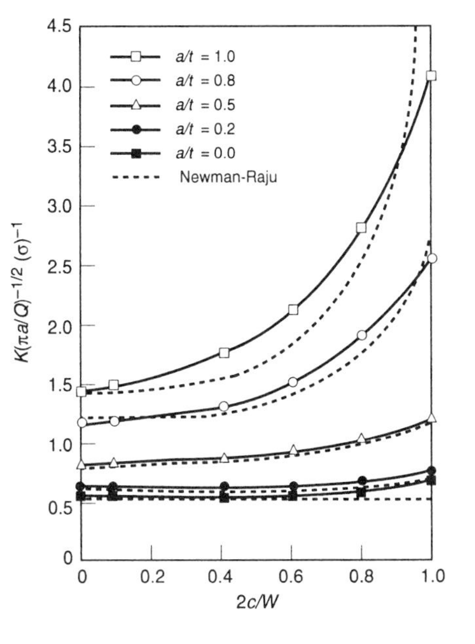

Fig. 5-21 New and old solutions for C-tip of a surface crack (Fig. 5-16) with $a/c = 0.2$ in tension. Source: Ref 5-19

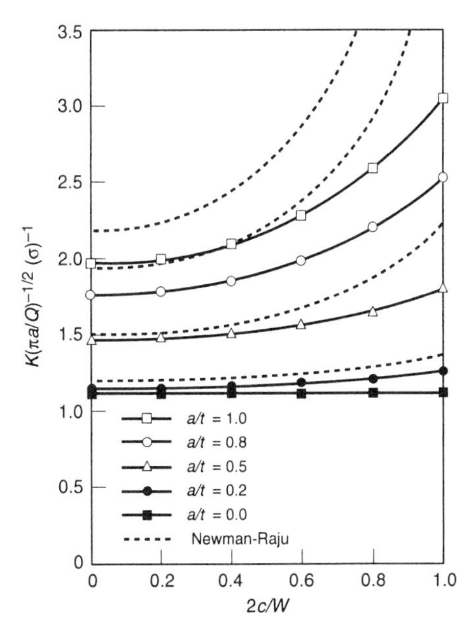

Fig. 5-22 New and old solutions for A-tip of a surface crack (Fig. 5-16) with $a/c = 0.2$ in tension. Source: Ref 5-19

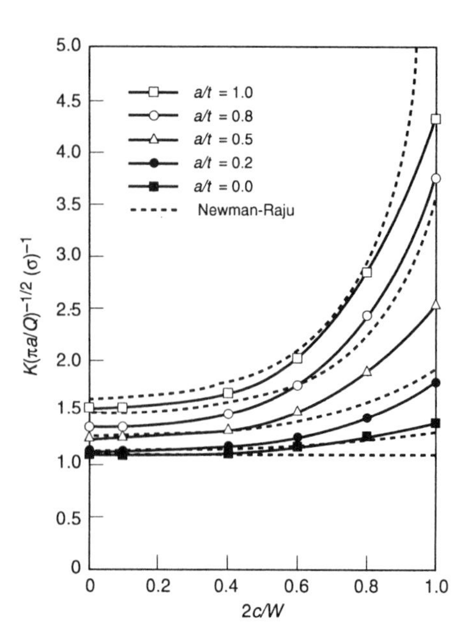

Fig. 5-23 New and old solutions for C-tip of a surface crack (Fig. 5-16) with $a/c = 1.0$ in tension. Source: Ref 5-19

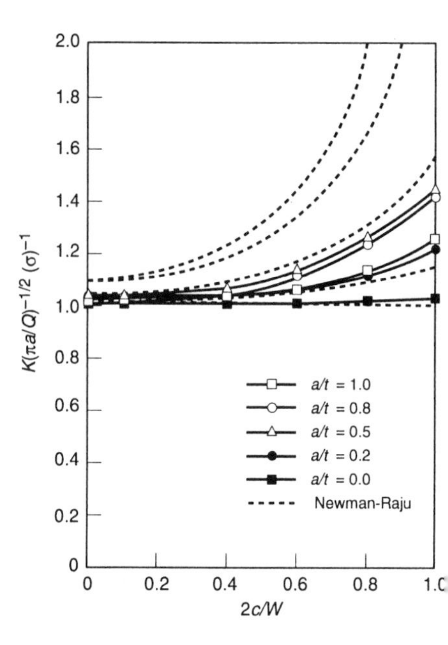

Fig. 5-24 New and old solutions for A-tip of a surface crack (Fig. 5-16) with $a/c = 1.0$ in tension. Source: Ref 5-19

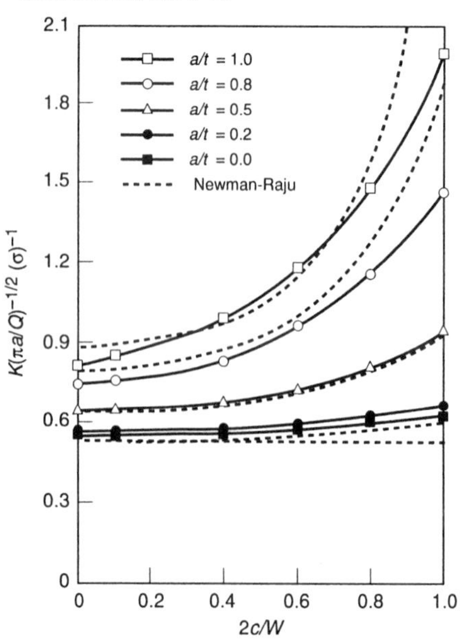

Fig. 5-25 New and old solutions for C-tip of a surface crack (Fig. 5-16) with $a/c = 0.2$ in bending. Source: Ref 5-19

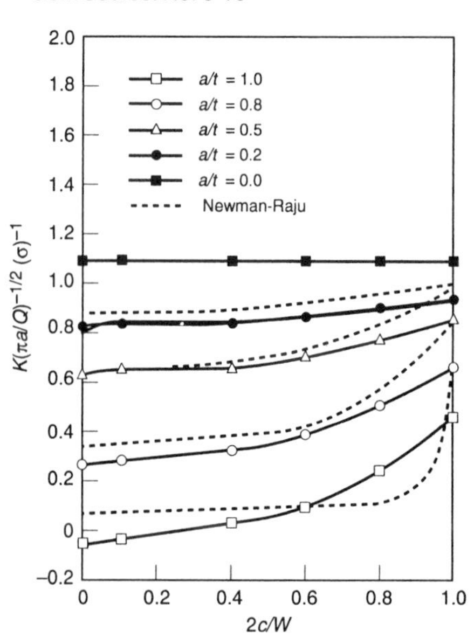

Fig. 5-26 New and old solutions for A-tip of a surface crack (Fig. 5-16) with $a/c = 0.2$ in bending. Source: Ref 5-19

parameter (the complete elliptical integral of the second kind presented in Fig. 5-17) and F_{ty} is the material tensile yield strength. As explained in Chapter 1, the term σ/F_{ty} is used for converting the physical crack length to the effective crack length (to include the effect of crack tip plasticity). If this term is deleted from Q, the elastic K is solely related to Θ. In the remainder of this chapter, Θ and $\sqrt{Q}$ are regarded as the same (without plasticity correction).

The angular function f_φ has the following form:

$$f_\varphi = [(a/c)^2 \cdot \cos^2 \varphi + \sin^2 \varphi]^{1/4} \qquad \text{(Eq 5-19a)}$$

for $a/c \leq 1$, and:

$$f_\varphi = [(c/a)^2 \cdot \sin^2 \varphi + \cos^2 \varphi]^{1/4} \qquad \text{(Eq 5-19b)}$$

for $a/c > 1$.

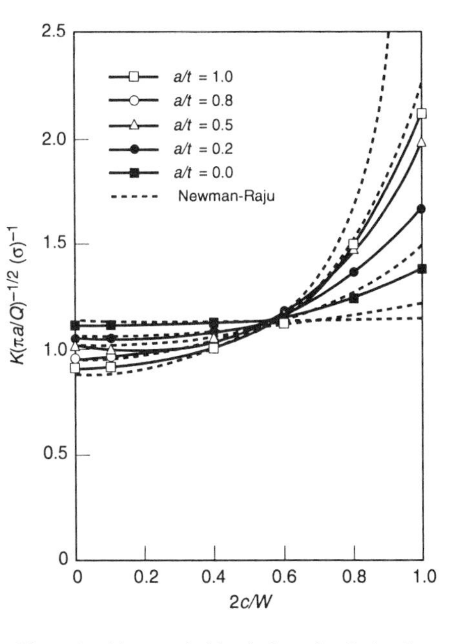

Fig. 5-27 New and old solutions for C-tip of a surface crack (Fig. 5-16) with $a/c = 1.0$ in bending. Source: Ref 5-19

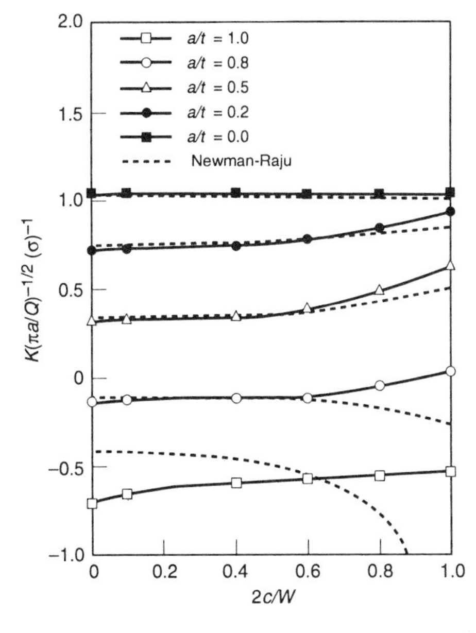

Fig. 5-28 New and old solutions for A-tip of a surface crack (Fig. 5-16) with $a/c = 1.0$ in bending. Source: Ref 5-19

In Eq 5-19(a) and (b), φ is a parametric angle measured from the plate surface toward the center of the crack (i.e., $\varphi = 0°$ is on the plate surface and $\varphi = 90°$ is at the maximum depth of the crack). This terminology is used in all open literature for defining the position of a point on the ellipse. However, φ is not an angle that actually connects the center of the ellipse to a specific point on the physical crack periphery. To translate φ to β (the angle between the plate surface and a specific point on the periphery of the ellipse, Fig. 5-18), the following relationship between φ and the geometric angle β can be used:

$$\beta = \tan^{-1}[(a/c) \cdot \tan \varphi] \qquad \text{(Eq 5-20)}$$

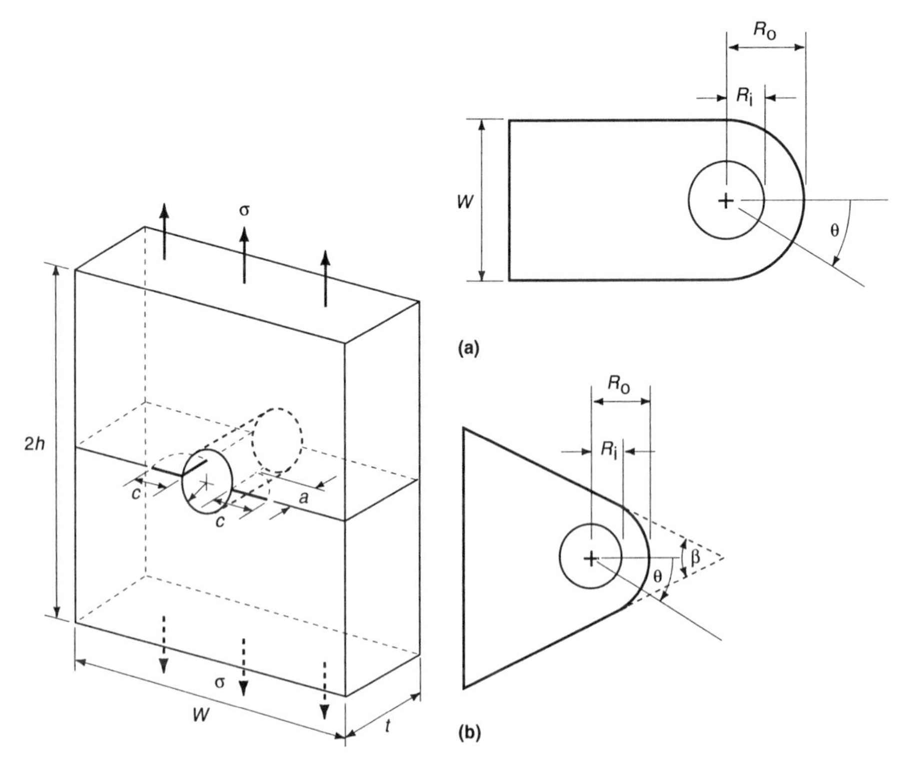

Fig. 5-29 Corner cracks at an open hole under uniform tension

Fig. 5-30 Attachment lug configurations. (a) Straight. (b) Tapered

The parameter α_f in Eq 5-18 is called the front face influence factor (dimensionless), accounting for the influence of the free surface coincident with the visible length of the crack. It is a function of a/c and φ. For $a/c = 0$, i.e.,

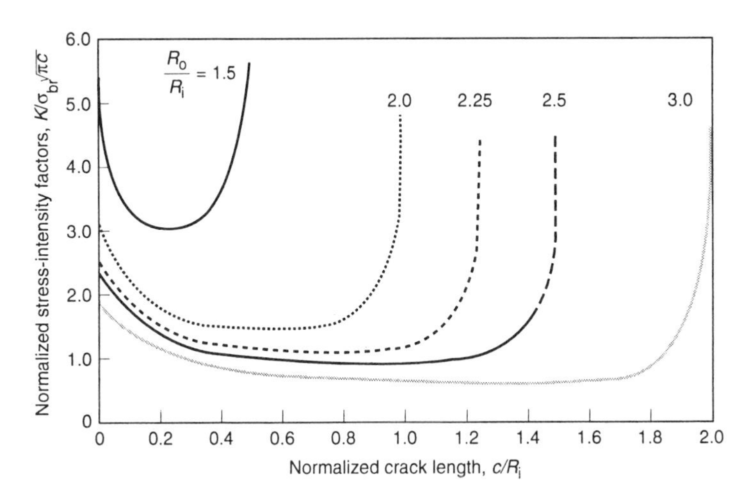

Fig. 5-31 Normalized stress intensity factors for single through-the-thickness cracks emanating from a straight lug subjected to a pin loading applied in the 0° loading direction. Source: Ref 5-24

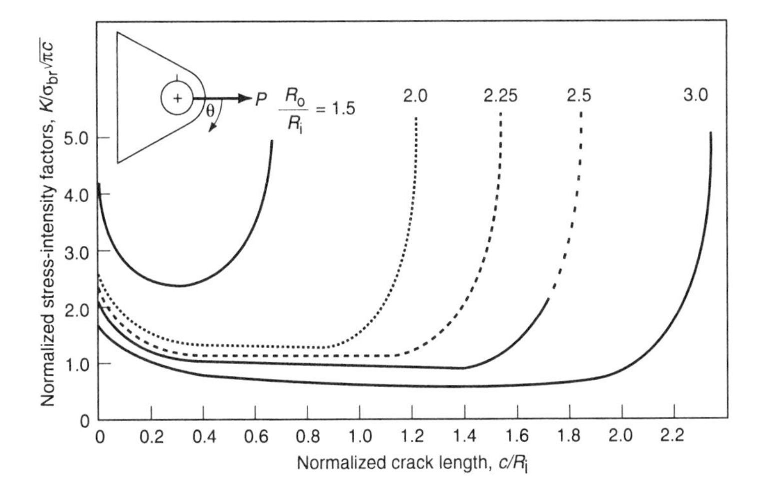

Fig. 5-32 Normalized stress intensity factors for single through-the-thickness cracks emanating from a tapered lug subjected to a pin loading applied in the 0° loading direction. Source: Ref 5-24

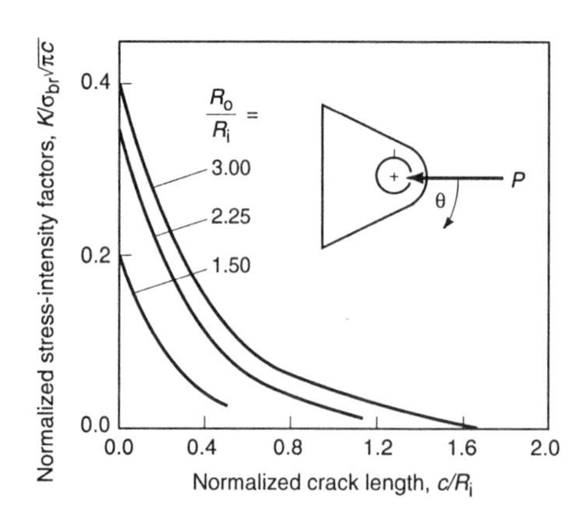

Fig. 5-33 Normalized stress intensity factors for single through-the-thickness cracks emanating from a tapered lug subjected to a pin loading applied in the 180° loading direction. Source: Ref 5-24

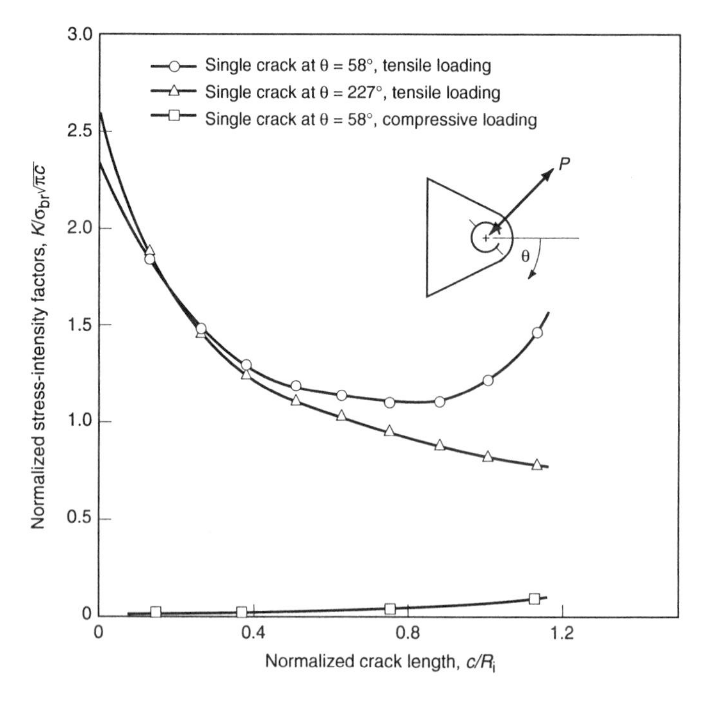

Fig. 5-34 Normalized stress intensity factors for single through-the-thickness cracks emanating from a tapered lug subjected to a pin loading applied in the –45° loading direction and its reversed direction, R_o/R_i = 2.25. Source: Ref 5-24

a scratch on the surface, one can idealize it as equivalent to an edge crack in a semi-infinite sheet. Therefore $\alpha_f = 1.122$ for all points along the entire crack front. For a semicircular crack ($a/c = 1$), the solution given by Smith (Ref 5-16) is shown in Fig. 5-19. The α_f values for all intermediate a/c ratios can be obtained by interpolation.

For a given a/c ratio, f_φ is a function of φ (radians). Therefore, the combination of α_f, f_φ, and $1/\sqrt{Q}$ is the source of the variance in K-values along the crack periphery. During crack propagation, each point along the crack front grows a different amount in a different direction. As a result, the crack shape continuously changes as the crack extends. In making structural life prediction, a minimum of two K-values (i.e., at the maximum depth, and on the surface) is required for each crack size and its corresponding aspect ratio. A demonstration of the fundamentals of K-variation as a function of a/c and φ

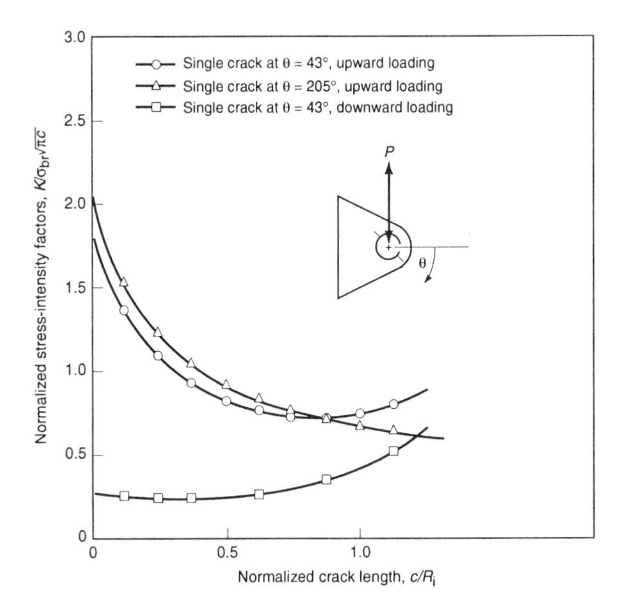

Fig. 5-35 Normalized stress intensity factors for single through-the-thickness cracks emanating from a tapered lug subjected to a pin loading applied in the −90° loading direction and its reversed direction, $R_o/R_i = 2.25$. Source: Ref 5-24

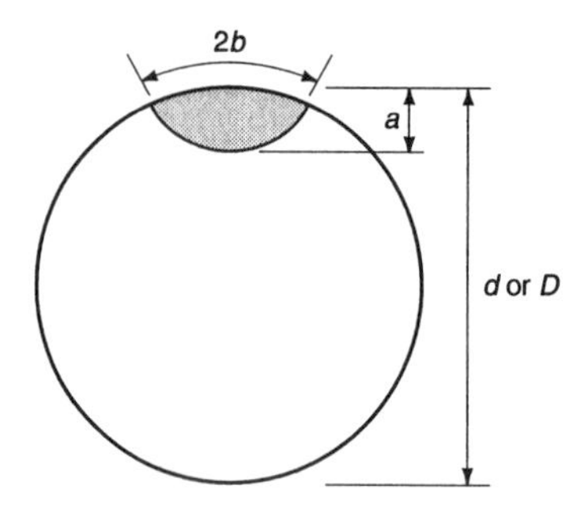

Fig. 5-36 Definition of crack dimensions for an almond-shaped crack in a solid cylinder

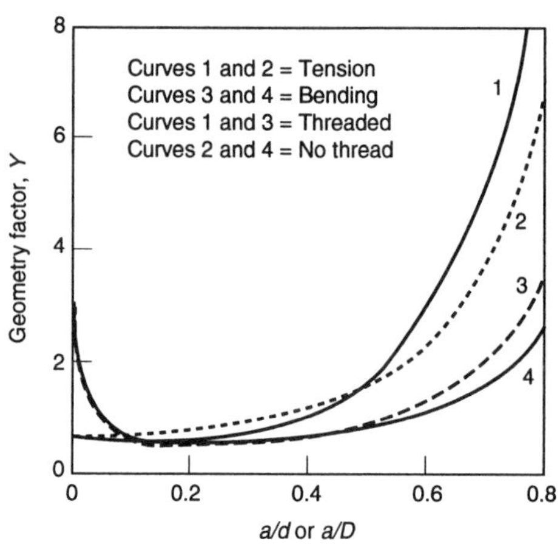

Fig. 5-37 Stress-intensity factors for threaded and unthreaded cylinders. Source: Ref 5-25

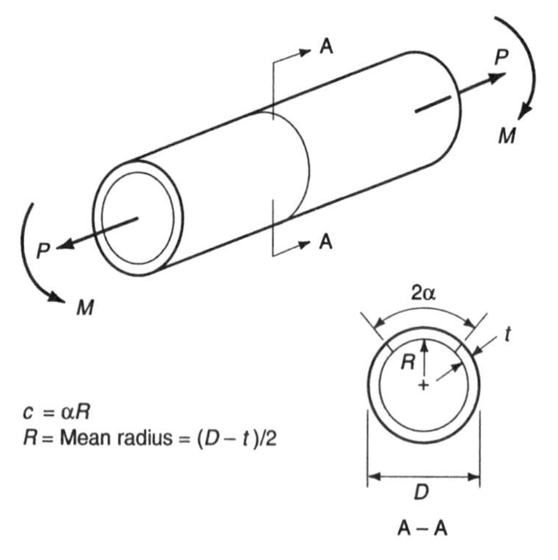

Fig. 5-38 A circumferential through-the-thickness crack in a hollow cylinder

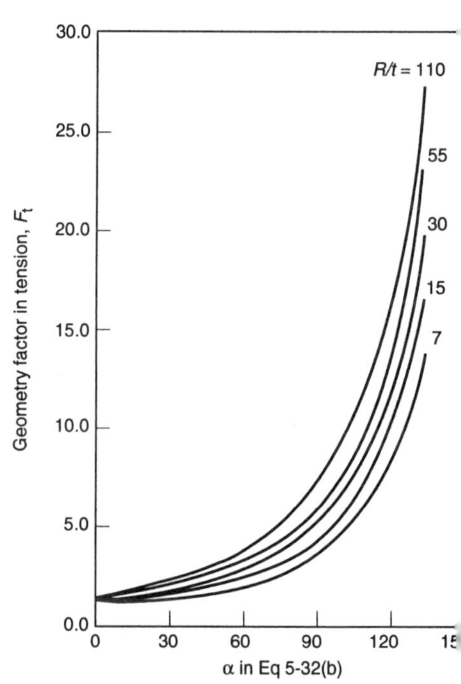

Fig. 5-39 Stress intensity factors for circumferential through crack in hollow cylinders subjected to tension (Eq 5-32a, b)

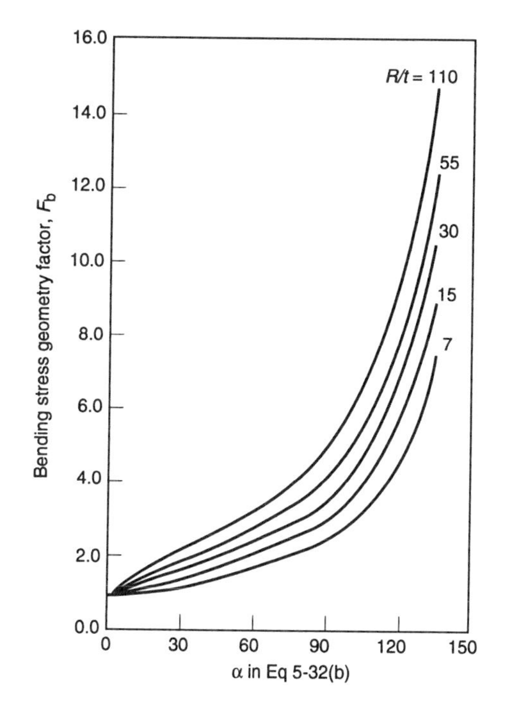

Fig. 5-40 Stress intensity factors for circumferential through crack in hollow cylinders subjected to bending (Eq 5-32a, b)

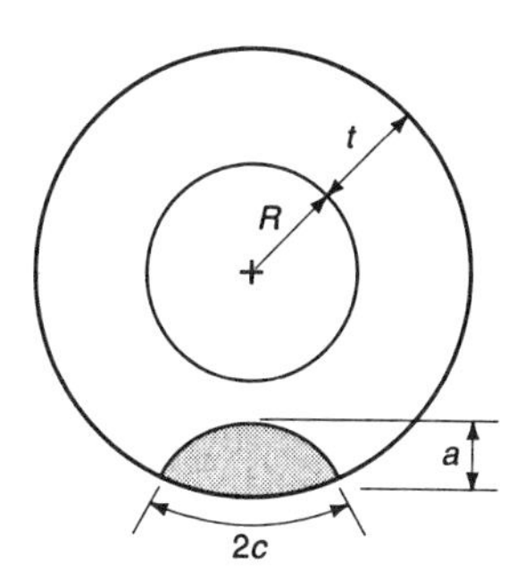

Fig. 5-41 Part-through crack on the circumferential plane in a hollow cylinder

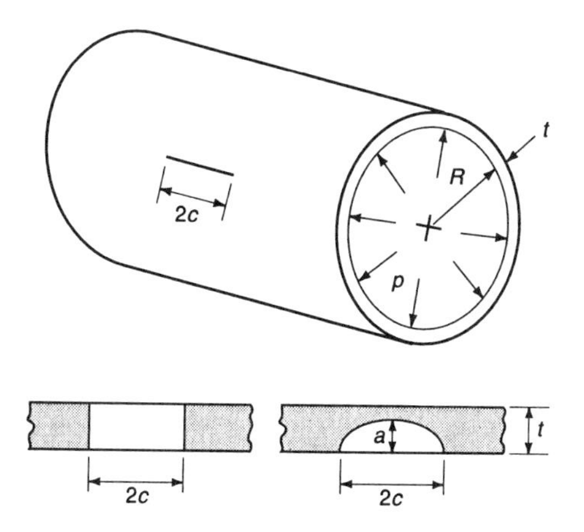

Fig. 5-42 Axial crack in a hollow cylinder

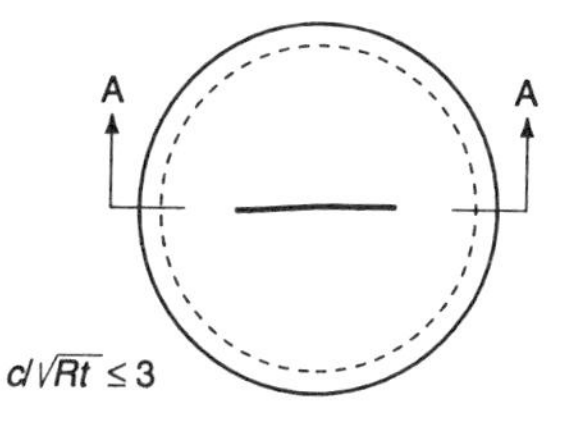

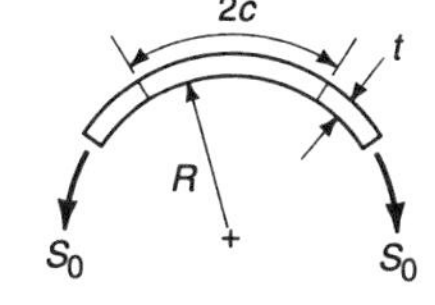

R = Mean radius = $(D - t)/2$

Fig. 5-43 Through-the-thickness crack in a spherical shell

is given in Fig. 5-20. Further discussion of Fig. 5-20, along with discussion of finite-element solutions, is presented in the following paragraphs.

Part-Through Crack in a Finite Plate

For a crack in a rectangular plate of finite thickness and width, the solution for K with a given crack size and shape should account for the influence of the width and the front and back faces of the plate. Finite-element solutions for cracks subjected to tension or bending have been developed by Newman and Raju (Ref 5-17, 5-18). The solution for tension loading was subsequently updated by Raju et al. (Ref 5-19). The new data (the values of $K/(\sigma \cdot \sqrt{\pi a/Q})$) are presented in Tables 5-1 and 5-2. These data have been built into the crack library in the NASA/FLAGRO computer program (Ref 5-20), with which interpolations are accomplished by using a nonlinear table look-up routine to obtain stress intensity factors that are not available

Table 5-3 Coefficients of closed-form stress-intensity equations for threaded and unthreaded cylinders

	Tension		Bending	
Coefficient	No thread(a)	Threaded(b)	No thread(a)	Threaded(c)
λ	...	2.4371	...	2.295
β	...	−36.5	...	−44.0
A	0.6647	0.5154	0.666	0.654
B	−1.2425	0.4251	−1.2628	−0.9
C	27.998	2.4134	10.737	0.8
D	−162.44	−15.4491	−50.539	10.5
E	472.23	36.157	139.29	−26.2
F	−629.63	...	−183.85	25.9
G	326.05	...	96.347	...

(a) Equivalent to the trigonometric equations of Ref 5-29. (b) Equation 8(b) of Ref 5-25 or Eq 9 of Ref 5-26. (c) Equation 8(d) of Ref 5-25

Table 5-4 Polynomial coefficients for Eq 5-32 (b) in tension

	R/t				
Coefficient	110	55	30	15	7
A	1.0679	1.0775	1.0996	1.0199	1.0203
B	0.0005	−0.0145	−0.0251	−0.015	−0.0145
C	0.0038	0.0037	0.0037	0.0025	0.0019
D	−0.0001	−0.0001	−0.0001	-8×10^{-5}	-6×10^{-5}
E	2×10^{-6}	2×10^{-6}	2×10^{-6}	1×10^{-6}	1×10^{-6}
F	-2×10^{-8}	-1×10^{-8}	-1×10^{-8}	-1×10^{-8}	-7×10^{-9}
G	5×10^{-11}	4×10^{-11}	4×10^{-11}	3×10^{-11}	2×10^{-11}

Note: Values are for circumferential through-the-thickness cracks in hollow cylinders.

in the tables. Comparisons of the new and old data are shown in Fig. 5-21 to 5-28. In some cases the differences are significant.

The old data were curve fitted, and a general expression for K, which included a group of correction factors for the width and the front and back faces of the plate, was developed by Newman and Raju (Ref 5-17, 5-18). These stress intensity correction factors are included here because the equations are in close form, covering a full range of geometric combinations, and have been used for some time by fracture mechanics analysts in the aircraft/aerospace industry:

$$K = \sigma \cdot F_s \cdot \sqrt{\pi a/Q} \qquad \text{(Eq 5-21)}$$

where $\sqrt{Q}$ is Θ (Fig. 5-17), as previously discussed, and F_s is a function of a/c, a/t, c/W, and φ such that:

$$F_s = [M_1 + M_2(a/t)^2 + M_3(a/t)^4] \cdot G_1 \cdot f_\varphi \cdot f_w \qquad \text{(Eq 5-22)}$$

For $a/c \leq 1$:

$$M_1 = 1.13 - 0.09(a/c) \qquad \text{(Eq 5-22a)}$$

$$M_2 = -0.54 + 0.89/[0.2 + (a/c)] \qquad \text{(Eq 5-22b)}$$

$$M_3 = 0.5 - 1/[0.65 + (a/c)] + 14.0(1 - a/c)^{24} \qquad \text{(Eq 5-22c)}$$

$$G_1 = 1 + [0.1 + 0.35(a/t)^2] \cdot (1 - \sin\varphi)^2 \qquad \text{(Eq 5-22d)}$$

Table 5-5 Polynomial coefficients for Eq 32(b) in bending

	R/t				
Coefficient	110	55	30	15	7
A	0.9524	0.9798	1.0158	1.0083	1.0107
B	0.0297	0.0101	−0.0039	−0.007	−0.008
C	0.0015	0.0019	0.0021	0.0017	0.0012
D	-6×10^{-5}	-6×10^{-5}	-6×10^{-5}	-5×10^{-5}	-4×10^{-5}
E	1×10^{-6}	1×10^{-6}	1×10^{-6}	7×10^{-7}	5×10^{-7}
F	-8×10^{-9}	-8×10^{-9}	-7×10^{-9}	-5×10^{-9}	-4×10^{-9}
G	2×10^{-11}	2×10^{-11}	2×10^{-11}	2×10^{-11}	1×10^{-11}

Note: Values are for circumferential through-the-thickness cracks in hollow cylinders.

Table 5-6 *F*-factors for internal thumbnail cracks on circumferential plane of a hollow cylinder

	$a/c = 0.2$					$a/c = 0.4$					$a/c = 0.6$					$a/c = 0.8$					$a/c = 1.0$				
R/t	$a/t = 0$	$a/t = 0.2$	$a/t = 0.5$	$a/t = 0.8$	$a/t = 1.0$	$a/t = 0$	$a/t = 0.2$	$a/t = 0.5$	$a/t = 0.8$	$a/t = 1.0$	$a/t = 0$	$a/t = 0.2$	$a/t = 0.5$	$a/t = 0.8$	$a/t = 1.0$	$a/t = 0$	$a/t = 0.2$	$a/t = 0.5$	$a/t = 0.8$	$a/t = 1.0$	$a/t = 0$	$a/t = 0.2$	$a/t = 0.5$	$a/t = 0.8$	$a/t = 1.0$
C-tip, uniform loading																									
1.0	0.580	0.593	0.610	0.846	1.117	0.630	0.650	0.665	0.841	1.041	0.670	0.688	0.702	0.831	0.976	0.695	0.709	0.722	0.817	0.919	0.700	0.713	0.726	0.796	0.872
2.0	0.600	0.617	0.671	0.824	0.975	0.660	0.669	0.714	0.837	0.956	0.695	0.703	0.741	0.838	0.930	0.715	0.721	0.752	0.828	0.898	0.710	0.722	0.747	0.806	0.860
4.0	0.613	0.633	0.726	0.898	1.049	0.664	0.681	0.746	0.894	1.014	0.698	0.712	0.772	0.880	0.974	0.716	0.727	0.774	0.858	0.930	0.718	0.727	0.762	0.827	0.883
10.0	0.591	0.644	0.785	1.000	1.178	0.651	0.689	0.797	0.967	1.108	0.692	0.718	0.799	0.930	1.041	0.714	0.732	0.791	0.891	0.975	0.717	0.730	0.774	0.849	0.913
300.0	0.538	0.583	0.747	1.075	1.398	0.601	0.679	0.818	1.023	1.199	0.668	0.722	0.829	0.969	1.074	0.700	0.739	0.817	0.919	0.996	0.726	0.736	0.785	0.878	0.960
C-tip, bending loading																									
1.0	0.337	0.265	0.111	0.080	0.050	0.358	0.308	0.216	0.150	0.120	0.370	0.338	0.293	0.253	0.230	0.375	0.355	0.340	0.343	0.354	0.371	0.360	0.359	0.378	0.400
2.0	0.400	0.403	0.410	0.420	0.430	0.430	0.443	0.450	0.465	0.493	0.460	0.470	0.482	0.520	0.559	0.480	0.485	0.503	0.548	0.590	0.482	0.486	0.505	0.547	0.587
4.0	0.498	0.510	0.569	0.678	0.775	0.539	0.550	0.602	0.698	0.782	0.567	0.577	0.622	0.704	0.775	0.581	0.590	0.628	0.696	0.756	0.583	0.590	0.620	0.675	0.723
10.0	0.544	0.595	0.722	0.915	1.072	0.605	0.637	0.732	0.888	1.019	0.646	0.664	0.734	0.858	0.965	0.668	0.677	0.728	0.824	0.909	0.670	0.675	0.713	0.786	0.851
300.0	0.538	0.583	0.747	1.075	1.398	0.601	0.679	0.818	1.023	1.199	0.668	0.722	0.829	0.969	1.074	0.700	0.739	0.817	0.919	0.996	0.726	0.736	0.785	0.878	0.960
A-tip, uniform loading																									
1.0	0.960	0.987	1.064	1.665	2.406	0.875	0.888	0.944	1.360	1.857	0.795	0.799	0.841	1.119	1.437	0.720	0.721	0.754	0.941	1.146	0.650	0.653	0.684	0.823	0.969
2.0	0.990	1.022	1.093	1.380	1.685	0.900	0.911	0.961	1.163	1.377	0.800	0.813	0.847	0.985	1.130	0.710	0.726	0.751	0.846	0.943	0.620	0.652	0.674	0.745	0.815
4.0	1.031	1.045	1.141	1.332	1.504	0.920	0.926	0.991	1.123	1.243	0.819	0.821	0.862	0.951	1.031	0.729	0.729	0.756	0.814	0.868	0.650	0.652	0.672	0.713	0.751
10.0	0.983	1.059	1.189	1.337	1.440	0.888	0.936	1.020	1.120	1.192	0.800	0.827	0.878	0.941	0.989	0.718	0.732	0.761	0.801	0.831	0.642	0.651	0.671	0.697	0.717
300.0	1.059	1.090	1.384	1.682	1.881	0.948	0.951	1.079	1.188	1.251	0.792	0.832	0.888	0.940	0.971	0.720	0.733	0.754	0.777	0.792	0.642	0.656	0.675	0.691	0.700
A-tip, bending loading																									
1.0	0.520	0.545	0.659	1.074	1.523	0.470	0.493	0.597	0.919	1.254	0.430	0.446	0.542	0.792	1.039	0.385	0.405	0.494	0.693	0.879	0.350	0.368	0.454	0.621	0.771
2.0	0.700	0.719	0.821	1.088	1.352	0.630	0.643	0.728	0.935	1.135	0.560	0.575	0.648	0.808	0.957	0.503	0.515	0.579	0.706	0.819	0.448	0.463	0.523	0.629	0.720
4.0	0.839	0.865	0.974	1.173	1.347	0.748	0.767	0.849	0.997	1.126	0.666	0.681	0.743	0.852	0.946	0.592	0.606	0.654	0.735	0.805	0.528	0.542	0.583	0.648	0.702
10.0	0.902	0.985	1.120	1.267	1.366	0.822	0.871	0.959	1.064	1.141	0.744	0.770	0.824	0.897	0.954	0.669	0.682	0.715	0.765	0.806	0.597	0.607	0.631	0.667	0.695
300.0	1.059	1.090	1.384	1.682	1.881	0.948	0.951	1.079	1.188	1.251	0.792	0.832	0.888	0.940	0.971	0.720	0.733	0.754	0.777	0.792	0.642	0.656	0.675	0.691	0.700

Source: Ref 5-32

Table 5-7 *F*-factors for external thumbnail crack on circumferential plane of a hollow cylinder

	$a/c = 0.2$					$a/c = 0.4$					$a/c = 0.6$					$a/c = 0.8$					$a/c = 1.0$				
R/t	$a/t = 0$	$a/t = 0.2$	$a/t = 0.5$	$a/t = 0.8$	$a/t = 1.0$	$a/t = 0$	$a/t = 0.2$	$a/t = 0.5$	$a/t = 0.8$	$a/t = 1.0$	$a/t = 0$	$a/t = 0.2$	$a/t = 0.5$	$a/t = 0.8$	$a/t = 1.0$	$a/t = 0$	$a/t = 0.2$	$a/t = 0.5$	$a/t = 0.8$	$a/t = 1.0$	$a/t = 0$	$a/t = 0.2$	$a/t = 0.5$	$a/t = 0.8$	$a/t = 1.0$
C-tip, uniform loading																									
1.0	0.590	0.672	0.893	1.249	1.552	0.664	0.713	0.871	1.138	1.368	0.712	0.739	0.846	1.039	1.209	0.734	0.747	0.818	0.954	1.075	0.731	0.739	0.788	0.882	0.966
2.0	0.560	0.660	0.876	1.177	1.416	0.643	0.706	0.859	1.086	1.271	0.699	0.734	0.838	1.006	1.148	0.727	0.744	0.814	0.938	1.046	0.728	0.737	0.787	0.881	0.964
4.0	0.540	0.653	0.873	1.162	1.383	0.630	0.701	0.858	1.081	1.257	0.691	0.731	0.839	1.006	1.145	0.722	0.742	0.815	0.940	1.046	0.725	0.735	0.786	0.880	0.962
10.0	0.542	0.646	0.867	1.172	1.414	0.630	0.697	0.855	1.087	1.275	0.689	0.728	0.838	1.010	1.153	0.720	0.741	0.815	0.941	1.049	0.722	0.734	0.785	0.879	0.961
300.0	0.538	0.583	0.747	1.075	1.398	0.601	0.679	0.818	1.023	1.199	0.668	0.722	0.829	0.969	1.074	0.700	0.739	0.817	0.919	0.996	0.726	0.736	0.785	0.878	0.960
C-tip, bending loading																									
1.0	0.592	0.643	0.742	0.870	0.967	0.659	0.690	0.761	0.861	0.940	0.704	0.720	0.768	0.844	0.908	0.727	0.731	0.760	0.819	0.871	0.729	0.724	0.740	0.785	0.829
2.0	0.552	0.645	0.798	0.972	1.092	0.632	0.691	0.801	0.939	1.040	0.687	0.720	0.795	0.902	0.987	0.716	0.731	0.780	0.863	0.932	0.720	0.724	0.757	0.820	0.876
4.0	0.545	0.645	0.835	1.075	1.254	0.624	0.690	0.827	1.014	1.158	0.678	0.717	0.814	0.956	1.069	0.706	0.728	0.794	0.899	0.987	0.710	0.722	0.767	0.845	0.912
10.0	0.524	0.633	0.850	1.136	1.357	0.612	0.684	0.840	1.057	1.229	0.672	0.715	0.823	0.984	1.115	0.703	0.727	0.801	0.918	1.016	0.705	0.721	0.772	0.859	0.932
300.0	0.538	0.583	0.747	1.075	1.398	0.601	0.679	0.818	1.023	1.199	0.668	0.722	0.829	0.969	1.074	0.700	0.739	0.817	0.919	0.996	0.726	0.736	0.785	0.878	0.960
A-tip, uniform loading																									
1.0	1.140	1.189	1.469	2.179	2.898	1.000	1.019	1.188	1.583	1.969	0.860	0.872	0.960	1.140	1.303	0.737	0.748	0.785	0.847	0.899	0.644	0.647	0.660	0.685	0.708
2.0	1.126	1.167	1.370	1.759	2.112	0.975	1.005	1.132	1.362	1.564	0.844	0.865	0.935	1.051	1.149	0.733	0.746	0.780	0.827	0.866	0.640	0.648	0.663	0.683	0.698
4.0	1.099	1.157	1.320	1.576	1.790	0.959	0.999	1.103	1.260	1.388	0.835	0.862	0.923	1.006	1.072	0.728	0.746	0.777	0.815	0.843	0.637	0.649	0.666	0.683	0.693
10.0	1.079	1.146	1.284	1.470	1.615	0.945	0.993	1.083	1.198	1.284	0.827	0.859	0.914	0.977	1.020	0.724	0.745	0.776	0.806	0.825	0.636	0.650	0.668	0.684	0.693
300.0	1.059	1.090	1.384	1.682	1.881	0.948	0.951	1.079	1.188	1.251	0.792	0.832	0.888	0.940	0.971	0.720	0.733	0.754	0.777	0.792	0.642	0.656	0.675	0.691	0.700
A-tip, bending loading																									
1.0	1.110	1.124	1.252	1.676	2.120	0.945	0.958	1.008	1.207	1.419	0.850	0.816	0.810	0.857	0.914	0.729	0.697	0.656	0.624	0.608	0.641	0.601	0.545	0.495	0.466
2.0	1.115	1.121	1.236	1.487	1.721	0.966	0.964	1.018	1.144	1.263	0.836	0.827	0.837	0.876	0.915	0.726	0.711	0.694	0.681	0.675	0.635	0.615	0.585	0.554	0.534
4.0	1.097	1.124	1.242	1.429	1.586	0.949	0.969	1.035	1.138	1.223	0.822	0.834	0.863	0.905	0.937	0.716	0.720	0.725	0.728	0.730	0.630	0.625	0.619	0.605	0.593
10.0	1.042	1.117	1.248	1.403	1.514	0.918	0.969	1.051	1.142	1.205	0.808	0.839	0.885	0.930	0.959	0.709	0.727	0.750	0.767	0.775	0.623	0.634	0.645	0.649	0.649
300.0	1.059	1.090	1.384	1.682	1.881	0.948	0.951	1.079	1.188	1.251	0.792	0.832	0.888	0.940	0.971	0.720	0.733	0.754	0.777	0.792	0.642	0.656	0.675	0.691	0.700

Source: Ref 5-32

$$f_W = \sqrt{\sec\left[(\pi c/W) \cdot \sqrt{(a/t)}\right]} \quad \text{(Eq 5-22e)}$$

and f_φ is given by Eq 5-19(a).

By inputting large values of W and t in Eq 5-22, the configuration of a shallow crack ($a/c \leq 1$) in a semi-infinite solid (i.e., without the presence of other boundaries such as width and back face) can be obtained. The stress intensity factor $F = K/\sigma\sqrt{\pi a}$, which is equal to $F_s/\sqrt{Q}$ (Eq 5-21), has been computed for several points on the crack periphery (for several a/c ratios) and is plotted in Fig. 5-20. Conceptually, the location of the highest K-value is at the maximum depth for $a < c$. However, the highest K-value is on the surface of the plate when $a \geq c$. As shown in Fig. 5-20, the switching actually takes place at $a/c \approx 0.8$. For this flaw shape (i.e., $a/c \approx 0.8$), all the F-values along the crack boundary are approximately the same. Thus, crack growth rates at each point along the crack boundary are approximately the same. Therefore, whether the crack starts as a scratch (having $a << c$) or a deep cavity (having $c << a$), given time its flaw shape will eventually stabilize at $a/c \approx 0.8$.

Table 5-8 *F*-factors for axial cracks in a pressurized cylinder (Fig. 5-42), *R*/*t* = 1

	$a/c = 0.2$		$a/c = 0.4$		$a/c = 0.6$		$a/c = 0.8$		$a/c = 1.0$	
a/t	F (90°)	F (0°)	F (90°)	F (0°)	F (90°)	F (0°)	F (90°)	F (0°)	F (90°)	F (0°)
Inside crack										
0	2.73	1.542	2.433	1.713	2.15	1.754	1.898	1.769	1.683	1.771
0.08	2.86	1.567	2.533	1.74	2.252	1.79	2.001	1.812	1.785	1.821
0.2	3.099	1.608	2.725	1.786	2.448	1.851	2.198	1.885	1.982	1.902
0.3	3.723	1.88	3.12	1.973	2.782	1.991	2.479	1.984	2.22	1.968
0.4	4.457	2.181	3.598	2.181	3.192	2.147	2.829	2.096	2.519	2.044
0.5	5.325	2.516	4.174	2.416	3.69	2.323	3.26	2.224	2.891	2.13
0.6	7.272	3.337	5.23	2.919	4.528	2.663	3.911	2.427	3.388	2.222
0.7	9.54	4.252	6.489	3.484	5.533	3.047	4.699	2.66	3.995	2.33
0.8	12.188	5.277	7.986	4.123	6.732	3.483	5.645	2.928	4.73	2.46
0.9	16.391	6.885	10.159	5.067	8.421	4.111	6.924	3.293	5.667	2.61
1	21.273	8.687	12.73	6.133	10.425	4.826	8.447	3.715	6.791	2.791
Outside crack										
0	0.789	0.479	0.663	0.514	0.577	0.508	0.501	0.498	0.437	0.487
0.08	0.807	0.441	0.687	0.5	0.598	0.505	0.52	0.503	0.454	0.5
0.2	0.836	0.382	0.726	0.476	0.633	0.499	0.552	0.512	0.482	0.52
0.3	0.957	0.396	0.81	0.506	0.698	0.528	0.6	0.54	0.517	0.547
0.4	1.086	0.411	0.901	0.537	0.768	0.558	0.653	0.57	0.555	0.576
0.5	1.224	0.427	0.997	0.569	0.844	0.59	0.71	0.601	0.598	0.606
0.6	1.593	0.581	1.188	0.695	0.981	0.684	0.804	0.668	0.655	0.652
0.7	1.985	0.742	1.391	0.827	1.129	0.783	0.905	0.739	0.717	0.699
0.8	2.402	0.91	1.608	0.964	1.287	0.886	1.014	0.813	0.785	0.749
0.9	3.091	1.232	1.924	1.2	1.51	1.053	1.159	0.923	0.868	0.812
1	3.823	1.568	2.261	1.446	1.75	1.228	1.316	1.039	0.957	0.879

For $a/c > 1$:

$$M_1 = [1.0 + 0.04(c/a)] \cdot (c/a)^{1/2} \qquad \text{(Eq 5-23a)}$$

$$M_2 = 0.2(c/a)^4 \qquad \text{(Eq 5-23b)}$$

$$M_3 = -0.11(c/a)^4 \qquad \text{(Eq 5-23c)}$$

$$G_1 = 1 + [0.1 + 0.35(c/a)(a/t)^2] \cdot (1 - \sin\varphi)^2 \qquad \text{(Eq 5-23d)}$$

and f_w is given by Eq 5-22(e), and f_φ is given by Eq 5-19(b).

In case the plate is subjected to bending load, Eq 5-21 becomes:

$$K = \sigma \cdot F_s \cdot H \cdot \sqrt{\pi a/Q} \qquad \text{(Eq 5-24)}$$

Table 5-9 F-factors for axial cracks in a pressurized cylinder, $R/t = 1.5$

	$a/c = 0.2$		$a/c = 0.4$		$a/c = 0.6$		$a/c = 0.8$		$a/c = 1.0$	
a/t	F (90°)	F (0°)	F (90°)	F (0°)	F (90°)	F (0°)	F (90°)	F (0°)	F (90°)	F (0°)
Inside crack										
0	2.097	1.194	1.93	1.326	1.713	1.395	1.52	1.436	1.355	1.462
0.08	2.189	1.212	1.974	1.352	1.757	1.416	1.563	1.453	1.396	1.475
0.2	2.343	1.241	2.055	1.397	1.836	1.452	1.639	1.481	1.47	1.497
0.3	2.692	1.411	2.255	1.522	1.999	1.543	1.771	1.544	1.575	1.536
0.4	3.078	1.592	2.48	1.656	2.185	1.641	1.922	1.612	1.698	1.579
0.5	3.505	1.784	2.736	1.801	2.397	1.747	2.097	1.685	1.842	1.626
0.6	4.402	2.237	3.223	2.078	2.775	1.945	2.383	1.818	2.05	1.704
0.7	5.391	2.719	3.769	2.373	3.201	2.157	2.708	1.96	2.291	1.788
0.8	6.485	3.234	4.381	2.69	3.682	2.385	3.077	2.113	2.568	1.88
0.9	8.152	4.019	5.259	3.135	4.356	2.702	3.578	2.323	2.925	2.002
1	9.995	4.859	6.242	3.614	5.113	3.044	4.144	2.55	3.333	2.134
Outside crack										
0	0.824	0.522	0.716	0.548	0.63	0.554	0.554	0.553	0.488	0.549
0.08	0.856	0.489	0.744	0.54	0.653	0.554	0.573	0.559	0.504	0.56
0.2	0.906	0.437	0.788	0.528	0.69	0.553	0.604	0.568	0.53	0.577
0.3	1.042	0.457	0.872	0.566	0.754	0.587	0.65	0.598	0.561	0.604
0.4	1.184	0.478	0.961	0.605	0.821	0.622	0.698	0.63	0.595	0.633
0.5	1.333	0.5	1.055	0.645	0.891	0.658	0.75	0.662	0.63	0.662
0.6	1.678	0.646	1.229	0.768	1.016	0.752	0.832	0.731	0.679	0.71
0.7	2.039	0.797	1.412	0.894	1.147	0.848	0.92	0.801	0.73	0.759
0.8	2.417	0.953	1.604	1.025	1.285	0.947	1.013	0.874	0.786	0.81
0.9	3.009	1.239	1.876	1.237	1.475	1.102	1.135	0.98	0.853	0.876
1	3.63	1.534	2.161	1.456	1.676	1.261	1.265	1.09	0.924	0.945

where $\sqrt{Q}$ is Θ (Fig. 5-17), F_s is given in Eq 5-22 for uniform tension, and σ is the applied bending stress. For $a/c \leq 1$, the function H has the form:

$$H = H_1 + (H_2 - H_1) \cdot \sin^{\rho} \varphi \qquad \text{(Eq 5-25a)}$$

where

$$\rho = 0.2 + a/c + 0.6(a/t) \qquad \text{(Eq 5-25b)}$$

$$H_1 = 1 - 0.34(a/t) - 0.11(a/c) \cdot (a/t) \qquad \text{(Eq 5-25c)}$$

$$H_2 = 1 + F_1(a/t) + F_2(a/t)^2 \qquad \text{(Eq 5-25d)}$$

In Eq 5-25(d):

$$F_1 = -1.22 - 0.12(a/c) \qquad \text{(Eq 5-25e)}$$

Table 5-10 F-factors for axial cracks in a pressurized cylinder, R/t = 2

	a/c = 0.2		a/c = 0.4		a/c = 0.6		a/c = 0.8		a/c = 1.0	
a/t	F (90°)	F (0°)	F (90°)	F (0°)	F (90°)	F (0°)	F (90°)	F (0°)	F (90°)	F (0°)
Inside crack										
0	1.796	1.027	1.68	1.141	1.495	1.216	1.33	1.264	1.188	1.296
0.08	1.873	1.043	1.705	1.165	1.518	1.231	1.35	1.271	1.207	1.297
0.2	1.996	1.066	1.749	1.208	1.558	1.257	1.387	1.283	1.241	1.298
0.3	2.249	1.197	1.883	1.308	1.665	1.329	1.471	1.332	1.304	1.328
0.4	2.52	1.332	2.03	1.413	1.782	1.405	1.563	1.384	1.376	1.359
0.5	2.81	1.474	2.19	1.524	1.911	1.485	1.665	1.438	1.456	1.392
0.6	3.378	1.798	2.508	1.723	2.156	1.634	1.848	1.545	1.587	1.465
0.7	3.988	2.137	2.853	1.931	2.422	1.79	2.048	1.658	1.733	1.541
0.8	4.644	2.491	3.226	2.148	2.713	1.953	2.268	1.775	1.894	1.621
0.9	5.6	3.019	3.762	2.442	3.122	2.176	2.571	1.938	2.108	1.733
1	6.63	3.572	4.341	2.75	3.567	2.41	2.902	2.108	2.345	1.852
Outside crack										
0	0.846	0.548	0.748	0.567	0.662	0.581	0.585	0.586	0.519	0.586
0.08	0.885	0.517	0.777	0.564	0.686	0.583	0.604	0.592	0.534	0.596
0.2	0.946	0.469	0.824	0.559	0.723	0.586	0.634	0.601	0.558	0.611
0.3	1.09	0.493	0.907	0.601	0.785	0.622	0.677	0.633	0.586	0.638
0.4	1.238	0.517	0.994	0.645	0.849	0.66	0.722	0.665	0.616	0.666
0.5	1.393	0.542	1.085	0.689	0.916	0.698	0.77	0.698	0.647	0.694
0.6	1.722	0.684	1.248	0.81	1.032	0.791	0.845	0.767	0.689	0.743
0.7	2.063	0.829	1.418	0.933	1.152	0.885	0.924	0.837	0.734	0.794
0.8	2.418	0.978	1.595	1.059	1.278	0.982	1.007	0.909	0.781	0.845
0.9	2.953	1.243	1.84	1.257	1.448	1.128	1.115	1.013	0.838	0.913
1	3.507	1.515	2.095	1.459	1.626	1.279	1.229	1.118	0.898	0.982

$$F_2 = 0.55 - 1.05(a/c)^{0.75} + 0.47(a/c)^{1.5} \quad \text{(Eq 5-25f)}$$

A final note on part-through crack growth behavior is directed at the estimate of stress intensity factors while the crack front is approaching the back face of the plate. Direct use of the above equations would lead to a discontinuity where the calculation of K is suddenly switched from the part-through crack solution to the through-crack solution. This is commonly known as the *transition phenomenon.* Literature that discusses techniques for making a smooth transition is listed at the end of this chapter.

Corner Crack(s) at a Circular Hole

The problem of corner crack(s) at a circular hole configuration (Fig. 5-29) has received considerable attention from the fracture mechanics community because of its effect on the life of aircraft structures. The first set of test data on fatigue and fracture strength of specimens containing a circular hole was presented at the Air Force Conference in 1969 (Ref 5-21). Since then,

Table 5-11 F-factors for axial cracks in a pressurized cylinder, $R/t = 3$

	a/c = 0.2		a/c = 0.4		a/c = 0.6		a/c = 0.8		a/c = 1.0	
a/t	F (90°)	F (0°)	F (90°)	F (0°)	F (90°)	F (0°)	F (90°)	F (0°)	F (90°)	F (0°)
Inside crack										
0	1.476	0.848	1.377	0.972	1.23	1.03	1.098	1.066	0.985	1.089
0.08	1.547	0.864	1.408	0.993	1.255	1.044	1.118	1.075	1.001	1.094
0.2	1.657	0.89	1.456	1.025	1.295	1.067	1.15	1.089	1.026	1.102
0.3	1.848	0.992	1.551	1.103	1.367	1.124	1.204	1.129	1.064	1.127
0.4	2.047	1.096	1.651	1.182	1.444	1.182	1.261	1.179	1.105	1.154
0.5	2.253	1.204	1.756	1.264	1.525	1.243	1.321	1.212	1.149	1.181
0.6	2.625	1.434	1.943	1.415	1.664	1.355	1.421	1.292	1.215	1.233
0.7	3.012	1.671	2.139	1.571	1.811	1.47	1.526	1.373	1.286	1.287
0.8	3.416	1.915	2.346	1.73	1.966	1.589	1.638	1.457	1.362	1.342
0.9	3.977	2.268	2.627	1.946	2.175	1.749	1.786	1.57	1.459	1.417
1	4.564	2.63	2.923	2.168	2.396	1.913	1.943	1.687	1.563	1.493
Outside crack										
0	0.85	0.516	0.763	0.577	0.681	0.6	0.607	0.612	0.544	0.619
0.08	0.904	0.51	0.801	0.584	0.71	0.608	0.629	0.622	0.56	0.629
0.2	0.986	0.502	0.859	0.596	0.755	0.621	0.664	0.636	0.585	0.644
0.3	1.134	0.544	0.939	0.646	0.813	0.663	0.703	0.669	0.609	0.671
0.4	1.287	0.586	1.022	0.698	0.873	0.705	0.744	0.703	0.634	0.698
0.5	1.443	0.629	1.106	0.75	0.935	0.747	0.786	0.737	0.66	0.725
0.6	1.737	0.752	1.238	0.865	1.026	0.835	0.844	0.803	0.691	0.771
0.7	2.04	0.877	1.374	0.982	1.121	0.925	0.904	0.869	0.722	0.818
0.8	2.35	1.004	1.514	1.102	1.218	1.016	0.966	0.936	0.756	0.866
0.9	2.792	1.204	1.693	1.276	1.341	1.146	1.042	1.029	0.792	0.928
1	3.246	1.407	1.878	1.454	1.468	1.279	1.12	1.124	0.831	0.991

considerable efforts have been directed at the development of analytical and semi-empirical solutions. A complete set of two-dimensional solutions was not available until 1983. Prior to that, engineers had to use a one-dimensional solution for residual strength and crack growth life analysis. Among the many engineering solutions available in the literature, Liu's semi-empirical equation (Ref 5-22) was widely adopted by fracture mechanics analysts in the aircraft/aerospace industry. It will be shown along with an application problem in Chapter 8. In this section we present the two-dimensional finite-element solution of Newman and Raju, in a form of curve-fitted equations (Ref 5-18). This group of equations is unique because it covers a full range of geometric variables needed for conducting life assessment.

The Double Crack Configuration

For the double symmetric crack configuration:

$$K = \sigma \cdot F_{ch} \cdot \sqrt{\pi a/Q} \qquad \text{(Eq 5-26)}$$

Table 5-12 F-factors for axial cracks in a pressurized cylinder, $R/t = 4$

	$a/c = 0.2$		$a/c = 0.4$		$a/c = 0.6$		$a/c = 0.8$		$a/c = 1.0$	
a/t	F (90°)	F (0°)	F (90°)	F (0°)	F (90°)	F (0°)	F (90°)	F (0°)	F (90°)	F (0°)
Inside crack										
0	1.325	0.762	1.235	0.889	1.105	0.939	0.988	0.969	0.888	0.989
0.08	1.393	0.78	1.267	0.907	1.13	0.952	1.008	0.979	0.903	0.995
0.2	1.496	0.806	1.317	0.936	1.17	0.974	1.038	0.994	0.925	1.006
0.3	1.661	0.895	1.397	1.003	1.229	1.024	1.08	1.03	0.953	1.029
0.4	1.83	0.986	1.479	1.072	1.29	1.075	1.124	1.066	0.983	1.053
0.5	2.004	1.078	1.564	1.142	1.354	1.127	1.17	1.103	1.014	1.078
0.6	2.3	1.27	1.702	1.273	1.455	1.224	1.24	1.171	1.058	1.122
0.7	2.604	1.466	1.845	1.406	1.56	1.322	1.313	1.24	1.104	1.167
0.8	2.918	1.666	1.994	1.542	1.669	1.423	1.389	1.311	1.153	1.213
0.9	3.339	1.95	2.187	1.726	1.809	1.558	1.485	1.405	1.212	1.273
1	3.773	2.24	2.387	1.914	1.956	1.696	1.585	1.501	1.275	1.334
Outside crack										
0	0.853	0.498	0.772	0.582	0.692	0.61	0.62	0.627	0.558	0.637
0.08	0.914	0.507	0.814	0.595	0.724	0.623	0.643	0.638	0.574	0.647
0.2	1.007	0.52	0.878	0.616	0.773	0.641	0.679	0.655	0.6	0.662
0.3	1.158	0.572	0.956	0.671	0.828	0.685	0.716	0.689	0.621	0.689
0.4	1.312	0.624	1.035	0.726	0.885	0.729	0.754	0.724	0.644	0.715
0.5	1.468	0.677	1.116	0.783	0.943	0.774	0.793	0.759	0.666	0.742
0.6	1.744	0.789	1.231	0.895	1.022	0.86	0.841	0.822	0.69	0.787
0.7	2.025	0.902	1.348	1.009	1.102	0.946	0.891	0.886	0.715	0.832
0.8	2.311	1.017	1.468	1.124	1.185	1.034	0.942	0.951	0.74	0.878
0.9	2.705	1.182	1.613	1.286	1.282	1.155	1.001	1.038	0.766	0.936
1	3.106	1.348	1.762	1.45	1.383	1.278	1.061	1.126	0.793	0.996

where σ is the applied stress (load over cross section, ignoring the hole and the crack) and $\sqrt{Q}$ is Θ (Fig. 5-17). If the panel length ($2h$) is sufficiently long, F_{ch} is a function of a/c, a/t, r/t, r/W, c/W, and φ:

$$F_{ch} = [M_1 + M_2(a/t)^2 + M_3(a/t)^4] \cdot G_1 \cdot G_2 \cdot G_3 \cdot f_\varphi \cdot F_w \qquad \text{(Eq 5-27a)}$$

For $a/c \leq 1$:

$$G_2 = (1 + 0.358\lambda + 1.425\lambda^2 - 1.578\lambda^3 + 2.156\lambda^4)/(1 + 0.13\lambda^2) \qquad \text{(Eq 5-27b)}$$

where

$$\lambda = \{1 + (c/r) \cdot \cos(0.85\varphi)\}^{-1} \qquad \text{(Eq 5-27c)}$$

The functions G_3 and F_w are given as follows:

$$G_3 = [0.1 + 0.04(a/c)] \cdot [1 + 0.1(1 - \cos\varphi)^2] \cdot [0.85 + 0.15(a/t)^{1/4}] \qquad \text{(Eq 5-27d)}$$

Table 5-13 F-factors for axial cracks in a pressurized cylinder, $R/t = 6$

	a/c = 0.2		a/c = 0.4		a/c = 0.6		a/c = 0.8		a/c = 1.0	
a/t	F (90°)	F (0°)	F (90°)	F (0°)	F (90°)	F (0°)	F (90°)	F (0°)	F (90°)	F (0°)
Inside crack										
0	1.179	0.691	1.097	0.801	0.986	0.847	0.885	0.874	0.799	0.892
0.08	1.247	0.706	1.133	0.82	1.012	0.861	0.904	0.885	0.811	0.9
0.2	1.351	0.728	1.187	0.849	1.052	0.883	0.932	0.902	0.829	0.912
0.3	1.491	0.806	1.253	0.909	1.1	0.929	0.964	0.935	0.849	0.935
0.4	1.634	0.884	1.32	0.97	1.148	0.974	0.997	0.968	0.869	0.957
0.5	1.779	0.963	1.388	1.032	1.198	1.021	1.031	1.001	0.89	0.98
0.6	2.004	1.129	1.484	1.143	1.267	1.103	1.077	1.059	0.917	1.018
0.7	2.232	1.296	1.582	1.256	1.337	1.187	1.124	1.118	0.945	1.056
0.8	2.464	1.466	1.682	1.371	1.409	1.272	1.172	1.178	0.973	1.095
0.9	2.76	1.707	1.804	1.524	1.495	1.384	1.23	1.256	1.007	1.145
1	3.062	1.951	1.928	1.678	1.584	1.497	1.289	1.335	1.014	1.195
Outside crack										
0	0.872	0.522	0.79	0.602	0.711	0.632	0.64	0.65	0.578	0.661
0.08	0.935	0.529	0.833	0.616	0.743	0.644	0.662	0.66	0.593	0.67
0.2	1.031	0.539	0.899	0.636	0.792	0.662	0.697	0.676	0.616	0.684
0.3	1.176	0.595	0.97	0.691	0.842	0.706	0.73	0.71	0.634	0.709
0.4	1.322	0.652	1.042	0.747	0.893	0.749	0.763	0.743	0.652	0.734
0.5	1.47	0.709	1.115	0.804	0.944	0.793	0.796	0.777	0.671	0.759
0.6	1.72	0.837	1.21	0.916	1.009	0.878	0.835	0.838	0.689	0.801
0.7	1.973	0.966	1.306	1.029	1.074	0.963	0.875	0.9	0.708	0.844
0.8	2.229	1.097	1.404	1.143	1.14	1.049	0.915	0.962	0.726	0.886
0.9	2.572	1.288	1.517	1.303	1.215	1.168	0.959	1.046	0.744	0.941
1	2.919	1.48	1.632	1.465	1.292	1.287	1.003	1.13	0.763	0.996

$$F_w = \sqrt{\sec[\pi(r+c)/W) \cdot \sqrt{(a/t)}] \cdot \sec(\pi r/W)} \qquad \text{(Eq 5-27e)}$$

The other functions (i.e., M_1, M_2, M_3, G_1, and f_φ) are the same as those previously given for the surface flaw.

For $a/c > 1$:

$$G_3 = [1.13 + 0.09(c/a)] \cdot [1 + 0.1(1 - \cos\varphi)^2] \cdot [0.85 + 0.15(a/t)^{1/4}] \qquad \text{(Eq 5-27f)}$$

The functions G_2 and F_w are given by Eq 5-27(b) and (e), respectively. The other functions (i.e., M_1, M_2, M_3, G_1, and f_φ) are the same as those previously given for the surface flaw.

Single Corner Crack

The stress intensity factors for a single corner crack at a hole can be estimated by using Eq 5-26 with a conversion factor given by Ref 5-23:

$$K_{\text{one crack}} = K_{\text{two cracks}} \cdot \sqrt{(4/\pi + ac/2tr)/(4\pi - ac/tr)} \qquad \text{(Eq 5-28a)}$$

Table 5-14 *F*-factors for axial cracks in a pressurized cylinder, *R/t* = 8

	$a/c = 0.2$		$a/c = 0.4$		$a/c = 0.6$		$a/c = 0.8$		$a/c = 1.0$	
a/t	F (90°)	F (0°)	F (90°)	F (0°)	F (90°)	F (0°)	F (90°)	F (0°)	F (90°)	F (0°)
Inside crack										
0	1.108	0.656	1.031	0.759	0.928	0.802	0.835	0.828	0.755	0.845
0.08	1.176	0.669	1.068	0.777	0.955	0.816	0.853	0.839	0.766	0.853
0.2	1.28	0.69	1.124	0.805	0.995	0.838	0.881	0.856	0.783	0.866
0.3	1.409	0.762	1.183	0.862	1.038	0.881	0.909	0.888	0.799	0.888
0.4	1.54	0.834	1.243	0.92	1.08	0.925	0.937	0.919	0.815	0.909
0.5	1.672	0.908	1.304	0.978	1.124	0.968	0.966	0.951	0.832	0.931
0.6	1.866	1.061	1.383	1.081	1.179	1.045	1.002	1.005	0.852	0.967
0.7	2.062	1.215	1.462	1.184	1.235	1.122	1.038	1.059	0.873	1.002
0.8	2.26	1.371	1.543	1.289	1.293	1.199	1.076	1.114	0.894	1.038
0.9	2.505	1.593	1.636	1.428	1.358	1.301	1.118	1.185	0.917	1.083
1	2.753	1.816	1.73	1.568	1.424	1.404	1.161	1.256	0.941	1.129
Outside crack										
0	0.883	0.534	0.8	0.613	0.721	0.644	0.65	0.662	0.588	0.673
0.08	0.946	0.54	0.844	0.627	0.753	0.656	0.672	0.672	0.603	0.382
0.2	1.043	0.55	0.91	0.647	0.802	0.673	0.706	0.688	0.624	0.696
0.3	1.185	0.608	0.977	0.702	0.849	0.716	0.736	0.72	0.64	0.72
0.4	1.327	0.667	1.045	0.758	0.897	0.76	0.767	0.753	0.657	0.744
0.5	1.471	0.726	1.114	0.814	0.945	0.804	0.798	0.787	0.674	0.768
0.6	1.707	0.862	1.198	0.926	1.001	0.887	0.847	0.744	0.688	0.809
0.7	1.946	0.999	1.284	1.039	1.059	0.972	0.907	0.774	0.704	0.85
0.8	2.186	1.138	1.37	1.153	1.117	1.057	0.968	0.805	0.719	0.891
0.9	2.504	1.342	1.467	1.312	1.181	1.174	1.05	0.838	0.733	0.943
1	2.824	1.548	1.565	1.472	1.245	1.292	0.974	1.132	0.747	0.996

In addition, the finite width correction factor should be adjusted for the single crack configuration:

$$F_w = \sqrt{\sec\left[(\pi(2r+c)/(2(W-c))\cdot\sqrt{(a/t)}\right]\cdot\sec(\pi r/W)} \quad \text{(Eq 5-28b)}$$

Crack at Pin Hole in a Lug

Two types of attachment lugs are considered here, the straight lug and the tapered lug (Fig. 5-30). The stress intensity factors for cracks at the bore of the pin hole are summarized below.

The Straight Lug

Figure 5-31 presents the finite-element solutions of stress intensity factors for an axially loaded straight attachment lug. The pin load is applied at the direction normal to the base of the lug ($\theta = 0°$). Either the bearing stress, σ_{br}, or the gross area stress, σ_0, can be used to compute stress intensity. That

Table 5-15 *F*-factors for axial cracks in a pressurized cylinder, *R/t* = 10

	a/c = 0.2		a/c = 0.4		a/c = 0.6		a/c = 0.8		a/c = 1.0	
a/t	F (90°)	F (0°)	F (90°)	F (0°)	F (90°)	F (0°)	F (90°)	F (0°)	F (90°)	F (0°)
Inside crack										
0	1.067	0.635	0.992	0.733	0.894	0.775	0.806	0.8	0.729	0.817
0.08	1.135	0.648	1.03	0.752	0.921	0.79	0.823	0.812	0.739	0.826
0.2	1.237	0.667	1.086	0.78	0.961	0.812	0.85	0.829	0.755	0.839
0.3	1.361	0.736	1.142	0.835	1.001	0.853	0.876	0.859	0.769	0.86
0.4	1.485	0.805	1.199	0.89	1.041	0.895	0.902	0.89	0.784	0.881
0.5	1.61	0.875	1.256	0.945	1.081	0.937	0.928	0.921	0.798	0.902
0.6	1.786	1.021	1.324	1.043	1.129	1.01	0.959	0.972	0.815	0.936
0.7	1.964	1.168	1.394	1.142	1.177	1.083	0.99	1.024	0.831	0.97
0.8	2.144	1.316	1.464	1.241	1.227	1.157	1.021	1.076	0.849	1.004
0.9	2.361	1.527	1.541	1.372	1.281	1.253	1.056	1.143	0.867	1.047
1	2.58	1.739	1.62	1.504	1.335	1.35	1.09	1.21	0.885	1.09
Outside crack										
0	0.889	0.542	0.806	0.62	0.728	0.651	0.657	0.669	0.595	0.681
0.08	0.953	0.547	0.85	0.633	0.76	0.663	0.678	0.679	0.609	0.689
0.2	1.051	0.556	0.917	0.653	0.808	0.68	0.712	0.695	0.629	0.703
0.3	1.19	0.616	0.981	0.709	0.853	0.723	0.74	0.727	0.644	0.726
0.4	1.33	0.676	1.047	0.765	0.899	0.766	0.769	0.76	0.659	0.75
0.5	1.471	0.736	1.113	0.821	0.945	0.81	0.798	0.792	0.675	0.774
0.6	1.699	0.878	1.191	0.933	0.997	0.893	0.829	0.852	0.688	0.814
0.7	1.929	1.02	1.27	1.045	1.05	0.977	0.86	0.912	0.701	0.854
0.8	2.16	1.163	1.35	1.159	1.103	1.062	0.891	0.972	0.714	0.894
0.9	2.462	1.375	1.436	1.317	1.16	1.178	0.923	1.053	0.726	0.945
1	2.766	1.589	1.524	1.477	1.217	1.295	0.956	1.134	0.738	0.996

is, for a single through-the-thickness crack on one side of the pin hole (perpendicular to the direction of loading, P), according to Ref 5-24:

$$K = \sigma_{br} \cdot F_{RB} \cdot \sqrt{\pi c} \quad \text{(Eq 5-29a)}$$

Because $\sigma_{br} = P/2R_i t$ and $\sigma_0 = P/2R_o t$, Eq 5-29(a) can be rewritten as:

$$K = \sigma_0 \cdot F_{RB} \cdot R_o/R_i \cdot \sqrt{\pi c} \quad \text{(Eq 5-29b)}$$

The values of F_{RB} for five R_o/R_i ratios are given in Fig. 5-31.

Equations 5-29(a) and (b) can be modified to analyze a corner crack. Two methods have been proposed (Ref 5-24). The method that involves analysis at two crack tips (i.e., points A and C of the corner crack) is rather complicated and incomplete, and it will not be discussed here. In the one-parameter method, the crack shape (i.e., a/c ratio) is assumed to be constant and equal to 1.33. The stress intensity factor at the lug surface point (i.e., point C) is computed using a corner crack correction factor:

Table 5-16 F-factors for axial cracks in a pressurized cylinder, $R/t = 20$

	a/c = 0.2		a/c = 0.4		a/c = 0.6		a/c = 0.8		a/c = 1.0	
a/t	F (90°)	F (0°)	F (90°)	F (0°)	F (90°)	F (0°)	F (90°)	F (0°)	F (90°)	F (0°)
Inside crack										
0	0.972	0.594	0.92	0.677	0.833	0.722	0.753	0.751	0.684	0.771
0.08	1.05	0.607	0.96	0.698	0.86	0.738	0.77	0.762	0.692	0.777
0.2	1.167	0.626	1.021	0.73	0.901	0.761	0.795	0.778	0.704	0.788
0.3	1.278	0.688	1.07	0.781	0.935	0.799	0.816	0.806	0.714	0.807
0.4	1.39	0.75	1.12	0.832	0.969	0.838	0.837	0.835	0.725	0.827
0.5	1.501	0.812	1.169	0.884	1.003	0.878	0.858	0.863	0.735	0.847
0.6	1.631	0.94	1.219	0.968	1.037	0.942	0.879	0.911	0.745	0.88
0.7	1.762	1.068	1.268	1.053	1.071	1.006	0.9	0.958	0.755	0.914
0.8	1.893	1.197	1.318	1.137	1.105	1.071	0.921	1.006	0.766	0.947
0.9	2.033	1.378	1.366	1.246	1.138	1.154	0.941	1.067	0.775	0.991
1	2.175	1.560	1.414	1.354	1.171	1.237	0.961	1.128	0.785	1.034
Outside crack										
0	0.884	0.546	0.829	0.62	0.751	0.659	0.679	0.684	0.617	0.701
0.08	0.958	0.555	0.971	0.683	0.78	0.673	0.697	0.694	0.627	0.708
0.2	1.069	0.569	0.933	0.655	0.823	0.693	0.725	0.709	0.641	0.718
0.3	1.185	0.627	0.985	0.716	0.858	0.732	0.746	0.738	0.651	0.738
0.4	1.3	0.684	1.037	0.766	0.894	0.771	0.768	0.766	0.662	0.759
0.5	1.416	0.742	1.09	0.817	0.93	0.81	0.79	0.795	0.673	0.779
0.6	1.565	0.866	1.143	0.905	0.965	0.877	0.811	0.845	0.682	0.814
0.7	1.715	0.99	1.196	0.994	1.001	0.944	0.832	0.894	0.691	0.849
0.8	1.865	1.114	1.25	1.082	1.037	1.012	0.854	0.944	0.7	0.884
0.9	2.039	1.293	1.302	1.2	1.071	1.101	0.873	1.009	0.708	0.929
1	2.214	1.473	1.354	1.317	1.106	1.19	0.893	1.074	0.715	0.975

$$\Theta_{71} = 1 - 0.2886/[1 + 2(a/c)^2 + (c/t)^2] \quad \text{(Eq 5-29c)}$$

and Eq 5-29(a) or (b) becomes:

$$K^C = K \cdot \Theta_{71} \quad \text{(Eq 5-29d)}$$

The superscript "C" stands for corner crack and K is the stress intensity factor for the through-the-thickness crack, given by Eq 5-29(a) or (b).

The Tapered Lug

All the equations listed above for the straight lug are also applicable to the tapered lug (with a new set of F_{RB} factors). The F_{RB} factors for the 0° loading are presented in Fig. 5-32. Limited finite-element data are available for tapered lugs loaded in other directions. The through-crack solutions for $\theta = 180°$, –45°, and –90° are presented in Fig. 5-33 to 5-35, respectively.

Table 5-17 F-factors for axial cracks in a pressurized cylinder, $R/t = 30$

	$a/c = 0.2$		$a/c = 0.4$		$a/c = 0.6$		$a/c = 0.8$		$a/c = 1.0$	
a/t	F (90°)	F (0°)	F (90°)	F (0°)	F (90°)	F (0°)	F (90°)	F (0°)	F (90°)	F (0°)
Inside crack										
0	0.941	0.581	0.897	0.659	0.812	0.705	0.735	0.735	0.669	0.755
0.08	1.022	0.593	0.938	0.681	0.84	0.72	0.752	0.745	0.676	0.761
0.2	1.144	0.612	0.999	0.713	0.881	0.744	0.777	0.761	0.688	0.771
0.3	1.251	0.672	1.046	0.763	0.913	0.782	0.796	0.788	0.697	0.79
0.4	1.358	0.732	1.094	0.813	0.946	0.82	0.816	0.816	0.705	0.809
0.5	1.466	0.792	1.141	0.864	0.968	0.858	0.835	0.844	0.714	0.828
0.6	1.582	0.914	1.185	0.944	1.007	0.919	0.853	0.89	0.723	0.862
0.7	1.698	1.036	1.229	1.024	1.037	0.981	0.871	0.936	0.731	0.895
0.8	1.814	1.159	1.272	1.104	1.067	1.043	0.889	0.983	0.739	0.928
0.9	1.932	1.331	1.311	1.205	1.093	1.122	0.905	1.042	0.747	0.972
1	2.05	1.503	1.35	1.306	1.12	1.2	0.921	1.102	0.755	1.015
Outside crack										
0	0.883	0.547	0.837	0.62	0.759	0.662	0.687	0.689	0.625	0.708
0.08	0.96	0.558	0.878	0.64	0.786	0.676	0.704	0.699	0.633	0.714
0.2	1.076	0.574	0.939	0.669	0.828	0.698	0.729	0.714	0.645	0.723
0.3	1.183	0.63	0.986	0.718	0.86	0.735	0.749	0.741	0.654	0.742
0.4	1.29	0.687	1.034	0.767	0.892	0.772	0.768	0.769	0.663	0.761
0.5	1.397	0.744	1.082	0.816	0.925	0.809	0.788	0.796	0.672	0.781
0.6	1.519	0.862	1.126	0.896	0.955	0.871	0.805	0.842	0.679	0.814
0.7	1.642	0.98	1.171	0.976	0.985	0.933	0.823	0.888	0.687	0.847
0.8	1.765	1.098	1.216	1.056	1.015	0.995	0.841	0.935	0.695	0.88
0.9	1.896	1.266	1.256	1.16	1.041	1.075	0.857	0.994	0.702	0.924
1	2.028	1.434	1.296	1.263	1.068	1.155	0.872	1.054	0.708	0.967

Crack in a Solid Cylinder

A crack in a solid cylinder usually has an almond shape, as shown schematically in Fig. 5-36. In the remainder of this section we will refer to it as a crack on the shank of a bolt, or at the thread of a bolt. However, one should keep in mind that this kind of crack can be found in any circumferential plane of a solid round bar. The stress intensity solutions presented herein are applicable to both.

The nomenclature for this crack type is defined in Fig. 5-36, where a is the crack depth, the point (point A) that travels through the diameter of the cylinder; b is the crack length (i.e., one-half of the crack tip-to-tip circumferential arc); d is the minor diameter at the thread of a bolt (or a notched round bar); and D is the diameter of a rod or the diameter at the unthreaded portion of a bolt, depending on the application.

Due to the geometric differences between a cylinder and a plate, this crack cannot be treated as an edge crack, or as a thumbnail crack in a rectangular cross section. It is in a class by itself. Many fracture mechanics

Table 5-18 F-factors for axial cracks in a pressurized cylinder, $R/t = 50$

	$a/c = 0.2$		$a/c = 0.4$		$a/c = 0.6$		$a/c = 0.8$		$a/c = 1.0$	
a/t	F (90°)	F (0°)	F (90°)	F (0°)	F (90°)	F (0°)	F (90°)	F (0°)	F (90°)	F (0°)
Inside crack										
0	0.917	0.57	0.878	0.645	0.796	0.691	0.721	0.722	0.657	0.743
0.08	1	0.583	0.919	0.667	0.824	0.707	0.738	0.732	0.664	0.748
0.2	1.125	0.601	0.982	0.7	0.866	0.73	0.762	0.747	0.674	0.757
0.3	1.229	0.659	1.027	0.749	0.896	0.767	0.781	0.774	0.682	0.776
0.4	1.333	0.717	1.073	0.798	0.927	0.805	0.799	0.802	0.69	0.795
0.5	1.438	0.776	1.119	0.848	0.958	0.842	0.817	0.829	0.698	0.814
0.6	1.543	0.893	1.158	0.924	0.984	0.902	0.833	0.874	0.705	0.847
0.7	1.648	1.011	1.197	1.001	1.011	0.961	0.849	0.919	0.712	0.88
0.8	1.753	1.129	1.237	1.078	1.037	1.021	0.864	0.964	0.719	0.913
0.9	1.854	1.294	1.269	1.173	1.059	1.096	0.877	1.022	0.725	0.957
1	1.955	1.459	1.301	1.269	1.081	1.172	0.89	1.08	0.731	1
Outside crack										
0	0.882	0.548	0.844	0.62	0.765	0.664	0.693	0.694	0.631	0.714
0.08	0.961	0.56	0.884	0.641	0.792	0.679	0.709	0.703	0.638	0.719
0.2	1.081	0.578	0.943	0.673	0.832	0.702	0.733	0.718	0.648	0.727
0.3	1.181	0.634	0.987	0.72	0.861	0.738	0.75	0.744	0.656	0.746
0.4	1.281	0.689	1.031	0.767	0.891	0.773	0.768	0.77	0.663	0.764
0.5	1.382	0.745	1.075	0.815	0.921	0.809	0.785	0.797	0.671	0.782
0.6	1.482	0.858	1.113	0.888	0.946	0.867	0.8	0.84	0.678	0.814
0.7	1.584	0.972	1.151	0.962	0.971	0.924	0.815	0.883	0.684	0.846
0.8	1.685	1.085	1.188	1.036	0.997	0.981	0.831	0.927	0.691	0.878
0.9	1.781	1.243	1.219	1.128	1.017	1.054	0.843	0.983	0.697	0.919
1	1.878	1.402	1.25	1.22	1.038	1.126	0.856	1.038	0.702	0.961

analysts believe that this type of crack has the shape of a circular arc, rather than one-half of an ellipse (Ref 5-25, 5-26). The center of the circle floats in between the free surface of the cylinder and a point infinitely far away from the cylindrical surface. As the crack front passes through the center of the rod, the crack front curvature will become flattened, approaching a straight crack front despite the fact that a circular shape (by geometric definition) had been maintained at all times. Experimental observations on crack growth in unthreaded and threaded rods (Ref 5-27 to 5-31) seem to support this claim.

Partly Circular Crack in a Bolt

For an almond-shaped crack in a bolt, a large portion of the crack growth activity will take place in only one-half of the cylindrical crack plane, that is, where $a \leq D/2$ (or $a \leq d/2$), before it becomes critical. Experimental data also show that the crack maintains a circular shape and a constant a/b ratio in this region. Referring back to the semi-elliptical flaw problem, a constant flaw shape during crack growth implies that the K-values along the crack periphery are nearly constant. However, this is not the case here because the crack tip on the surface, which travels along the circumference of the cylindrical cross section, has to travel a longer distance than the crack middle point (which propagates in the depth direction), in order to keep up with the crack shape aspect ratios. That is, $db/dN > da/dN$ in each increment of crack extension. This means that the K-value at point B would be higher than that at point A. Therefore, a two-dimensional crack growth scheme is more suitable for this type of crack geometry. However, if the relationship between a/D and a/b is clearly established, and the K-values (or the geometric factor of β or Y) at locations on the crack propagation path are specifically determined for the expected crack geometries, crack growth can be predicted by monitoring only one point on the crack periphery (i.e., by treating the crack configuration as if it were a one-dimensional crack). In this section the geometric factor Y will be expressed as a function of a/D (or a/d), for stress intensity factors at point A. Justifications for using this approach have been made (Ref 5-26) by showing good correlations between the stress intensity equations presented below and available test data. A bibliography documenting the other approaches published in the open literature is given in a review paper by Reemsnyder, which is listed at the end of this chapter.

Crack on Bolt Shank or at the Thread

For a crack on the circumferential surface of an unthreaded (or threaded) cylinder, the one-dimensional crack tip stress intensity, K, can be defined as:

$$K = S \cdot Y \cdot \sqrt{\pi c} \qquad \text{(Eq 5-30)}$$

where S is the applied stress [$4P/(\pi D^2)$, $4P/(\pi d^2)$, $32M/(\pi D^3)$, or $32M/(\pi d^3)$, depending on the application]. The equation below is suitable for cracks on the shank of a bolt, or at the root of a screw thread, subjected to either tension or bending:

$$Y = \lambda \cdot \exp[\beta x] + A + Bx + Cx^2 + Dx^3 + Ex^4 + Fx^5 + Gx^6 \qquad \text{(Eq 5-31)}$$

Here, x equals a/D, or a/d, whichever is appropriate. In the case of an unthreaded cylinder the first two terms in Eq 5-31 do not exist, because these terms cover the local stress concentration caused by the screw thread. The magnitude of the exponential term decreases rapidly as crack length increases. The values of the coefficients are given in Table 5-3. A graphical presentation of Eq 5-31 is shown in Fig. 5-37.

It should be noted that λ and β are associated with the stress concentration factor and the local stress distribution near the notch root (or screw thread). It is expected that their values may vary, depending on the depth, the root radius, and the pitch angle of the thread. Therefore, it would be conceptually impossible to have a single curve for Y to cover a wide range of geometric details. The values for λ and β listed in Table 5-3 represent a mixture of geometries that were used in the development of analytical and experimental data. The finite-element models had a 45° pitch angle with root radii in a range of 0 to 0.125 mm. The test data were generated from notched specimens that either contained machined grooves that simulated the 12-UNF-3A screw thread (having a root radius of 0.38 mm) or had an unknown geometry. Although these values for λ and β certainly will not be universally suitable for all the local geometries encountered in practice, it is conceivable that they are applicable to most thread geometries as long as the crack size is not too small. In any event, Table 5-3 is a guide to making crack growth life estimates.

Crack on the Circumferential Plane of a Hollow Cylinder

In this category of cracks we consider the through-the-thickness crack and the thumbnail crack originated on the inner or outer wall of a hollow cylinder (a tube, a pipe, or a pressure vessel) that is subjected to axial loads (i.e., uniform tension, bending, or both).

Through-the-Thickness Crack

The nomenclature for a through-the-thickness crack is defined in Fig. 5-38, where M is the bending moment, P is the tensile load, α is the angle that represents one-half of the crack tip-to-tip circumferential arc, t is the thickness of the cylinder wall, R is the average radius of the inner and outer cylinder wall, c is the half-crack length $(R \cdot \alpha)$, S_t is the tension stress $(P/(2\pi Rt))$, and S_b is the bending stress $(M/(\pi tR^2))$. The stress intensity equation is:

$$K = S \cdot F \cdot \sqrt{\pi c} \qquad \text{(Eq 5-32a)}$$

where S can be designated as S_t or S_b, depending on the loading condition. Likewise, F is designated as F_t or F_b. The dimensionless factors F_t and F_b are given in Ref 5-20 as a group of close-form solutions. However, application of these equations may be cumbersome because of their complexity. To help readers cut through the lengthy calculations, these factors have been computed for several specific R/t ratios (with an assumed Poisson's ratio of 0.3). The results are presented in Fig. 5-39 and 5-40. Each of these curves was fitted by a polynomial equation:

$$F = A + B\alpha + C\alpha^2 + D\alpha^3 + E\alpha^4 + F\alpha^5 + G\alpha^6 \qquad \text{(Eq 5-32b)}$$

The coefficients are listed in Tables 5-4 and 5-5. For R/t ratios not included in these figures or tables, interpolated values for F_t or F_b can be used.

The solutions for F_t and F_b were developed on the basis of the thin-wall theory (for $R/t \geq 30$). Figures 5-39 and 5-40 and Tables 5-4 and 5-5 can be used for tubes having thicker walls or smaller curvatures (i.e., $R/t < 30$). Although $R/t = 7$ is included in these figures and tables, it is recommended that $R/t = 15$ be considered as the limit of applicability.

Part-Through Crack

For a thumbnail crack on the inner or outer wall of the hollow cylinder (Fig. 5-41), the expression for K is given by Ref 5-32 as:

$$K = S \cdot F \cdot \sqrt{\pi a} \qquad \text{(Eq 5-33)}$$

where S can be S_t or S_b, depending on the loading condition. In either case, F is a function of a/c, R/t, and a/t. The values of F for internal and external cracks subjected to tension or bending are given in Tables 5-6 and 5-7.

Pressurized Cylinder and Sphere

Axial Cracks in a Cylinder

In this section we consider a long cylinder subjected to internal pressure. A through-the-wall-thickness crack or a part-through crack is placed along the length of the cylinder (Fig. 5-42). The cylinder is sufficiently long as compared to the length of the crack, and the end effect on crack tip stress intensity is not considered.

Through-the-Thickness Crack

In a pressurized thin-wall cylinder (Fig. 5-42), the elastic stress intensity factor of Folias/Erdogan was given by Ref 5-33 as:

$$K = S \cdot Y \cdot \sqrt{\pi c} \qquad \text{(Eq 5-34a)}$$

where

$$Y = (1 + 0.52\lambda + 1.29\lambda^2 - 0.074\lambda^3)^{1/2} \qquad \text{(Eq 5-34b)}$$

where $\lambda = c/\sqrt{(Rt)}$, S is the pR/t stress, and R is the inner radius of the cylinder. Equation 5-34(a) accounts for the effects of shell curvature on stress intensity. Poisson's ratio was assumed to be 1/3. The applicable range is $0 \leq \lambda \leq 10$.

Part-Through Crack

Now consider a pair of thumbnail cracks symmetrically located on the inner or outer wall of a hollow cylinder (Fig. 5-42). The expression for K can be written as:

$$K = S \cdot \alpha \cdot \sqrt{\pi a/Q} \qquad \text{(Eq 5-35a)}$$

where $S = pR/t$, and α is a function of R/t, a/c, a/t, φ, and the nonuniform tangential stresses acting on the crack plane. Having the nonuniform stress distribution normalized to the pR/t stress, the α function can be written as (Ref 5-34):

$$\alpha_i = (t/R) \cdot r^2/(r^2 - R^2) \cdot [2H_0 - 2H_1\,(a/R) + 3H_2(a/R)^2 - 4H_3(a/R)^3] \qquad \text{(Eq 5-35b)}$$

for an internal crack, or as:

$$\alpha_0 = (t/R) \cdot R^2/(r^2 - R^2) \cdot [2G_0 + 2G_1(a/r) + 3G_2(a/r)^2 + 4G_3(a/r)^3] \quad \text{(Eq 5-35c)}$$

for an external crack.

Here r and R are the outer and inner radius of the cylinder, respectively. The H and G values are functions of R/t, a/c, a/t, and φ. Each H or G value corresponds to a particular loading distribution. The subscript 0 corresponds to uniform tension, subscript 1 to linear distribution, subscript 2 to quadratic, and subscript 3 to cubic. Using the H and G values given in Ref 5-32, α_i and α_0 values for the pressurized cylinder were computed for 11 R/t ratios (1, 1.5, 2, 3, 4, 6, 8, 10, 20, 30, and $\geq$ 50).

Let $F = \alpha/\sqrt{Q}$ so that Eq 5-35(a) is reduced to:

$$K = S \cdot F \cdot \sqrt{\pi a} \quad \text{(Eq 5-35d)}$$

The computed F values are presented in Tables 5-8 to 5-18. In Ref 5-32, the stress intensity factor tables list H and G values for only five R/t ratios (1, 2, 4, 10, and $\geq$ 50), in combination with three a/c ratios (0.2, 0.4, and 1.0) and five a/t ratios (0, 0.2, 0.5, 0.8, and 1.0). For other R/t ratios listed in Tables 5-8 to 5-18, the F values have been obtained by linear interpolation. Likewise, interpolated F values can be used for other parametric combinations not included in these tables.

Although the α factors were originally derived for the double crack configuration, they are also applicable to the single crack configuration. When Eq 5-35(a) or (d) is used for the single crack configuration, the maximum error is 4% (higher than the actual value), depending on the a/c ratio.

Crack in a Pressurized Sphere

Currently, the Erdogan solution for a through-the-thickness crack (Ref 5-20) is the only solution available for the pressurized spherical shell (Fig. 5-43):

$$K = S \cdot Y \cdot \sqrt{\pi c} \quad \text{(Eq 5-36a)}$$

where

$$Y = (1 + 3\lambda^{1.9})^{0.4} \quad \text{(Eq 5-36b)}$$

where $\lambda = c/\sqrt{(Rt)}$, S is the $pR/2t$ stress, and R is the average of the inner and outer radii. This equation accounts for the effects of shell curvature on stress intensity. The Poisson's ratio was assumed to be 1/3. The applicable range is $0 \leq \lambda \leq 3$.

REFERENCES

5-1. M. Isida, Stress Intensity Factor of the Tension of an Eccentrically Cracked Strip, *J. Appl. Mech., (Trans. ASME),* Series E, Vol 33, 1966, p 674–675

5-2. M. Isida, Effect of Width and Length on Stress Intensity Factors of Internally Cracked Plate under Various Boundary Conditions, *Int. J. Fract. Mech.,* Vol 7, 1971

5-3. W.F. Brown, Jr. and J.E. Srawley, *Plane Strain Crack Toughness Testing of High Strength Metallic Materials,* STP 410, American Society for Testing and Materials, 1966, p 77–79

5-4. E.P. Phillips, The Influence of Crack Closure on Fatigue Crack Growth Thresholds in 2024-T3 Aluminum Alloy, *Mechanics of Fatigue Crack Closure,* STP 982, American Society for Testing and Materials, 1988, p 515

5-5. H. Tada, P.C. Paris, and G.R. Irwin, *Stress Analysis of Cracks Handbook,* Del Research Corp., Hellertown, PA, 1973

5-6. O.L. Bowie, Analysis of an Infinite Plate Containing Radial Cracks Originating at the Boundaries of an Internal Circular Hole, *J. Math. Physics,* Vol 35, 1956, p 60

5-7. P.C. Paris and G.C. Sih, Stress Analysis of Cracks, *Symposium on Fracture Toughness Testing and Its Applications,* STP 381, American Society for Testing and Materials, 1965, p 30–83

5-8. J.C. Newman, Jr., "Predicting Failure of Specimens with Either Surface Cracks or Corner Cracks at Holes," Report TN D-8244, National Aeronautics and Space Administration, Washington, D.C., June 1976

5-9. J.C. Newman, Jr., "An Improved Method of Collocation for the Stress Analysis of Cracked Plates with Various Shaped Boundaries," Report TN D-6376, National Aeronautics and Space Administration, Washington, D.C., Aug 1971

5-10. Metals Test Methods and Analytical Procedures, *Annual Book of ASTM Standards,* Vol 03.01, *Metals—Mechanical Testing; Elevated and Low Temperature Tests; Metallography,* American Society for Testing and Materials, 1995

5-11. C.C. Poe, Jr., "Stress-Intensity Factor for a Cracked Sheet with Riveted and Uniformly Spaced Stringers," Report TR-R-358, National Aeronautics and Space Administration, Langley Research Center, Hampton, VA, May 1971

5-12. C.C. Poe, Jr., "The Effect of Broken Stringers on the Stress Intensity Factor of a Uniformly Stiffened Sheet Containing a Crack," Report TM X-71947, National Aeronautics and Space Administration, Langley Research Center, Hampton, VA, 1973

5-13. W.J. Crichlow, "The Optimum Design of Shell Structure for Static Strength, Stiffness, Fatigue and Damage Tolerance Strength," paper presented at AGARD Symposium on Structural Optimization (Istanbul, Turkey), 6–8 Oct 1969

5-14. C.C. Poe, Jr., Fatigue Crack Propagation in Stiffened Panels, *Damage Tolerance in Aircraft Structures,* STP 486, American Society for Testing and Materials, 1971, p 79–97

5-15. G.R. Irwin, Crack-Extension Force for a Part-Through Crack in a Plate, *J. Appl. Mech., (Trans. ASME),* Series E, Vol 84, 1962, p 651–654
5-16. F.W. Smith, The Elastic Analysis of the Part-Circular Surface Flaw Problem by the Alternating Method, *The Surface Crack: Physical Problems and Computational Solutions,* The American Society of Mechanical Engineers, 1972, p 125–152
5-17. J.C. Newman, Jr. and I.S. Raju, An Empirical Stress-Intensity Factor Equation for the Surface Crack, *Eng. Fract. Mech.,* Vol 15, 1981, p 185–192
5-18. J.C. Newman, Jr. and I.S. Raju, Stress-Intensity Factor Equations for Cracks in Three-Dimensional Finite Bodies, *Fracture Mechanics: Fourteenth Symposium,* Vol I, *Theory and Analysis,* STP 791, American Society for Testing and Materials, 1983, p I-238 to I-265
5-19. I.S. Raju, S.R. Mettu, and V. Shivakumar, Stress Intensity Factor Solutions for Surface Cracks in Flat Plates Subjected to Nonuniform Stresses, *Fracture Mechanics: Twenty-Fourth Volume,* STP 1207, American Society for Testing and Materials, 1994, p 560–580
5-20. R.G. Forman, V. Shivakumar, and J.C. Newman, Jr., "Fatigue Crack Growth Computer Program NASA/FLAGO Version 2.0," Report JSC-22267A, National Aeronautics and Space Administration, Washington, D.C., May 1994
5-21. L.R. Hall and R.W. Finger, "Fracture and Fatigue Growth of Partially Embedded Flaws," Proceedings of Air Force Conference on Fatigue and Fracture of Aircraft Structures and Materials, Report AFFDL-TR-70–144, Air Force Flight Dynamics Laboratory, Wright-Patterson Air Force Base, Dayton, OH, 1970, p 235–262
5-22. A.F. Liu, Stress Intensity Factor for a Corner Flaw, *Eng. Fract. Mech.,* Vol 4, Pergamon Press, 1972, p 175–179
5-23. R.C. Shah, Stress Intensity Factors for Through and Part-Through Cracks Originating at Fastener Holes, *Mechanics of Crack Growth,* STP 590, American Society for Testing and Materials, 1976, p 429–459
5-24. K. Katherisan, T.M. Hsu, and T.R. Brussat, "Advanced Life Analysis Methods—Crack Growth Analysis Methods for Attachment Lugs," Report AFWAL-TR-84-3080, Vol II, Air Force Flight Dynamics Laboratory, Wright-Patterson Air Force Base, Dayton, OH, Sept 1984
5-25. A.F. Liu, Evaluation of Current Analytical Methods for Crack Growth in a Bolt, *Durability and Structural Reliability of Airframes—ICAF 17,* Vol 2, EMAS Ltd., West Midlands, U.K., 1993, p 1141–1155
5-26. A.F. Liu, Behavior of Fatigue Cracks in a Tension Bolt, *Structural Integrity of Fasteners,* STP 1236, American Society for Testing and Materials, 1994, p 124–140
5-27. T.L. Mackay and B.J. Alperin, Stress Intensity Factors for Fatigue Cracking in High Strength Bolts, *Eng. Fract. Mech.,* Vol 21, 1985, p 391–397
5-28. D. Wilhem, J. Fitzgerald, J. Carter, and D. Dittmer, An Empirical Approach to Determining K for Surface Cracks, *Adv. Fract. Res.,* Vol 1, 1981, p 11–21
5-29. R.G. Forman and V. Shivakumar, Growth Behavior of Surface Cracks in the Circumferential Plane of Solid and Hollow Cylinders, *Fracture Mechanics: Seventeenth Volume,* STP 905, American Society for Testing and Materials, 1986, p 59–74

5-30. R.G. Forman and S.R. Mettu, Behavior of Surface and Corner Cracks Subjected to Tensile and Bending Loads in Ti-6Al-4V Alloy, *Fracture Mechanics: Twenty-Second Symposium,* Vol I, STP 1131, American Society for Testing and Materials, 1992, p 519–546
5-31. R.R. Cervay, "Empirical Fatigue Crack Growth Data for a Tension Loaded Threaded Fastener," Report AFWAL-TR-88-4002, Air Force Flight Dynamics Laboratory, Wright-Patterson Air Force Base, Dayton, OH, Feb 1988
5-32. S.R. Mettu, I.S. Raju, and R.G. Forman, "Stress Intensity Factors for Part-Through Surface Cracks in Hollow Cylinders," JSC Report 25685/LESC Report 30124, NASA Lyndon B. Johnson Space Center/Lockheed Engineering and Sciences Co. Joint Publication, July 1992
5-33. J.C. Newman, Jr., "Fracture Analysis of Surface and Through Cracks in Cylindrical Pressure Vessels," TN D-8325, National Aeronautics and Space Administration, Washington, D.C., Dec 1976
5-34. I.S. Raju and J.C. Newman, Jr., Stress-Intensity Factors for Internal and External Surface Cracks in Cylindrical Vessels, *J. Pressure Vessel Technol., (Trans. ASME),* Vol 104, 1982, p 293–298

SELECTED REFERENCES

Handbooks

- J.P. Gallagher, F.J. Giessler, A.P. Berens, and R.M. Engle, Jr., "USAF Damage Tolerant Design Handbook: Guidelines for the Analysis and Design of Damage Tolerant Aircraft Structures," Report AFWAL-TR-82-3073, Flight Dynamics Laboratory, Air Force Wright Aeronautical Laboratories, Wright-Patterson Air Force Base, Dayton, OH, May 1984
- A.F. Liu, Summary of Stress-Intensity Factors, *ASM Handbook,* Vol 19, *Fatigue and Fracture,* ASM International, 1996, p 980–1000
- Y. Murakami et al., Ed., *Stress-Intensity Factors Handbook,* Vol 1 and 2, Pergamon Press, 1987
- P.C. Paris and G.C. Sih, Stress Analysis of Cracks, *Symposium on Fracture Toughness Testing and Its Applications,* STP 381, American Society for Testing and Materials, 1965, p 30–83
- D.P. Rooke and D.J. Cartwright, *Compendium of Stress Intensity Factors,* Her Majesty's Stationery Office, London, U.K., 1976
- G. Shih, *Handbook of Stress-Intensity Factors,* Institute of Fracture and Solid Mechanics, Lehigh University, Bethlehem, PA, 1973
- H. Tada, P.C. Paris, and G.R. Irwin, *Stress Analysis of Cracks Handbook,* Del Research Corp., Hellertown, PA, 1973
- H. Tada, P.C. Paris, and G.R. Irwin, *Stress Analysis of Cracks Handbook,* Paris Productions, Inc., Saint Louis, MO, 1985

Mechanically Fastened Joints

- D.L. Ball, The Development of Mode I, Linear-Elastic Stress Intensity Factor Solutions for Cracks in Mechanically Fastened Joints, *Eng. Fract. Mech.,* Vol 27, 1987, p 653–681

- T.R. Brussat, S.T. Chiu, and M. Creager, "Flaw Growth in Complex Structure," Vol I, "Technical Discussion," Report AFFDL-TR-77-79, Air Force Flight Dynamics Laboratory, Wright-Patterson Air Force Base, Dayton, OH, 1977
- H. Reemsnyder, Fatigue of Mechanically Fastened Joints, *ASM Handbook,* Vol 19, *Fatigue and Fracture,* ASM International, 1996, p 287–294
- J.M. Waraniak and A.F. Liu, Fatigue and Crack Propagation Analysis of Mechanically Fastened Joints, Paper 83–0839, presented at the AIAA/ASME/ASCE/AHS 24th Structures, Structural Dynamics and Materials Conf. (Lake Tahoe, Nevada), 2–4 May 1983, synopsis appears in *J. Airc.,* Vol 21, 1984, p 225–226

Transition of Part-Through Crack or Corner Crack to Through-Crack

- R.M. Ehret, "Fracture Control Design Methods for Space Vehicles," Vol II, "Assessment of Fracture Mechanics Technology for Space Shuttle Applications," Report CR-134597, National Aeronautics and Space Administration, Lewis Research Center, 1974
- J.C. Ekvall, T. R. Brussat, A.F. Liu, and M. Creager, "Engineering Criteria and Analysis Methodology for the Appraisal of Potential Fracture Resistant Primary Aircraft Structure," Report AFFDL-TR-72-80, Air Force Flight Dynamics Laboratory, Wright-Patterson Air Force Base, Dayton, OH, 1972
- A.F. Grandt, Jr., J.A. Harter, and B.J. Heath, The Transition of Part-Through Cracks at Holes into Through-The-Thickness Flaws, *Fracture Mechanics: Fifteenth Symposium,* STP 833, American Society for Testing and Materials, 1984, p 7–23
- K. Katherisan, T.M. Hsu, and T.R. Brussat, "Advanced Life Analysis Methods—Crack Growth Analysis Methods for Attachment Lugs," Report AFWAL-TR-84-3080, Vol II, Air Force Flight Dynamics Laboratory, Wright-Patterson Air Force Base, Dayton, OH, Sept 1984
- J.M. Waraniak and A.F. Liu, "Fatigue and Crack Propagation Analysis of Mechanically Fastened Joints," Paper 83–0839, presented at the AIAA/ASME/ASCE/AHS 24th Structures, Structural Dynamics and Materials Conference (Lake Tahoe, Nevada), 2–4 May 1983, synopsis appears in *J. Airc.,* Vol 21, 1984, p 225–226

Chapter 6

Environmentally Assisted Crack Growth

This chapter discusses crack growth and fracture in environments or temperatures other than room-temperature air. Cyclic frequency, wave form, and hold time are closely associated with environmental and temperature effects. Discussions of these are included here. Emphasis is placed on understanding the crack growth mechanisms and methods for structural life assessment.

Stress-Corrosion Cracking

The conventional definition of *stress corrosion* applies to the chemical interaction between the corrosive media and the metal, whether it is a test coupon or a structural member. It simply means that the part is simultaneously subjected to stress and corrosion, except that the stress comes from a constant load over a substantial time period. While the part is submerged in the corrosive agent (e.g., salt water) and under a sustained load, intergranular corrosion pits generally are formed on the free surface of the part. Intergranular fracture will follow. Because grain boundary is a logical source and sink for chemical diffusion, the general conception always associates an intergranular fracture surface with stress corrosion. In fact, most engineering alloys do exhibit intergranular cracking in the case of stress corrosion. However, cracks can propagate along other form of dislocations inside the grain, e.g., a twin plane. Therefore transgranular fracture also can occur occasionally.

Becoming aware of the fracture mode, through metallurgical failure analysis, will certainly help engineers in developing new alloys, improving an existing alloy, or selecting an alloy for a certain application. References 6-1 and 6-2 and the *ASM Handbook,* Volume 19, contain a number of review papers on this subject. Because fracture mechanics analysts are only interested in the analytical aspects of predicting the residual strength of their structural components, it will make little difference to them whether the fracture path is intergranular or transgranular for a certain material in a certain environment. However, it is significant that cracks propagated by different modes progress at different rates. The state-of-the-art fracture mechanics analysis methodology does not have a good handle on this problem. Therefore, this chapter only touches on a few key points that are relevant to fracture mechanics analysis.

In fracture mechanics terminology, *stress-corrosion cracking* is not the same as stress corrosion. In the case of stress-corrosion cracking, it is assumed that the structural part contains a pre-existing crack. Certain environments can have a very pronounced effect on subcritical crack growth. The effect may be thought of as the promotion of time-dependent crack extension at stress intensity levels below K_{Ic}, or K_c. Processes commonly referred to as electrochemical and hydrogen embrittlement are examples of this. The scenario involves a cracked structure (or a test specimen containing a crack) that is submerged in a corrosive environment and subjected to a sustained load for a prolonged period. At some point the crack will start to grow. When the crack grows to a critical length, rapid fracture will occur. The final absolute value of the fracture toughness obtained in this manner (computed using final crack length and the sustained load level) will be the same as the fracture toughness (based on final crack length and fracture load) obtained from monotonically increasing load tests conducted in a nonaggressive environment.

Figure 6-1 illustrates the sustained load environmental crack growth behavior. This type of curve can be obtained by loading a series of specimens to various percentages of the monotonic fracture stress. In order words, each specimen is loaded to a certain percentage of the baseline critical stress intensity factor value for the same crack size. These loads are maintained until fracture (or retired after a very long time period, in the case of no failure). The initial stress intensity for a given specimen is calculated based on the initial precrack size and the sustained load applied to the specimen. The corresponding stress intensity factor is usually called K_i (or K_{Ii}). The time required to fracture depends on the applied stress level and the properties of the material. The times to failure for all specimens are recorded, and plotted against their corresponding K_i values. The apparent threshold level, indicating no failure below this K-level, is called the K_{ISCC} (for plane strain) or K_{SCC} (for plane stress).

Some investigators believe that K_{SCC} does not exist. One possible reason is that thin sheet is not likely to promote flat fracture surfaces that are intrinsically associated with stress-corrosion cracking. Figure 6-2 indicates that the susceptibility to stress-corrosion cracking decreases with decreasing thickness and that there is a critical thickness below which stress-corrosion cracking does not occur. This critical thickness is dependent on the alloy composition, heat treatment, orientation, and loading rate. Apparently, the critical thickness relates to the transition from plane strain to plane stress. There is only a small quantity of test data available, and it cannot definitely prove or disprove the existence of K_{SCC}.

Results of tests conducted on various specimen geometries indicate that K_{ISCC} is a material property. Thus, K_{ISCC} may be used as a measure of the susceptibility of a material to stress-corrosion cracking in a given environment. Some actual examples (for Ti-6Al-4V and D6AC steel) are shown in Fig. 6-3. Certainly, K_{ISCC} depends on material processing. For a given material composition, thermomechanical treatment, and environment, K_{ISCC} may be different for different orientations of the test specimen. For example, in aluminum alloys, susceptibility to stress-corrosion cracking is much greater in the short-transverse direction than in any other grain direction. Therefore, extreme care should be taken to use the correct value of K_{ISCC} for a specific application.

During an environmental sustained-load test, the increments of the crack extension may be recorded so that a curve for crack length versus time is

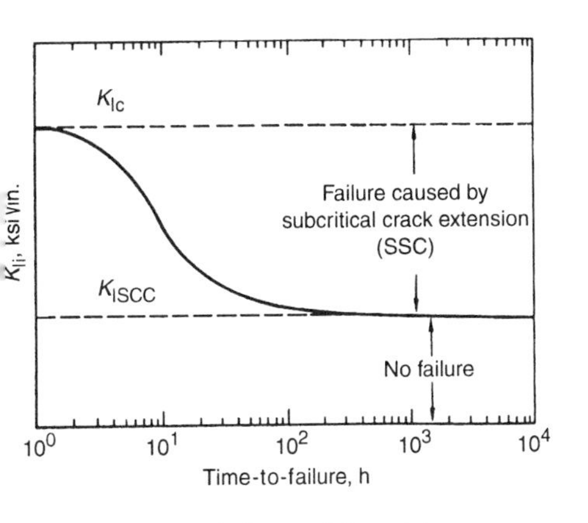

Fig. 6-1 Schematic time-to-failure data

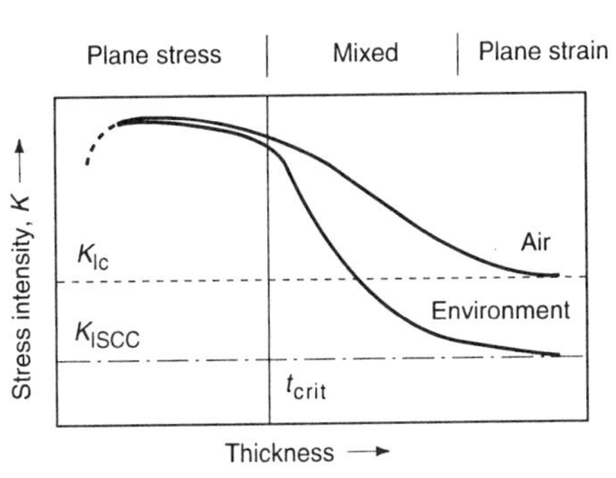

Fig. 6-2 Schematic plot showing the effect of specimen thickness on the stress-corrosion cracking susceptibility of titanium alloys. Note: Stress-corrosion cracking does not occur when $t < t_{crit}$. Source: Ref 6-3

obtained. A typical curve for crack extension versus time is shown in Fig. 6-4. From the curve of crack length versus time obtained from the same test, a crack growth rate curve (*da/dt* versus *K*) can be plotted. It has been shown that this curve is a basic material property. The relationship between *da/dt* and *K* may be divided into three regions (see Fig. 6-5). In stage 1, *da/dt* is highly sensitive to changes in *K*, exhibiting a threshold below which cracks do not propagate. This threshold value of *K* corresponds to K_{ISCC}. In this region, crack growth is controlled by the interactive effects of the mechanical and chemical driving forces. Stage 2 represents crack growth above K_{ISCC}. Crack growth behavior in this region is bounded by the solid and the dotted line shown in the figure. In many alloy/environment systems, rates of crack growth appear to be independent of *K;* that is, *da/dt* is constant regardless of *K* (the dotted line behavior shown in the figure). In such cases the primary driving force for crack growth is no longer mechanical but is related to the chemical corrosion processes occurring at the crack tip. In stage 3, crack growth rate is again strongly dependent on *K*, increasing

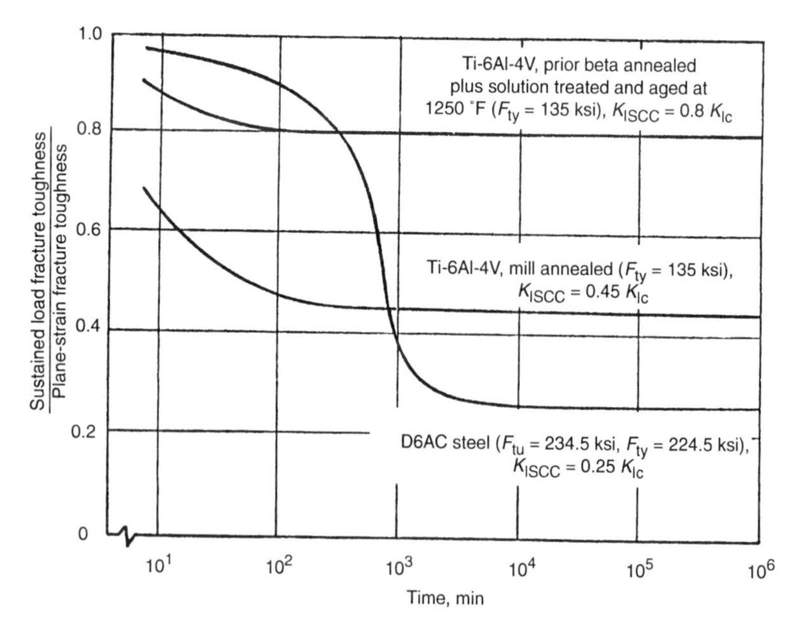

Fig. 6-3 Effect of environment (3.5% salt solution) and time at sustained load on fracture toughness for titanium alloys and steel. Source: Ref 6-3, 6-4

rapidly as K approaches K_{Ic}. In this stage, crack growth is mechanically controlled, unaffected by environment.

Corrosion Fatigue Crack Propagation

Corrosion fatigue crack propagation is no different from ordinary fatigue crack growth, except that the test is run in a corrosive environment (e.g., salt water, high-humidity air, etc.). Therefore, it is common practice to conduct tests and record and implement the *da/dN* data as discussed in Chapter 3. Comprehensive documentation of corrosion fatigue testing is available (Ref 6-7). However, keep in mind that corrosion fatigue crack growth testing uses a precracked specimen, and the pitting process required in a conventional corrosion fatigue test is skipped. As far as the microscopic crack propagation path and fracture surface appearance are concerned, it is generally agreed that the fracture path in a traditional corrosion fatigue specimen (repeated cyclic loading in a corrosive medium) is transgranular. There are exceptions. For example, after several specimens of the same alloy were

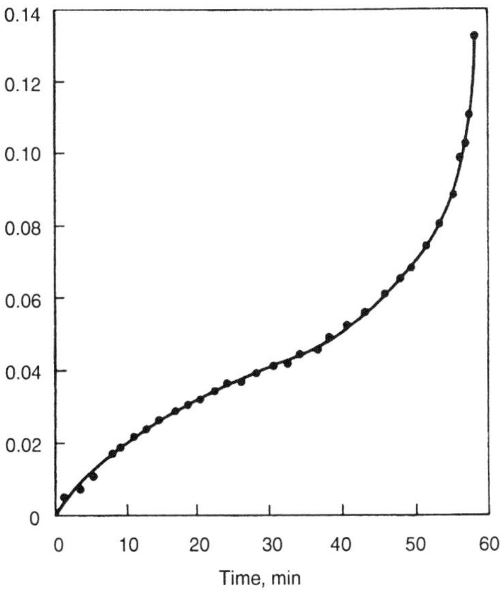

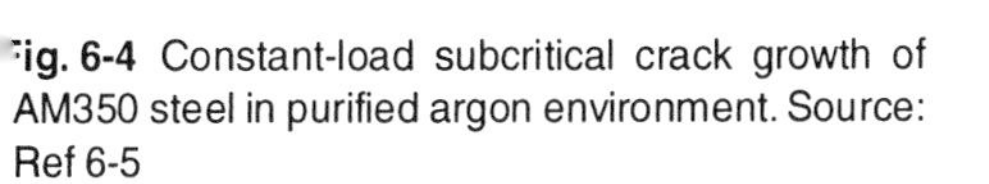
Fig. 6-4 Constant-load subcritical crack growth of AM350 steel in purified argon environment. Source: Ref 6-5

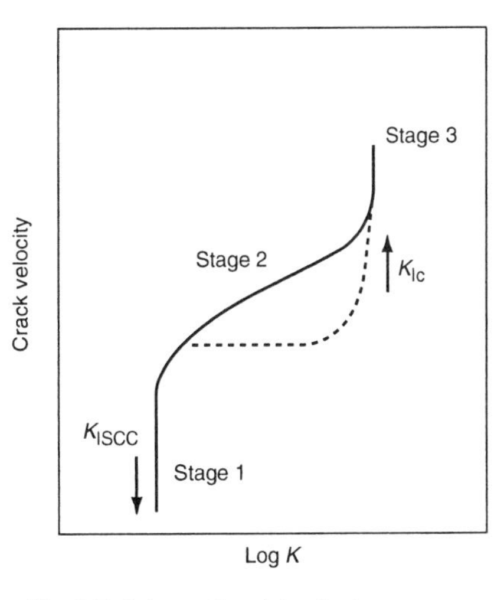

Fig. 6-5 Schematic plot of stress-corrosion cracking velocity against stress intensity factor. Source: Ref 6-6

tested in the same environment (7079-T651 tested in NaCl solution), the fracture paths were either transgranular, intergranular, or mixed (Ref 6-8). These disturbing results are probably attributable to variation in the degree of stress-corrosion cracking component in an individual test. When a specimen exhibits time-dependent behavior (because it is tested at a very low frequency or subjected to a long hold time), its fracture path may have a tendency toward intergranular cracking.

Changes in macroscopic fracture surface appearance due to an aggressive environment have been reported by Vogelesang (Ref 6-9). He showed that the fracture surface of 7075-T6 aluminum ($t = 6$ mm) tested in vacuum is primarily a shear mode fracture. In increasingly aggressive environments, a transition to flat fracture is observed (Fig. 6-6). Vogelesang has suggested that the mode of failure depends on whether the applied shear stress or the applied normal stress first reaches a critical value at the crack tip. An aggressive environment may reduce critical normal stress, because decohesion is facilitated by such mechanisms as adsorption, dissolution, and oxidation. Thus, fracture planes appear to be perpendicular to the tensile stress axis. In the absence of aggressive environment, crack propagation can only occur by slip deformation, leading to a shear lip fracture surface. In the

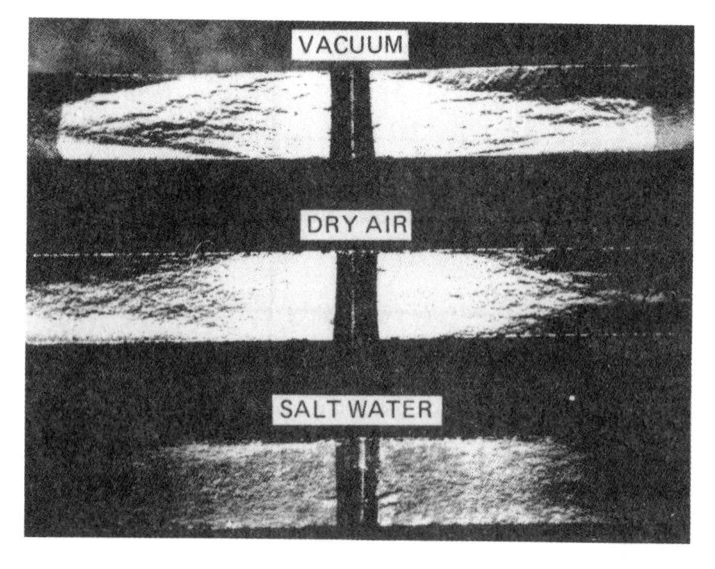

Fig. 6-6 The transition from the tensile mode fracture (salt water) to the shear mode fracture (vacuum) in three environments. Source: Ref 6-9

following sections we will concentrate on the discussion of crack growth mechanisms and methods to handle the effects of frequency, wave form, hold time, and environment on constant-amplitude crack growth rate.

Crack Growth Mechanism

According to McEvilly and Wei (Ref 6-10), corrosion fatigue crack propagation exhibits three basic types of crack growth rate behavior (Fig. 6-7). True corrosion fatigue describes the behavior when fatigue crack growth rates are enhanced by the presence of aggressive environment at levels of K below K_{ISCC} (Fig. 6-7a). This behavior is characteristic of materials that do not exhibit stress corrosion, i.e., $K_{ISCC} = K_{Ic}$. Stress-corrosion fatigue describes corrosion under cyclic loading that occurs whenever the stress in the cycle is greater than K_{ISCC}. This is characterized by a plateau in crack growth rate (Fig. 6-7b) similar to that observed in stress-corrosion cracking (Fig. 6-5). The most common type of corrosion fatigue behavior, shown in Fig. 6-7(c), is characterized by stress-corrosion fatigue above K_{ISCC}, superimposed on true corrosion fatigue at all stress intensity levels.

For most of the commonly used alloys, the effect of frequency on constant-amplitude crack growth rate is probably negligible in dry air environ-

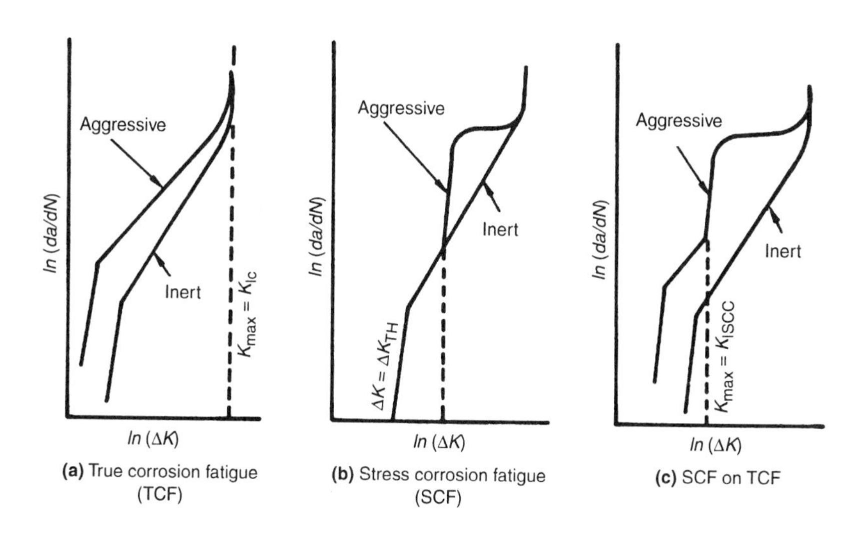

Fig. 6-7 Schematic diagrams showing three basic types of corrosion fatigue crack growth behavior. Source: 6-10

ments. A lightly corrosive environment (humid air) gives rise to higher crack growth rates than a dry environment. The cyclic crack growth rate varies significantly at elevated temperature and in chemically aggressive environments. The effects of frequency, wave form, and hold time on constant-amplitude crack growth rate are generally magnified by the presence of a corrosive medium. The faster the testing speed, the slower the fatigue crack growth rate (or the longer the hold time, the faster the growth rate). Although there is no concurrence of opinion as to the reason for the environmental effect, it is certainly due to corrosive action. As a result, the influence of the environment is time- and temperature-dependent.

In studying the environmental fatigue crack growth behavior in titanium, Dawson and Pelloux have classified environments into three groups according to the possible influence of loading frequency (Ref 6-11). These models are shown schematically in Fig. 6-8. In normally inert environments, such as vacuum, helium, argon, or air, fatigue crack growth rates in titanium exhibit no effect of frequency (Fig. 6-8a). In liquids such as methanol, a "normal" frequency effect is found, in that higher fatigue crack growth rates are found at lower frequencies (Fig. 6-8b). In halide-containing solutions such as salt water, "cyclic stress-corrosion cracking" with a characteristic discontinuity in the da/dN versus ΔK curve is found (Fig. 6-8c). The lower the loading frequency, the lower the ΔK value at which the discontinuity is observed,

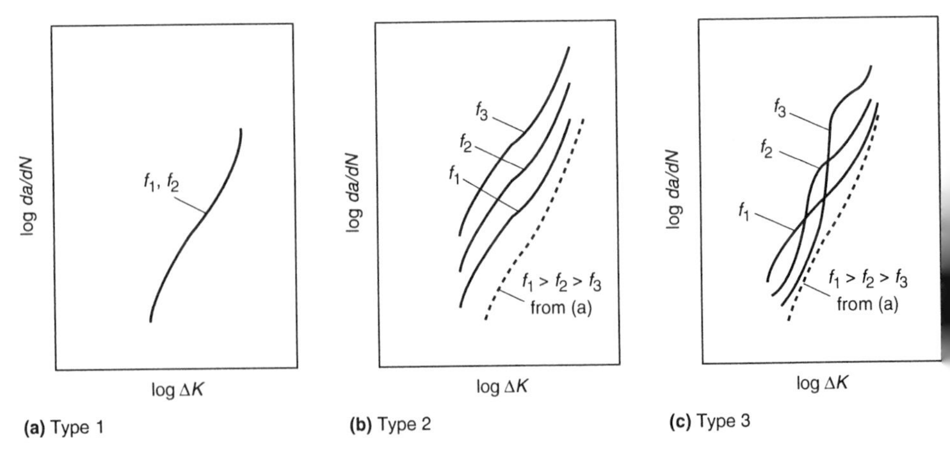

Fig. 6-8 Influence of loading frequency on fatigue crack growth for three classes of environments (schematic). (a) Little or no effect of frequency, as in vacuum, inert gas, or air. (b) Fatigue crack growth increases with decreasing frequency, as in methanol. (c) Cyclic stress-corrosion cracking effect, as in salt water. Source: Ref 6-11

where the limiting value is such that $K_{max} = K_{ISCC}$. Figure 6-9 illustrates the corrosion fatigue behavior for X-65 line pipe steel tested in salt water with a superimposed cathodic potential. This behavior is somewhat close to type 3 of Fig. 6-8. When the same data are compared to the generalized model of McEvilly and Wei, they fall somewhere between types B and C behavior in Fig. 6-7.

Diffusion is perhaps the one important mechanism that has not received close attention from fracture mechanics analysts. Penetration of chemical agent into a localized crack tip region would severely damage the bulk of material near the crack tip. The damaged material no longer represents the original bulk material and exhibits accelerated crack growth rates as compared to its baseline behavior. This *damaged zone* is called the *environment-affected zone (EAZ)* by some investigators. Theoretically, the size of the EAZ can be calculated using a textbook formula, i.e., the width of the EAZ (the diffusion distance) is expressed as:

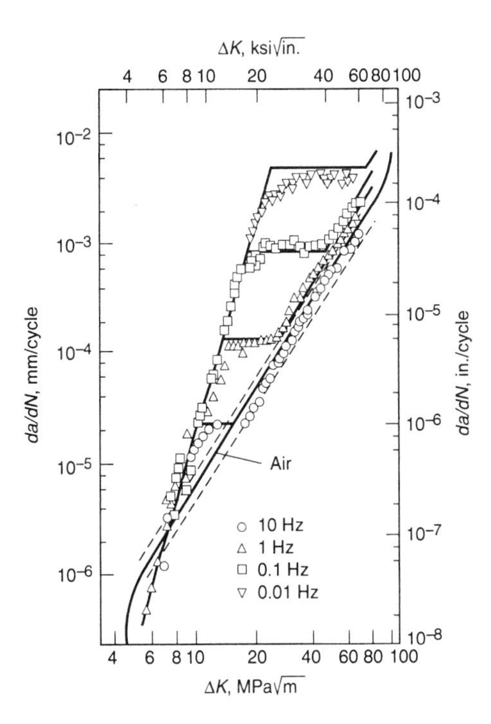

Fig. 6-9 Crack growth in X-65 line pipe steel exposed to air and 3.5% salt water with a superimposed catholic potential. Source: Ref 6-12

CATHODIC

THE POPE DIDN'T PONTIFICATE THE POTENTIAL

HARTBELLE 8/2015

$$r_e = \sqrt{Dt} \qquad \text{(Eq 6-1)}$$

where t is hold time, and the diffusion coefficient D is a function of the activation energy Q and the gas constant R (8.314 J/K), and is given by:

$$D = w^2 \cdot v' \cdot \exp(-Q/RT) \qquad \text{(Eq 6-2)}$$

where T is the absolute temperature in K, w is the jump distance, and v' is the jump frequency. The value for Q (in J or cal/mole) is sometimes available in the open literature. It is easy to understand that Eq 6-2 is rather academic, not meant for day-to-day use in solving engineering problems.

At the present time, experimental testing is the only means for determining the EAZ size and the crack growth behavior inside the EAZ. In the case of high-temperature crack growth testing, oxidation due to penetration of hot air into a localized crack tip region is considered the primary source for creating the damaged zone. A typical example, taken from Ref 6-13, is shown in Fig. 6-10. The data were developed from the Inco 718 alloy, tested at 590 °C. The loading profile consists of one trapezoidal stress cycle (i.e., with hold time) and many triangular or sinusoidal stress cycles of the same magnitude, applied before and after the trapezoidal load. The test was conducted in such a way that a constant K-level was maintained at each stress cycle. A record of crack lengths versus stress cycles was taken (Fig. 6-10). This type of data can be reduced to a plot of crack growth rate versus crack lengths, as shown in Fig. 6-11. The purpose of making this plot is to show the influence of the sustained load on crack growth rate, so the crack length immediately following the sustained load cycle is set to 0. It appears that the data points are divided into two parts, each of which can be approximated by a straight line. The first part characterizes the crack growth behavior inside the EAZ, where the crack growth rate is initially higher than normal because the material has been damaged by the hot air. The crack growth rates corresponding to those post-sustained-load cycles gradually decrease to a level that is otherwise normal for an undamaged material. Thus, the horizontal line implies that the crack traveled through the EAZ and resumed a stable crack growth behavior. The intersection of these two lines determines the size of the EAZ. Many test data of this type are reported in Ref 6-13, covering a range of combinations of temperature, hold time, and stress intensity level. Figures 6-10 and 6-11 are typical of the data reported.

Analytical Methods

There are many analytical methods that offer a handle on estimating environmental fatigue crack growth rates as functions of frequency and hold time. Several commonly known models are listed here as Ref 6-14 to 6-19. The Wei-Landes superposition principle (Ref 6-14) has been universally adopted by the fracture mechanics community. The Saxena model (Ref 6-19) works well with tests that are associated with variation of frequency alone. The other models are more or less variations of the Wei-Landes superposition model, but heavily involved with empirical constants. In this section, only the Wei-Landes and the Saxena models are presented.

The Principle of Superposition

In the Wei-Landes method, the rate of fatigue crack growth in an aggressive environment is considered to be equal to the algebraic sum of the rate in an inert reference environment and that of an environmental component, computed from sustained-load crack growth data obtained in an identical aggressive environment and the load profile represented by $K(t)$. The total

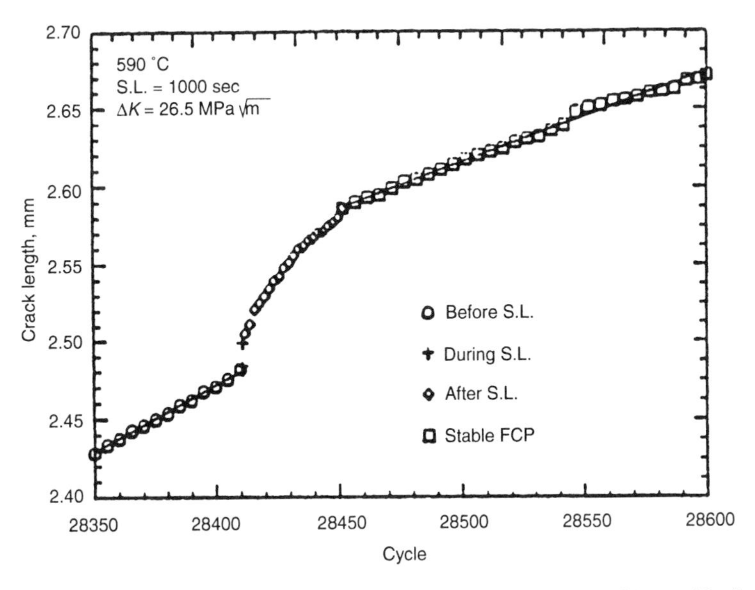

Fig. 6-10 Effect of long hold time on constant-amplitude crack growth of Inco 718 alloy. Source: Ref 6-13

crack growth rate in an aggressive environment ($\Delta a/\Delta N$) can be expressed as (Ref 6-14):

$$\frac{\Delta a}{\Delta N} = \left(\frac{\Delta a}{\Delta N}\right)_{\mathrm{i}} + \int \frac{da}{dt} \cdot K(t) \cdot dt \qquad \text{(Eq 6-3)}$$

where $(\Delta a/\Delta N)_{\mathrm{i}}$ is the crack growth rate in an inert reference environment, usually room-temperature air. The term da/dt is the sustained load crack growth rate in an identical aggressive environment. It can be determined from test data obtained from sustained-load tests. The function $K(t)$ represents the load profile as a function of time. The effects of frequency and loading variable are incorporated through $K(t)$. In a constant-amplitude, sinusoidal loading condition, $K(t)$ can be written as:

$$\mathrm{K}(t) = \frac{K_{\max}}{2}\left[(1+R) + (1-R)\cdot \cos \omega t\right] \qquad \text{(Eq 6-4)}$$

This expression accounts for the effects of frequency, mean load, range of cyclic loads, and the hold time wave form on the sustained-load growth

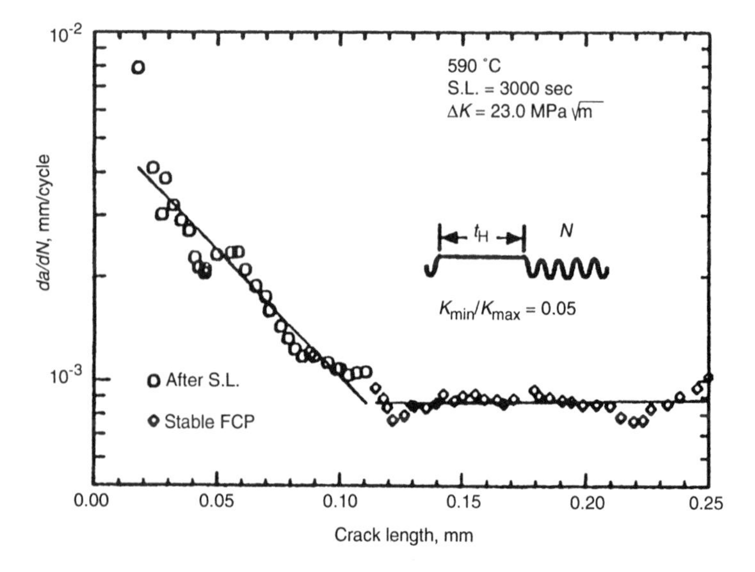

Fig. 6-11 Measured *da/dN* after a sustained load cycle. Source: Ref 6-13

component. The computational procedure is illustrated schematically in Fig. 6-12. An example comparing the predictions with actual test data is shown in Fig. 6-13. It is seen that except in one test, excellent correlations have been obtained. The case that showed a poor correlation was tested at 1 Hz (the fastest among all the tests), indicating the limit of this analytical model. As will be discussed in a later section of this chapter, a similar limitation also occurs on correlating the prediction with test data of the Inco 718 alloy at high temperature. It is also noteworthy that the crack growth behavior of Fig. 6-13 seems to match the type 3 behavior of Fig. 6-8.

In any event, it is thought that this approach allows predictions of corrosion fatigue crack growth rates simply by adding the inert fatigue crack growth rate per cycle and the crack extension due to stress corrosion. Therefore it eliminates the work needed to conduct fatigue crack growth tests in a corrosive environment.

Later, Wei modified Eq 6-3 to a general form (Ref 6-15):

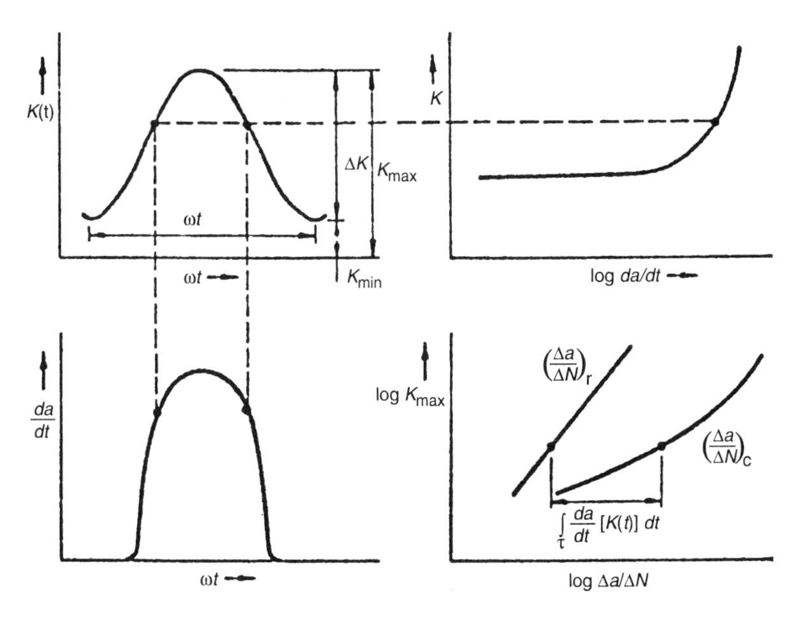

Fig. 6-12 Schematic diagram illustrating the suggested method of analysis. (a) Stress intensity spectrum (top left). (b) Rate of crack growth under sustained load in an aggressive environment (top right). (c)Environmental contribution to crack growth in fatigue (bottom left). (d) Integrated effects on environment and K_{max} on fatigue crack growth rate (bottom right). Source: Ref 6-14

$$\frac{\Delta a}{\Delta N} = \left(\frac{da}{dN}\right)_{\mathrm{i}} + \left(\frac{da}{dN}\right)_{\mathrm{e}} + \left(\frac{da}{dN}\right)_{\mathrm{SCC}} \quad \text{(Eq 6-5)}$$

or:

$$\frac{\Delta a}{\Delta N} = \left(\frac{da}{dN}\right)_{\mathrm{i}} + \left(\frac{da}{dN}\right)_{\mathrm{e}} + \int \frac{da}{dt}\, dt \quad \text{(Eq 6-6)}$$

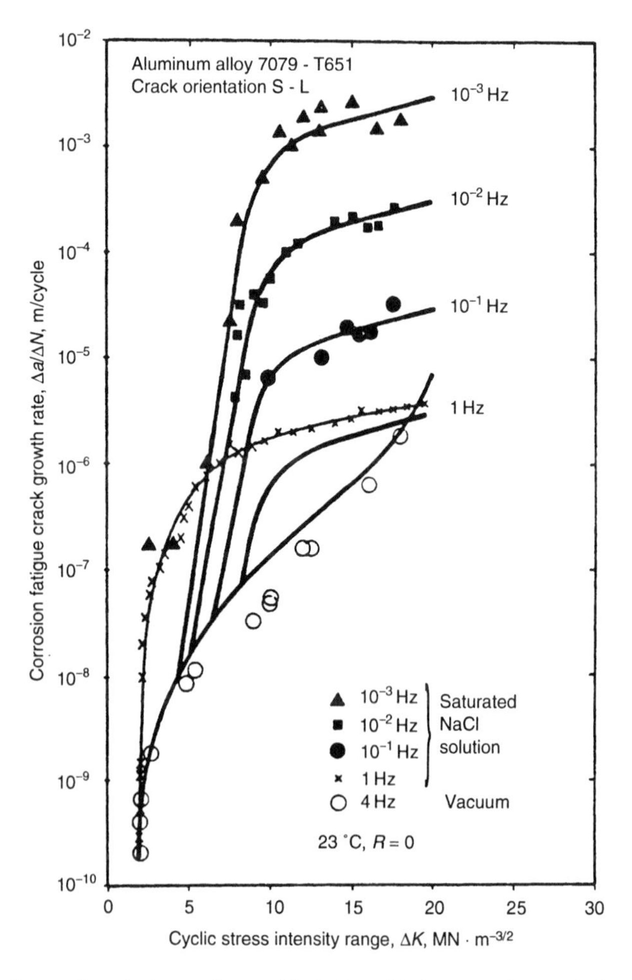

Fig. 6-13 Comparison between theory (Eq 6-3) and fatigue crack growth data of 7079-T651 aluminum alloy tested in vacuum and saturated NaCl solution. Source: Ref 6-8

where $(da/dN)_i$ is for "pure" fatigue in inert environment, $(da/dN)_e$ accounts for the contribution from cyclic loading in environmental exposure, and $(da/dN)_{SCC}$ is the contribution by sustained-load crack growth at K levels above K_{ISCC}. In other words, the integral term signifies the crack growth due to sustained load, which is a function of K_{max}.

Most of the test data used by Wei and Landes for developing Eq 6-3 and 6-5 were from constant-amplitude tests. On the basis of the data gathered, the steady-state response of fatigue crack growth to environments may be grouped into three basic types in relation to K_{ISSC}, as previously shown in Fig. 6-7:

1. The type A behavior is typified by the aluminum-water system. Environmental effects result from interaction of fatigue and environmental attack. The integral term in Eq 6-5 vanishes for this type of behavior.
2. The type B behavior is typified by the hydrogen-steel system. Environmental crack growth is directly related to sustained-load crack growth, with no interaction effects. The second term of Eq 6-5 vanishes for this type of behavior.
3. The type C behavior is the behavior of most alloy-environment systems. Above K_{ISSC} the behavior approaches that of type B, whereas below K_{ISSC} the behavior tends toward type A, with the associated interaction effects. The transition between the two types of behavior is not always sharply defined. In this case the second term on the right side of Eq 6-5 vanishes when K_{max} is above K_{ISSC}. The third term becomes zero when K_{max} is below K_{ISSC}.

The Effect of Load Frequency

Although the Saxena model was originally designed for high-temperature applications, it is conceivable that it is equally applicable to corrosive environments because temperature is, after all, an aspect of environment. In Ref 6-19 the model for handling the frequency effects has three regions: a frequency-independent region, a mixed region, and a frequency-dependent region (Fig. 6-14). There is no frequency effect in region I. Thus, crack growth rate can be expressed in a conventional way:

$$\frac{da}{dN} = \left(\frac{da}{dN}\right)_0 \qquad \text{(Eq 6-7)}$$

for $f \geq f_0$, where f_0 is the characteristic frequency that separates the cycle-dependent and mixed regions. The subscript "0" in Eq 6-7 stands for the reference frequency, f_0.

In region II, i.e., the mixed region ($f_0 > f \geq f_1$), the time dependence of *da/dN* becomes significant. The crack growth rate at a frequency f can be expressed as:

$$\frac{da}{dN} = \left(\frac{da}{dN}\right)_0 + \left[\int_0^{1/f} \left(\frac{da}{dt}\right) dt - \int_0^{1/f_0} \left(\frac{da}{dt}\right) dt\right] \qquad \text{(Eq 6-8)}$$

The third term on the right side (the second term inside the bracket) is added to Eq 6-8 because there is no time-dependent effect at and above a frequency of f_0, and the term inside the bracket should reduce to zero. According to Ref 6-20, Eq 6-8 eventually reduces to:

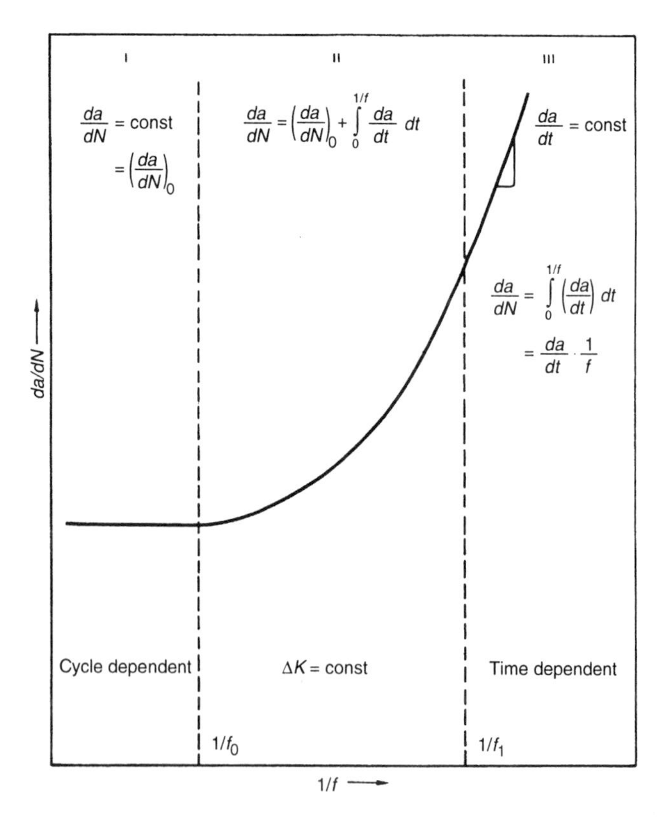

Fig. 6-14 Schematic representation of fatigue crack growth behavior with frequency. Source: Ref 6-19

$$\frac{da}{dN} = \left(\frac{da}{dN}\right)_0 + C_4 \cdot \Delta K^{\alpha} \cdot [1/\sqrt{f} - 1/\sqrt{f_0}] \qquad \text{(Eq 6-9)}$$

To determine the constants in Eq 6-9, *da/dN* versus ΔK data for two frequencies are required. The steps involved in determining these constants are shown in Fig. 6-15. In Fig. 6-15(a), the data for two frequencies, f_0 and f_2, are plotted schematically. The best-fit straight line on a log-log plot for the reference frequency f_0 is obtained by conducting a simple linear regression analysis of the *da/dN* versus ΔK data. This line defines the first term on the right side of Eq 6-9, the conventional Paris power law regime. Next, the differences in growth rates for the two frequencies are determined at several ΔK-levels, as shown in Fig. 6-15(a). These data points are then plotted as shown in Fig. 6-15(b). Knowing the slope α and the intercept C' for this new regression line, the value of C_4 can be calculated using:

$$C_4 = C' \cdot [1/\sqrt{f_2} - 1/\sqrt{f_0}]^{-1} \qquad \text{(Eq 6-10)}$$

At low frequency, i.e., $f < f_1$, *da/dt* reaches a steady value and remains constant. This is the time-dependent region (i.e., region III). Crack growth rate in this region can be described by:

$$\frac{da}{dN} = \frac{da}{dt} \cdot \frac{1}{f} \qquad \text{(Eq 6-11)}$$

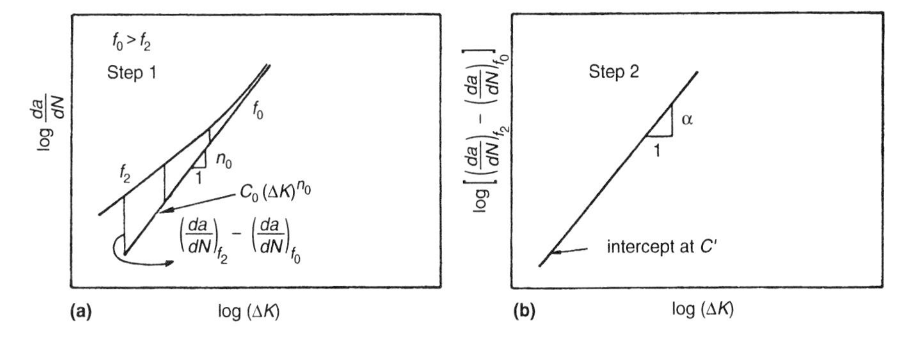

Fig. 6-15 Schematic representation of the steps involved in determining the empirical constants in Eq 6-9. Source: Ref 6-19

Load and Environment Interaction

Depending on service application, predicting the spectrum crack growth life of a structural member may become an important element within the general scope of life assessment. In Chapter 3 we hypothesized that the crack tip plastic zone is responsible for the crack growth behavior under spectrum loading. Consider two identical tests, one in dry air and another in a corrosive medium. For simplicity, assume that the loading profile consists of one high load followed by a group of lower load cycles of the same magnitude. By definition, the plastic zone size at a given K-level does not change regardless of environment, i.e., plastic zone size is a function of K only. However, the amount of retardation will be less in the corrosive environment than in dry air for the same stress intensity range and overload ratio. In other words, in the corrosive environment fewer cycles are required for a fatigue crack to propagate through the overload plastic zone, because crack growth rates are higher. Thus, the difference between the amount of retardation in dry air and in a corrosive environment appears to be caused simply by a faster crack growth rate in a corrosive environment. However, published test data have shown mixed results. Some of these phenomena are discussed below.

When aluminum alloys are tested in a corrosive medium (as opposed to room-temperature air or vacuum), the number of delay cycles, N_D, caused by an overload will be decreased (Ref 6-21, 6-22). As for titanium (Ti-6Al-4V, mill annealed), N_D will be increased in 3.5% NaCl solution, or nearly unchanged in moist air with 30 to 60% relative humidity or in distilled water as compared to dehumidified argon (Ref 6-23).

Cyclic frequency and hold time may further magnify the environmental effect due to the time factor involved. For tests conducted in corrosive environment, the hold time would have helped to create an EAZ and consequently would have altered the baseline da/dN property of the bulk of material inside the EAZ. The scenario is that da/dN is even higher in the EAZ (as compared to the da/dN for the undamaged material). Therefore, it would be logical to assume that retardation would be further reduced owing to hold time. On the basis of limited data, cyclic frequency has an insignificant effect on single-overload crack growth retardation. However, the multiple-overload test results of titanium in salt water strongly indicate that N_D is frequency dependent (Ref 6-23). It seems that the slow frequencies associated with the multiple-overload cycles could have caused an effect equivalent to that of hold time. That is, decreasing the frequency of a group of cycles or increasing the hold time of a single cycle will result in an increased number of delay cycles. As previously mentioned in Chapter 3, when titanium is tested in room-temperature air without corrosive agent, N_D may be increased or decreased, depending on where the hold time is being applied. Unfortunately, test data that would show the effect due to the combination of hold time and corrosive medium are not available.

According to Ref 6-22, hold time at zero load (below K_{min}) would cause significant reduction of N_D as compared to no hold time. For aluminum alloys, such as 2219 and 7075 (peak aged or overaged), tested in humid air,

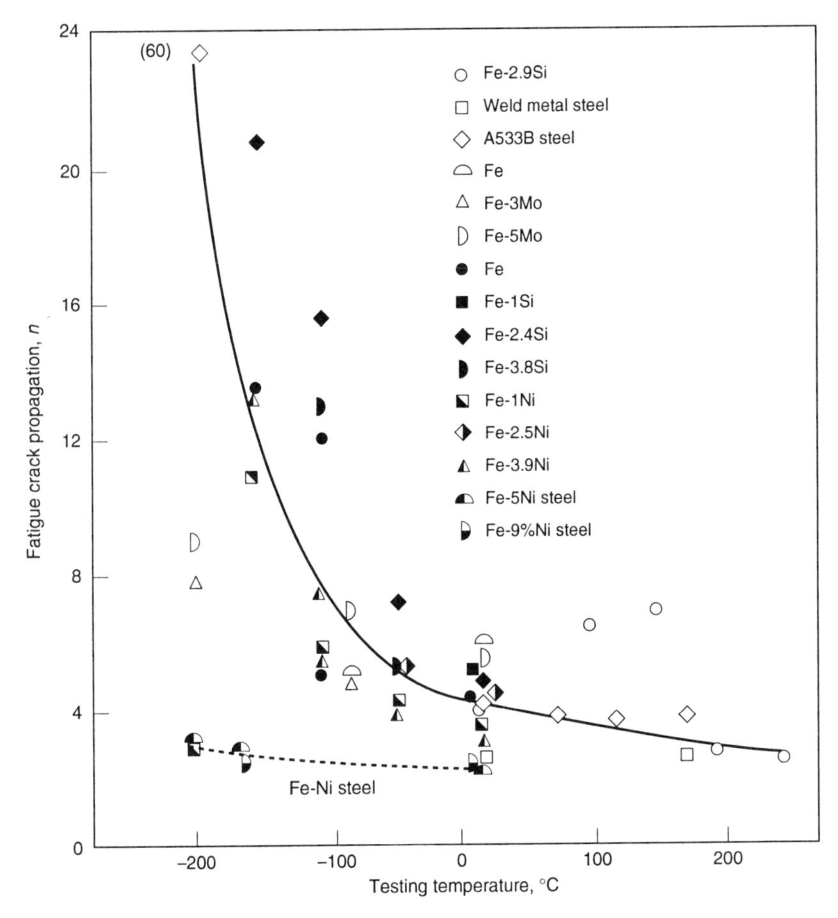

Fig. 6-16 Influence of testing temperature on fatigue crack propagation exponent for iron-base alloys. Source: Ref 6-24

N_D decreases as hold time increases. Holding the constant load level at K_{mean} does not alter any crack growth retardation characteristics in aluminum (Ref 6-21). Again, unfortunately, there are no data showing the effect of hold time at the peak of an overload for aluminum.

In summary, the information gathered here is by no means complete; more research is needed in developing a quantitative database to systematically identify the post-sustained-load phenomena. In any event, it is being realized that suitable models for delay must not only incorporate the effects of loading variables, but also account for material property variations. This type of information is essential for the development of a reliable crack growth life assessment method that can handle load and environment interaction.

Crack Growth at Low Temperature

So far, gathering low-temperature fatigue crack growth rates data and trying to understand exactly what happened (in a certain way) is the only type of documentation available in the open literature. At low temperatures the reaction kinetics are slower and the air can contain less water vapor. This may reduce crack growth rates in certain alloys. Sometimes the effect of temperature on fatigue crack growth in the low-temperature range is very small. However, for some alloys (e.g., steels in particular) that exhibit ductile-to-brittle transition at low temperature, crack growth rates can be very high when brittleness causes "cleavage" during crack propagation. Figure 6-16 gives an example, demonstrating the increase in the crack growth rate exponent with decreasing temperature for a number of iron-base alloys and steels. In any event, using the actual da/dN and fracture toughness data for material evaluation and selection and performing life assessment is the only way to deal with low-temperature crack growth and fracture.

Crack Growth at High Temperature

Current fracture mechanics theory treats cyclic crack growth as a linear elastic phenomenon. The residual strength of a test coupon, or a structural component, is frequently computed based on linear elastic fracture indices. Elastic-plastic or fully plastic analysis such as the J-integral approach is used when large-scale yielding occurs. All the existing crack growth analysis methods for spectrum life prediction basically deal with using material constant-amplitude crack growth rate data to compute the crack growth history of a structural element. For crack growth at high temperature, the conventional crack growth methodology that was based on material room-temperature behavior will no longer be applicable.

The need for an updated fracture mechanics technology that can handle the combined effects of temperature, stress amplitude, cyclic frequency, and dwell time was first recognized by researchers in the nuclear and aircraft engine industries and government agencies. Substantial research efforts have been made since the mid-1970s. The accomplishments have been documented in a number of review papers and books; some of them are listed here as Ref 6-25 to 6-35. Testing techniques and data reduction procedures are given in Ref 6-34, 6-36, and 6-37. Application of these new technologies to damage tolerance analysis of aircraft structures is discussed in Ref 6-38 to 6-41.

The key products developed from mid-1970 to 1996 include:

- New fracture mechanics indices for characterizing material residual strength and sustained-load crack growth at high temperature, namely, the steady-state creep parameter, C^* (Ref 6-42), and the transient creep parameter, C_t (Ref 6-43)
- A large quantity of test data revealing the various aspects of high-temperature crack growth behavior
- A comprehensive document on the application of the data in life prediction (Ref 6-44)
- A comprehensive document of life assessment methods on a component-specific basis for boilers, turbines, pressure vessels, and advanced steam plants (Ref 6-45)
- Computer models for constant-amplitude crack growth at high temperature (Ref 6-46, 6-47)
- An updated computer code for spectrum crack growth life prediction (Ref 6-48)

This section discusses the variables that affect the material crack growth rate behavior and the essential elements in making life prediction. In accord with the aforementioned review articles, creep crack growth can be characterized in terms of fracture mechanics crack growth by the steady-state parameter C^* (in the large-scale crack growth regime) and the transient parameter C_t. Under small-scale creep, C_t is designated as $(C_t)_{SSC}$ and is theoretically equivalent to K (the crack tip stress intensity factor). It becomes C^* when the amount of creep deformation approaches a steady-state creep condition. Therefore, C_t can be used for the entire range of creep deformations. Because K is a more convenient parameter and all the crack growth analysis methodologies have been developed based on K, it is better to use K in lieu of C_t when the situation permits.

If a material is creep resistant or environment sensitive, its creep zone will be smaller than the creep zones in the creep ductile materials. After reviewing and analyzing a vast amount of creep crack growth data (Table 6-1), it can be concluded that most high-temperature superalloys are creep resistant and environment sensitive. The linear elastic index K may, in some cases, adequately characterize the high-temperature crack growth behavior in these alloys.

In the following paragraphs both the C_t-based and the K-based approaches will be presented. There is also discussion of some elements that are required in formulating a cumulative damage model to account for the effect of variable-amplitude/variable-shape stress cycles.

Characterization Parameters

Depending on material, temperature, time, and environment, crack growth at high temperature can exhibit various degrees of creep deformation at the crack tip region. Consider that creep resistance of a material is characterized by the coefficient and exponent of the relationship of minimum creep rate to stress:

$$\dot{\varepsilon} = A \cdot \sigma^n \quad \text{(Eq 6-12)}$$

The degree of creep deformation can be described as *small-scale creep*, *transient creep*, or *steady-state creep*. Under small-scale creep, the creep zone is small in comparison to the crack length, and the size of the body and its growth is constrained by the surrounding elastic material. Under steady-state creep, the creep zone engulfs the entire untracked ligament of the material (e.g., between the crack tip and the free edge of the test coupon). The transient creep condition represents an intermediate level of creep deformation.

Theoretically, these three levels of creep deformation are analogous to the three levels of crack tip plasticity in the subcreep temperature regime: the small-scale yielding level in which the load-carrying behavior of the specimen (or component) is linear, the fully plastic level, and the intermediate (elastic-plastic) level. Therefore, under small-scale creep the driving force for crack growth will be K. Under steady-state creep, crack growth occurs essentially by creep. The fracture index will be the energy rate line integral C^*, which is considered to be the counterpart of J_p for a fully plastic body. Generally speaking, the controlling parameter depends on the level of creep

Table 6-1 Controlling parameters for creep crack growth analysis

Alloy	Temperature, °C	Controlling parameter	Other parameters attempted	Remarks	Reference
Aluminums					
RR58	150–200	C^*	K	...	6-62
2219-T851	150	K	...	...	6-63
2219-T851	175	K	σ_{nom}, σ_{net} σ_{ref}, C^*	...	6-70
Low-alloy steels					
A470 Class 8	482, 538	C^*, C_t	K	...	6-43, 6-64
1.0Cr-0.5Mo	535	C^*	K	...	6-66
1.0Cr-0.5Mo	525	C^*	K, J, σ_{ref}	...	6-68
1.0Cr-1.0Mo-.25V	427–538	C_t	K, $C(t)$, C^*	...	6-54
0.5Cr-0.5Mo-.25V	540	C^*, C_t	$C(t)$	(a)	6-71
2.25Cr-1.0Mo	540	C^*, C_t	$C(t)$	(a)	6-71
1.25Cr-0.5Mo	482, 538	C_t	...	(a)	6-67
0.16C	400, 500	C^*	σ_{net}	(b)	6-65
Stainless Steels					
800H	535	C^*	K, σ_{net}	...	6-58
316	740	σ_{net}	K	...	6-51
316	593	C^*	K, σ_{ref}	...	6-52, 6-53
316	593	J	C^*	(c)	6-64
316	600, 650	C^*	...	...	6-65
304	593	C^*	...	...	6-49, 6-50
304	650	C^*	K, σ_{net}, σ_{ref}	...	6-69
304	593	C^*	C_t	...	6-64
304	650	C^*	σ_{net}	(d)	6-65
Superalloys					
Udimet 700	850	K	J, C^*	...	6-57
Discaloy	649	C^*	K, σ_{nom}	...	6-42
Astroloy	704	K	...	...	6-55
Waspaloy	604	K	...	...	6-55
Nimonic 80A	650	K	C^*	...	6-66
Nimonic 115	704	K	...	...	6-55
René 95	704	K	...	...	6-55
Inco 718	538	K	C^*	...	6-56, 6-61
Inco 718	538	K	...	(e)	6-60
Inco 718	593	K	C^*	...	6-56, 6-61
Inco 718	649	K	C^*	...	6-56, 6-61
Inco 718	649	K	...	(e)	6-60
Inco 718	704	K	...	...	6-55
Inco 718	704	K	...	(e)	6-55
IN 100	732	K, C^*, σ_{net}	...	(f)	6-59

(a) Both as-processed and used materials were used. (b) Tested in vacuum and in air. (c) Net section failure per Ref 6-26. (d) Also correlated with C_t (Ref 6-64). (e) Tested in vacuum. (f) C^* and σ_{net} correlated with CT specimens only deformation, which is indicated by the size of a creep zone under load (Ref 6-72). The size of creep zone r_{cr} at the tip of a stationary crack is a function of *K* (the elastic stress intensity factor), *E* (the material elastic modulus), *A* and *n* (the minimum creep strain rate coefficients), and, most importantly, *t* (time at load). As explained in Chapter 1, the size of r_{cr} is proportional to $t^{2/(n-1)}$. The rate of creep zone expansion $\dot{r}_{cr}$ (i.e., the derivation of r_{cr} with respect to *t*) is proportional to $t^{-(n-3)/(n-1)}$. The estimated creep zone size will ultimately provide a measure of the degree of creep deformation, thereby hinting what would be the appropriate fracture mechanics index for a given case under consideration. However, there are no clear boundaries, or criteria, to divide the r_{cr} into small-scale creep and transient creep categories. The Saxena parameter C_t (discussed in Chapter 1), which is embedded with a term for the size of the creep zone, provides a universal characterization for the entire range of creep deformations.

For the purpose of finding out which fracture index is applicable to characterize material behavior at high temperature, a correlation check is traditionally made during the course of material evaluation and material data development. The majority of the published work in the 1970s was devoted to data correlation. Fracture and crack growth rate data were correlated by using *K, J, C**, *C*(*t*), C_t, δ (the crack tip displacement rate), and non-fracture mechanics parameters such as σ_{net} (net section stress), σ_{ref} (reference stress), and σ_{eq} (equivalent stress). Table 6-1 presents a sampling of test data for four types of engineering alloys tested at temperature ranges near their individual limits for high-temperature application. It is shown that these alloys exhibit a wide range of creep crack growth behavior, from environment sensitive to creep ductile. Evidently the majority of the superalloys listed in Table 6-1 correlate with *K*. For alloys that are either creep resistant (e.g., Udimet 700) or sensitive to environment (e.g., Inco 718) the crack growth rates (*da/dt,* the amount of crack growth increment per unit time) are controlled by *K* (i.e., *K* can be used as an independent variable for data correlation). Aluminum alloys might or might not correlate with *K* (probably because of their ductility). Low-alloy steels and stainless steels generally correlate with C_t and *C**.

On two separate occasions, the crack growth rate, *da/dt,* for 316 stainless steel has been shown to correlate with σ_{net}, or *J.* According to Riedel (Ref 6-26), a *da/dt* data set may correlate with *K, C**, or (the fully plastic) *J,* depending on the combination of σ_{net} and time at load. Therefore, high-net-section stresses would have caused those 316 stainless steel specimens to correlate with σ_{net}, or *J.* Keep in mind that these data were creep rupture test data; each test had been run for a long period of time. In the case of fatigue crack growth testing, the hold time of a given stress cycle is usually short, so that the creep zone size is small. It is conceivable that the linear elastic K would be the controlling parameter for most cases.

Creep Crack Growth

To perform a remaining-life assessment of a component under creep crack growth conditions, two principal ingredients are needed: (a) an appropriate expression for relating the driving force K, C^*, or C_t to the nominal stress, crack size, material constants, and geometry of the component being analyzed; and (b) a correlation between this driving force and the crack growth rate in the material, which has been established on the basis of prior data or by laboratory testing of samples from the component. Once these two components are available, they can be combined in order to derive the crack size as a function of time. The general methodology for doing this is illustrated below, assuming C_t to be the driving crack tip parameter.

In general, C_t and C^* are determined based on graphical interpretation of the laboratory test record. The load line deflection rate and specimen geometry are the key elements in the calibration of a generalized expression for calculating the resulting C_t or C^* of a test. Reference 6-34 provides a detailed account of this type of procedure. However, this type of analysis has presented considerable difficulties to fracture mechanics analysts in performing structural crack growth life predictions.

Another analytical expression for calculating C_t has been given by Saxena (Ref 6-25), discussed in Chapter 1. Recalling Eq 1-53, we have:

$$C_t = (C_t)_{ssc} + C^* \quad \text{(Eq 6-13)}$$

In this equation, the first term denotes the contribution from small-scale creep and the second term denotes the contribution from steady-state, large-scale creep. The first term is time variant, whereas the second term is time invariant. In the limit of $t \to 0$, approaching small-scale creep conditions, the first term dominates, implying that K is the controlling parameter in crack growth, with time also explicitly entering the relationship. In the limit $t \to \propto$, the first term becomes zero and C_t becomes identical with C^*.

The term $(C_t)_{ssc}$ for the contribution of small-scale creep has been defined as:

$$(C_t)_{ssc} = \frac{4\alpha\chi\tilde{F}_{cr}(\theta, n)\cdot\eta}{E(n-1)}\cdot\frac{K^4}{W}\cdot\frac{F'}{F}\cdot(EA)^{2/(n-1)}\cdot(t)^{-(n-3)/(n-1)} \qquad \text{(Eq 6-14)}$$

where $\eta = (1 - \nu^2)$ for plane strain (1 for plane stress) and χ is a scaling factor that is approximately equal to 1/3 (for plane strain) as determined by finite-element analysis (Ref 6-25). F is a K-calibration factor, a function of a/W. It is given by $KB\sqrt{W}/P$ where K is the elastic stress intensity factor, a function of a/W, geometry, and loading condition. F' is the derivative of F, that is, $dF/d(a/W)$. The term α is given by:

$$\alpha = \frac{1}{2\pi}\left[\frac{(n+1)\,I_n}{2\pi\eta}\right]^{2/(n-1)} \qquad \text{(Eq 6-15)}$$

The material properties A and n can be obtained from creep tests. On the basis of another finite-element analysis (Ref 6-25), the term $\chi\tilde{F}_{cr}(\theta, n)$ in Eq 6-14 was found to have a value of approximately 1/7.5 (for plane strain). Therefore, Eq 6-14 can be reduced to:

$$(C_t)_{ssc} = \frac{4\alpha\beta(1-\nu^2)}{E(n-1)}\cdot\frac{K^4}{W}\cdot\frac{F'}{F}\cdot(EA)^{2/(n-1)}\cdot(t)^{-(n-3)/(n-1)} \qquad \text{(Eq 6-16a)}$$

where $\beta \cong 1/7.5$. Alternatively, taking advantage of Fig. 1-14, after proper substitution of Eq 1-52 and 6-15 into Eq 6-14, Eq 6-14 can be simplified as:

$$(C_t)_{ssc} = \frac{2\chi\eta}{\pi(n-1)}\cdot\frac{K^4}{W}\cdot\frac{F'}{F}\cdot A^{2/(n-1)}\cdot(E\cdot t)^{-(n-3)/(n-1)}\cdot F_{cr}(\theta, n) \qquad \text{(Eq 6-16b)}$$

where $\chi \cong 1/3$ (for plane strain). Whichever is convenient, either Eq 6-16(a) or (b) can be used as the first term of Eq 6-13.

The steady-state creep crack growth parameter C^* is analogous to the J-integral in the fully plastic condition (Ref 6-42). Therefore, the close-form solution for C^* takes the same form as that for J_p. In other words, an expression for C^* can be derived by adapting an appropriate J_p solution for a geometry under consideration. The expression for C^* is just the same as for J_p, except that the material stress-strain coefficients in the J_p equation are replaced by the material creep strain rate coefficients. Taking the center-cracked panel as an example, in Chapter 4 we stated that:

$$J_p = \alpha\sigma_0\varepsilon_0 a \cdot (1 - 2a/W) \cdot h_1 \cdot \left(\frac{P}{P_0}\right)^{n+1} \qquad \text{(Eq 6-17)}$$

After making proper substitutions of the creep coefficients, Eq 6-17 becomes (Ref 6-49):

$$C^* = \alpha\dot{\varepsilon}_0\sigma_0 a \cdot (1 - 2a/W) \cdot h_1 \left(\frac{P}{P_0}\right)^{n+1} \qquad \text{(Eq 6-17a)}$$

or:

$$C^* = A \cdot a \cdot (1 - 2a/W)^{-n} \cdot h_1 \cdot \left(\frac{P}{W}\right)^{n+1} \qquad \text{(Eq 6-17b)}$$

for plane stress, and:

$$C^* = A \cdot a \cdot (1 - 2a/W)^{-n} \cdot h_1 \cdot \left(\frac{\sqrt{3}P}{2W}\right)^{n+1} \qquad \text{(Eq 6-17c)}$$

for plane strain. Here, the physical unit for C^* is (in. · lb/in.^{2})/h, or MPa · m/h.

For the compact specimen:

$$C^* = \alpha\dot{\varepsilon}_0\sigma_0 \cdot (W - a) \cdot h_1 \cdot \left(\frac{M}{M_0}\right)^{n+1} \qquad \text{Eq (6-18a)}$$

or:

$$C^* = A \cdot (W - a)^{-n} \cdot h_1 \cdot \left(\frac{P \cdot (W + a)/2}{0.2679 \cdot (W - a)^2}\right)^{n+1} \qquad \text{(Eq 6-18b)}$$

for plane stress, and:

$$C^* = A \cdot (W - a)^{-n} \cdot h_1 \cdot \left(\frac{P \cdot (W + a)/2}{0.364 \cdot (W - a)^2}\right)^{n+1} \qquad \text{(Eq 6-18c)}$$

for plane strain.

Equations 6-13 to 6-16(b) can be used to estimate C_t from an applied load (stress) and from a knowledge of the elastic and creep behavior of the material, the *K*-calibration expression, and the C^* expression for the geometry of interest. The *K*-expression can be found in handbooks. The J_p-expressions (which are required to derive C_t) are not as abundantly available for different geometries as the *K*-expressions. Reference 6-73 is one of the few sources that provides engineering J_p solutions. At the present time, this is viewed as a limitation of the technology. More detailed descriptions of the derivations of the C_t and C^* expressions, and the manner of obtaining some of the constants and calculating their values, are given in Ref 6-34. When the engineering J_p solutions are not available, an alternative method can be used. The procedure allows one to use the linear elastic K in combination with the reference stress (that is related to the Von Mises yield stress) to estimate C^*. This approach is described in Ref 6-33 and 6-74.

Once C_t is known, it can be correlated to the crack growth rate through the constants b and m in the following relation:

$$\dot{a} = bC_t^m \quad \text{(Eq 6-19)}$$

Values of the constants b and m for all the materials analyzed by Saxena et al. are listed in Table 6-2. It has been shown (from Eq 6 and 7 of Ref 6-34) that m should have the approximate value $n/(n+1)$, where n is the creep rate exponent.

Crack growth calculations are performed with the current values of a and the corresponding values da/dt to determine the time increment required for increasing the crack size by a small amount Δa, that is, $\Delta t = \Delta a/\dot{a}$. This provides new values of a, t, and C_t, and the process is then repeated. When

Table 6-2 Creep crack growth constants *b* and *m* for various ferritic steels

	b				*m*	
	Upper scatter line		Mean		Upper	
Material	BU(a)	SI(b)	BU(a)	SI(b)	scatter	Mean
All base metal	0.094	0.0373	0.022	0.00874	0.805	0.805
2¼Cr-1Mo weld metal	0.131	0.102	0.017	0.0133	0.674	0.674
1¼Cr-½Mo weld metal	(c)	(c)	(c)	(c)	(c)	(c)
2¼Cr-1Mo and 1¼Cr–½Mo heat-affected-zone/fusion-line material	0.163	0.0692	0.073	0.031	0.792	0.792

(a) BU = British units: *da/dt* in in./h; C_t in in. · lb/in. · h × 10^3. (b) SI = Système International units: *da/dt* in mm/h; C_t in kJ/m^2 · h. (c) Insufficient data; creep crack growth rate behavior comparable to that of base metal. Source: Saxena, Han, and Banerji, "Creep Crack Growth Behavior in Power Plant Boiler and Steam Pipe Steels," EPRI Project 2253-10, published in Ref 6-45

the value of a reaches the critical size a_c as defined by K_{Ic}, J_{Ic}, or any other appropriate failure parameter, failure is deemed to have occurred.

Ainsworth et al. (Ref 6-74) have recently described a unified approach for structures containing defects. This approach incorporates structural failure by rupture, incubation behavior preceding crack growth, and creep crack growth in a single framework. Service life is governed by a combination of time to rupture, time of incubation, and time of crack growth. All of these quantities are calculated using a reference stress that is specifically applicable to the geometry of the component and is derived analytically or based on scale-model tests. If the desired service life exceeds the calculated rupture time, retirement may be necessary. In the opposite situation, further analysis is carried out to calculate the incubation time during which no crack growth is expected to occur. If the calculation indicates that the incubation time is less than the desired service life, then a crack growth analysis is performed to calculate the crack growth life. If the sum of the two is less than the desired service life, operation beyond that point would be considered unsafe.

High-Temperature Fatigue Crack Growth

As mentioned in the foregoing paragraphs, the use of the linear elastic stress intensity factor K may be adequate for analyzing high-temperature fracture resistance of creep-resistant superalloys. This section briefly summarizes the factors affecting high-temperature fatigue crack growth in the context of traditional K-factor analysis.

If cyclic crack growth testing at high temperature is done in a traditional way (i.e., with a sinusoidal or symmetrically triangular wave form at a moderately high frequency), the crack growth rates are functions of ΔK and R. Thus, the phenomenon is similar to those at room temperature, with the following exceptions: (a) for a given R, the value of ΔK_{th} is higher at higher temperature; and (b) for a given R, the terminal ΔK-value is higher at higher temperature, because K_c is usually higher at a higher temperature (due to the fact that the material tensile yield strength is lower at a higher temperature).

A schematic representation of temperature influence on da/dN is shown in Fig. 6-17. The crossover phenomenon of the ΔK_{th} values seems real in high-temperature fatigue crack growth, because it has been experimentally observed in a number of materials, i.e., Inco 718, René 95, titanium alloys, and titanium aluminides (Ref 6-75). The crossover of the terminal ΔK values is due to the fact that K_c usually increases with temperature, so that $(1 - R) \cdot K_c$ also increases with temperature. Crack growth rates in the linear range are not always higher at temperatures higher than room temperature, as implied in Fig. 6-17. Depending on frequency and ΔK range, some materials (particularly those sensitive to environment) may exhibit slower crack growth rates at intermediate temperatures. Moisture, which might have acted as a corrosive medium, was vaporized by heat. Therefore, the magnitude of the environmental fatigue component that is attributed to moisture would be reduced. An example showing a three-dimensional representation of effects of temperature and frequency on *da/dN* for an air-cast Cr-Mo-V rotor steel is presented in Fig. 6-18.

In the power-law (Paris equation) crack growth regime, the effects of temperature, stress ratio R, and hold time have been investigated for many high-temperature alloys. Typical behavior and crack growth results for specific alloys are covered elsewhere (e.g., *ASM Handbook,* Volume 19). A general comparison of temperature effects on the fatigue crack growth of several different high-temperature alloys is shown in Fig. 6-19. Because the reported data were obtained at various ΔK ranges and temperature ranges, the general comparison is based on a constant ΔK, arbitrarily chosen as 30 MPa$\sqrt{m}$ (27 ksi$\sqrt{in.}$). A clear trend of crack growth rate increase with increasing temperature can be seen in Fig. 6-19. At temperatures up to about

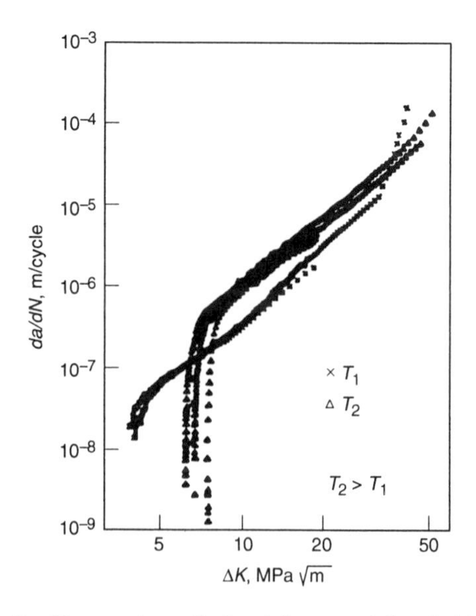

Fig. 6-17 Schematic of temperature effect on fatigue crack threshold and growth rates

50% of the melting point (550 to 600 °C, or 1020 to 1110 °F), the growth rates are relatively insensitive to temperature, but the sensitivity increases rapidly at higher temperatures. The crack growth rates for all the materials at temperatures up to 600 °C, relative to the room temperature rates, can be estimated by a maximum correlation factor of 5 (2 for ferritic steels).

Cyclic frequency (or duration of a stress cycle, e.g., with hold time) also plays an important role in high-temperature crack growth. At high frequency, i.e., fast loading rate with short hold time (or no hold time), the crack growth rate is cycle dependent and can be expressed in terms of da/dN (the amount of crack growth per cycle). At low frequency (or with long hold time), however, the crack growth rate is time dependent, i.e., da/dN is in proportion to the total time span of a given cycle. For the tests with different cycle times, all crack growth rate data points are collapsed into a single curve, of which da/dt is the dependent variable. A mixed region exists between the two extremes. The transition from one type of behavior to

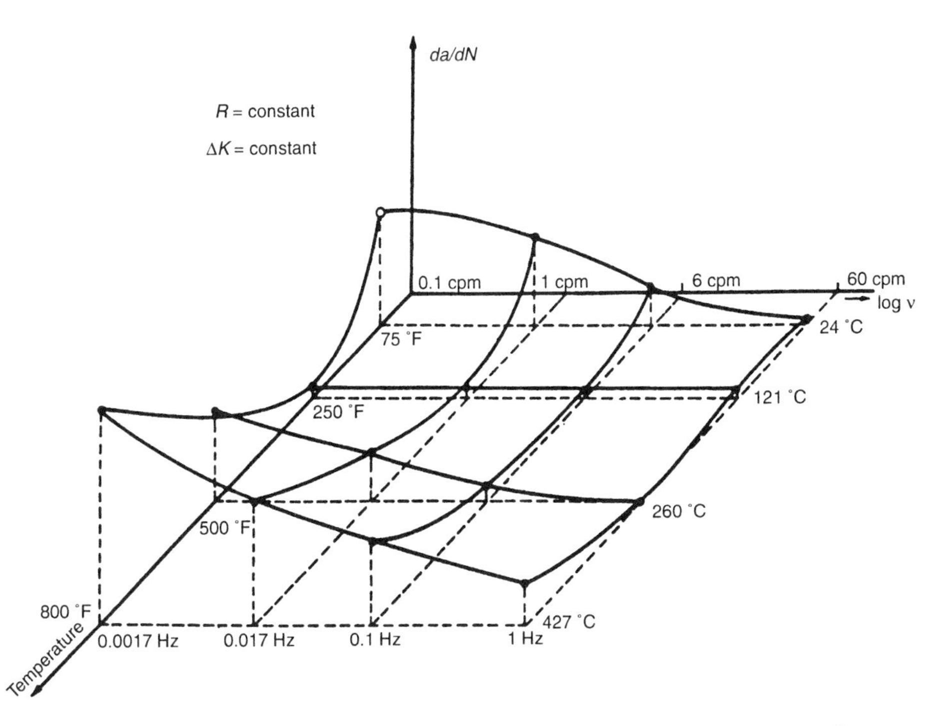

Fig. 6-18 Three-dimensional representation of effects of frequency and temperature on fatigue crack growth rate in air. Source: Ref 6-76

another depends on material, temperature, frequency, and R (Ref 6-77). For a given material and temperature combination, the transition frequency is a function of R. The frequency range at which the crack growth rates remain time dependent increases as R increases (Ref 6-77). The limiting case is R approaches unity. It is equivalent to crack growth under sustained load, for which the crack growth rates at any frequency will be totally time dependent.

To further explore the complex interaction mechanisms of stress, temperature, time, and environmental exposure, a vast amount of experimental and analytical data was compiled from a bibliography of 42 references. Crack growth behavior was examined for 36 types of loading profiles and more than 60 combinations of material, temperature, frequency and time. A compilation of the results is presented in Ref 6-40. The evaluation method was to classify the data into groups representing a variety of isolated loading events. In this way, the phenomenological factors that influence load/environment interaction mechanisms can be determined. This classification also enables micromechanical modeling to be made to account for the contribution of each variable to the total behavior of crack growth. A composite of these load segments provides a basis for spectrum loading simulation.

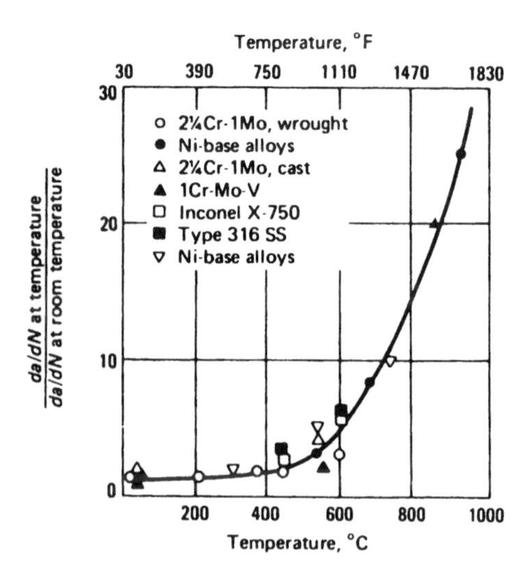

Fig. 6-19 Variation of fatigue crack growth rates as a function of temperature at $\Delta K = 30$ MPa$\sqrt{m}$. Source: Ref 6-45

At room temperature, cyclic frequency and the shape of a stress cycle have insignificant effects on either constant-amplitude or spectrum crack growth behavior. The magnitude and sequential occurrences of the stress cycles are the only key variables that affect room-temperature crack growth behavior. Therefore, an accurate representation of the material crack growth rate data as a function of the stress amplitude ratio (i.e., the so-called crack growth law or crack growth rate equation), and a load interaction model for monitoring the load sequence effects on crack growth (commonly called the crack growth retardation/acceleration model) are the only two essential elements in cycle-dependent crack growth life predictive methodology. However, the time factor in a given stress cycle (whether it is associated with hold time or low frequency), which promotes time-dependent crack growth behavior, might play a significant role in high-temperature crack growth. Therefore, the section below briefly reviews cycle-dependent versus time-dependent crack growth.

Cycle-Dependent Versus Time-Dependent Crack Growth

Research conducted on conventional high-temperature superalloys, Inco 718 in particular, has shown that sustained-load creep crack growth rate data can be used to predict cyclic crack growth in the time-dependent regime (Ref 6-78, 6-79). In regions in which cycle-dependent and time-dependent phenomena are both present, implementation of a semi-empirical technique may be required. A summary of how to formulate a collective procedure to predict crack growth in the time-dependent regime is given below.

Applying the Wei-Landes superposition principle for subcritical crack growth in an aggressive environment (Ref 6-14), the crack growth rate for a given stress cycle can be treated as the sum of three parts:

- The uploading part (i.e., the load rising portion of a cycle)
- The hold time
- The down loading (unloading) portion of a cycle

Therefore:

$$\frac{da}{dN} = \left(\frac{da}{dN}\right)_{r} + \left(\frac{da}{dN}\right)_{H} + \left(\frac{da}{dN}\right)_{d} \quad \text{(Eq 6-20)}$$

The subscripts "r," "H," and "d" stand for uploading (the uprising phase of a cycle), hold time, and downloading, respectively. Unlike the original Wei-

Landes model (i.e., Eq 6-3 and 6-5), the crack growth rate of Eq 6-20 has nothing to do with the reference (inert) environment. The test is conducted at temperature.

It has been shown frequently by experimental tests that the amount of *da* for the unloading part is negligible unless the stress profile is asymmetric, and the ratio of uploading time to unloading time is significantly small (that is, the unloading time compared to the uploading time is sufficiently long). For simplicity, we will limit our discussion to those cases based on two components only, i.e., by setting $(da/dN)_d$ equal to zero. However, the $(da/dN)_r$ term may be cycle dependent, time dependent, or mixed. This term consists of two parts: one part accounts for the cyclic wave contribution and the other part accounts for the time contribution. Consequently, Eq 6-20 can be rewritten as:

$$\frac{da}{dN} = \left(\frac{da}{dN}\right)_c + \left(\frac{da}{dN}\right)_t + \left(\frac{da}{dN}\right)_H \qquad \text{(Eq 6-21)}$$

The first term on the right side of Eq 6-21 represents the cycle-dependent part of the cycle. It comes from the conventional crack growth rate data at high frequency, i.e., it follows the *crack growth laws* cited in the literature (such as the Paris and Walker equations). A comprehensive review of this

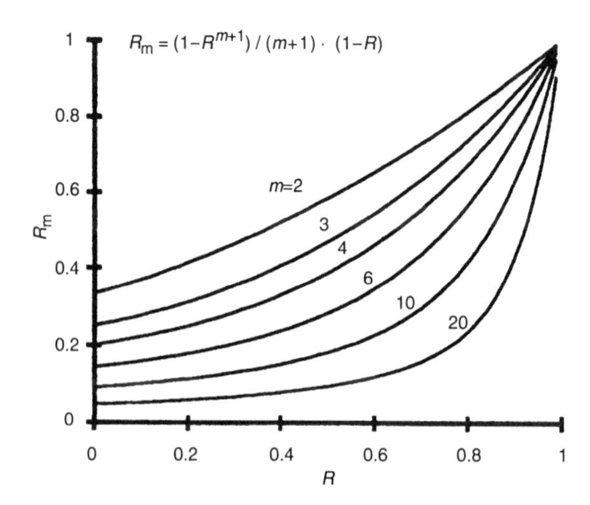

Fig. 6-20 A plot of the coefficient R_m versus R-ratio

subject is given in Chapter 3. In reality, when a stress cycle is totally cycle dependent, the magnitude of the second term on the right side of Eq 6-21 will be negligibly small. On the other hand, when a stress cycle is totally time dependent, the contribution of $(da/dN)_c$ to the total da/dN is negligible; thereby the validity of Eq 6-21 in respect to full frequency range is maintained.

When a crack growth rate component exhibits time-dependent behavior, it is equivalent to crack growth under a sustained load of which the crack growth rate description is defined by da/dt (instead of da/dN) as:

$$\frac{da}{dt} = C \cdot (K_{max})^m \qquad \text{(Eq 6-22)}$$

This quantity is obtained from a sustained-load test. To express the second term on the right side of Eq 6-21 in terms of da/dt, consider a generalized function that can describe K at any given time in a valley-to-peak cycle:

$$\mathrm{K}(t) = R \cdot K_{max} + 2K_{max} \cdot (1 - R) \cdot t_r \cdot f \qquad \text{(Eq 6-23)}$$

where t_r is the time required for ascending the load from valley to peak and f is the frequency of the cyclic portion of a given load cycle. For symmetric loading (i.e., $t_r = f/2$), Eq 6-23 gives $\mathrm{K}(t) = K_{min}$ at $t_r = 0$, and $\mathrm{K}(t) = K_{max}$ at $t_r = 1/2f$. The amount of crack extension over a period t_r can be obtained by replacing the K_{max} term of Eq 6-22 by $\mathrm{K}(t)$ and integrating:

$$\left(\frac{da}{dN}\right)_t = C\int_0^{t_r} [\mathrm{K}(t)]^m dt \qquad \text{(Eq 6-24)}$$

For any positive value of m, Eq 6-24 yields:

$$\left(\frac{da}{dN}\right)_t = C \cdot (K_{max})^m \cdot t_r \cdot R_m \qquad \text{(Eq 6-25)}$$

where

$$\mathrm{R_m} = (1 - R^{m+1})/[(m + 1) \cdot (1 - R)] \qquad \text{(Eq 6-26)}$$

A parametric plot for R_m as a function of R is shown in Fig. 6-20. It is seen that R_m increases as R increases. Therefore, for a given K_{max}, $(da/dN)_r$ increases as R increases in the time-dependent regime. This trend is the reverse of those customary observed in the high-frequency, cycle-dependent regime.

The third term on the right side of Eq 6-21 simply equals to *da/dt* times the time at load. Recognizing that the first term on the right side of Eq 6-25 is actually equal to *da/dt,* Eq 6-21 can be expressed as:

$$\frac{da}{dN} = \left(\frac{da}{dN}\right)_{\mathrm{c}} + \frac{da}{dt} \cdot t_{\mathrm{r}} \cdot R_{\mathrm{m}} + \frac{da}{dt} \cdot t_{\mathrm{H}} \qquad \text{(Eq 6-27)}$$

where t_H is the hold time.

The applicability of Eq 6-27 is demonstrated in Fig. 6-21 and 6-22. In these figures, the test data were generated from an Inco 718 alloy, at 649 °C, having various combinations of ΔK, R, t_H, and f. The test data, which were extracted from the open literature (Ref 6-78, 6-79), are presented in the figures along with the predictions.

One of the two data sets in Fig. 6-21 was generated by using trapezoidal stress cycles of which $f = 1$ Hz (i.e., 0.5 s for uploading and 0.5 s for

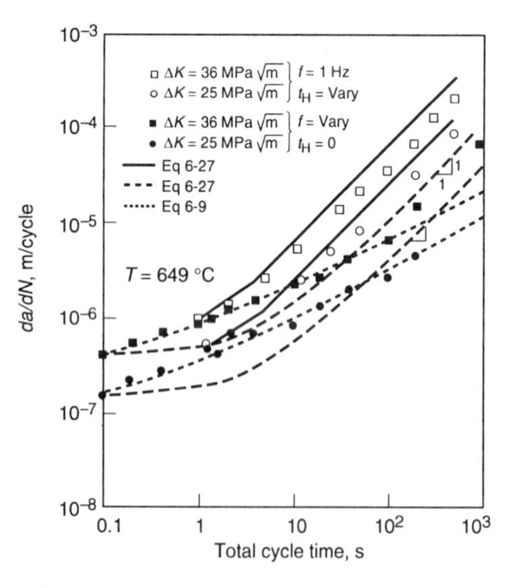

Fig. 6-21 High-temperature fatigue crack growth rates of Inco 718 (actual and predicted values, $R = 0.1$)

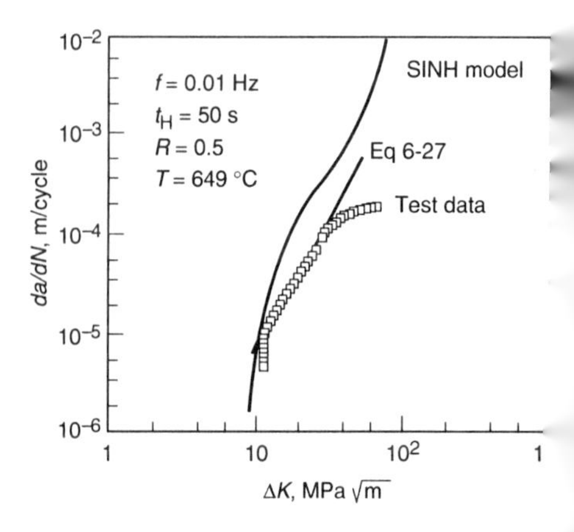

Fig. 6-22 High-temperature fatigue crack growth rates of Inco 718 (actual and predicted values, $R = 0.5$)

unloading) and by varying hold time (i.e., $t_H = 1$ to 500 s). The data points for the other data set were obtained by conducting tests at various frequencies without hold time. A constant ΔK level, either 25 MPa$\sqrt{m}$ or 36 MPa$\sqrt{m}$ with $R = 0.1$, was applied to all the tests. Crack growth rate per cycle was plotted as a function of total time per cycle. For example, for a total cycle time of 100 s, it would mean that the test was conducted at a frequency of 1 Hz with $t_H = 99$ s, or $f = 0.01$ Hz without hold time. The predictions were made by using Eq 6-22 and 6-27 with $C = 2.9678 \times 10^{-11}$ m/s and $m = 2.65$ (the separately determined material *da/dt* constants). The value for the $(da/dN)_c$ term was set to those experimental data points for $f = 10$ Hz. The correlation between Eq 6-27 and the trapezoidal load test data is quite good.

It is also shown in Fig. 6-21 that Eq 6-27 correlates with the triangular load test data in the time-dependent region ($f \leq 0.02$ Hz, or total time ≥ 50 s) but fails to predict the crack growth rates in the mixed region (0.02 Hz $< f <$ 10 Hz). For this group of data, a better correlation was obtained by using the latest version of the Saxena equation, i.e., Eq 6-9. For the data sets in Fig. 6-21, the value for f_0 was assumed to be 10 Hz. Following the procedure described in the section on corrosion fatigue crack growth, it was determined that $C_4 = 1.075 \times 10^{-10}$, $\alpha = 2.35$.

The sample case shown in Fig. 6-22 involves all three loading variables, t_H, t_r, and R (as compared to those data sets shown in Fig. 6-21, where the crack growth rates were functions of t_H and R, or f and R). The test conditions for this data set were $R = 0.5$, $f = 0.01$ Hz (i.e., $t_r = 50$ s), $t_H = 50$ s, and $K_{max} = 20$ to 140 MPa$\sqrt{m}$ ($\Delta K = 10$ to 70 MPa$\sqrt{m}$). A very good match was obtained up to $\Delta K = 35$ MPa$\sqrt{m}$. It thus appears that Eq 6-27 is superior to the other crack growth models. A comparison with the SINH model (Ref 6-46), a model that is widely used by the engine industry, is shown in Fig. 6-22.

In conclusion, the crack growth behavior of a stress cycle having a trapezoidal wave form can be predicted by using the combination of conventional high-frequency *da/dN* data, sustained load data (*da/dt*), and Eq 6-27. For stress cycles having a triangular wave form, test data for a specific frequency in question may be required. Otherwise, a set of test data containing several frequencies is needed for developing the empirical constants in the Saxena equation. Equation 6-9 is basically an empirical function for curve fitting and data interpolation; it is not a scientific rule that dictates the frequency effect on crack growth behavior. Therefore, although not essential, it is desirable to have an all-around method that can describe the *da/dN* behavior in the mixed region.

In summary, as long as load/environment interactions are absent, the total crack growth rate for a loading block containing both triangular and trapezoidal stress cycles will be:

$$\left(\frac{da}{dN}\right)_{\text{Total}} = \sum_i \left(\frac{da}{dN}\right)_i \qquad \text{(Eq 6-28)}$$

where i denotes the ith loading step in the entire group of loads under consideration. The amount of da for each loading step is determined by using Eq 6-9 or 6-27.

The methods presented in the foregoing sections are applicable to isothermal conditions (i.e., constant temperature). Crack growth may be subjected to thermal-mechanical loading conditions (i.e., both mechanical loads and temperatures are fluctuating). In addition, neither type of these input cycles need be in-phase. Research on this type of crack growth phenomena is limited (Ref 6-61, 6-80) and will not be discussed here. The methods presented here have been demonstrated with data obtained from creep-resistant and environment-sensitive material. One investigation has shown that the method of superposition may be unconservative for creep ductile material such as the low-alloy steels (Ref 6-81).

Load and Environment Interaction

Some load/environment interaction phenomena of corrosion fatigue were discussed in an earlier section of this chapter. It seems that depending on the material and service application, the coexistence of the crack tip plastic zone and the EAZ may bring about conflicting results for the delay behavior in general. In the case of crack growth at high temperature, we just learned that there is a third type of crack tip deformation zone, namely the creep zone. It is conceivable that the existence of the creep zone could add more confusion to the subject matter.

In preceding sections of this chapter, the methods for determining the size of the EAZ and the creep zone were discussed. In the following, we will demonstrate how to determine the size of a crack tip plastic zone at high temperature. Because crack tip plasticity is a function of mechanical load and material tensile yield strength, we can simply use the material tensile yield strength at temperature for F_{ty}. The equation for r_p as a function of K and F_{ty} is given in Chapter 3. This concept can be extended to determine the size of r_p for a load cycle with hold times at the peak. That is, r_p can be calculated by using an effective yield strength that is a function of temperature and time. Such an effective yield strength can be obtained from a creep rupture test in which the combined effect of temperature and time is accounted for. The notion is that, given time, the dwell time in a trapezoidal stress cycle promotes creep deformation at the crack tip. The F_{ty} term in Eq 3-15 may be considered as an effective yield stress required to achieve a 0.2% strain in a creep test. The ratio of effective yield stress to room-temperature tensile yield strength can be plotted against the Larsen-Miller parameter, which takes into account the combined effect of temperature and time at load. Figure 6-23 is a plot showing such a relationship for two high-temperature alloys: the Inco 718 alloy and the super-$\alpha 2$ titanium aluminide forging (Ti-25Al-10Nb-3V-1Mo in atomic percentage, which is equivalent to Ti-14A1-20Nb-3V-2Mo in weight percentage).

Some alloy that is normally susceptible to environment may become susceptible to creep at higher temperatures. When an alloy is somewhat susceptible to creep (even though crack extension is still controlled by K), crack growth behavior inside an EAZ may not be the same as shown in Fig. 6-10 and 6-11. The crack growth rates for those load cycles following a sustained load would have been decreased instead of increased (Ref 6-82). An actual example of such a phenomenon is shown in Fig. 6-24. In this case, the IN 100 alloy was tested at 732 °C with a loading profile very much the same as shown in Fig. 6-11. The stress levels for the trapezoidal and the sinusoidal

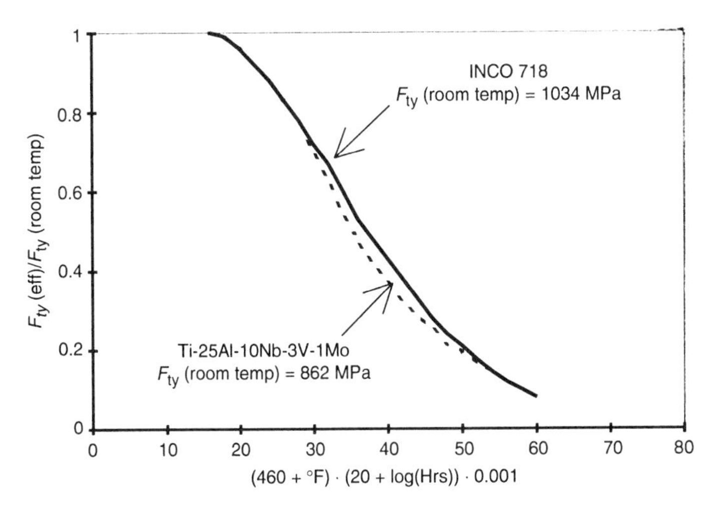

Fig. 6-23 Effective tensile yield strength variations as a function of temperature and hold time represented by the Larsen-Miller parameter

stress cycles were the same ($K_{max} = 38.5\ MPa\sqrt{m}$), so crack growth retardation was not supposed to happen. According to the theory of the EAZ, the post-sustained-load crack growth should have behaved as shown in Fig. 6-10. However, the crack growth behavior of this alloy subsequent to the sustained load was very much the same as those coming from a high/low load sequence profile. That is, the post-sustained-load crack growth rates were retarded instead of accelerated. The mechanism for such behavior is not fully understood, because this is the only data set of this type reported in the literature. Referring to Table 6-1, there is a gray area such that elastic stress intensity factor K is not the only crack growth controlling parameter for the IN 100 alloy. Besides K, two other fracture indices, i.e., C^* and σ_{net}, also correlate with the sustained load fracture test results. Thus, one possible explanation for this phenomenon is that the crack tip deformation zone here resembles a creep zone in which stress relaxation is prominent.

In summary, it is hypothesized that two out of three crack tip deformation models can co-exist in a material/temperature system: r_p and r_{cr}, or r_p and

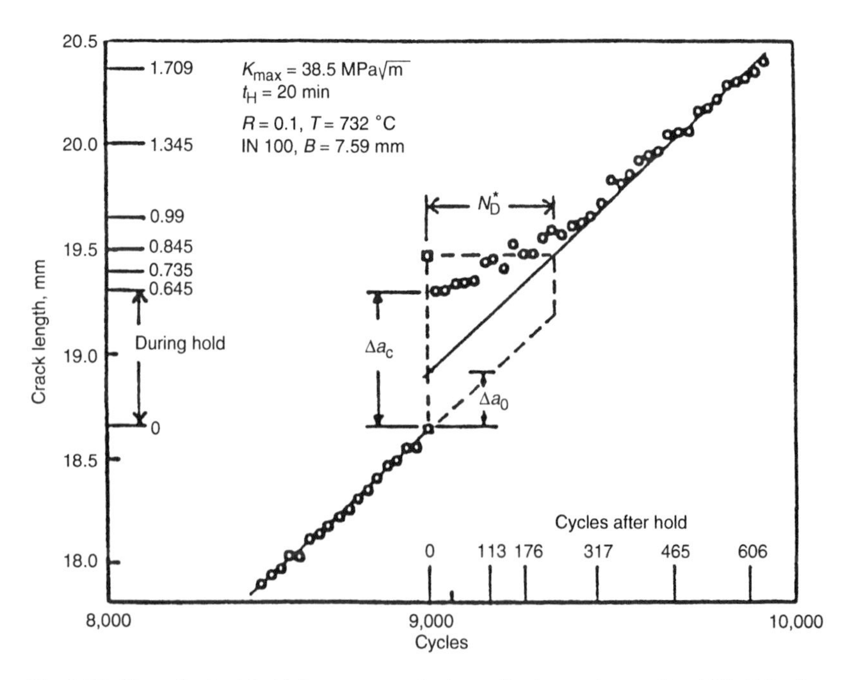

Fig. 6-24 The effect of hold time on constant-amplitude crack growth of IN 100 alloy. Source: Ref 6-82

EAZ. It may even be possible that all three types of crack tip deformation zones co-exist at a given time. Each of them has a unique way of controlling the crack growth rates for the current and the subsequent cycles. The plastic zone is the source for altering the crack tip stress field during crack propagation. The size of a creep zone indicates the relative degree of creep deformation. It closely associates with the change of crack-growth-controlling parameters with time. If an alloy is sensitive to environment, an EAZ will exist at the crack tip.

The analytical methods described in this chapter are capable of handling constant-amplitude stress cycles involving variations in frequency (with or without hold time in a given cycle). At present, they are inadequate in dealing with loading spectra that contain both trapezoidal and sinusoidal stress cycles. The characterizing elements discussed only provide some insight to the problem that is faced. More research is needed in order to develop a reliable model for predicting spectrum crack growth at high temperature.

REFERENCES

6-1. H. Arup and R.N. Parkins, Ed., *Stress Corrosion Research*, Sijthoff and Noordhoff, Alphen aan den Rijn, The Netherlands, 1979

6-2. R.H. Jones, Ed., *Stress-Corrosion Cracking*, ASM International, 1992

6-3. D.E. Piper, S.H. Smith, and R.V. Carter, *Metall. Eng. Quarterly*, Vol 8, 1968, p 50

6-4. C.E. Feddersen and D.P. Moon, "A Compilation and Evaluation of Crack Behavior Information on D6AC Steel Plate and Forging Materials for the F-111 Aircraft," Defense Metals Information Center, Battelle Columbus Laboratories, Columbus, OH, 25 June 1971

6-5. H.H. Johnson, Environmental Cracking in High-Strength Materials, *Fracture*, Vol III, H. Liebowitz, Ed., Academia Press, 1971, p 679–720

6-6. R.P. Wei, S.R. Novak, and D.P. Williams, *Material Research and Standards*, TMRSA, Vol 12, American Society for Testing and Materials, 1972, p 25

6-7. P. Anderson, Corrosion Fatigue Testing, *ASM Handbook,* Vol 19, *Fatigue and Fracture*, ASM International, 1996, p 193–209

6-8. M.O. Speidel, Stress Corrosion and Corrosion Fatigue Crack Growth in Aluminum Alloys, *Stress Corrosion Research*, Sijthoff and Noordhoff, Alphen aan den Rijn, The Netherlands, 1979, p 117–175

6-9. L.B. Vogelesang, "Some Factors Influencing the Transition from Tensile Mode to Shear Mode Under Cyclic Loading," Report LR-222, Delft University of Technology, The Netherlands, Aug 1976

6-10. A.J. McEvily, Jr. and R.P. Wei, *Corrosion Fatigue: Chemistry, Mechanics and Microstructure*, National Association of Corrosion Engineers, 1972, p 381–395

6-11. D.B. Dawson and R.M.N. Pelloux, Corrosion Fatigue Crack Growth Rates of Titanium Alloys Exposed in Aqueous Environments, *Metall. Trans.*, Vol 5, 1974, p 723–731

6-12. O. Vosikovsky, Fatigue-Crack Growth in an X-65 Line-Pipe Steel at Low Cyclic Frequencies in Aqueous Environments, *J. Eng. Mater. Technol.*, (*Trans ASME*), Series H, Vol 97, 1975, p 298–305
6-13. K.M. Chang, Elevated Temperature Fatigue Crack Propagation after Sustained Loading, *Effects of Load and Thermal Histories on Mechanical Behavior of Materials,* TMS, 1987, p 13–26
6-14. R.P. Wei and J.D. Landes, Correlation between Sustained-Load and Fatigue Crack Growth in High-Strength Steels, *Material Research and Standards*, TMRSA, Vol 9, American Society for Testing and Materials, 1969, p 25–28
6-15. R.P. Wei, On Understanding Environment Enhanced Fatigue Crack Growth—A Fundamental Approach, *Fatigue Mechanisms*, STP 675, American Society for Testing and Materials, 1979, p 816
6-16. H.D. Dill and C.R. Saff, "Environment-Load Interaction Effects on Crack Growth," Report AFFDL-TR-78-137, Air Force Flight Dynamics Laboratory, Wright-Patterson Air Force Base, Dayton, OH, 1980
6-17. E.J. Imhof and J.M. Barsom, *Progress in Flaw Growth and Fracture Toughness Testing*, STP 743, American Society for Testing and Materials, 1973, p 182–205
6-18. A. Saxena, R.S. Williams, and T.T. Shih, A Model for Representing and Predicting the Influence of Hold Time on the Fatigue Crack Growth Behavior at Elevated Temperature, *Fracture Mechanics: Thirteenth Conf.*, STP 743, American Society for Testing and Materials, 1981, p 89–99
6-19. A. Saxena, "A Model for Predicting the Effect of Frequency on Fatigue Crack Growth Behavior at Elevated Temperature," Scientific Paper 79-1D3-EVFLA-P2, Westinghouse Research and Development Center, Pittsburgh, PA, Sept 1979
6-20. A. Saxena and J.L. Basani, Time-Dependent Fatigue Crack Growth Behavior at Elevated Temperature, *Fracture: Interactions of Microstructure, Mechanisms and Mechanics*, AIME, 1984, p 357–383
6-21. G.R. Chanani, "Fundamental Investigation of Fatigue Crack Growth Retardation in Aluminum Alloys," Report AFML-TR-76–156, Air Force Materials Laboratory, Wright-Patterson Air Force Base, Dayton, OH, 1976
6-22. W.P. Slagle, D. Mahulikar, and H.L. Marcus, Effect of Hold Times on Crack Retardation in Aluminum Alloys, *Eng. Fract. Mech.*, Vol 13, 1980, p 889–895
6-23. T.T. Shih and R.P. Wei, Load and Environment Interactions in Fatigue Crack Growth, *Prospects of Fracture Mechanics*, Noordhoff International Publishing, 1974, p 231–248
6-24. W.W. Gerberich and N.R. Moody, A Review of Fatigue Fracture Topology Effects on Threshold and Growth Mechanisms, *Fatigue Mechanisms*, STP 675, American Society for Testing and Materials, 1979, p 292–341
6-25. A. Saxena, Mechanics and Mechanism of Creep Crack Growth, *Fracture Mechanics: Microstructure and Micromechanisms*, ASM International, 1989
6-26. H. Riedel, Creep Crack Growth, *Flow and Fracture at Elevated Temperatures*, American Society for Metals, 1983, p 149–177
6-27. H. Ghonem, T. Nicholas, and A. Pineau, "Analysis of Elevated Temperature Fatigue Crack Growth Mechanisms in Alloy 718," paper presented at ASME Winter Annual Meeting, 1991
6-28. J.M. Larsen and T. Nicholas, Cumulative-Damage Modeling of Fatigue Crack Growth in Turbine Engine Materials, *Eng. Fract. Mech.*, Vol 22, 1985, p 713–730

6-29. T. Nicholas, J.H. Laflen, and R.H. Van Stone, in *Proc.: Conf. on Life Prediction for High Temperature Gas Turbine Materials,* Syracuse University Press, 1986, p 4.1–4.61
6-30. H.P. Van Leeuwen, "The Application of Fracture Mechanics to the Growth of Creep Cracks," AGARD Report 705, 56th Meeting of the Structures and Materials Panel, North Atlantic Treaty Organization, London, April 1983
6-31. K. Sadananda and P. Shahinian, Review of the Fracture Mechanics Approach to Creep Crack Growth in Structural Alloys, *Eng. Fract. Mech.*, Vol 15, 1981, p 327–342
6-32. K. Sadanada and P. Shahinian, Creep Crack Growth Behavior and Theoretical Modeling, *Metal Science*, Vol 15, 1981, p 425–432
6-33. G.A. Webster and A.R. Ainsworth, *High Temperature Component Life Assessment*, Chapman & Hall, London, 1994
6-34. R. Norris, P.S. Crover, B.C. Hamilton, and A. Saxena, Elevated-Temperature Crack Growth, *ASM Handbook,* Vol 19, *Fatigue and Fracture*, ASM International, 1996, p 507–519
6-35. A.F. Liu, High-Temperature Life Assessment, *ASM Handbook,* Vol 19, *Fatigue and Fracture*, ASM International, 1996, p 520–526
6-36. D.C. Maxwell and T. Nicholas, High-Temperature Fatigue Crack Growth Testing, *ASM Handbook,* Vol 19, *Fatigue and Fracture*, ASM International, 1996, p 181–183
6-37. ASTM Standard E 1457, Metals Test Methods and Analytical Procedures, *Annual Book of ASTM Standards,* Vol 03.01, *Metals—Mechanical Testing; Elevated and Low Temperature Tests; Metallography*, American Society for Testing and Materials, 1995
6-38. A. Nagar, "A Review of High Temperature Fracture Mechanics for Hypervelocity Vehicle Application," AIAA Paper 88-2386, presented at AIAA/ASME/ASCE/AHS/ASC 29th Structures, Structural Dynamics and Materials Conf. (Williamsburg, VA), 1988
6-39. D.M. Harmon, C.R. Saff, and J.G. Burns, "Development of an Elevated Temperature Crack Growth Routine," AIAA Paper 88–2387, AIAA/ASME/ASCE/AHS/ASC 29th Structures, Structural Dynamics and Materials Conf. (Williamsburg, VA), 1988
6-40. A.F. Liu, "Element of Fracture Mechanics in Elevated Temperature Crack Growth," AIAA Paper 90–0928, presented at AIAA/ASME/ASCE/AHS/ASC 31st Structures, Structural Dynamics and Materials Conf. (Long Beach, CA), 2–4 April 1990, p 981–994
6-41. A.F. Liu, Assessment of a Time Dependent Damage Accumulation Model for Crack Growth at High Temperature, *Proc. 1994,* Vol 3, Int. Council of the Aeronautical Sciences, 1994, p 2625–2635
6-42. J.D. Landes and J.A. Begley, A Fracture Mechanics Approach to Creep Crack Growth, *Mechanics of Crack Growth,* STP 590, American Society for Testing and Materials, 1976, p 128–148
6-43. A. Saxena, Creep Crack Growth under Non-Steady State Conditions, *Fracture Mechanics: Seventeenth Volume*, STP 905, American Society for Testing and Materials, 1986, p 185–201
6-44. A. Saxena, "Life Assessment Methods and Codes," Report EPRI TR-103592, Electric Power Research Institute, Jan 1996

6-45. R. Viswanathan, *Damage Mechanisms and Life Assessment of High-Temperature Components*, ASM International, 1989
6-46. J.M. Larsen, B.J. Schwartz, and C.G. Annis, Jr., "Cumulative Damage Fracture Mechanics under Engine Spectra," Report AFML-TR-79-4159, Air Force Materials Laboratory, Wright-Patterson Air Force Base, Dayton, OH, 1980
6-47. A. Utah, "Crack Growth Modeling in Advanced Powder Metallurgy Alloy," Report AFWAL-TR-80-4098, Air Force Materials Laboratory, Wright-Patterson Air Force Base, Dayton, OH, 1980
6-48. D.M. Harmon and C.R. Saff, "Damage Tolerance Analysis for Manned Hypervelocity Vehicles," Vol I, Final Technical Report, WRDC-TR-89-3067, Flight Dynamics Laboratory, Wright Research and Development Center, Air Force Systems Command, Wright-Patterson Air Force Base, Dayton, OH, Sept 1989
6-49. A. Saxena, Evaluation of C* for the Characterization of Creep-Crack-Growth Behavior in 304 Stainless Steel, *Fracture Mechanics: Twelfth Conf.*, STP 700, American Society for Testing and Materials, 1980, p 131–151
6-50. T.T. Shih, A Simplified Test Method for Determining the Low Rate Creep Crack Growth Data, *Fracture Mechanics: Fourteenth Symposium,* Vol II, *Testing and Applications*, STP 791, American Society for Testing and Materials, 1983, p II-232 to II-247
6-51. R.D. Nicholson and C.L. Formby, *Int. J. Fract.*, Vol 11, 1975, p 595–604
6-52. K. Sadananda and P. Shahinian, Evaluation of J* Parameter for Creep Crack Growth in Type 316 Stainless Steel, *Fracture Mechanics: Fourteenth Symposium,* Vol II, *Testing and Applications*, STP 791, American Society for Testing and Materials, 1983, p II-182 to II-196
6-53. K. Sadananda and P. Shahinian, Parametric Analysis of Creep Crack Growth in Austenitic Stainless Steel, *Elastic-Plastic Fracture: Second Symposium,* Vol I, *In-Elastic Crack Analysis*, STP 803, American Society for Testing and Materials, 1983, p I-690 to I-707
6-54. A. Saxena and B. Gieseke, Transients in Elevated Temperature Crack Growth, *Proc. of MECAMAT, Int. Seminar on High Temperature Fracture Mechanisms and Mechanics,* 1987, p III/19 to III/36
6-55. S. Floreen, The Creep Fracture of Wrought Nickel-Base Alloys by a Fracture Mechanics Approach, *Metall. Trans. A*, Vol 6, 1975, p 1741–1749
6-56. K. Sadananda and P. Shahinian, Creep Crack Growth in Alloy 718, *Metall. Trans. A*, Vol 8, 1977, p 439–449
6-57. K. Sadananda and P. Shahinian, *Metall. Trans. A*, Vol 9, 1978, p 79–84
6-58. M. Welker, A. Rahmel, and M. Schutze, Investigations on the Influence of Internal Nitridation on Creep Crack Growth in Alloy 800 H, *Metall. Trans. A*, Vol 20, 1989, p 1553–1560
6-59. R.C. Donath, T. Nicholas, and S.L. Fu, *Fracture Mechanics: Thirteenth Conf.*, STP 743, American Society for Testing and Materials, 1981, p 186–206
6-60. M. Stucke, M. Khobaib, B. Majumdar, and T. Nicholas, Environmental Aspects in Creep Crack Growth in Nickel Base Superalloy, *Advances in Fracture Research*, Vol 6, Pergamon Press, 1986, p 3967–3975
6-61. G.K. Haritos, D.L. Miller, and T. Nicholas, Sustained-Load Crack-Growth in Inconel 718 Under Nonisothermal Conditions, *J. Eng. Mater. Technol.*, (*Trans. ASME*), Series H, Vol 107, 1985, p 172–179

6-62. K.M. Nikbin and G.A. Webster, Temperature Dependence of Creep Crack Growth in Aluminum Alloy RR58, *Micro and Macro Mechanics of Crack Growth*, 1982, p 137–147
6-63. J.G. Kaufman, K.O. Bogardus, D.A. Mauney, and R.C. Malcolm, Creep Cracking in 2219-T851 Plate at Elevated Temperatures, *Mechanics of Crack Growth*, STP 590, American Society for Testing and Materials, 1976, p 149–168
6-64. A. Saxena and J.D. Landes, Characterization of Creep Crack Growth in Metals, *Advances in Fracture Research*, Vol 6, Pergamon Press, 1986, p 3977–3988
6-65. S. Taira, R. Ohtani, and T. Kitamura, Application of J-Integral to High-Temperature Crack Propagation, Part I: Creep Crack Propagation, *J. Eng. Mater. Technol.*, (*Trans. ASME*), Series H, Vol 101, 1979, p 154–161
6-66. H. Riedel and W. Wagner, Creep Crack Growth in NIMONIC 80A and in a 1Cr-1/2Mo Steel, *Advances in Fracture Research*, Vol 3, Pergamon Press, 1986, p 2199–2206
6-67. S. Jani and A. Saxena, Influence of Thermal Aging on the Creep Crack Growth of a Cr-Mo Steel, *Effects of Load and Thermal Histories on Mechanical Behavior of Materials*, TMS/AIME, 1987, p 201–220
6-68. H.P. van Leeuwen and L. Schra, Fracture Mechanics and Creep Crack Growth of 1%Cr-1/2%Mo Steel with and without Prior Exposure to Creep Conditions, *Eng. Fract. Mech.*, Vol 127, 1987, p 483–499
6-69. R. Koterazawa and T. Mori, Applicability of Fracture Mechanics Parameters to Crack Propagation under Creep Condition, *J. Eng. Mater. Technol.*, (*Trans. ASME*), Series H, Vol 99, 1977, p 298–305
6-70. P.L. Bensussan and R.M.N. Pelloux, Creep Crack Growth in 2219-T851 Aluminum Alloy: Applicability of Fracture Mechanics Concepts, *Advances in Fracture Research*, Vol 3, Pergamon Press, 1986, p 2167–2179
6-71. H. Riedel and V. Detampel, Creep Crack Growth in Ductile, Creep-Resistant Steels, *Int. J. Fract.*, Vol 33, 1987, p 239–262
6-72. H. Riedel and J.R. Rice, Tensile Cracks in Creeping Solids, *Fracture Mechanics: Proc. of the Twelfth National Symposium on Fracture Mechanics,* STP 700, American Society for Testing and Materials, 1980, p 112–130
6-73. V. Kumar, M.D. German, and C.F. Shih, "An Engineering Approach for Elastic-Plastic Fracture Analysis," Report NP 1931, Electric Power Research Institute, Palo Alto, CA, 1981
6-74. R.A. Ainsworth, G.G. Chell, M.C. Coleman, I.W. Goodal, D.J. Gooch, J.R. Haigh, S.T. Kimmins, and G.J. Neate, CEGB Assessment Procedure for Defects in Plant Operating in the Creep Range, *Fatigue Fract. Eng. Mater. Struct.*, Vol 10, 1987, p 115–127
6-75. J.E. Allison and J.C. Williams, *Scripta Metall.*, Vol 19, 1985, p 773–778
6-76. T.T. Shih and G.A. Clarke, Effect of Temperature and Frequency on the Fatigue Crack Growth Rate Properties of a 1950 Vintage Cr-Mo-V Rotor Material, *Fracture Mechanics*, STP 677, American Society for Testing and Materials, 1979, p 125–143
6-77. T. Nicholas and N.E. Ashbaugh, Fatigue Crack Growth at High Load Ratios in the Time-Dependent Regime, *Fracture Mechanics: Nineteenth Symposium*, STP 969, American Society for Testing and Materials, 1988, p 800–817

6-78. G.K. Haritos, T. Nicholas, and G.O. Painter, Evaluation of Crack Growth Models for Elevated-Temperature Fatigue, *Fracture Mechanics: Eighteenth Symposium*, STP 945, American Society for Testing and Materials, 1988, p 206–220
6-79. T. Nicholas and T. Weerasooriya, Hold-Time Effects in Elevated Temperature Fatigue Crack Propagation, *Fracture Mechanics: Seventeenth Volume*, STP 905, American Society for Testing and Materials, 1986, p 155–168
6-80. L.A. James, Environmentally Assisted Cracking Behaviour of a Low Alloy Steel under Non-Isothermal Conditions, *Fatigue and Crack Growth: Environmental Effects, Modeling Studies, and Design Considerations*, PVP 306, ASME, 1995, p 19–27
6-81. R. Ohtani, Substance of Creep-Fatigue Interaction Examined from the Point of View of Crack Propagation Mechanics, *Low Cycle Fatigue and Elasto-Plastic Behaviour of Materials,* Elsevier, London, 1987, p 211–222
6-82. J.M. Larsen and T. Nicholas, Load Sequence Crack Growth Transients in a Superalloy at Elevated Temperature, *Fracture Mechanics, Fourteenth Symposium,* Vol II, *Testing and Applications,* STP 791, American Society for Testing and Materials, 1983, p II-536 to II-552

Chapter 7

Fracture Mechanics for Mixed Crack Tip Displacement Modes

When a cracked plate is subjected to arbitrary loading, the stress field near the crack tip can be divided into three basic types (Fig. 7-1). The tensile mode, or opening mode (mode 1), is associated with local displacement in which the crack surfaces move directly apart. The shear mode, or sliding mode (mode 2), is characterized by displacements in which the crack surfaces slide over one another in the direction perpendicular to the leading edge of the crack. The tearing mode, or torsion mode (mode 3), results in the crack surfaces sliding with respect to one another in the direction parallel to the leading edge of the crack. The stress and displacement fields near the crack tip are characterized by the stress intensity factor K_1 (for mode 1), K_2 (for mode 2), or K_3 (for mode 3), or the combination of any two (or all three) of these modes. The linear superposition of these three modes is sufficient to describe the most general case of crack tip stress fields.

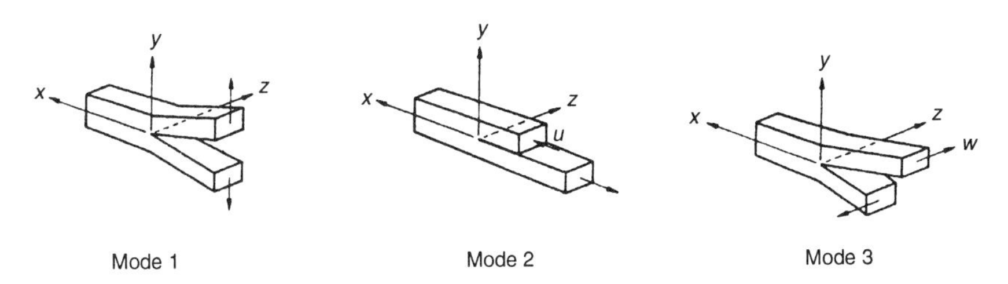

Fig. 7-1 Basic modes of crack (surface) displacements for isotropic materials

So far, we have centered our discussions around the tensile mode configuration—the applied loading was perpendicular to the crack. In the following paragraphs we will first introduce the crack tip stress and displacement fields for mode 2 and mode 3 (taken from Ref 7-1 and 7-2), as supplements to the mode 1 solutions given in Chapter 1. Afterward, we will discuss a generalized configuration, a stationary crack subjected to a generalized in-plane loading condition. That is, the crack may lie on an oblique angle with respect to the loading direction. The loads may be uniaxial tension, pure shear, or in a biaxial stress field. Depending on the situation, the crack tip displacement field would be pure mode 1, pure mode 2, or combined modes 1 and 2. The problem for mode 3 (which involves out-of-plane loading) is not considered here.

Stress and Displacement Fields at Crack Tip

The stresses and displacements for a mode 2 crack can be written as:

$$\sigma_y = \frac{K_2}{\sqrt{2\pi r}} \sin\frac{\theta}{2} \cos\frac{\theta}{2} \cos\frac{3\theta}{2}$$

$$\sigma_x = -\frac{K_2}{\sqrt{2\pi r}} \sin\frac{\theta}{2}\left[2 + \cos\frac{\theta}{2} \cos\frac{3\theta}{2}\right]$$

$$\tau_{xy} = \frac{K_2}{\sqrt{2\pi r}} \cos\frac{\theta}{2}\left[1 - \sin\frac{\theta}{2} \sin\frac{3\theta}{2}\right] \qquad \text{(Eq 7-1)}$$

for plane strain:

$$\sigma_z = \nu(\sigma_x + \sigma_y)$$

$$\tau_{xz} = \tau_{yz} = 0$$

$$v = \frac{K_2}{G}\left(\frac{r}{2\pi}\right)^{1/2} \cos\frac{\theta}{2}\left[-1 + 2\nu + \sin^2\frac{\theta}{2}\right]$$

$$u = \frac{K_2}{G}\left(\frac{r}{2\pi}\right)^{1/2} \sin\frac{\theta}{2}\left[2 - 2\nu + \cos^2\frac{\theta}{2}\right]$$

$$w = 0 \qquad \text{(Eq 7-2)}$$

For plane stress:

$$\sigma_z = \tau_{xz} = \tau_{yz} = 0$$

$$v = \frac{K_2}{G}\left(\frac{r}{2\pi}\right)^{1/2} \cos\frac{\theta}{2}\left[-1 + 2\left(\frac{\nu}{1+\nu}\right) + \sin^2\frac{\theta}{2}\right]$$

$$u = \frac{K_2}{G}\left(\frac{r}{2\pi}\right)^{1/2} \sin\frac{\theta}{2}\left[2 - 2\left(\frac{\nu}{1+\nu}\right) + \cos^2\frac{\theta}{2}\right]$$

$$w = \frac{K_2}{G} \cdot \frac{z}{\sqrt{2\pi r}} \cdot \left(\frac{\nu}{1-\nu}\right) \cdot \sin\frac{\theta}{2} \qquad \text{(Eq 7-3)}$$

For a mode 3 crack, which is neither plane stress nor plane strain, the stresses and displacements are:

$$\sigma_x = \sigma_y = \sigma_z = \tau_{xy} = 0$$

$$\tau_{xz} = -\frac{K_3}{\sqrt{2\pi r}} \sin\frac{\theta}{2}$$

$$\tau_{yz} = \frac{K_3}{\sqrt{2\pi r}} \cos\frac{\theta}{2}$$

$$v = u = 0$$

$$w = \frac{K_3}{G}\sqrt{\frac{2r}{\pi}} \sin\frac{\theta}{2} \qquad \text{(Eq 7-4)}$$

Equations 7-1 to 7-4 and those stated in Chapter 1 for uniaxial tension are approximations. Some nonsingular terms (the so-called higher-order terms) have been left out. These nonsingular terms account for the effect of the load applied parallel to the plane of the crack. Therefore, in the case of biaxial loading (with a stress component parallel to the crack) the nonsingular terms should not be omitted. The updated crack tip stress and displacement fields (which combine modes 1 and 2 by superposition) are (Ref 7-2 and 7-3):

$$\sigma_y = \frac{K_1}{\sqrt{2\pi r}} \cos\frac{\theta}{2}\left[1 + \sin\frac{\theta}{2}\sin\frac{3\theta}{2}\right] + \frac{K_2}{\sqrt{2\pi r}} \sin\frac{\theta}{2}\cos\frac{\theta}{2}\cos\frac{3\theta}{2}$$

$$\sigma_x = \frac{K_1}{\sqrt{2\pi r}} \cos\frac{\theta}{2}\left[1 - \sin\frac{\theta}{2}\sin\frac{3\theta}{2}\right] - \frac{K_2}{\sqrt{2\pi r}} \sin\frac{\theta}{2}\left[2 + \cos\frac{\theta}{2}\cos\frac{3\theta}{2}\right] + S_y(1-\lambda)\cos 2\beta$$

$$\tau_{xy} = \frac{K_1}{\sqrt{2\pi r}} \sin\frac{\theta}{2}\cos\frac{\theta}{2}\cos\frac{3\theta}{2} + \frac{K_2}{\sqrt{2\pi r}} \cos\frac{\theta}{2}\left[1 - \sin\frac{\theta}{2}\sin\frac{3\theta}{2}\right] \quad \text{(Eq 7-5)}$$

$$v = \frac{K_1}{G}\sqrt{\frac{r}{2\pi}} \sin\frac{\theta}{2}\left[\frac{1}{2}(\kappa+1) - \cos^2\frac{\theta}{2}\right] + \frac{K_2}{G}\sqrt{\frac{r}{2\pi}} \cos\frac{\theta}{2}\left[\frac{1}{2}(1-\kappa) + \sin^2\frac{\theta}{2}\right]$$

$$+ \frac{(1-\lambda)S_y}{8G}\{r[\sin(2\beta-\theta) + \kappa\sin(2\beta+\theta) - 2\sin\theta\cos 2\beta] + (\kappa+1)a\sin 2\beta\}$$

$$u = \frac{K_1}{G}\sqrt{\frac{r}{2\pi}} \cos\frac{\theta}{2}\left[\frac{1}{2}(\kappa-1) + \sin^2\frac{\theta}{2}\right] + \frac{K_2}{G}\sqrt{\frac{r}{2\pi}} \sin\frac{\theta}{2}\left[\frac{1}{2}(\kappa+1) + \cos^2\frac{\theta}{2}\right]$$

$$+ \frac{(1-\lambda)S_y}{8G}\{r[\cos(\theta+2\beta) + \kappa\cos(\theta-2\beta) - 2\sin\theta\sin 2\beta] + (\kappa+1)a\cos 2\beta\}$$

(Eq 7-6)

where $\lambda = S_x/S_y$, $\kappa = (3 - 4\nu)$ for plane strain and $(3-\nu)/(1+\nu)$ for generalized plane stress, and r and θ are polar coordinates defining the relationship between a point and the crack tip (see Fig. 1-5). For the crack orientation, β

is denoted as an angle between the crack plane and a plane of reference (i.e., the y-z plane, see Fig. 7-2 to 7-4).

Stress Intensity Factors

As shown in Fig. 7-2 to 7-4, a mixed-mode condition occurs whenever there are both normal-stress and shear-stress components acting on the crack. The applied in-plane stresses can be uniaxial, biaxial, or pure shear. The stress intensity factors are:

$$K_I = \sigma\sqrt{\pi a} \cdot \phi_i$$

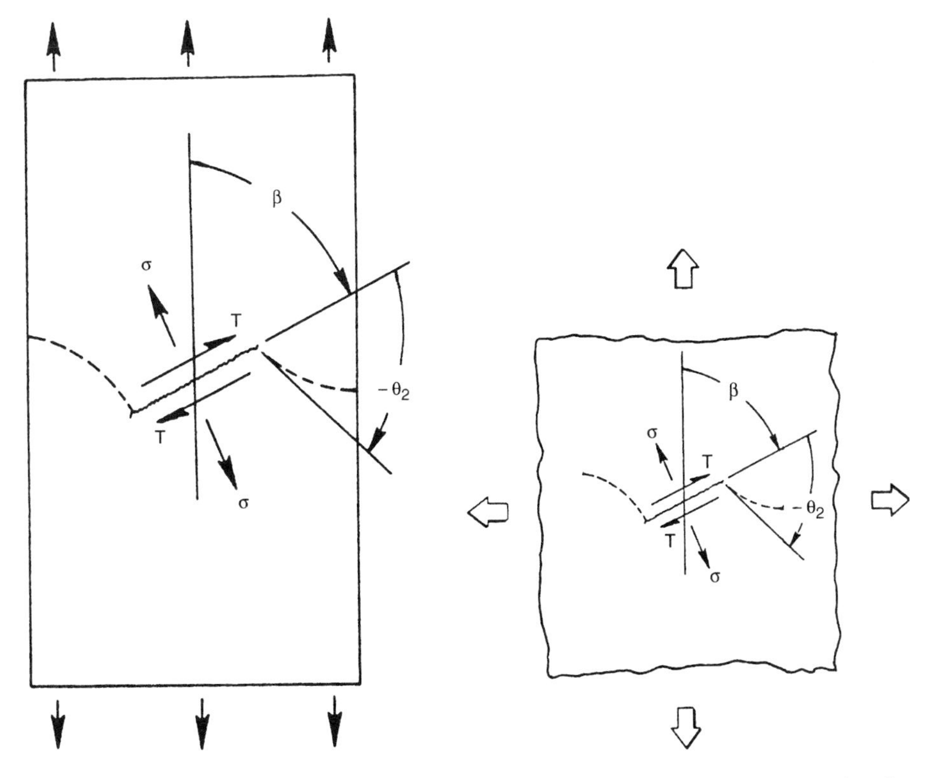

Fig. 7-2 Stress components on a crack subjected to uniaxial tension

Fig. 7-3 Stress components on a crack subjected to biaxial load

$$K_2 = \tau\sqrt{\pi a} \cdot \phi_{ii} \quad \text{(Eq 7-7)}$$

where ϕ_i and ϕ_{ii} are dimensionless functions accounting for the effect of geometry and loading conditions. The tension and shear stresses (σ and τ) can be obtained by using simple, straightforward stress analysis procedures, e.g., the Mohr circle or a finite-element computer code.

In the literature (particularly in textbooks or handbooks for stress intensity solutions), often a minus sign appears in front of the stress intensity expression for K_2. Most finite-element computer codes also print out the K_2 as a negative value. This minus sign should not be confused as a negative value. For a mode 2 crack the sign of K_2 will be consistent with the shear stress in the region near the crack tip. The minus sign is a convention used to indicate the relative movement of crack surfaces at a given crack tip (one surface slides forward while another surface slides backward). A plus sign would indicate an opposite situation when applicable. Therefore, the absolute value of K_2 is always used for engineering fracture mechanics analysis.

Letting ϕ_i and ϕ_{ii} be equal to unity (i.e., for an infinitely wide sheet), the stress intensity factors K_1 and K_2 for the configuration shown in Fig. 7-3 can

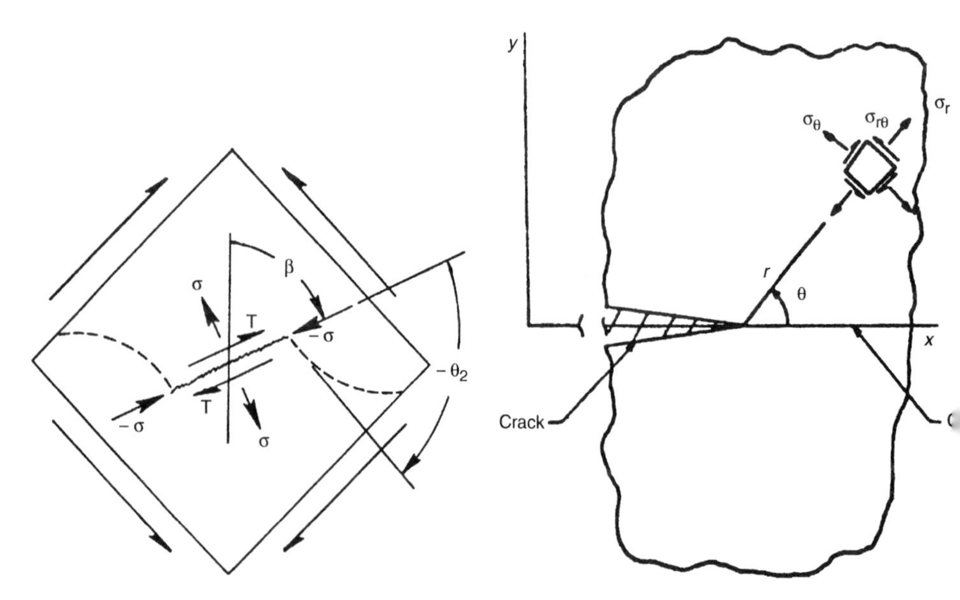

Fig. 7-4 Stress components on a crack subjected to shear load

Fig. 7-5 Coordinates used to describe stresses near a crack tip

be expressed in terms of the remote reference stresses as (Ref 7-3, 7-4):

$$K_1 = \frac{S_y\sqrt{\pi a}}{2}[(1+\lambda) - (1-\lambda)\cos 2\beta]$$

$$K_2 = \frac{S_y\sqrt{\pi a}}{2}(1-\lambda)\sin 2\beta \qquad \text{(Eq 7-8)}$$

where $\lambda = S_x/S_y$ with S_y being a positive quantity (i.e., always in tension), S_x can be either positive (tension) or negative (compression).

For a flat crack, i.e., $\beta = \pi/2$, the crack is normal to S_y and parallel to S_x, Eq 7-8 reduces to:

$$K_1 = S_y\sqrt{\pi a}$$

$$K_2 = 0 \qquad \text{(Eq 7-8a)}$$

i.e., independent of the biaxial stress ratio . However, for an inclined crack the effect of the horizontal load appears directly in the expression for K_1 and K_2. For example, in the case of uniaxial tension (i.e., $\lambda = 0$), Eq 7-8 reduces to:

$$K_1 = S_y\sqrt{\pi a}\cdot\sin^2\beta$$

$$K_2 = S_y\sqrt{\pi a}\cdot\sin\beta\cdot\cos\beta \qquad \text{(Eq 7-8b)}$$

For $\lambda = -1$:

$$K_1 = -S_y\sqrt{\pi a}\cdot\cos 2\beta$$

$$K_2 = S_y\sqrt{\pi a}\cdot\sin 2\beta \qquad \text{(Eq 7-8c)}$$

If the crack is subjected to in-plane shear in lieu of tension and compression, the stress intensity factors can simply be written as:

$$K_1 = -S_s\sqrt{\pi a}\cdot\cos 2\beta$$

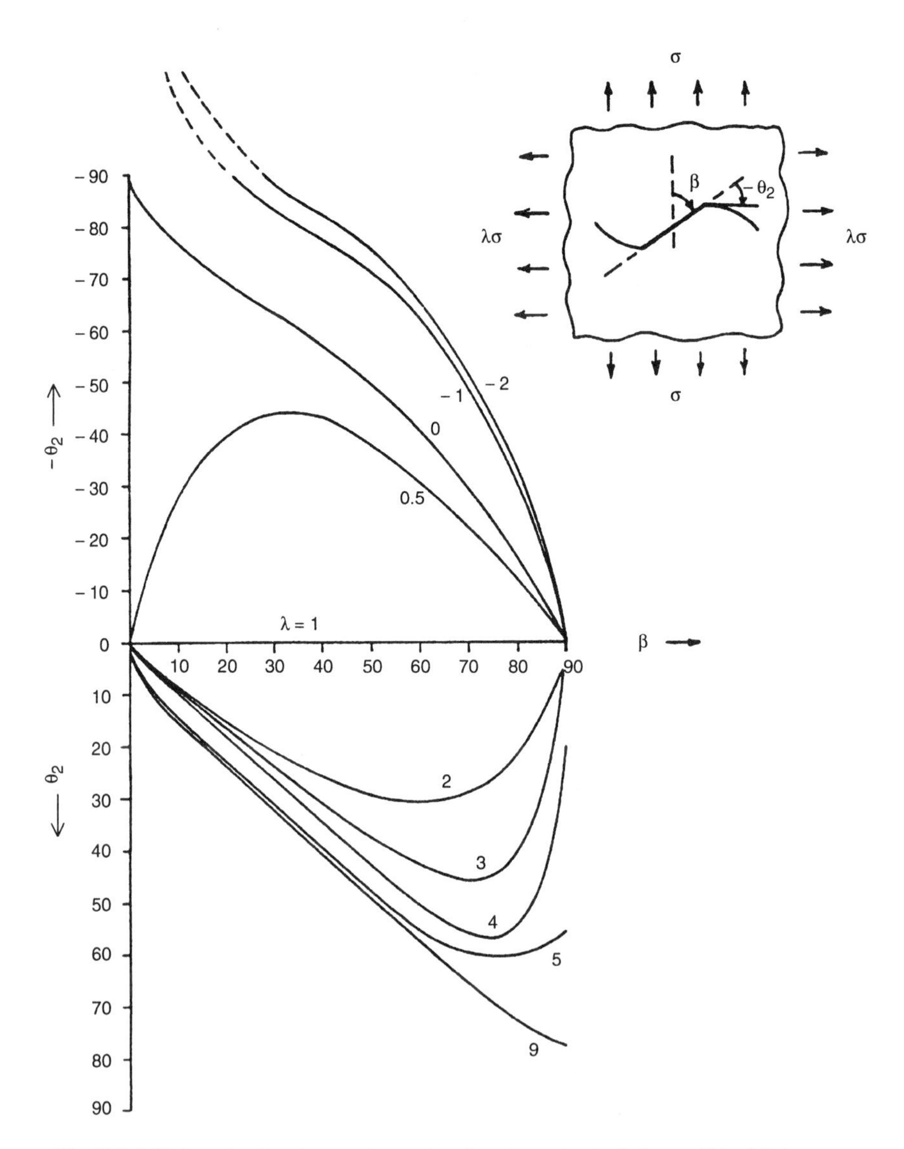

Fig. 7-6 Initial crack extension angle as a function of crack orientation and biaxial stress ratio. Source: Ref 7-4

$$K_2 = S_s\sqrt{\pi a} \cdot \sin 2\beta \qquad \text{(Eq 7-8d)}$$

In either case, pure shear is attended at $\beta = \pi/4$, where $K_1 = 0$ and $K_2 = S_y\sqrt{\pi a}$ (or $S_s\sqrt{\pi a}$, depending on loading condition). For an opposite situation, i.e., $\lambda = 1$, the effect of S_x disappears and K_1 and K_2 are once again given by Eq 7-8(a), with K_1 being independent of β (Ref 7-4).

Direction of Crack Extension

In Chapter 1, we presented the crack tip stress field with a rectangular coordinate system. After converting the stress terms in Eq 1-18 and 7-1 from a rectangular coordinate system to a cylindrical coordinate system (see Fig. 7-5 for notations), we have (Ref 7-5):

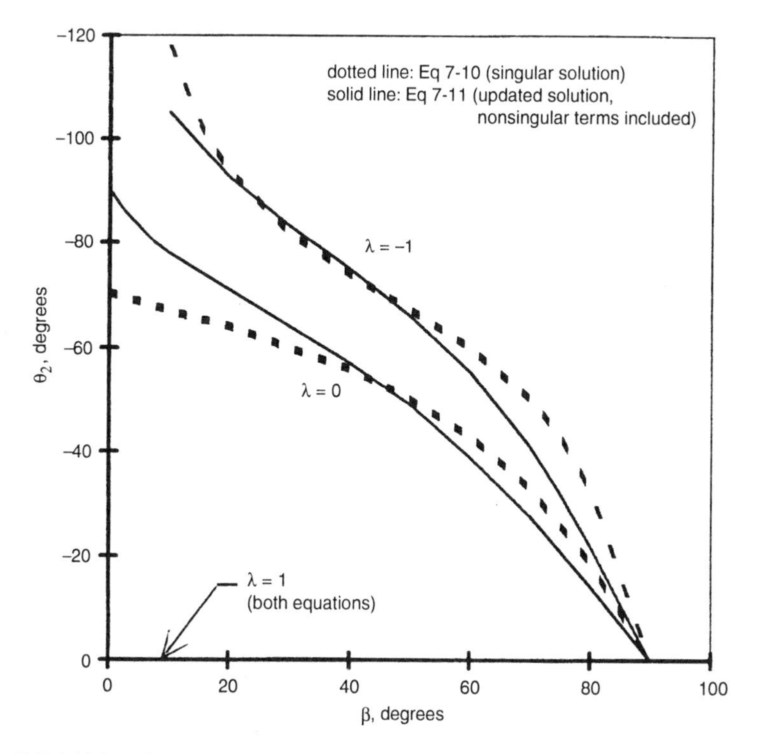

Fig. 7-7 Initial crack extension angle as a function of crack orientation and biaxial stress ratio. Comparison of two analytical solutions

$$\sigma_r = \frac{1}{\sqrt{2\pi r}} \cos\frac{\theta}{2}\left[K_1\left(1 + \sin^2\frac{\theta}{2}\right) + K_2\left(\frac{3}{2}\sin\frac{\theta}{2} - 2\tan\frac{\theta}{2}\right)\right]$$

$$\sigma_\theta = \frac{1}{\sqrt{2\pi r}} \cos\frac{\theta}{2}\left[K_1 \cos^2\frac{\theta}{2} + \frac{3}{2}K_2 \sin\theta\right]$$

$$\tau_{r\theta} = \frac{1}{\sqrt{2\pi r}} \cos\frac{\theta}{2}\left[K_1 \sin\theta + K_2\left(3\cos\theta - 1\right)\right] \quad \text{(Eq 7-9)}$$

On the basis of the maximum stress criterion for brittle fracture, the crack will move along a path normal to the direction of greatest tension. Thus, the shear stress on a line of expected path of crack extension is zero. By setting $\tau_{r\theta}$ to zero, and defining θ_2 as the angle at which the circumferential stress σ_θ is a maximum, we have:

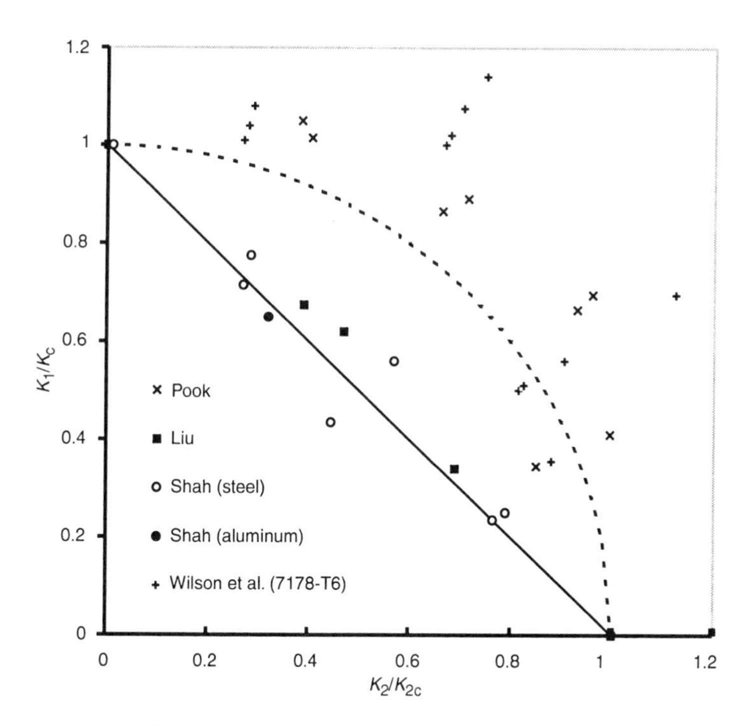

Fig. 7-9 Behavior of K_1 and K_2 interactions

$$K_1 \sin \theta_2 + K_2 (3 \cos \theta_2 - 1) = 0 \qquad \text{(Eq 7-10)}$$

where K_1 and K_2 are functions of β and are given by Eq 7-8 to 7-8(d).

Repeating the same procedure using Eq 7-5, we can have another equation in which the effect of biaxial stresses is included:

$$1.75 \cdot [D_1 (3 \cos \theta_2 - 1) - 3D_2 \sin \theta_2] + 4D_3 \sin \frac{\theta_2}{2} \cos \theta_2 = 0 \qquad \text{(Eq 7-11)}$$

where $D_1 = -3(1 - \lambda)\sin \beta \cos \beta$, $D_2 = \lambda + (1 - \lambda)\sin 2\beta$, and $D_3 = (1 - \lambda)\cos 2\beta$.

A graphical presentation of Eq 7-11 is given in Fig. 7-6, showing the angle at onset of crack extension (θ_2) as a function of original crack inclination angle (β) and biaxial stress ratio (λ).

Figure 7-7 shows comparisons between the two solutions (Eq 7-10 and 7-11) for three selected biaxial stress ratios ($\lambda = 0$, -1, and $+1$). The curve for the "singular solution" with $\lambda = 0$ is the familiar one originally proposed by Erdogan and Sih (Ref 7-5). The curve for the "updated solution" with $\lambda = 0$ is equivalent to that of Williams and Ewing (Ref 7-6). Similar analytical results, which tend to agree with the updated solution, have also been reported in the literature (Ref 7-7). For $\lambda = 1$, the solution for θ_2 is unaffected

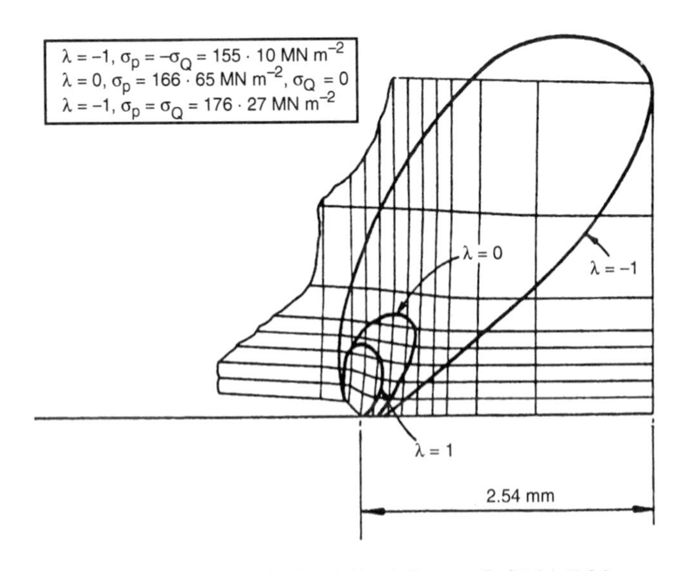

Fig. 7-10 Crack tip plastic zone under biaxial load. Source: Ref 7-31, 7-34

by the nonsingular terms in the crack tip stress field equations. As stated earlier, the initial crack angle β plays no role in the direction of crack propagation. The crack will grow straight and stay on its original orientation. Laboratory test results are available in substantiation of this (Ref 7-8, 7-9).

The curves in the upper part of Fig. 7-6 (for $\lambda \leq 1$) merge to a point at $\beta = 90°$, $\theta_2 = 0$. This means that while S_y is the most dominating stress component a flat crack ($\beta = 90°$) will stay on its original plane. As shown in the lower half of Fig. 7-6, when S_x is greater than S_y the crack will turn toward the y-axis whether the original crack plane is parallel, or inclined to S_x. The crack extension angle θ_2 will be positive (counter-clockwise) and the crack will end up normal to S_x, the most dominating stress in the system.

Experimental evidence for an initially flat crack subjected to $\lambda = 1.5$ biaxial stress ratio is shown in Fig. 7-8. More data of this kind (including $\lambda =$

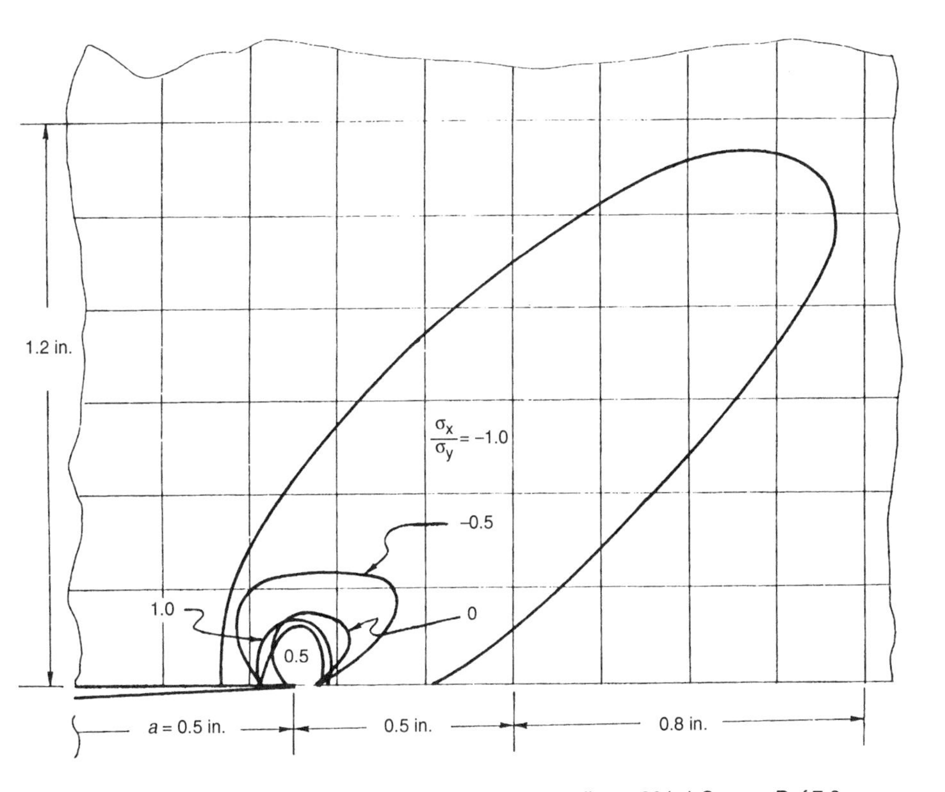

Fig. 7-11 Crack tip plastic zone for 2024-T351 aluminum alloy at 30 ksi. Source: Ref 7-9

1.25 and 1.75) are also available (Ref 7-9). Although this curved crack validates, at least qualitatively, the physical aspect of crack growth under high biaxial stress ratios, Eq 7-11 cannot be used for predicting the entire crack propagation path. The crack tip stress, displacements, and stress intensity solutions were derived for straight cracks. The "initial take-off angle" θ_2 is only a means for indicating the direction in which the crack will start propagating. Subsequently, a new θ_2 will be needed for the slightly bifurcated crack. Then, another θ_2 will be needed for the next crack increment, and so forth. Eventually the crack will turn to a plane normal to the most dominating tensile stress direction (as shown in Fig. 7-8). Thus, an analytical crack growth methodology for predicting the crack propagation path is needed. Unfortunately, such a method is currently unavailable.

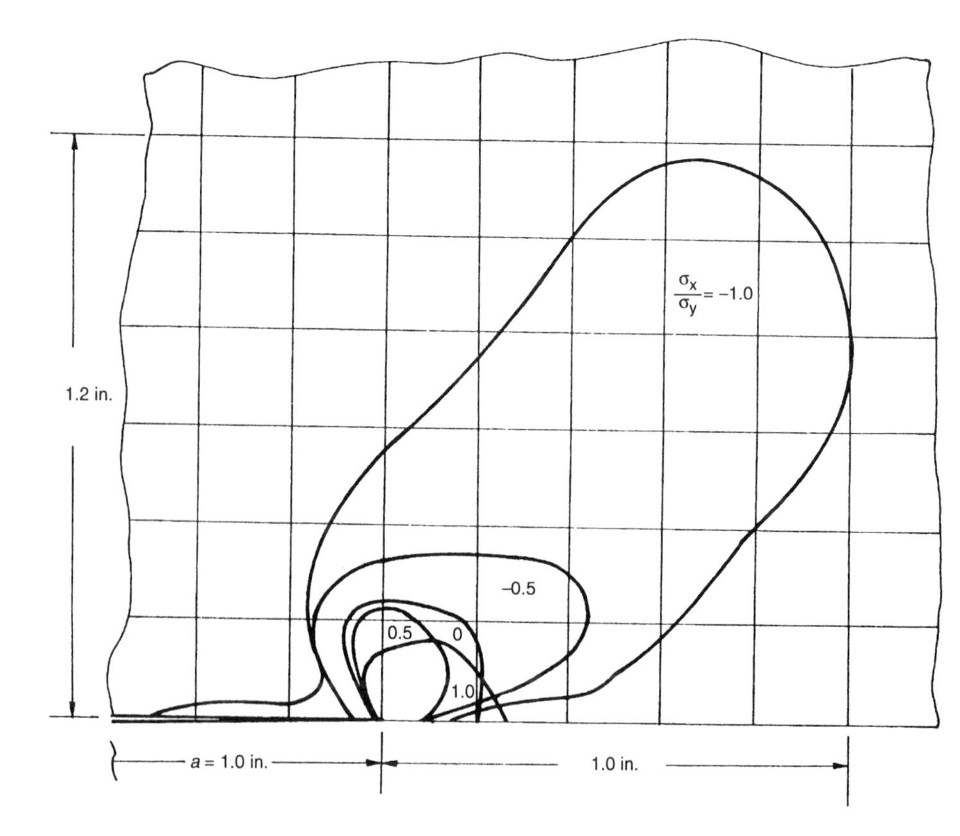

Fig. 7-12 Crack tip plastic zone for 7075-T7351 aluminum alloy at 30 ksi. Source: Ref 7-9

Iida and Kobayashi (Ref 7-10) have pointed out that the K_2 component will vanish very rapidly once the crack starts to propagate. Consequently, the crack will quickly change its direction while crack propagation is in progress, turning normal to the loading direction. Their work was based on an inclined crack subjected to uniaxial loading (i.e., a situation normally associated with Eq 7-10). Their observation is supported by a number of fracture test data from other sources (Ref 7-11, 7-12). However, the crack behavior shown in Fig. 7-8 seems to indicate otherwise. Though the crack did finally end up perpendicular to the most dominating loading direction, the entire course of crack propagation involved many fatigue cycles. It is conceivable that the K_2 component of the curved crack vanished during each fatigue cycle but rapid fracture did not occur until the crack had reached its critical length. In each consecutive crack growth increment, the crack re-established its mixed-mode status and was associated with a different set of K_1 and K_2 combinations. An approximate method proposed by Leevers et al.

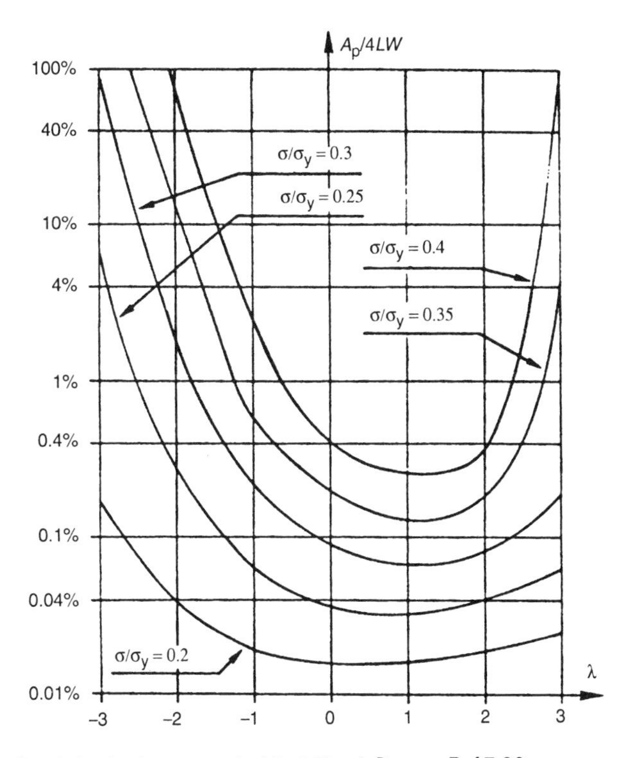

Fig. 7-13 Crack tip plastic zone under biaxial load. Source: Ref 7-30

(Ref 7-13) has been used to make correlation with the curved crack test data, and moderate success was achieved (Ref 7-9). In any event, development of a more sophisticated methodology is required.

Failure Criteria

Now we will expand Eq 1-24 to cover a general case that includes all three crack tip displacement modes, on the basis of the superposition principle. According to Ref 7-14, the total strain energy release rate can be expressed as:

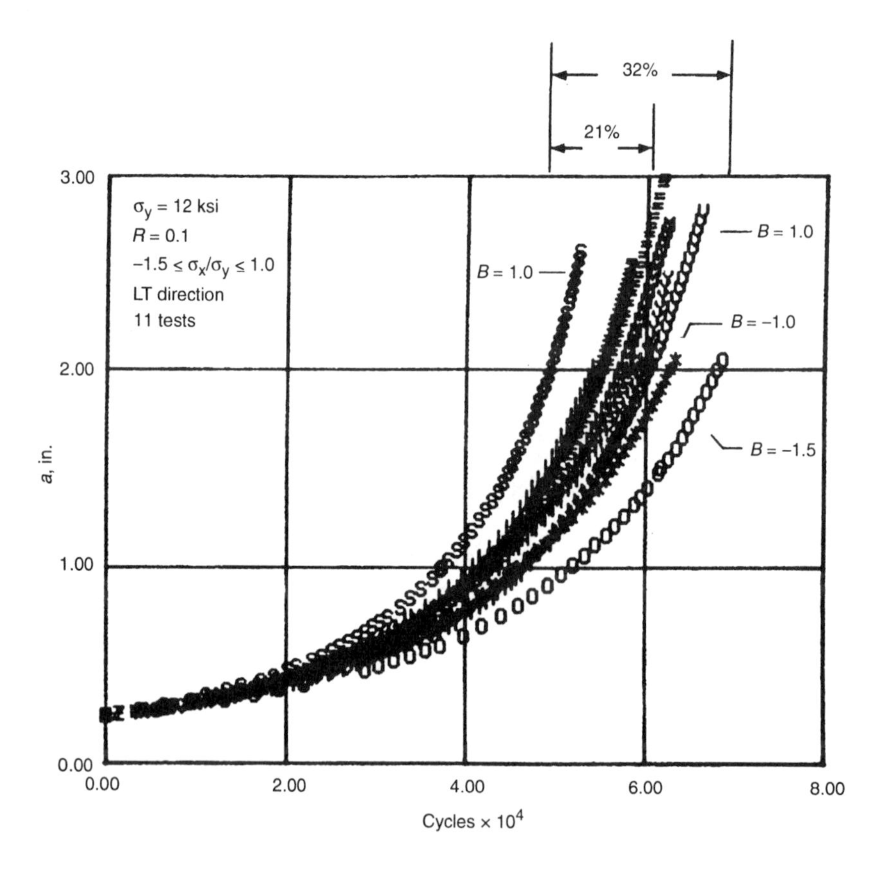

Fig. 7-14 Crack growth histories of the 7075-T7351 aluminum alloy cruciform specimens. Source: Ref 7-9

$$\mathcal{G} = \mathcal{G}_1 + \mathcal{G}_2 + \mathcal{G}_3$$

$$= \lim_{\alpha \to 0} \frac{2}{\alpha} \int_0^{\alpha} \left(\frac{\sigma_y v}{2} + \frac{\tau_{yx} u}{2} + \frac{\tau_{yz} w}{2} \right) dr \qquad \text{(Eq 7-12)}$$

To make a connection between $\mathcal{G}$ and K, we can use a technique demonstrated in Chapter 1. When the stresses in Eq 7-12 are appropriately obtained with $r = \alpha - r$ and $\theta = 0$, and the displacements are obtained with r and $\theta = \pi$, and the values are substituted into Eq 7-12, we have:

$$\mathcal{G}_1 = K_1^2 \cdot E'$$

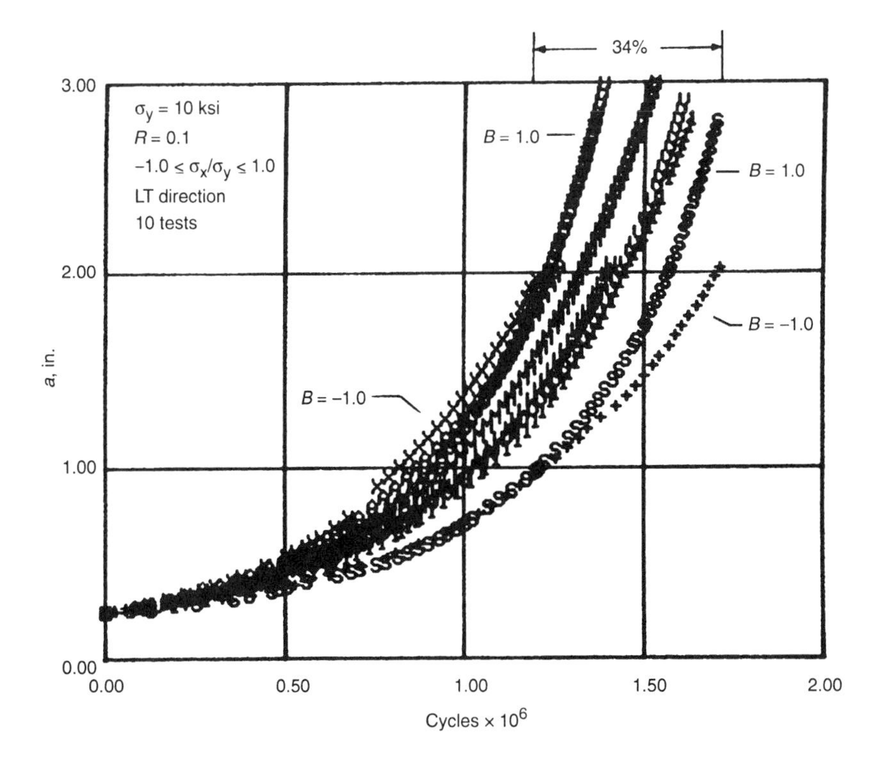

Fig. 7-15 Crack growth histories of the 2024-T351 aluminum alloy cruciform specimens. Source: Ref 7-9

$$\mathcal{G}_2 = K_2^2 \cdot E'$$

$$\mathcal{G}_3 = K_3^2/2G = \frac{1+\nu}{E} K_3^2 \qquad \text{(Eq 7-13)}$$

where $E' = 1/E$ for plane stress and $(1 - \nu^2)/E$ for plane strain. $\mathcal{G}_3$ is neither plane stress nor plane strain. Note that Eq 7-12 and 7-13 were derived with the higher-order terms omitted. Thus, having the nonsingular terms included, the results of the integrand for $\mathcal{G}$ in Eq 7-12 will be altered, as well as the $\mathcal{G}$ and K relationship, i.e., Eq 7-13. The biaxial ratio λ will appear in these equations to account for the effect of S_x on fracture loads.

Now we can derive a failure criterion for the mixed mode 1 and mode 2 from Eq 7-12 and 7-13:

$$K^2 = K_1^2 + K_2^2$$

or:

$$K = \sqrt{K_1^2 + K_2^2} \qquad \text{(Eq 7-14)}$$

Here, K can be considered an effective value equivalent to the total contribution of K_1 and K_2. This is simply stating that the plate will fail when K reaches a critical value while the interaction between K_1 and K_2 satisfies the relationship of Eq 7-12 (or Eq 7-14). But what is the critical value? Rewriting Eq 7-14, we have:

$$1 = \left(\frac{K_1}{K}\right)^2 + \left(\frac{K_2}{K}\right)^2 \qquad \text{(Eq 7-14a)}$$

To satisfy Eq 7-14 or Eq 7-14(a), K_{1c} must equal K_{2c}. Therefore, while no known test data are available, it is commonly assumed that K_{2c} equals K_{1c}. In reality, K_{1c} may or may not equal K_{2c}. In fact, the maximum stress criterion used by Erdogan and Sih (Ref 7-5) has predicted that $K_{2c} = 0.866$ K_{1c}. On the other hand, on the basis of the strain energy density criterion (Ref 7-15), Sih and MacDonald (Ref 7-16) have predicted that:

$$\frac{K_{2c}}{K_{1c}} = \sqrt{\frac{3(1-2\nu)}{2(1-\nu)-\nu^2}} \qquad \text{(Eq 7-15)}$$

which gives $K_{2c} = 0.905K_{1c}$ for $\nu = 1/3$. A compilation of available fracture test data is presented in Table 7-1. The K_{1c} and K_{2c} values shown in the table are reportedly based on initial crack length and fracture load. On the basis of these data, the K_{2c} to K_{1c} ratio varies from 0.75 to 1.1. The test data listed in Table 7-1 include materials in four categories: polymethylmethacrylate, high- and moderate-strength aluminum alloys, and steel. For a given material, the experimentally developed K_{2c} to K_{1c} ratios have been consistent, even though different specimen types were used by different investigators. Extrapolated K_{2c} values have been obtained from tensile-loaded specimens. Therefore, it seems that the K_{2c} to K_{1c} ratios compiled in Table 7-1 are material dependent but specimen geometry and investigator independent. Many other investigators (Ref 7-22 to 7-26) have developed various types of specimens and devices for determining the pure K_{2c} value, but they did not offer any information as to the K_{2c} to K_{1c} ratio for any material. With the knowledge gained from the test data listed in Table 7-1, it seems logical for us to modify Eq 7-14(a) to become:

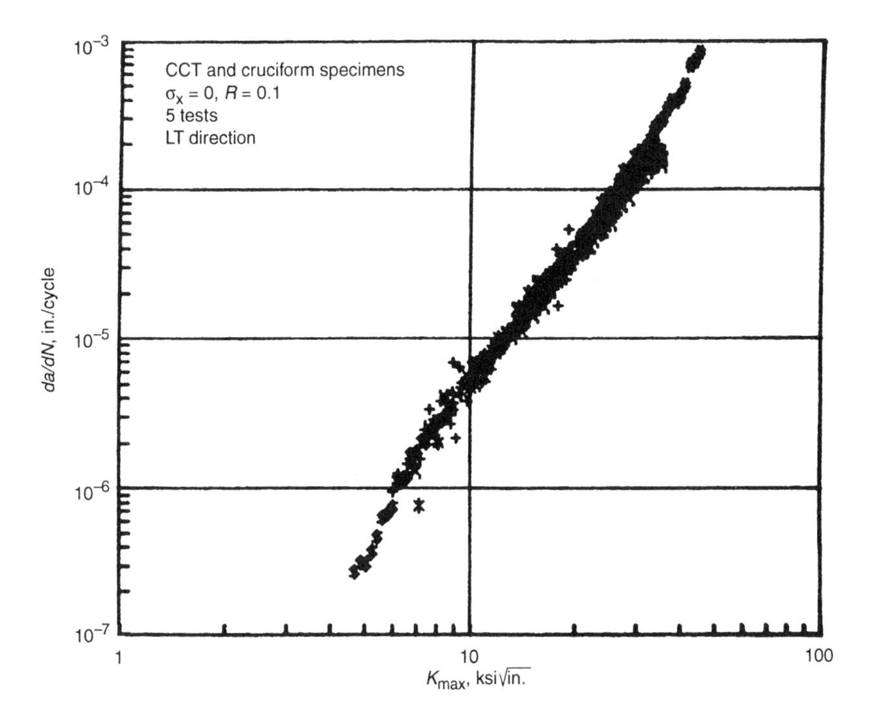

Fig. 7-16a Cyclic crack growth rate behavior of 7075-T7351 aluminum alloy. Source: Ref 7-9

$$\left(\frac{K_1}{K_{1c}}\right)^2 + \left(\frac{K_2}{K_{2c}}\right)^2 = 1 \qquad \text{(Eq 7-16)}$$

or:

$$\left(\frac{K_1}{K_{1c}}\right)^2 + \left(\frac{K_2}{\alpha K_{1c}}\right)^2 = 1 \qquad \text{(Eq 7-16a)}$$

where $\alpha = K_{2c}/K_{1c}$.

To validate Eq 7-16 as a failure criterion by empirical correlation, test data developed from testing of angle-cracked specimens are plotted in Fig. 7-9 for evaluation. Although they are limited and by no means conclusive, the data of Liu (Ref 7-12) and Shah (Ref 7-17) show the following relationship:

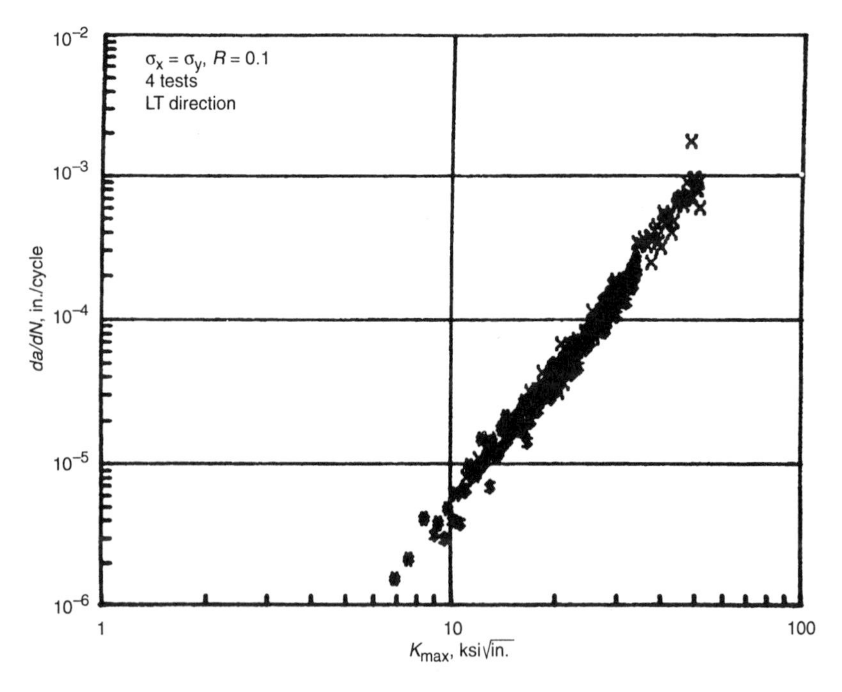

Fig. 7-16b Cyclic crack growth rate behavior of 7075-T7351 aluminum alloy. Source: Ref 7-9

$$\frac{K_1}{K_{1c}} + \frac{K_2}{K_{2c}} = 1 \qquad \text{(Eq 7-17)}$$

or:

$$\frac{K_1}{K_{1c}} + \frac{K_2}{\alpha K_{1c}} = 1 \qquad \text{(Eq 7-17a)}$$

On the other hand, as shown in Fig. 7-9, there is no clear correlation between K_1/K_{1c} and K_2/K_{2c} among the test data reported by Pook (Ref 7-18) or Wilson (Ref 7-11). Therefore, it is anticipated that the real interaction behavior falls in between Eq 7-16 and 7-17. Thus, we can write a generalized interaction criterion as:

$$\left(\frac{K_1}{K_{1c}}\right)^{\xi} + \left(\frac{K_2}{K_{2c}}\right)^{\zeta} = 1 \qquad \text{(Eq 7-18)}$$

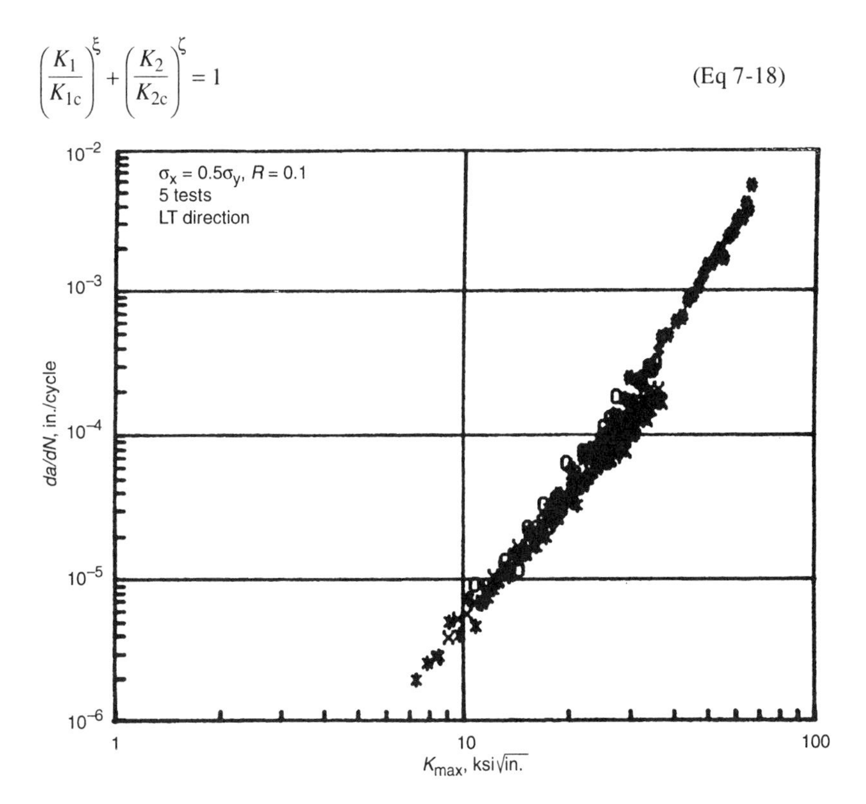

Fig. 7-16c Cyclic crack growth rate behavior of 7075-T7351 aluminum alloy. Source: Ref 7-9

Table 7-1 Mode 1 and mode 2 fracture toughness for various materials

Material	Thickness, mm	K_{1c} (MPa√m)	K_{2c} (MPa√m)	K_{2c}/K_{1c}	Specimen type (for mode 2)	Loading for mode 2	Source Reference
7075-T7651 aluminum (5.6 Zn, 2.5 Mg)	7.62	89.2	89.2	1.0	Picture frame	Shear	7-12
7075-T651 aluminum (5.6 Zn, 2.5 Mg)	7.62	70.0	70.0	1.0	Edge-angle crack	Tension	7-17
2024-T351 aluminum (4.5 Cu, 1.5 Mg)	7.95	99.4	74.2	0.75	Picture frame	Shear	7-12
2024-T3 aluminum (4.5 Cu, 1.5 Mg)	(a)	(a)	(a)	0.75	Center-angle crack	Tension	7-21
DTD 5050 aluminum (5.5 Zn)	10.2, 12.7	31.5	25.0(b)	0.79	Center-angle crack	Tension	7-18
4340 steel (@−200 °F)	6.35	45.7	49.7	1.08	Thin-walled tube	Torsion	7-17
AISI 01 tool steel	7.11	26.4	27.4	1.04	Compact shear	Shear	7-19
Polymethylmethacrylate	10.0	1.06(c)	0.94	0.89	Compact shear	Shear	7-20
Polymethylmethacrylate	3.0	0.52	0.46	0.89	Center crack	Shear forces	7-5

Note: Unless otherwise noted, all K_{1c} values are conventional center-crack K_c values. (a) Not reported in references. (b) Extrapolated value. (c) Compact specimen

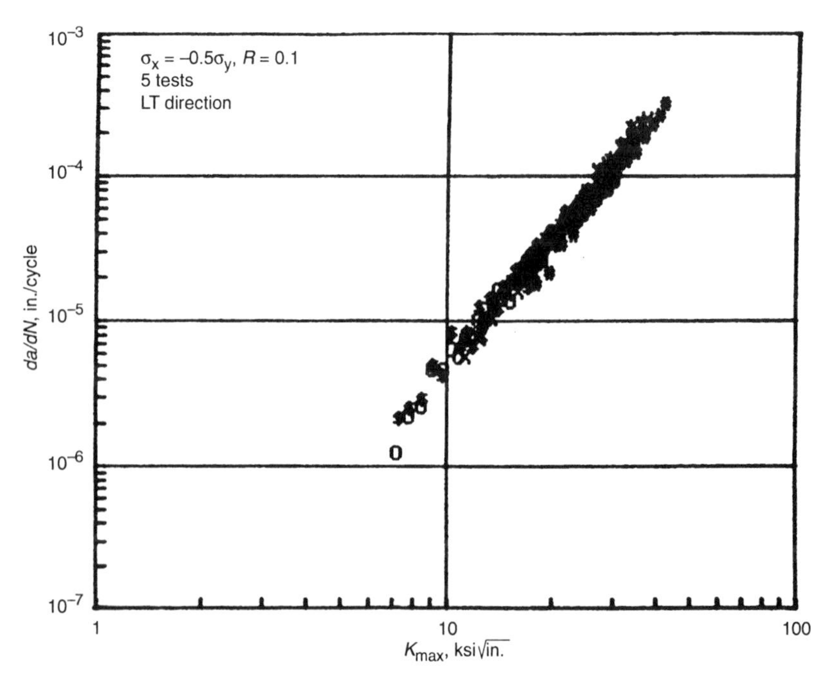

Fig. 7-16d Cyclic crack growth rate behavior of 7075-T7351 aluminum alloy. Source: Ref 7-9

or:

$$\left(\frac{K_1}{K_{1c}}\right)^{\xi} + \left(\frac{K_2}{\alpha K_{1c}}\right)^{\zeta} = 1 \qquad \text{(Eq 7-18a)}$$

and treat ξ and ζ as experimentally determined coefficients. This equation also provides flexibility in making experimental data correlation, because ξ and ζ need not be the same value for a given material. This concept was first introduced by Wu for establishing failure criteria for an anisotropic material (Ref 7-27). Equation 7-17 provides a more conservative fracture strength estimate than Eq 7-16 and 7-18, but the analytically derived failure criterion, i.e., Eq 7-16 and Eq 7-16(a), is a universally known failure criterion for mixed-mode failure.

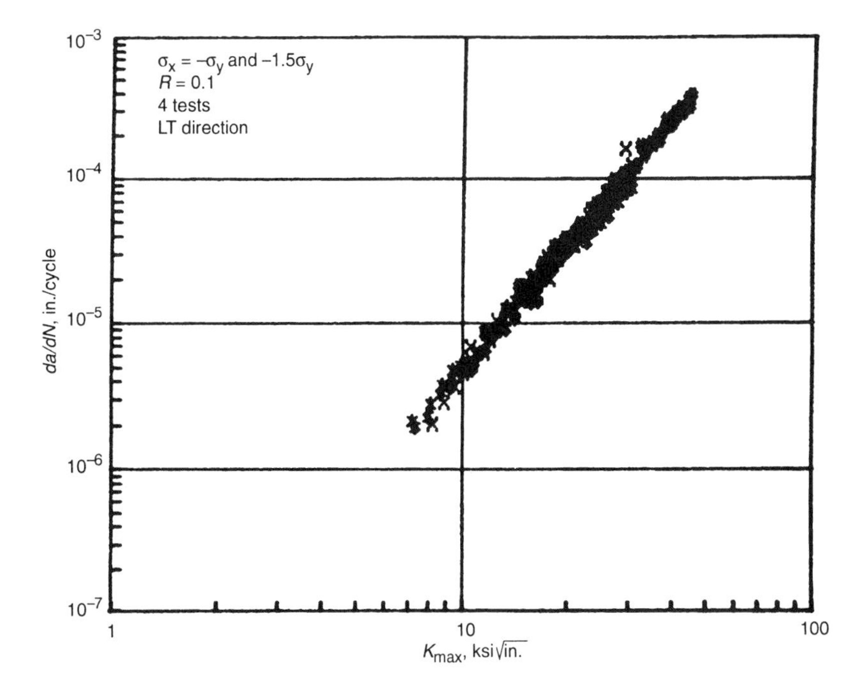

Fig. 7-16e Cyclic crack growth rate behavior of 7075-T7351 aluminum alloy. Source: Ref 7-9

Crack Tip Plastic Zone

According to Rice (Ref 7-28), all the plastic zone equations listed in Chapter 1 for mode 1 loading are adaptable to mode 2 and mode 3 loading. This is easily done by appropriately replacing the *K, S,* and F_{ty} terms with the proper values corresponding to a given mode. The crack tip plastic zone radius r_y for a crack subjected to a loading condition that involves mixed mode 1 and mode 2 has been given by Pook (Ref 7-18) as:

$$r_y = \frac{1}{2\pi} \cdot \frac{K_1^2 + 3K_2^2}{F_{ty}^2} \quad \text{(Eq 7-19)}$$

Thus, for a given stress level, the crack tip plastic zone size for pure shear is three times larger than that for the case of pure tension.

Crack tip plastic zone size under biaxial stresses can be determined analytically, by determining the maximum shear stress contours (Ref 7-4, 7-

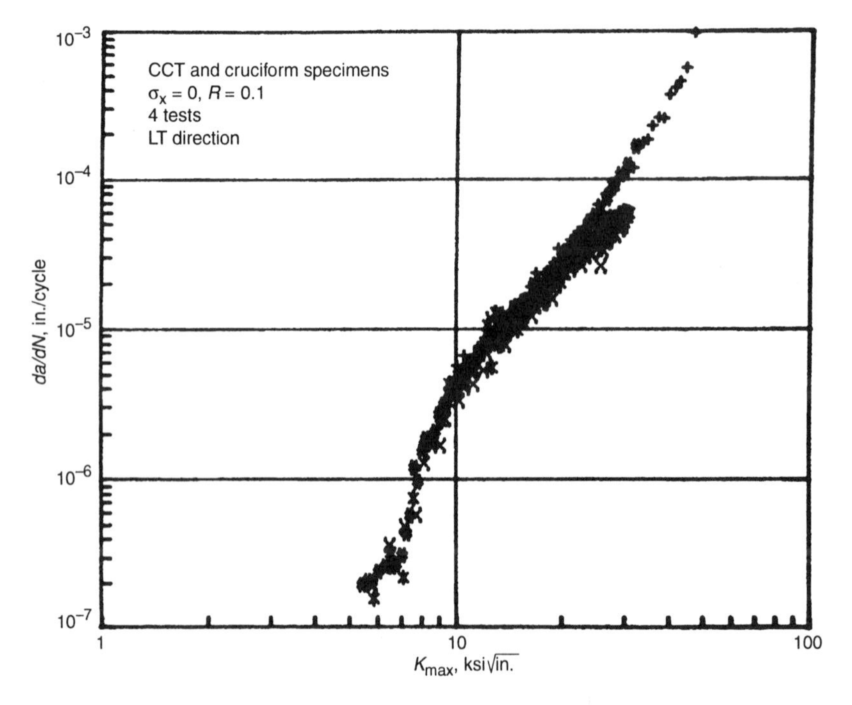

Fig. 7-17a Cyclic crack growth rate behavior of 2024-T351 aluminum alloy. Source: Ref 7-9

29), or by elastic-plastic finite-element analysis (Ref 7-9, 7-30 to 7-34). These works hold that the compression load parallel to the crack causes the enlargement in plastic zone size. However, plastic zone size decreases with tension load parallel to the crack. As compared to uniaxial tension, the plastic zone size is much larger for negative biaxial stress ratios but slightly smaller for positive biaxial stress ratios (Fig. 7-10). Examples of finite-element results for two aluminum alloys (2024-T351 and 7075-T7351) are shown in Fig. 7-11 and 7-12. These plastic zone sizes have been determined using a stepwise linear routine that comes with the NASTRAN structural analysis computer code. Figure 7-13 shows yet another set of elastic-plastic finite-element analysis results (Ref 7-30). It shows that crack tip plastic zone size decreases with increase of λ and reaches a minimum at $\lambda \cong 1$.

There is general agreement among the data in Fig. 7-10 to 7-13. However, the work of Liebowitz et al. has pushed the envelope of biaxialty to the territory that includes high positive biaxial stress ratios ($\lambda >> 1$). Figure 7-13 shows that plastic zone size will increase again as the biaxial stress ratio

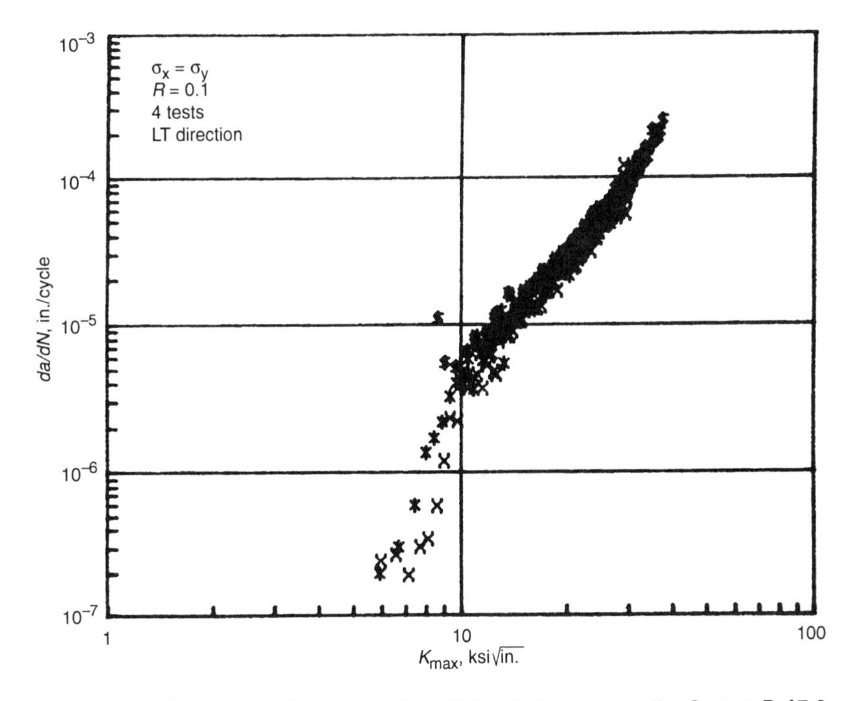

Fig. 7-17b Cyclic crack growth rate behavior of 2024-T351 aluminum alloy. Source: Ref 7-9

increases further (from $\lambda > 1$). In Fig. 7-13, σ_y is the 0.2% offset yield stress; plastic zone size is presented as the ratio of the area of the plastic zone (A_p) to the total area of the sheet ($4LW$). Details regarding the finite-element model and the analytical method used are reported in Ref 7-30 and will not be elaborated here.

Concerning the nature of the fracture path, Cotterell (Ref 7-35) has written that for a stable crack propagation path to prevail, the crack tip plastic zone (the theoretical isochromatic pattern) must not tilt backward with respect to the direction of crack propagation. Here the term *stable growth* simply means that the advancing crack tip will stay on its original crack plane during crack propagation. On the other hand, *unstable growth* means that the crack will turn away from its original crack plane during crack propagation. Now we can apply this phenomenon as a supplemental supporting argument to explain why the crack in Fig. 7-8 turns away in such a manner from its original crack plane. Examining the theoretical plastic zone patterns reported in Ref 7-4, we can make the following conclusions. The

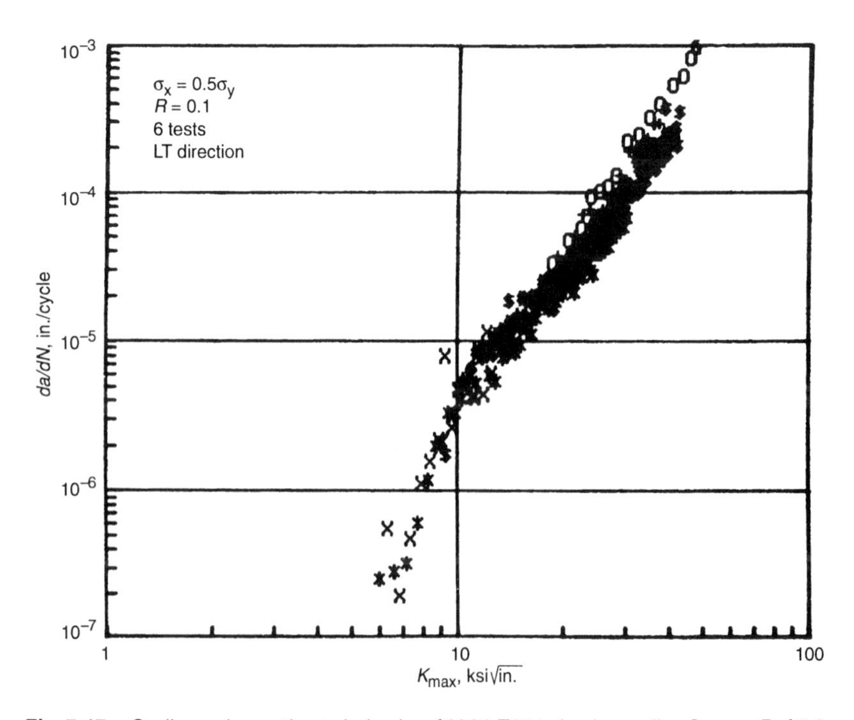

Fig. 7-17c Cyclic crack growth rate behavior of 2024-T351 aluminum alloy. Source: Ref 7-9

tip of the butterfly-shaped plastic zone of a biaxially loaded crack is located right above (and below) the crack tip when the crack is subjected to equibiaxial loading. Otherwise, the tip of the plastic zone will either be in front of, or fall behind, the crack tip. When the biaxial stress ratio is lower than 1, including 0 (uniaxial tension) and negative biaxial stress ratios, the plastic zone will tilt forward. Thus, the crack propagation path is stable. When the biaxial stress ratio is greater than 1, the shape of the plastic zone tilts backward. Therefore, the crack in Fig. 7-8, which was subjected to a 1.5 biaxial stress ratio, turned away from its original crack plane.

The Effect of Biaxial Stresses on Fatigue Crack Growth

For a crack oriented perpendicular to S_y, the addition of S_x (parallel to the crack) will not alter the mode 1 stress intensity factor (Ref 7-14, 7-33). The

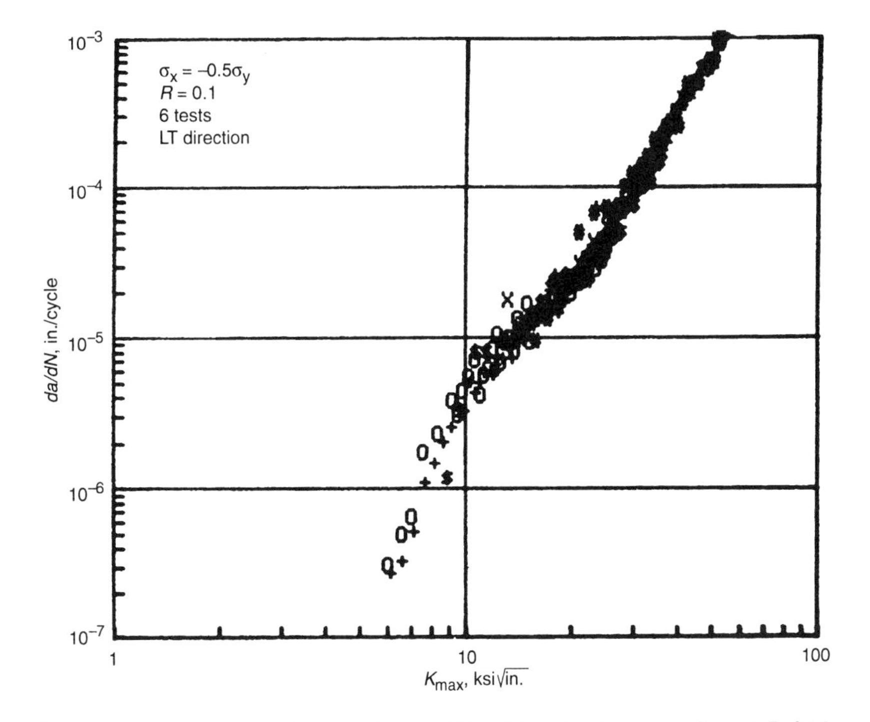

Fig. 7-17d Cyclic crack growth rate behavior of 2024-T351 aluminum alloy. Source: Ref 7-9

only requirement is that the crack is not associated with any geometric stress concentration. In Chapter 3, we considered that K is the only driving force that controls constant-amplitude da/dN. Therefore, we can expect that biaxial stress state will not affect the constant-amplitude crack growth behavior. However, we also stated that crack tip plastic zone size is a function of biaxial stress ratio. Consequently, we anticipate that biaxial stress state may affect the variable-amplitude crack growth behavior. In the following sections, we will demonstrate these phenomena with a set of experimental test data that cover a wide range of biaxial stress ratios and fatigue loading conditions.

Constant-Amplitude Fatigue Crack Growth

An experimental test program that was carried out at Northrop (Ref 7-8, 7-9, 7-36, 7-37) consisted of many types of fatigue crack growth testing to explore the effect of biaxial stresses on crack growth. Cruciform specimens

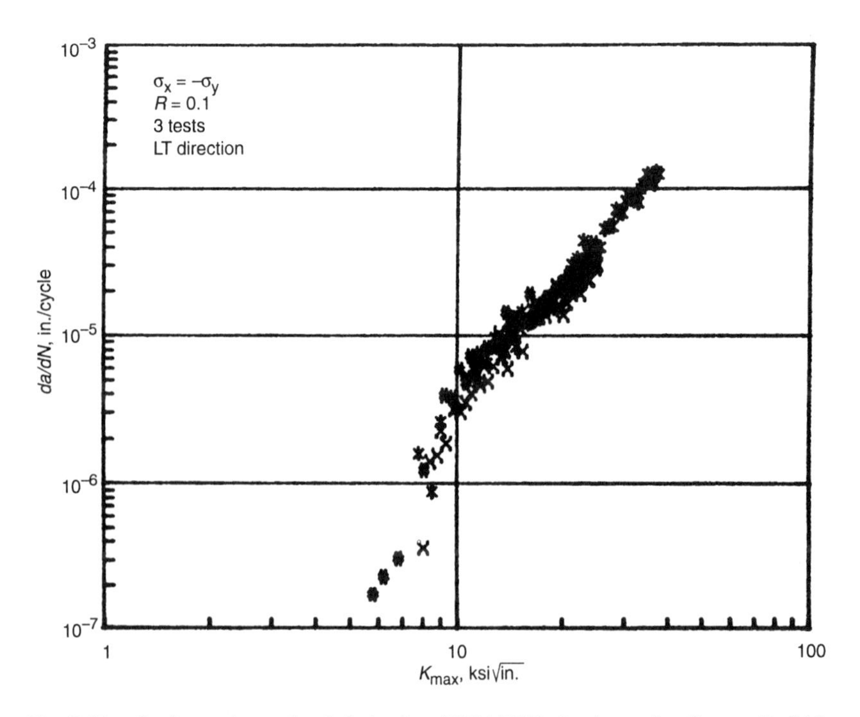

Fig. 7-17e Cyclic crack growth rate behavior of 2024-T351 aluminum alloy. Source: Ref 7-9

containing a straight through-the-thickness crack have been used for fatigue crack growth testing. Forty-six specimens of 2024-T351 and 7075-T7351 aluminum alloys were tested under various biaxial stress ratios (–1.5, ±1.0, ±0.5, 0) at $R = 0.1$, and at various applied stress levels ($0.2 \le S_y/F_{ty} \le 0.6$). Details about the test setup and experimental procedures are given in Ref 7-9. Details concerning specimen design, stress analysis, and cracked finite element analysis are given in Chapter 8.

The effect of biaxial loading conditions on constant-amplitude crack growth rate can be evaluated from the standpoint of experimental scatter of the *a* versus *N* curves (Fig. 7-14, 7-15). There the scatter (on *N*) for growing a crack from 12.7 mm (0.5 in.) to 101.6 mm (4.0 in.) was only ±16% (32% total scatter band) for the 7075-T7351 specimens and ±17% (34% total scatter band) for the 2024-T351 specimens. As identified in Fig. 7-14, the scatter band for the 7075-T7351 specimens at $\lambda = 1.0$ was already 21% (as compared to the 32% total scatter band). For the 2024-T351 specimens, the 34% scatter band actually represents the scatter band for specimens subjected to a –1.0 biaxial ratio (Fig. 7-15). It is also shown in these figures that *a* versus *N* curves for the remaining tests are randomly sandwiched between

Table 7-2 Loading spectrum for biaxial stress fatigue crack growth testing

Loading step	P_{max}(a)	P_{min}(a)	Cycles
1	84.92	5.62	3
2	54.60	7.41	103
3	77.78	7.41	6
4	44.60	11.67	372
5	91.75	5.62	1
6	68.25	7.41	20
7	38.89	11.67	811
8	58.25	0	203
9	35.40	7.41	656
10	61.90	16.03	11
11	49.84	11.67	279
12	60.63	7.41	36
13	47.78	0	2
14	79.68	16.03	1
15	41.43	7.41	421
16	100.00	5.62	1
17	60.79	11.67	97
18	47.46	16.03	61
19	54.12	11.67	139
20	37.46	0	8
21	74.29	11.67	7
22	48.10	7.41	343

(a) Percent of peak stress in the spectrum (100% = 30 ksi). Source: Ref 7-37

these scatter bands. No definite trend in the biaxialty effects is shown in either one of these figures.

The *a* versus *N* record for all 46 tests was reduced to a *da/dN* versus K_{max} format. Composite graphs for $\lambda = 0$, ±1.0, and −1.5 for each material are presented in Fig. 7-16 (a–e) and 7-17 (a–e). The 2024 alloy exhibited more experimental scatter than the 7075 alloy, but the test results revealed that the crack growth rate curves are almost identical. That is, for the same material and cyclic stress amplitude there is no effect on fatigue crack growth rate due to differing biaxial stress ratios. It is significant that each of these composite curves represents many sets of test data points and that each set of those data was generated from different combinations of stress levels and cyclic frequencies. Six technicians were involved at different times in recording *a* versus *N* data. Even so, it is very evident that the crack growth rate behavior remained consistent. Similar test results have been obtained for $R = 0.7$ (Ref 7-9). These data also show no biaxial stress effect on fatigue crack growth rates.

Variable-Amplitude Fatigue Crack Growth

As discussed in Chapter 3, for crack growth under uniaxial, mode 1 loading, the role of the plastic zone is to influence the spectrum crack growth behavior. In the case of biaxial loading it is anticipated that the crack tip plastic zone will play a similar role. A quantitative evaluation has been

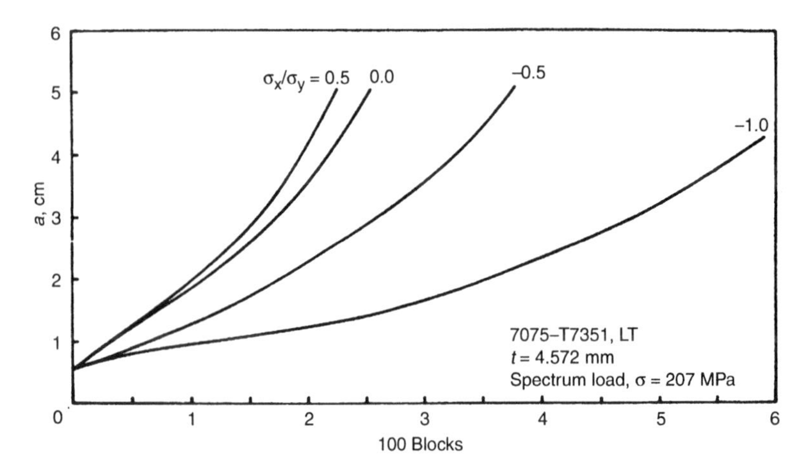

Fig. 7-18 Effect of biaxial stress ratio on variable-amplitude crack growth. Source: Ref 7-37

carried out by conducting spectrum load tests on cruciform specimens at four biaxial stress ratios ($\kappa = -1.0, -0.5, 0, 0.5$). The spectrum profile is given in Table 7-2. The test results (crack length versus number of loading blocks) are presented in Fig. 7-18. These data show that cracks grow faster at positive biaxial stress states but slower at negative biaxial stress states, as compared to uniaxial tension ($S_x = 0$). For a spectrum that consists of very few cycles of high loads, the retardation behavior will prevail. Therefore, a stress state associated with a larger plastic zone will result in a longer period of crack growth retardation, and vice versa, as indicated in Fig. 7-18. As previously stated, the crack tip plastic zone is larger for a crack subjected to a negative biaxial stress ratio, but slightly smaller for a crack subjected to a positive biaxial stress ratio. Thus, crack growth retardation correlates with biaxial stress ratios. However, on the basis of the constant-amplitude test results, it appears that crack tip plastic zone size variation with biaxial stress ratio contributes no effect to constant-amplitude fatigue crack growth.

REFERENCES

7-1. G.R. Irwin, Analysis of Stresses and Strains Near the End of a Crack Transversing a Plate, *J. Appl. Mech., (Trans. ASME),* Series E, Vol 24, 1957, p 361

7-2. G.R. Irwin, Fracture, *Hanbuch der Physik*, Vol VI, Springer-Verlag, Berlin, 1958, p 551-590

7-3. J. Eftis, N. Subramonian, and H. Liebowitz, Crack Border Stress and Displacement Equations Revisited, *Eng. Fract. Mech.*, Vol 9, 1977, p 189–210

7-4. J. Eftis and N. Subramonian, The Inclined Crack under Biaxial Load, *Eng. Fract. Mech.*, Vol 10, 1978, p 43–67

7-5. F. Erdogan and G.C. Sih, On the Crack Extension in Plates under Plane Loading and Transverse Shear, *J. Basic Eng., (Trans. ASME),* Series D, Vol 85, 1963, p 519–527

7-6. J.G. Williams and P.D. Ewing, Fracture under Complex Stress—The Angle Crack Problem, *Int. J. Fract. Mech.*, Vol 8, 1972, p 441–446

7-7. K.J. Chang, On the Maximum Strain Criterion—A New Approach to the Angled Crack Problem, *Eng. Fract. Mech.*, Vol 14, 1981, p 107–124

7-8. A.F. Liu and J.R. Yamane, Crack Growth under Equibiaxial Tension, *Res Mech.*, Vol 5, 1982, p 1–11

7-9. A.F. Liu and D.F. Dittmer, "Effect of Multiaxial Loading on Crack Growth," Report AFFDL-TR-78-175 (in three volumes), Air Force Flight Dynamics Laboratory, Wright-Patterson Air Force Base, Dayton, OH, Dec 1978

7-10. S. Iida and A.S. Kobayashi, Crack Propagation Rate in 7075-T6 Plates under Cyclic Tensile and Transverse Shear Loadings, *J. Basic Eng., (Trans. ASME),* Series D, Dec 1969, p 764–769

7-11. W.K. Wilson, W.G. Clark, Jr., and E.T. Wessel, "Fracture Mechanics Technology for Combined Loadings and Low-to-Intermediate Strength Metals," Technical Report 10276 (Final), Vehicular Components and Materials Laboratory, U.S. Army Tank Automotive Command, Warren, MI, Nov 1968

7-12. A.F. Liu, Crack Growth and Failure of Aluminum Plate under In-Plane Shear, *AIAA J.*, Vol 12, 1974, p 180–185
7-13. P.S. Leevers, J.C. Radon, and L.E. Culver, *J. Mech. Phys. Solids*, Vol 24, 1976, p 381–395
7-14. H. Tada, P.C. Paris, and G.R. Irwin, "Stress Analysis of Cracks Handbook," Del Research Corp., Hellertown, PA, 1973
7-15. G.C. Sih, A Special Theory of Crack Propagation, *Methods of Analysis and Solutions to Crack Problems*, G.C. Sih, Ed., Noordhoff International Publishing, Leyden, The Netherlands, 1973, p 21
7-16. G.C. Sih and B. MacDonald, Fracture Mechanics Applied to Engineering Problems—Strain Energy Density Fracture Criterion, *Eng. Fract. Mech.*, Vol 6, 1974, p 361–386
7-17. R.C. Shah, Fracture under Combined Modes in 4340 Steel, *Fracture Analysis*, STP 560, American Society for Testing and Materials, 1974, p 29–52
7-18. L.P. Pook, The Effect of Crack Angle on Fracture Toughness, *Eng. Fract. Mech.*, Vol 3, 1971, p 483–486
7-19. L. Banks-Sills, M. Arcan, and H. Gabay, A Mode II Fracture Specimen—Finite Element Analysis, *Eng. Fract. Mech.*, Vol 19, 1984, p 739–750
7-20. L. Banks-Sills and M. Arcan, A Compact Mode II Fracture Specimen, *Fracture Mechanics: Seventeenth Volume*, STP 905, American Society for Testing and Materials, 1986, p 347–363
7-21. B.C. Hoskin, D.G. Graff and P.J. Foden, see: D. Broek, *Elementary Engineering Fracture Mechanics*, Noordhoff International Publishing, Leyden, The Netherlands, 1974, p 332, or A.P. Parker, *The Mechanics of Fracture and Fatigue*, E.&F.N. Spon Ltd., London, 1981, p 90
7-22. D.L. Jones and D.B. Chisholm, An Investigation of Edge Sliding Modes in Fracture Mechanics, *Eng. Fract. Mech.*, Vol 7, 1975, p 261–270
7-23. A. Otsuka, K. Mori, T. Okshima, and S. Tsuyama, Mode II Fatigue Crack Growth in Aluminum Alloys and Mild Steel, *Advances in Fracture Research: Proc., Fifth Int. Conf. on Fracture*, Pergamon Press, 1981, p 1851–1858
7-24. E.W. Smith and K.J. Pasoe, The Behavior of Fatigue Cracks Subjected to Applied Biaxial Stress: A Review of Experimental Evidence, *Fatigue Eng. Mater. Struct.*, Vol 6, 1983, p 201
7-25. R. Roberts and J.J. Kibler, Mode II Fatigue Crack Propagation, *J. Basic Eng.*, (*Trans. ASME*), Series D, Vol 93, 1971, p 671–680
7-26. M. Truchon, M. Amestoy, and K. Dang-Van, Experimental Study of Fatigue Crack Growth under Biaxial Loading, *Advances in Fracture Research: Proc., Fifth Int. Conf. on Fracture*, Pergamon Press, 1981, p 1841–1849
7-27. E.M. Wu, Application of Fracture Mechanics to Anisotropic Plates, *J. Appl. Mech., (Trans. ASME),* Series E, Vol 34, 1967
7-28. J.R. Rice, Mechanics of Crack Tip Deformation and Extension by Fatigue, *Fatigue Crack Propagation*, STP 415, American Society for Testing and Materials, 1967, p 247–311
7-29. S.H. Smith, in *Prospect of Fracture Mechanics*, G.C. Sih, H.C. Van Elst, and D. Broek, Ed., Noordhoff International Publishing, Leyden, The Netherlands, 1974, p 367–388
7-30. H. Liebowitz, J.D. Lee, and J. Eftis, Biaxial Effects in Fracture Mechanics, *Eng. Fract. Mech.*, Vol 10, 1978, p 315–335

7-31. P.D. Hilton, Plastic Intensity Factors for Cracked Plates Subjected to Biaxial Loading, *Int. J. of Fract.*, Vol 9, 1973, p 149–156
7-32. N.J.I. Adams, *Eng. Fract. Mech.*, Vol 5, 1973, p 983–991
7-33. K.J. Miller and A.P. Kfouri, An Elastic-Plastic Finite Element Analysis of Crack Tip Fields under Biaxial Loading Conditions, *Int. J. Fract.*, Vol 10, 1974, p 393–404
7-34. A.P. Kfouri and K.J. Miller, The Effect of Load Biaxialty on the Fracture Toughness Parameters J and G, *Fracture 1977*, Vol 3, 1977, p 241–245
7-35. B. Cotterell, Notes on the Paths and Stability of Cracks, *Int. J. Fract. Mech.*, Vol 2, 1966, p 526–533
7-36. A.F. Liu, J.E. Allision, D.F. Dittmer, and J.R. Yamane, Effect of Biaxial Stresses on Crack Growth, *Fracture Mechanics*, STP 677, American Society for Testing and Materials, 1979, p 5–22
7-37. A.F. Liu, Crack Growth under Variable Amplitude Biaxial Stresses, *Proc. of the 1982 Joint Conf. on Experimental Mechanics,* Part 2, Society for Experimental Stress Analysis, 1982, p 907–912

Chapter 8

Application

The current trend in structural life assessment is to determine the damage tolerance capability of a structure by assuming that the structure contains a pre-existing flaw in a critical location of the structure. The first logical step in structural life assessment is to identify the potentially fracture-critical parts and the critical location(s) in each part. Fracture mechanics analysis is then conducted on these parts to determine the actual criticality of each part by checking the analysis results against the design criteria. At the same time, trade studies can be conducted on the same structural part to attain an optimized design.

In preliminary design (or sometimes, in redesign), it is often convenient to conduct a parametric analysis involving the design stress level and missions to failure as a function of initial flaw size. A schematic example of this type of investigation is given in Fig. 8-1. Here the calculated number of missions to failure is plotted against the operating stress levels, which are interpreted as fractions of the baseline design stress level. These data can be converted into a display of structural weight increase as a function of initial flaw size or operating stress level for various service life requirements (Fig. 8-2). A reference value for initial flaw size may be selected, based on expected detection capability. The effect on structural weight of applying a safety factor either to the initial flaw size or to the required service life may be quickly determined.

A more thorough analysis, such as those presented in Ref 8-1, can be performed. In Ref 8-1, requirements were selected for static strength, fatigue strength, damage tolerance, inspectability, and in-service inspection intervals. Several candidate design configurations and materials were studied. Each structural configuration and material combination was analyzed to determine whether any one of the designs would satisfy all the criteria. The structures will be further compared on the basis of weight and cost.

Crack Growth Predictive Analysis Techniques

As mentioned in Chapter 4, a sophisticated crack growth analysis procedure contains at least five key elements. They are pictorially presented in Fig. 8-3 as a recap. Analytical results are sensitive to many input variables, including:

- Operational usage (including stress level)
- Fracture mechanics methodology
- Structural geometry
- Initial crack length
- Material properties

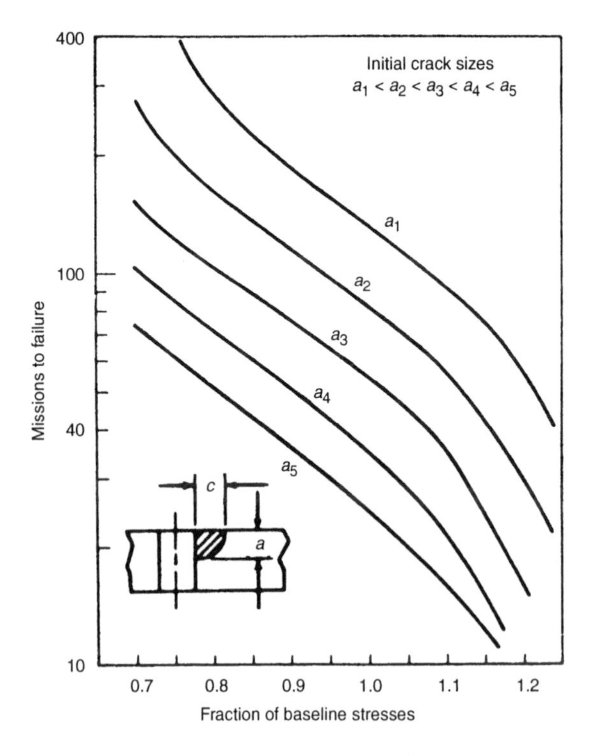

Fig. 8-1 Missions to failure as a function of applied stress levels and initial crack sizes (a schematic illustration, not data)

These variables interact with each other, and it is difficult to pinpoint the significance of the sensitivity attributed to each one. However, the sensitivity of most of these variables lies in their effect on the stress intensity factor. For example, when the initial stress intensity factor in one case is below or slightly above the threshold value in the *da/dN* curve and the initial stress intensity factor in another case is relatively higher, the difference in calculated life will be substantial. Therefore, starting from a given initial crack length, a realistic and accurate crack growth history can be predicted, depending on how all the elements shown in Fig. 8-3 are being handled. Many fundamental considerations, concepts, and techniques have been discussed in depth in the preceding chapters. The techniques that are required to determine damage accumulation (i.e., how to do the final integration in Fig. 8-3) are discussed below.

Damage Accumulation Techniques

So far we have shown that crack length, or the increment of crack growth, is related to number of stress cycles (or flight time) through a driving force K (or ΔK). To calculate a crack growth history, one may choose to determine how much a crack will grow in each applied load cycle or a group of load cycles. Or, one may find it more appropriate to calculate the number of load cycles (or flights, or segments of flight) it takes to grow a crack from a prescribed crack length to the next prescribed crack length.

Most crack growth accumulation analysis routines are structured around fatigue crack propagation laws. As discussed in Chapter 3, a crack subjected

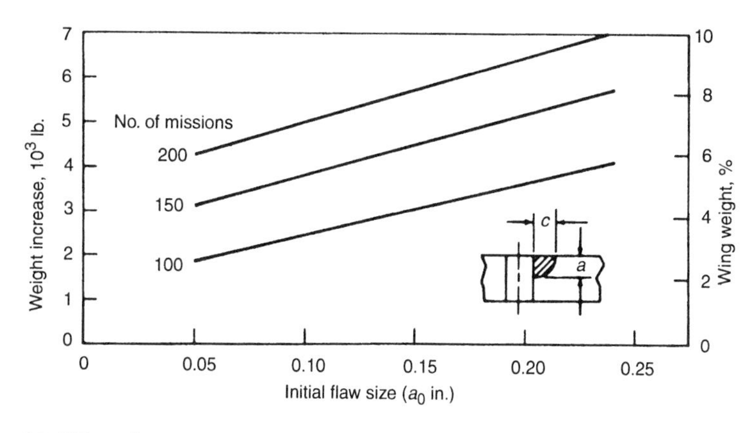

Fig. 8-2 Effect of initial flaw size on structural weight for safe life (a schematic illustration, not data)

to a constant-amplitude applied differential stress intensity, ΔK, has been observed to propagate at a material-dependent rate:

$$\frac{da}{dN} = f(\Delta K, C, n, K_c, \Delta K_0) \quad \text{(Eq 8-1)}$$

To estimate the crack extension over a discrete number of constant-amplitude load cycles requires integration of Eq 8-1 along *a*, or:

$$N - N_0 = \int_{a_0}^{a_n} \frac{da}{f(\Delta K, \ldots)} \quad \text{(Eq 8-2)}$$

Numerical integration of Eq 8-2 is required in most cases due to the complexity of the crack growth rate expression and/or the stress intensity solu-

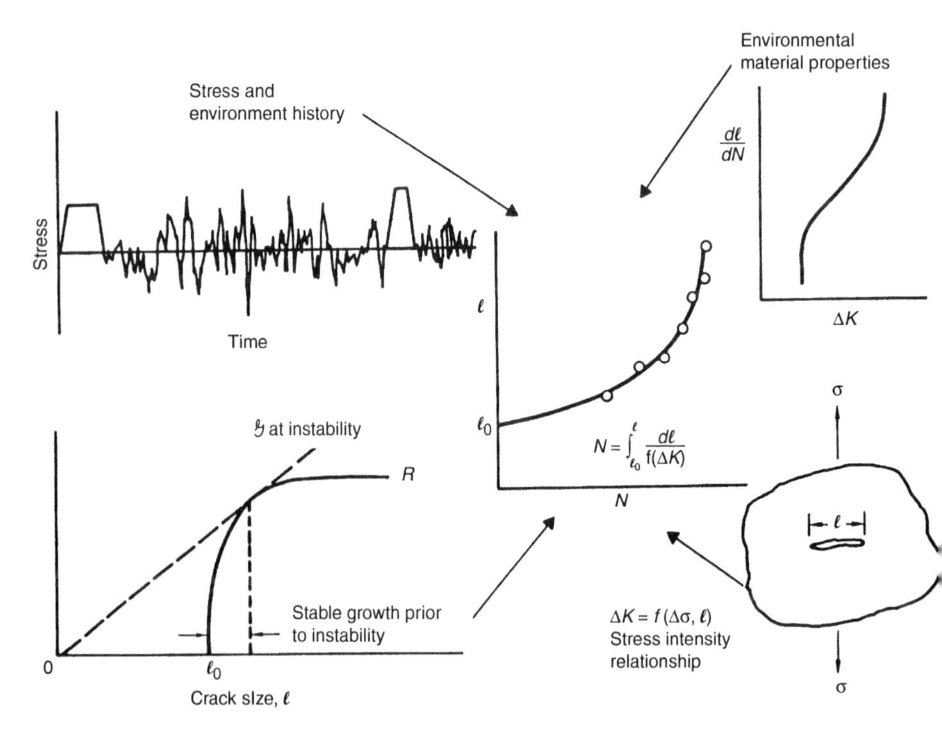

Fig. 8-3 Required elements in a life assessment methodology

tion. Two approaches can be taken to this integration: incrementing either crack size or cycles at a chosen fixed rate.

The number of cycles required to grow a crack a discrete distance, Δa, can be calculated by simply evaluating ΔK at the average crack length $(a + \Delta a/2)$ and calculating the average growth rate, da/dN, from Eq 8-1 and solving:

$$\Delta N = \Delta a/(da/dN) = \Delta a \,/\, |f(\Delta K, \ldots)| \qquad \text{(Eq 8-3)}$$

For a crack configuration that experiences large stress intensity gradients, dK/da, the Δa increment must be small to preserve accuracy. By summing the number of cycles calculated from Eq 8-3, an output of a versus N can be generated and arbitrarily terminated at a specific stress intensity, crack size, or cyclic life, or the crack can be grown to failure.

Conversely, an increase in crack size resulting from a discrete number of load cycles can be evaluated by:

$$\Delta a = \Delta N/(da/dN) = \Delta N \,/\, |f(\Delta K, \ldots)| \qquad \text{(Eq 8-4)}$$

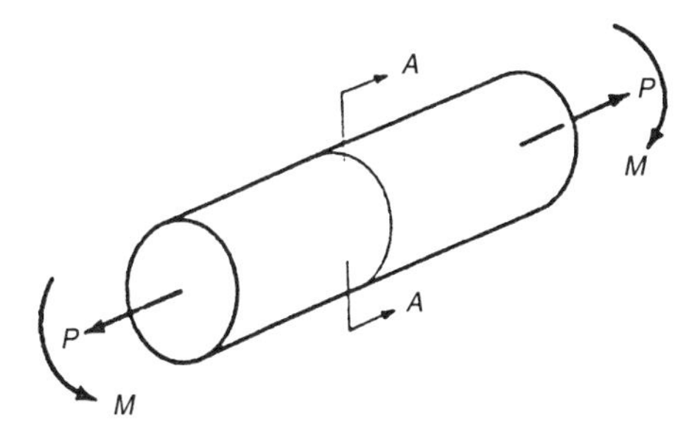

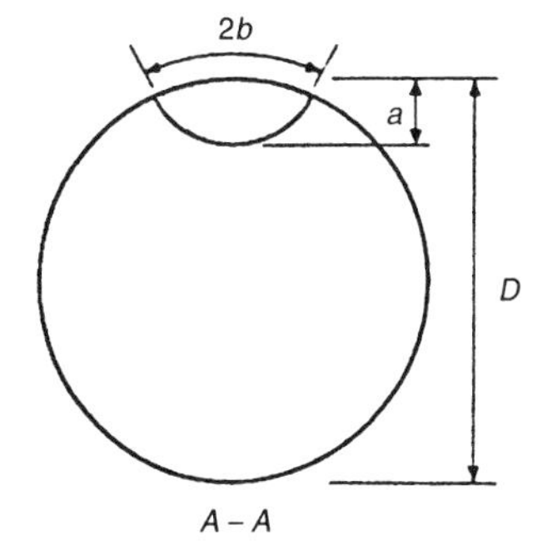

Fig. 8-4 Definitions of a cracked rod

The average growth rate for this calculation is slightly more difficult to estimate; however, several integration techniques are available in the literature to provide an accurate evaluation of Eq 8-4.

The third method for calculating damage accumulation is the cycle-by-cycle summation technique. This technique, in conjunction with the deterministic input of the loads spectra, is best suited for incorporation with crack growth interaction models to account for load interaction effects. In general, the procedure consists of the following steps:

1. The initial crack size follows from the damage tolerance assumption as a_1. The stress range in the first cycle is $\Delta\sigma_1$. Then determine ΔK_1.
2. Determine da/dN at ΔK_1 from the tabulated da/dN versus ΔK data (or an equation such as those presented in Chapter 3). Take into account the appropriate R-ratio at each loading step in the load spectrum.
3. The crack extension Δa_1 in cycle 1 is $\Delta a_1 = (da/dN)_1 \cdot 1$.
4. The new crack length will be $a_2 = a_1 + \Delta a_1$.
5. Repeat steps 1 to 4 for every following cycle, in the ith cycle replacing a_1 by a_i and a_2 by a_{i+1}.
6. If incorporation of a load interaction model is desired, follow the procedures given in Chapter 3 for determining K_{eff}, a_{eff}, etc.

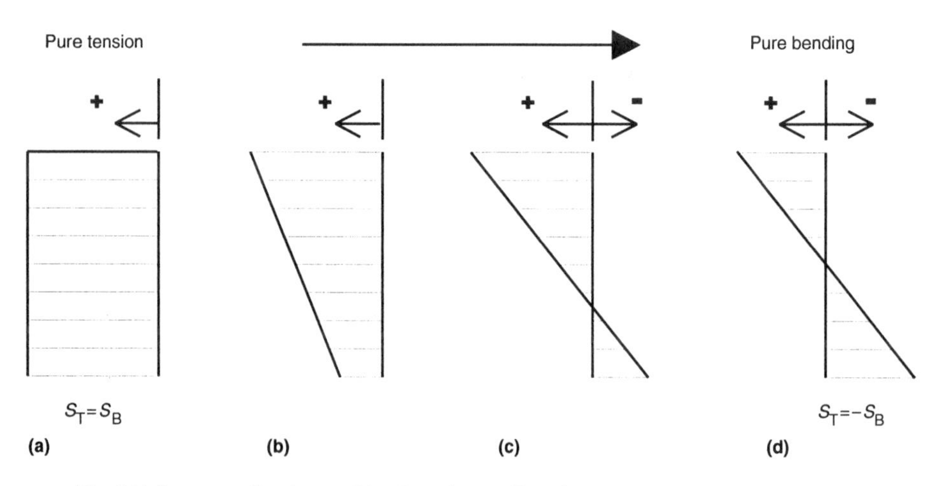

Fig. 8-5 Stress profiles for combined tension and bending

Since stress intensity factor varies along the periphery of a part-through crack, a two-dimensional crack growth scheme is an essential item in a crack growth computing routine. More important, the two-dimensional crack growth computing routine must be able to calculate crack growth increments at each point (e.g., on the surface and at maximum depth) independently. The reason for this is that in some stress level and stress intensity factor combinations, the stress intensity at one point on the crack may be lower than the ΔK-threshold. But another point on the same crack may have a stress intensity value significantly higher than the ΔK-threshold. This situation implies that one point on the crack may stop growing while another point on the same crack is still growing. However, this situation may be only temporary, because the crack shape continuously changes, even though the crack is growing in one direction. The stress intensity level that is originally below the ΔK-threshold may become high enough to cause crack extension due to the new flaw geometry, or simply due to changes in applied stress level (in the subsequent loading step), or both.

Examples

Several examples are presented in this section to demonstrate how life assessment methods are effectively implemented. Illustrations are not lim-

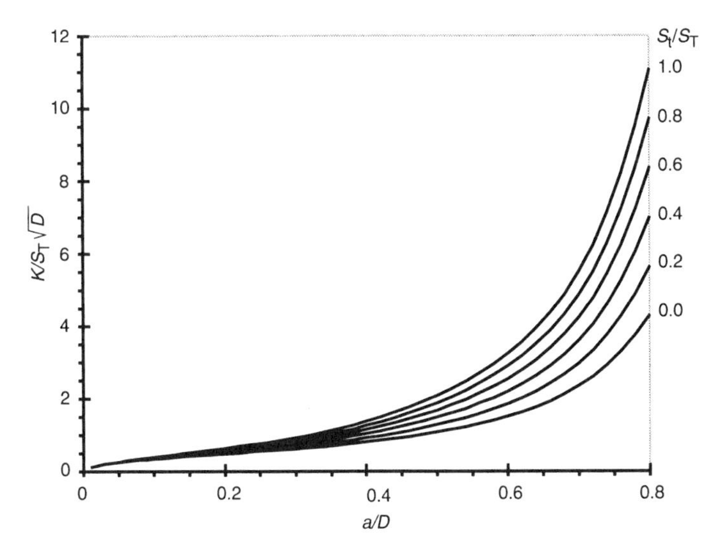

Fig. 8-6 Mode 1 stress intensity factors for a rod under combined tension and bending

ited to structural sizing alone. Emphasis is also placed on application of these techniques to fracture mechanics and structural life assurance research.

Example 1: Fracture Analysis of a Rod Subjected to Combined Tension and Bending

Description of the Problem

Consider a solid round bar containing a part-through crack subjected to either pure tension, pure bending, or combined tension and bending. The overall configuration of the bar is shown in Fig. 8-4. The objective of this problem is to construct a generalized fracture map to be used as a convenient tool for determining the allowable stress level or critical crack length.

The Approach

1. Determine the cross-sectional stress distribution by means of finite-element analysis, or by using any other stress analysis tool. The resulting stress

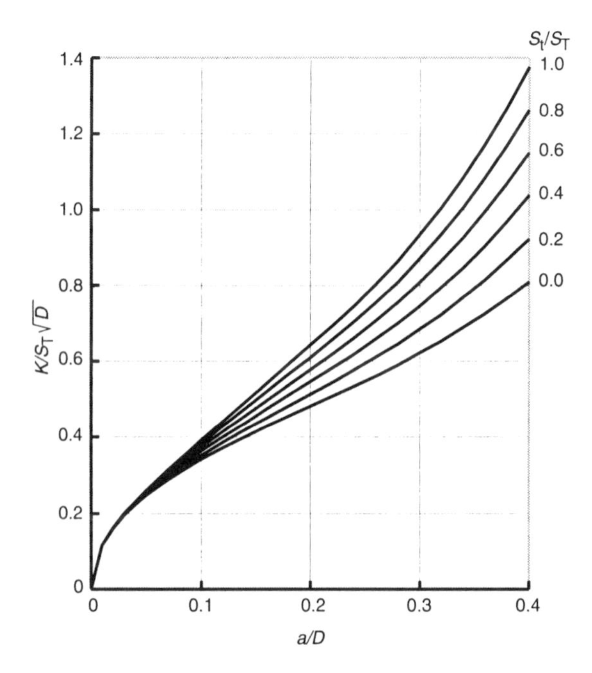

Fig. 8-7 A magnified view of Fig. 8-6

distribution usually has a shape like one of those shown in Fig. 8-5(a) to (d), starting from pure tension for Fig. 8-5(a) with an increasing amount of bending toward Fig. 8-5(d).

2. The stress intensity factors given in Chapter 5 are for pure tension or pure bending, not for combined tension and bending. In accord with the superposition principle, we can break down the total stress distribution (which has the shape of Fig. 8-5b or 8-5c) into two parts. Then, we can make use of the existing stress intensity factors to solve the problem of combined tension and bending.

Now we need to define the following terminology: S_T for the stress of the top extreme fiber of the rod, S_B for the stress of the bottom extreme fiber of the rod, S_t for the tension component, and S_b for the bending component. Both S_T and S_B are determine by stress analysis, but S_t and S_b are unknowns.

On the basis of a rule of algebra, we need two equations to solve two unknowns:

$$S_T = S_t + S_b \qquad \text{(Eq 8-5)}$$

and

$$S_B = S_t - S_b \qquad \text{(Eq 8-6)}$$

For the purpose of showing how to implement these equations, here are two hypothetical examples.

2(a). For the case of Fig. 8-2(b), for $S_T = 16.32$ MPa and $S_B = 9.71$ MPa, we have:

$$16.32 = S_t + S_b$$

$$9.71 = S_t - S_b$$

After solving these two equations, we have $S_t = 13.015$ MPa and $S_b = 3.305$ MPa. Comparing these values with S_T, we have $S_t \approx 0.8\ S_T$ and $S_b \approx 0.2\ S_T$. When the maximum stress S_T is used as the reference stress level, we can say that this loading condition is a mixture of 80% tension and 20% bending. Therefore, the total stress intensity factor for a given crack length will be:

$$K = S_T\sqrt{\pi a}\,(0.8Y_t + 0.2Y_b) \qquad \text{(Eq 8-7)}$$

where Y_t and Y_b are the geometric factors for tension and bending, respectively. Their values are given in Chapter 5.

2(b). For the case of Fig. 8-2(c), for $S_T = 26.786$ MPa and $S_B = -9.595$ MPa, we have:

$$26.786 = S_t + S_b$$

$$-9.595 = S_t - S_b$$

After solving these two equations, we have $S_t = 8.596$ MPa and $S_b = 18.19$ MPa. Therefore, this loading condition is equivalent to a mixture of 32% tension and 68% bending, and the total stress intensity factor for a given crack length will be:

$$K = S_T\sqrt{\pi a}\,(0.32Y_t + 0.68Y_b) \qquad \text{(Eq 8-8)}$$

3. A family of parametric curves plotting $K/(S_T\sqrt{D})$ versus a/D is presented in Fig. 8-6. Figure 8-7 is a magnified version of the lower a/D region of Fig. 8-6. For a given material for which the value of K_{Ic} is known, and for which the combination of tension and bending (in percentage) has been determined by stress analysis, the allowable stress (in terms of S_T) as a function of the critical crack length, or vice versa, can be easily determined from these figures. For example, using the first example given above, the maximum stress S_T at the assumed crack origin is 16.32 MPa, which is a

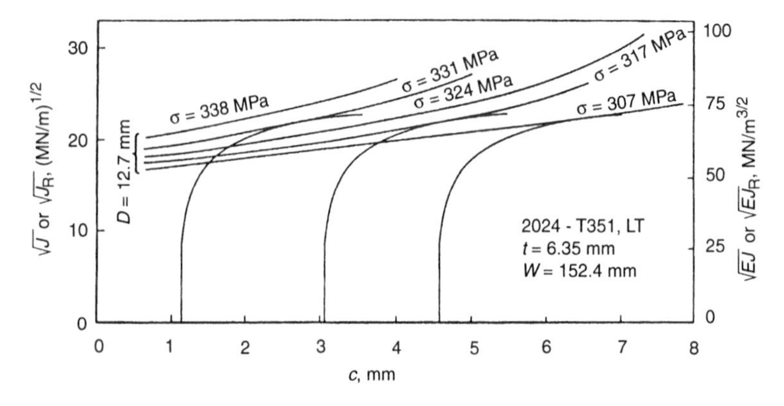

Fig. 8-8 Determination of fracture stress and critical crack length for three crack-at-a-hole specimens. Source: Ref 8-2

composite of 13.015 MPa in tension and 3.305 MPa in bending. The diameter of the rod under consideration is 10 mm. The rod is made of steel, which has a K_{Ic} value of 80 MPa$\sqrt{m}$. To determine the critical crack length for this stress level, we simply compute the value for $K/(S_T\sqrt{D})$, which is 1.55. For $K/(S_T\sqrt{D}) = 1.55$ with $S_t/S_T = 0.8$, we located $a/D = 0.43$ in Fig. 8-6. Therefore, the critical crack length would be 4.3 mm. If the maximum S_T at the assumed crack origin is 50 MPa instead of 16.32 MPa, the computed value

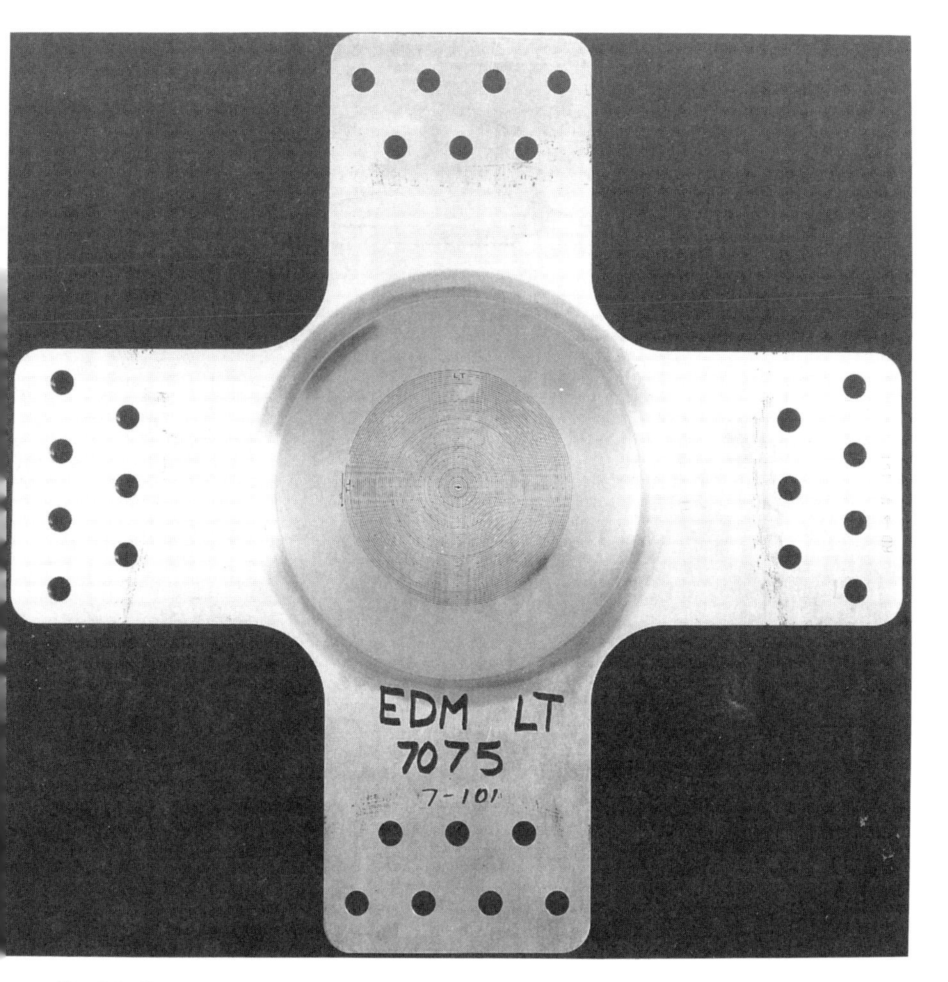

Fig. 8-9 Photograph of a cruciform specimen. Source: Ref 8-3

for $K/(S_T\sqrt{D})$ would be 0.506. From Fig. 8-7, we find $a/D = 0.15$ for the case of 80% tension and 20% bending. Therefore, the critical crack length would be 1.5 mm.

Example 2: Application of the Crack Growth Resistance Curve Technique

Description of the Problem

Residual strength tests have been conducted using specimens containing a very small through-the-thickness crack at the edge of a circular hole. The objective of this example is to predict the fracture strengths of these specimens and compare the analytical predictions with test results.

Description of the Crack-at-a-Hole Tests

Four specimens were fabricated from 2024-T351 aluminum alloy, 6.53 mm thick, 15.24 cm wide, and 45.72 cm long. A circular hole was placed at the center of each specimen. The hole diameter was 12.7 mm for three of the specimens and 19.05 mm for the other. An electrical discharge machining slot was placed on one side of the hole, parallel to the width, at the

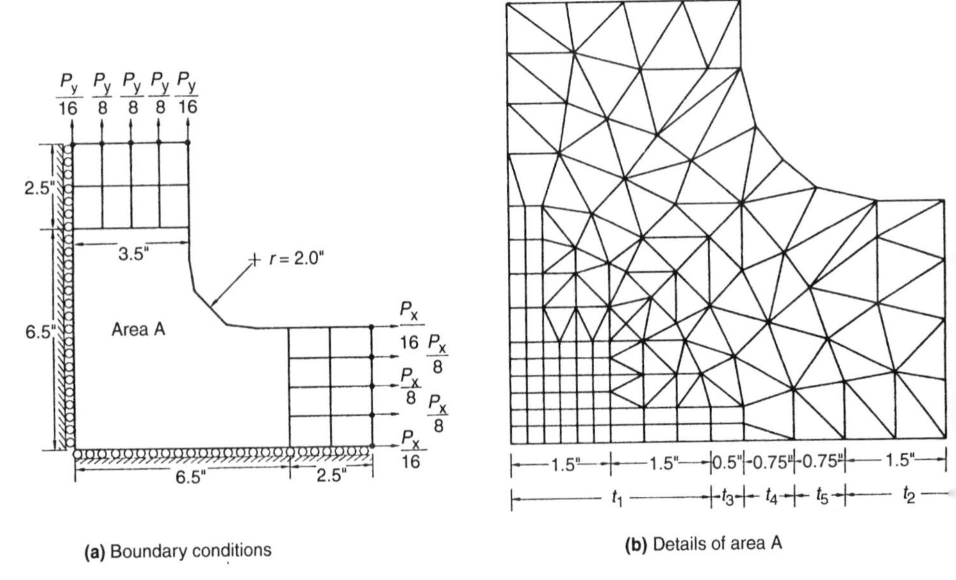

Fig. 8-10 Finite-element model of one-quarter of a cruciform specimen. Source: Ref 8-3, 8-4

midlength of the specimen. Each specimen was subsequently precracked to a desired initial crack length, ranging from 1 to 4.5 mm, approximately. The precracking stress level was kept below 100 MPa, with $R = 0.1$, for all four specimens. Then, each specimen was subjected to monotonically increasing load to failure. Anti-buckling guides were used during each test (see Fig. 2-7). During each test, the crack length extension record (as a function of increasing load level) was monitored. The crack lengths were measured visually with reference to photographically prepared gridlines on the specimen surface. The accuracy of the readings was enhanced by use of a zoom stereomicroscope (×7 to ×30) equipped with a fine-scale (0.0254 mm)

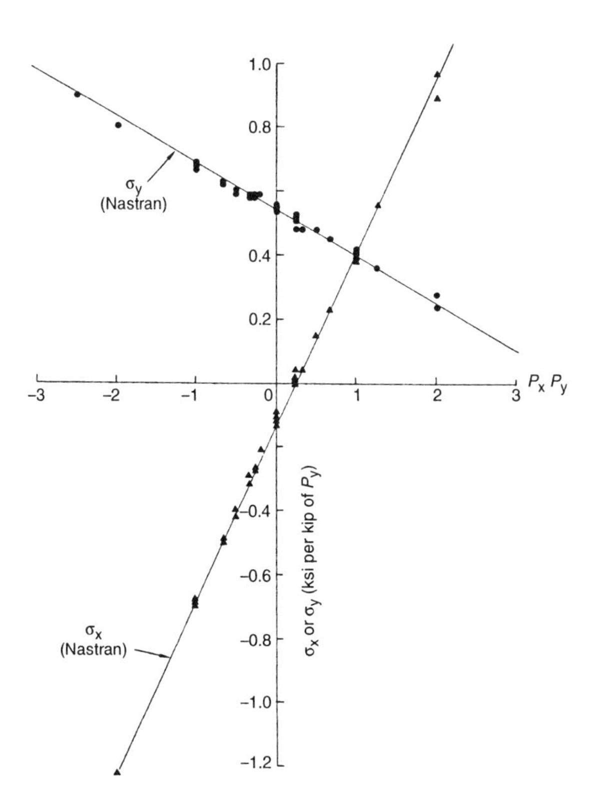

Fig. 8-11 Load-stress relationship at the center of a cruciform specimen. Data symbols are rosette data points. Source: Ref 8-3, 8-4

reticle for fractional grid readings. The crack growth records are presented in Table 8-1.

The Approach

As mentioned in Chapter 2, there is a great chance that the fracture behavior of a very short crack at the edge of a hole will be disturbed by gross plastic deformation at the hole, caused by geometric stress concentration. The J-integral approach would be the proper way to tackle this problem. Shows et al. have used this approach in combination with the crack growth resistance curve to predict the residual strength of a cracked hole that exhibits excessive local plasticity (Ref 8-2). Their rationale was based on the assumption that the R-curve for K and the R-curve for J are analogous. Their approach includes the following elements and procedures:

1. Obtain a generic R-curve for the same material and thickness as the crack-at-a-hole specimens. For this purpose we tested a center-cracked panel that had the same material and overall geometry as the others, except that it had no hole. The test setup and the technique for crack growth measurement were the same as in the other tests. The Δc versus load record was reduced to the K_R-curve format.

Table 8-1 Fracture test results of specimens containing a crack at a hole

Specimen No.	A2-22		A2-26		A2-35		A2-36		A2-7	
r	19.05		12.7		12.7		12.7		No hole	
t	6.528		6.553		6.553		6.528		6.553	
	P	c	P	c	P	c	P	c	P	c
Slow	0	0.978	0	1.143	0	3.048	0	4.559	0	26.13
stable	127.7	1.092	133.5	1.219	133.5	3.099	133.5	4.572	91.2	26.13
crack	163.7	1.194	169.0	1.219	155.7	3.124	177.0	4.699	121.4	26.18
growth	192.6	1.308	195.7	1.219	175.7	3.239	198.0	4.699	138.8	26.23
record	210.8	1.346	218.0	1.219	198.0	3.264	219.3	4.699	157.9	26.23
	234.9	1.486	244.7	1.448	221.5	3.289	248.2	4.699	170.4	26.28
	254.0	1.511	262.5	1.473	241.1	3.302	266.9	4.890	182.8	26.49
	273.1	1.562	280.2	1.473	263.8	3.302	284.7	5.017	191.3	26.66
	289.6	1.676	293.1	1.473	270.9	2.302	297.6	5.080	200.2	26.79
	299.8	1.892	298.9	1.549	284.2	3.315	306.9	5.080	210.0	27.07
	306.0	2.045	310.5	1.575	306.9	3.493	314.9	5.461	217.1	27.32
	309.2	2.743	313.2	1.778	321.6	3.645	319.4	5.461	219.3	27.71
	306.0	6.312	316.7	1.956	324.7	3.734	321.6	5.842	223.8	28.37
	301.6	9.716	323.8	2.083	330.5	4.445	324.3	6.287	...	...
	...	...	332.7	2.540	329.2	5.537	324.7	6.985	...	...
	...	...	332.7	2.870	328.7	9.017	...	...	...	...
	...	...	330.5	4.826	...	...	...	...	...	...
	...	...	328.3	9.373	...	...	...	...	...	...

Note: Hole radius, r, in mm. Specimen thickness, t, in mm. Monotonic load, P, in 10^3 N. Crack length, c, in mm. Source: Ref 8-2

2. Place the R-curve at the scale of an initial crack length for a given crack-at-a-hole test.
3. Following the R-curve procedure described in Chapter 2, i. e., the descriptions that go along with Fig. 2-16, determine the fracture stress for each crack-at-a-hole specimen. For each applied stress level, it is necessary to compute an applied J-curve in lieu of the applied K-curve. (The method of calculating the individual J-values was presented in Chapter 4.) Then, graphically determine the fracture stress level for that test. In other words, the fracture stress of a given crack-at-a-hole specimen is determined by superimposing the baseline K_R-curve onto a family of applied J-curves of the specimen. As shown in Fig. 8-8 (which shows the results for three tests of small-hole specimens), the fracture stress and the critical crack length of each test is defined by the tangential point of the baseline K_R-curve and the applied J-curve.

The actual fracture stress of a test is defined as the maximum load divided by the gross cross-sectional area of the specimen. The critical crack length

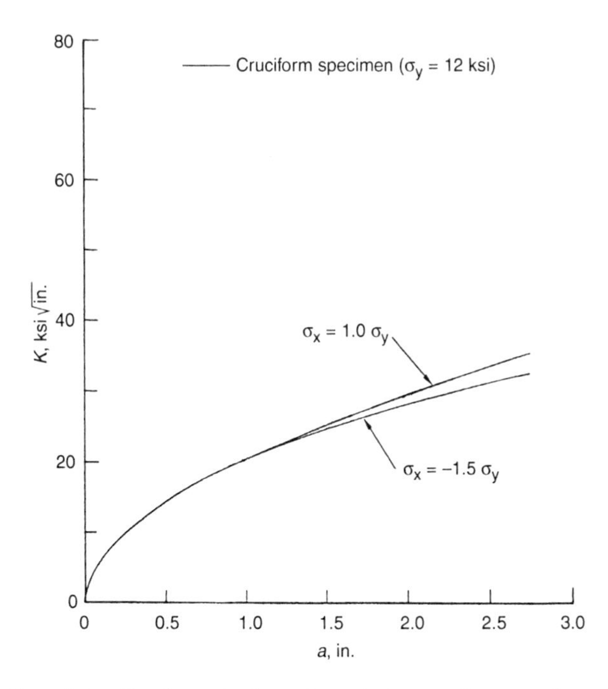

Fig. 8-12 Stress intensities for center through-cracks in a cruciform specimen. Source: Ref 8-3, 8-4

for a test is the measured crack length corresponding to the maximum load. The predicted fracture strengths and critical crack lengths for all four tests are compared with actual test results in Table 8-2. In this analysis, we are actually superimposing a J-curve (which includes both the elastic and plastic components) onto an elastic K_R-curve. Because there is no gross area yielding in the center-cracked specimen, and because there is only the elastic component in a J_R value in this case, J_R and K_R are the same. Therefore, the approach we just used is validated.

Example 3: Specimen Design Procedure—A Case History

Description of the Problem

There is no standard specimen configuration and test procedure for generating fatigue crack growth data under biaxial loading conditions. To ensure a successful test program, a specimen design/evaluation plan must include several important steps:

- Specimen requirement considerations
- Specimen shape and size optimization
- Stress analysis (of uncracked specimens) to determine what input loads might be needed in order to obtain a desired stress level and biaxial stress ratio
- Determination of stress intensity factors for these crack configurations and biaxial stress ratios

These issues are discussed below.

Specimen Design Considerations

- The specimen should be designed to avoid fatigue damage at the grip or in any area other than that containing the crack.

Table 8-2 Comparison of predicted fracture stresses and critical crack lengths with test results

Specimen		Fracture stress, MPa		Critical crack length, mm	
I.D.	D, mm	Predicted	Actual	Predicted	Actual
A2-22	19.05	331	310.8	2.616	2.743
A2-26	12.7	331	332.7	2.794	2.870
A2-35	12.7	317	330.5	4.572	4.445
A2-36	12.7	307	324.7	6.731	6.985

Note: 2024-T351 aluminum, LT. t = 6.53 mm, W = 152.4 mm. Source: Ref 8-2

- The specimen should be capable of taking compression load.
- The size of the specimen should be large enough to minimize boundary effects on crack tip stress intensity, but it should not be too large, so that the required load levels can be kept within the capacity of the testing machine.
- The stress distribution across the specimen width should be fairly uniform.
- The specimen configuration should be relatively simple in order to minimize machining costs.

The Approach for Specimen Optimization

After considering many options, a cruciform specimen configuration was chosen. An overall view of the specimen is shown in Fig. 8-9. A closeup photograph of the center region (with a crack) is shown in Fig. 7-8. The NASTRAN computer code has been used to accomplish the following:

- Optimization of specimen configuration
- Establishment of the load and stress relationship in the uncracked specimen
- Determination of crack tip stress intensity factors

A finite-element model representing one-quarter of the specimen is shown in Fig. 8-10. The thickness of the center region (t_1) and the thickness and width of the loading arms (t_2 and W) are the three primary variables affect-

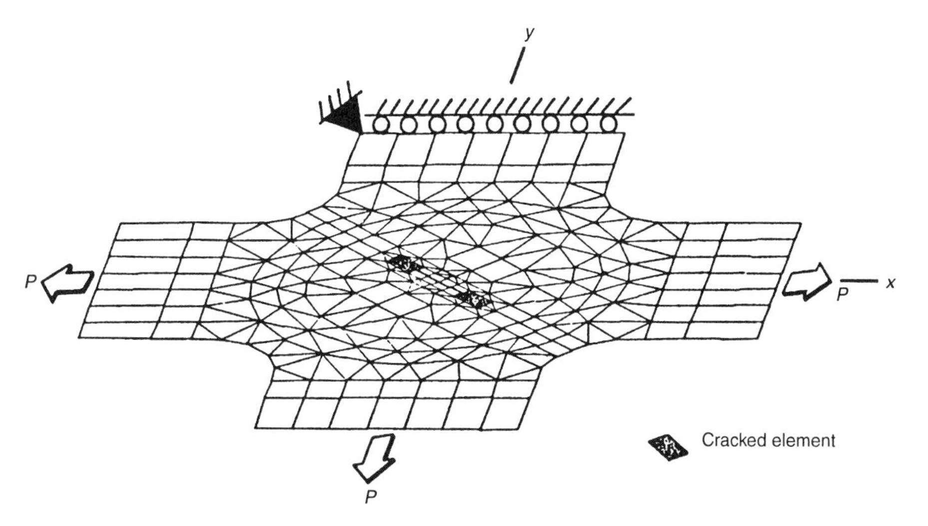

Fig. 8-13 Finite-element model for an angle crack in a cruciform specimen. Source: Ref 8-3, 8-5

ing the stress distribution. A 12.7 mm (0.5 in.) thick loading arm was preselected for t_2 to eliminate one of the variables and also to minimize material and machining costs. After a series of iterations, it was decided that the specimen would have an overall size of 600 by 600 mm, machined from a 12.7 mm thick aluminum plate. The t_1 and W dimensions have been optimized by conducting stress analysis on a dummy panel configuration (without a crack). As shown in Fig. 8-10, the optimized dimensions are $W =$ 177.8 mm (7 in.) and $t_1 = 4.57$ mm (0.18 in.). Also shown in Fig. 8-10 is the center region of the specimen, which contains a flat circular area, 152.4 mm (6 in.) in diameter, for housing the crack. The tapered region, which symmetrically connects the center region to the top and bottom surfaces of the loading arms, is represented by three rings of triangular elements of different intermediate thicknesses (t_3, t_4, and t_5) to simulate the curvature connecting t_1 and t_2. A radius of 50.8 mm (2 in.) has been used to minimize stress concentration at each corner where the loading arms are connected to each other.

Load and Stress Relationship

While optimizing the specimen configuration using the NASTRAN model shown in Fig. 8-10, various combinations of input loads (P_x and P_y) were

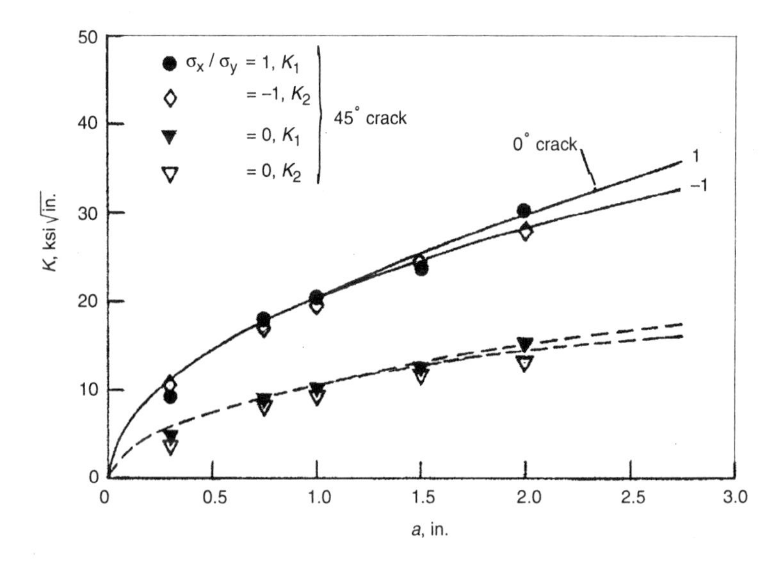

Fig. 8-14 Comparison of stress intensities for a 45° crack and a 0° crack in a cruciform specimen made of 7075-T7351 aluminum, $t = 0.18$ in., $\sigma_y = 12$ ksi. Source: Ref 8-3, 8-5

applied, and stresses (σ_x and σ_y) at the very center of the specimen and along the presumed crack line were recorded. To establish the load and stress relationship in an uncracked specimen, the magnitude of σ_y and σ_x (per 1000 lb of P_y) are plotted as functions of the P_x to P_y ratio. The graphs shown in Fig. 8-11 represent the stresses at the center of the cruciform specimen. A pair of equations representing the load and stress relationship (the straight lines in Fig. 8-11) can then be obtained:

$$\sigma_y = \frac{P_y}{12}[6.55 - 1.73(P_x/P_y)] \qquad \text{(Eq 8-9)}$$

and

$$\sigma_x = \frac{P_y}{12}[6.57(P_x/P_y) - 1.75] \qquad \text{(Eq 8-10)}$$

For an actual test, the required P_x and P_y values corresponding to any desirable σ_x and σ_y combinations can be determined by solving these equations.

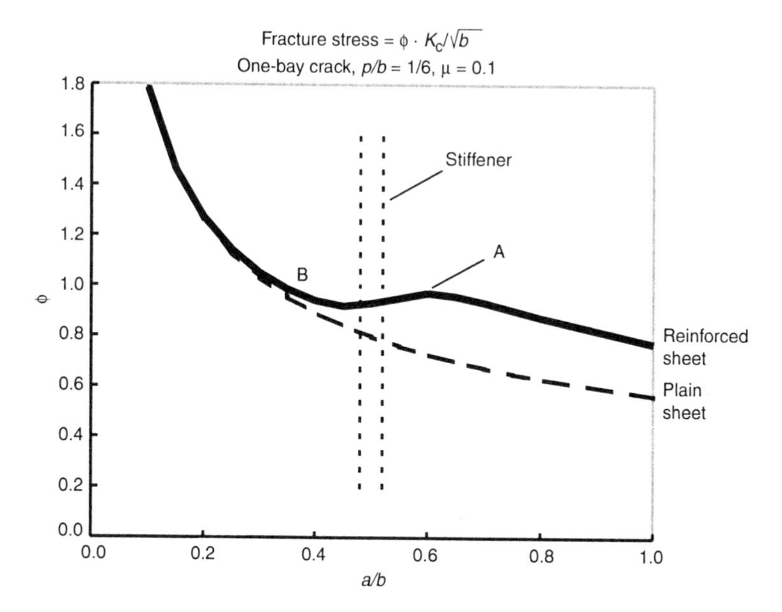

Fig. 8-15 Effect of reinforcement on residual strength

The straight lines in Fig. 8-11 are analytical data that were determined by finite-element analysis. The individual data points are stress survey data taken from back-to-back rosettes at the center of the cruciform specimen. Back-to-back rosettes were also placed across the x- and y-axes to ensure stress uniformity. The stress survey records are available in Ref 8-3.

Stress Intensity Factors

Finite-element analyses were performed to determine stress intensity factors for the straight-crack and the angle-crack configurations under various biaxial stress ratios.

The Straight Crack. Elastic K_1-values of the normal crack (i.e., $\beta = 90°$ with β being the angle between the crack line and the y-axis) of various sizes have been determined by using the same quarter-specimen model and special five-node "crack tip" element shown in Fig. 4-6 and 4-7. For each crack length analyzed, the cracked element was placed among the conventional isoparametric elements to represent the crack tip region of the specimen with a preselected crack length. The crack face was assumed to be traction free, and the crack length was represented by relieving the boundary constraints.

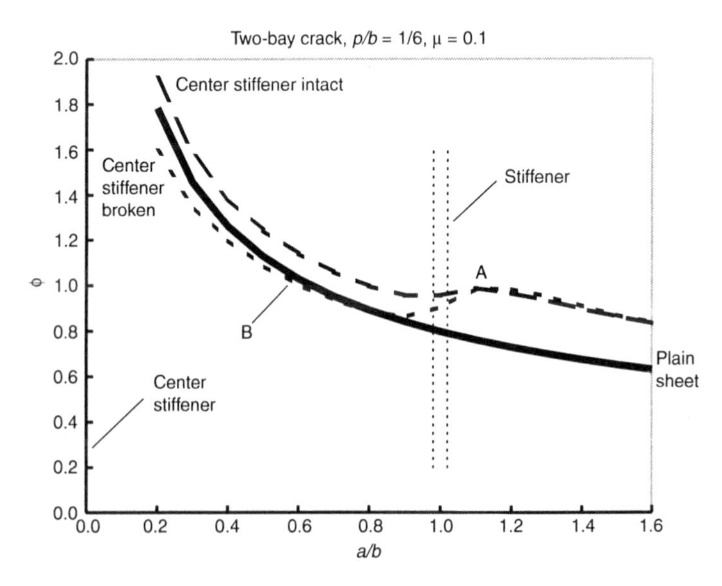

Fig. 8-16 Effect of broken stiffener on residual strength

The analyses were performed for eight different half-crack lengths, varying from 6.35 to 69.85 mm (0.25 to 2.75 in.). Each crack length was loaded to a stress level of 82.7 MPa (12 ksi) in the y-direction in combination with various stress levels in the x-direction, covering a range of biaxial stress ratios ($-1.5 \le \sigma_x/\sigma_y \le 1.0$). The results are plotted in Fig. 8-12. It is seen that the effects of loading conditions on elastic K-values are negligible. The apparent deviations in K, for a crack length greater than 38.1 mm (1.5 in.), were mainly due to the effect of specimen geometry rather than the effect of biaxial stress ratios. That is, the tension-compression loading cases exhibited more reduction in σ_y in the area near the rim.

The Angle Crack. Finite-element analyses were performed on cruciform specimens containing a 45° crack. Three loading cases were analyzed: $\sigma_x = 0$, $\sigma_x = \sigma_y$, and $\sigma_x = -\sigma_y$. In all three cases, the input stress for σ_y was 82.7 MPa (12 ksi). Due to the asymmetric nature of the problem, a full-specimen model was used. Two nine-node "crack tip" elements (Fig. 4-6) were incor-

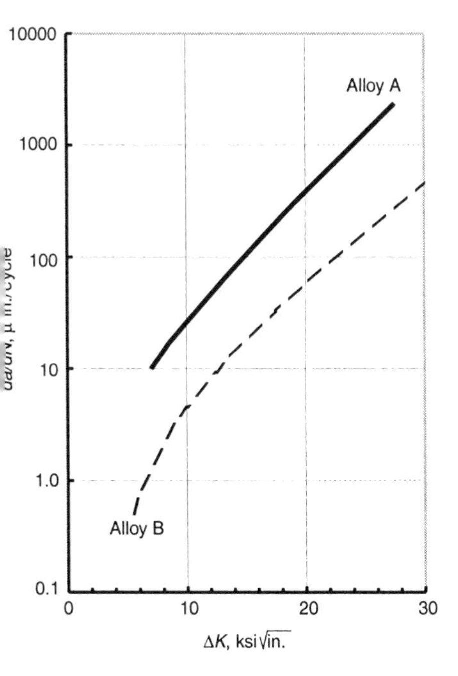

Fig. 8-17 Fatigue crack growth rates for alloys A and B (not factual data)

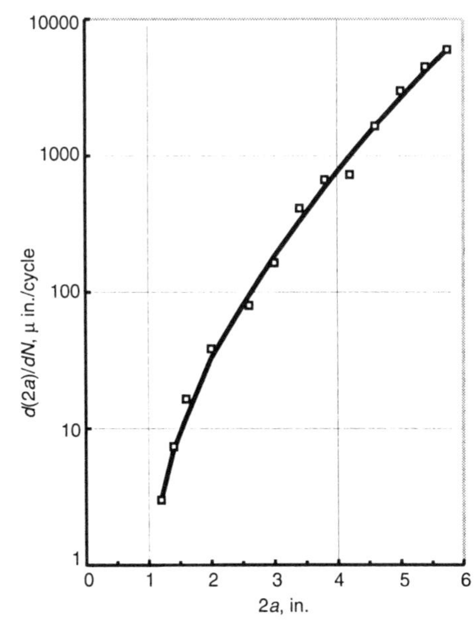

Fig. 8-18 Fatigue crack growth record of panel A (not factual data)

porated into the model (Fig. 8-13). For each biaxial ratio, elastic K_1 and K_2 values for five crack lengths ($a = 7.62$ to 50.8 mm) were computed for each crack tip (there are two crack tips in each model).

Theoretically, in an infinite sheet, the $\sigma_x = \sigma_y$ loading condition will be pure mode 1 and the $\sigma_x = -\sigma_y$ loading condition will be pure mode 2. The appropriate K-values corresponding to each loading condition are plotted in Fig. 8-14. In the upper part of the graph, the K_1 and K_2 values for the 1 to 1 and the 1 to –1 biaxial ratios, respectively, are identical to the K_1 values for the straight cracks (shown as solid lines here). As for the uniaxial loading condition ($\sigma_x = 0$), the K_1 and K_2 for each 45° crack were approximately equal to one-half of the corresponding pure mode 1 and mode 2 stress intensity values. The expected values, according to the angle-crack equations given in Chapter 7, are indicated by the dotted lines in Fig. 8-14. A very good agreement between finite-element analysis and theory has been obtained.

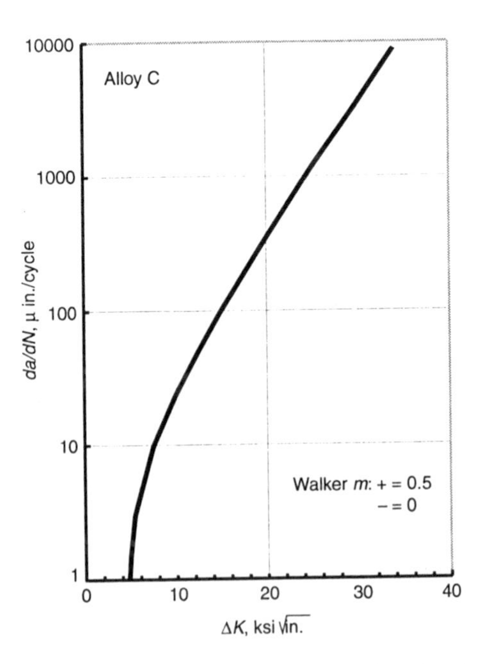

Fig. 8-19 Fatigue crack growth rates for alloy C (not factual data)

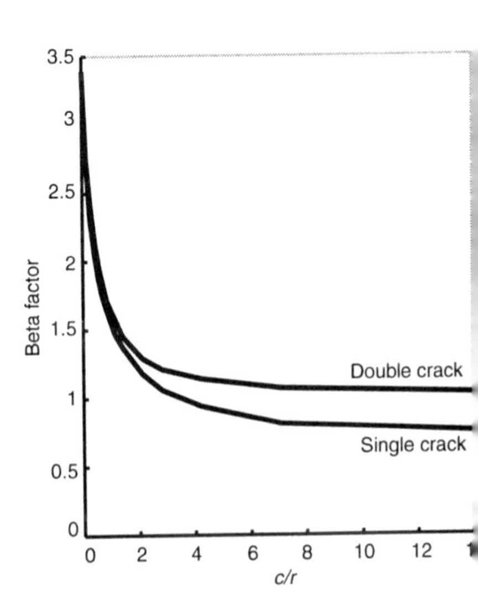

Fig. 8-20 One-dimensional K-factors for corner cracks at a circular hole. Source: Ref 8-6

Concluding Remarks

After the characteristics of a newly designed specimen are thoroughly checked, it can be put to use for generating crack growth rates data at various biaxial ratios. The predetermined stress intensity factors can be used for making data correlation (Ref 8-3 to 8-5). Some of the test results are discussed in Chapter 7. The lesson learned from this case history is that careful planning is an essential part of experimental research. This is the only way to get meaningful test results.

Example 4: Construction of an Operating Stress Map for Reinforced Skin Panels

Description of the Problem

Two parametric diagrams for determining fracture stresses of reinforced skin panels will be constructed. One diagram is for a panel containing a one-bay crack (i.e., the crack is in the middle of the bay between two stiffeners). The other compares the fracture strengths of two panels, one containing a two-bay crack located beneath an intact center stiffener, and one with the center stiffener broken. The specifications of the panels are:

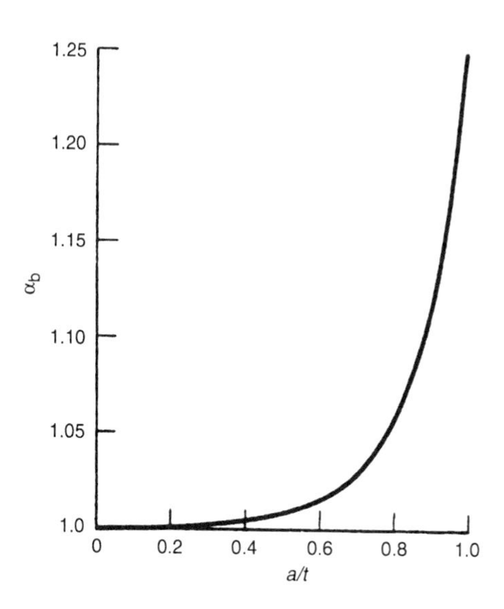

Fig. 8-21 Back face correction factor. Source: Ref 8-7

- The reinforcements are made of flat straps.
- The bay width times the skin thickness is nine times the cross-sectional area of the strap.
- The elastic modulus for the skin is the same as that for the stiffener.
- The ratio of attachment spacing to stiffener spacing is one-sixth.

The Approach

The fracture stress for a flat panel reinforced with attached stiffeners is given by Eq 5-13 as:

$$\sigma = K_c/(\sqrt{\pi a} \cdot C) \qquad \text{(Eq 8-11)}$$

where the stress intensity modification factor C is the ratio of the stress intensity factor for the stiffened panel to the stress intensity factor for the plain sheet (without stiffener). The values for the C-factors are given in Fig. 5-10 and 5-11. In these figures the C-factor is given as a function of p/b, a/b, and μ, where p is the attachment spacing (rivet pitch) and b is the stiffener spacing (same as the bay width), and the relative stiffness parameter μ is defined as:

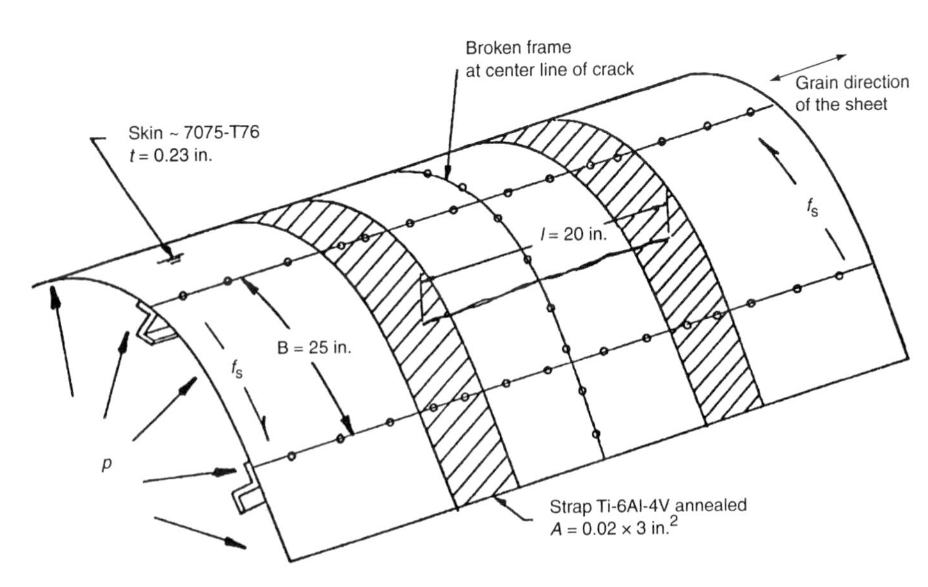

Fig. 8-22 Sketch of a 20 in. crack in an aircraft fuselage

$$\mu = \left[1 + \frac{b \cdot t}{A} \cdot \frac{E}{E_s}\right]^{-1} \qquad \text{(Eq 8-12)}$$

where t is the skin thickness, A is the cross-sectional area of the strap, and E and E_s are the elastic modulus of the skin and the strap, respectively. For the present example, $\mu = 0.1$ because $(b \cdot t)/A = 9$ and $E = E_s$.

To plot σ versus a/b, we can rearrange Eq 8-11 as:

$$\sigma_s = (K_c/\sqrt{b})/(\sqrt{\pi a/b} \cdot C) \qquad \text{(Eq 8-13)}$$

for the stiffened sheet, and:

$$\sigma_c = (K_c/\sqrt{b})/(\sqrt{\pi a/b}) \qquad \text{(Eq 8-14)}$$

for the plain sheet.

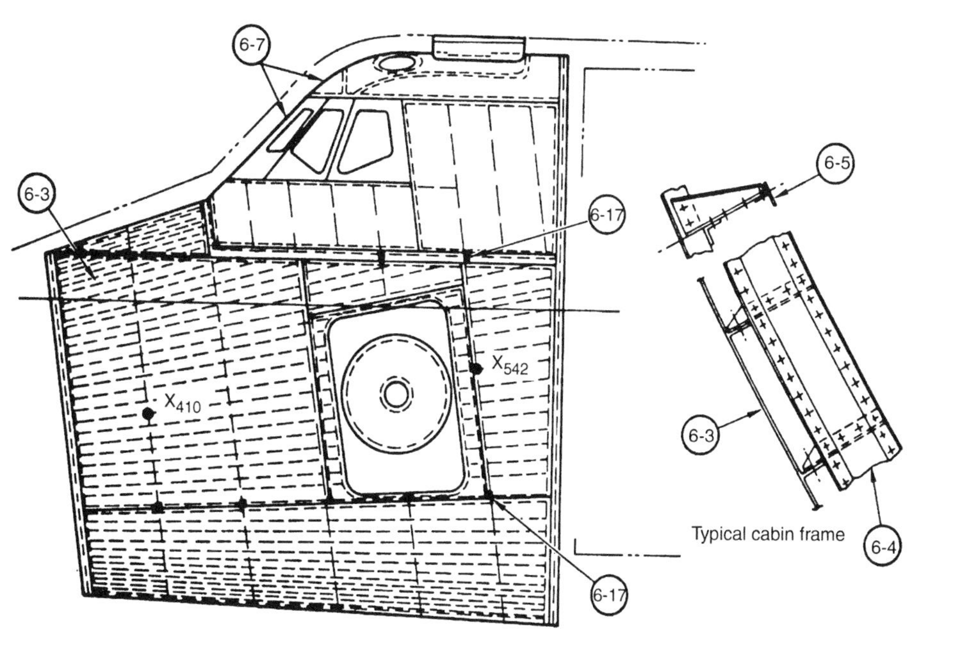

Fig. 8-23 Crew compartment configuration

Letting $\phi_s = 1/(\sqrt{\pi a/b} \cdot C)$ and $\phi_c = 1/(\sqrt{\pi a/b}\ )$, we can plot ϕ_s versus a/b for the stiffened sheet, and ϕ_c versus a/b for the plain sheet. For any given b and K_c, the fracture stresses σ_s and σ_c corresponding to an a under consideration can be determined from Eq 8-13 and 8-14.

The One-Bay Crack. Plotting ϕ_s or ϕ_c versus a/b for this crack configuration is quite straightforward. Just obtain a series of C-values from Fig. 5-10 and calculate ϕ_s and ϕ_c. The results are shown in Fig. 8-15. Evidently the biggest gain in residual strength in a reinforced sheet, compared with a plain sheet, is where the crack tip is slightly past the reinforcement (at point A). If rapid fracture should occur at any crack length that is associated with a ϕ_s value lower than that at point A, the sheet actually can take more load. The sheet will not fail until the applied stress level reaches point A. In other words, the residual strength curve for the reinforced sheet has an imaginary flat portion connecting points A and B. A reinforced sheet in which rapid fracture starts at any crack length beyond point B ($a/b \geq 0.37$) will not actually fail until the stress level reaches point A. This is a special "crack-arrest" feature that is inherent to a reinforced sheet.

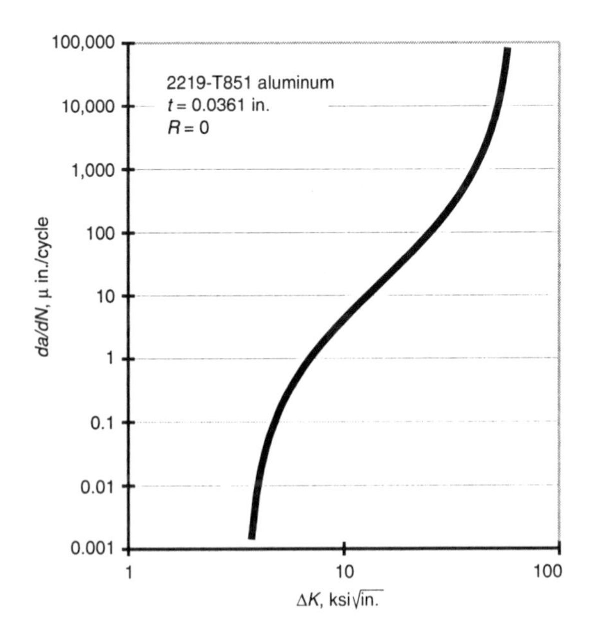

Fig. 8-24 Fatigue crack growth rates for 2219-T851 aluminum

A Two-Bay Crack with an Intact Center Stiffener. For a crack originating at a point beneath a stiffener, simply repeat the same procedure as before, but take the C-values from Fig. 5-11. The results are presented in Fig. 8-16. Similarly, there is a crack-arrest region where $0.8 \leq a/b \leq 1.1$.

A Two-Bay Crack with a Broken Stiffener. This situation can be handled by compounding the C-factors of two separate cases: a panel without a center stiffener and a panel with a broken stiffener.

For a panel with a broken stiffener, Eq 8-13 can be written as:

$$\sigma_{bs} = (K_c\sqrt{b})/\sqrt{\pi a/b} \cdot C \cdot C_b) \qquad \text{(Eq 8-15)}$$

where C_b is the broken stiffener factor given in Fig. 4-32. The parameter λ in Fig. 4-32 is modified to be the function of a/b. After multiplying b/b to the product of λ, λ becomes $(a/b) \cdot [(1/\mu) - 1]$. Then:

a/b	C_b
0.2	1.12
0.3	1.08
0.4	1.055
0.5	1.045
0.6	1.04
0.7	1.038
0.8	1.036
0.9	1.034
1.0	1.032
1.1	1.03
1.2	1.03
1.3	1.03
1.5	1.03
1.6	1.03
1.8	1.03
1.9	1.03
2.0	1.03

The second part of the problem involves obtaining the C-factors for an imaginary configuration, that is, a double bay without a middle stiffener. The C-factors corresponding to this configuration can be obtained by adjusting the reinforcement parameters to suit the present situation. That is, the stiffener spacing (b) for the imaginary configuration is actually $2b$ and therefore $p/b = 1/6$ is treated as $p/b = 1/12$. Consequently, $\mu \cong 0.05$.

The graphs in Fig. 8-16 indicate that the fracture stresses are lower (as compared to those for the plain sheet) for shorter cracks having their crack tips not far away from the broken stiffener. However, the effect of a broken stiffener vanishes at $a/b = 0.8$, and the two residual strength curves (the curves for panels with and without a broken stiffener) eventually merge at $a/b = 1.1$. The graph also implies that for crack lengths larger than $a/b = 0.6$ (point B) the panel with a broken stiffener will be able to sustain a higher stress level (during slow stable crack growth) until it reaches point A at $a/b = 1.1$.

Example 5: Examples of Fatigue Crack Growth Calculations

Three step-by-step examples are given in this section to demonstrate some simple procedures for computing cumulative fatigue crack growth damages. Both constant-amplitude and block spectrum fatigue crack growth are included. In these examples we use the actual *da/dN* curve instead of a crack growth rate equation for each material involved. The Walker *m* will be used to take care of the *R*-ratio effect when necessary.

Constant-Amplitude Crack Growth

A flat panel made of alloy A, for which *da/dN* data is shown in Fig. 8-17, is subjected to zero-to-tension ($R = 0$) constant-amplitude loading. The

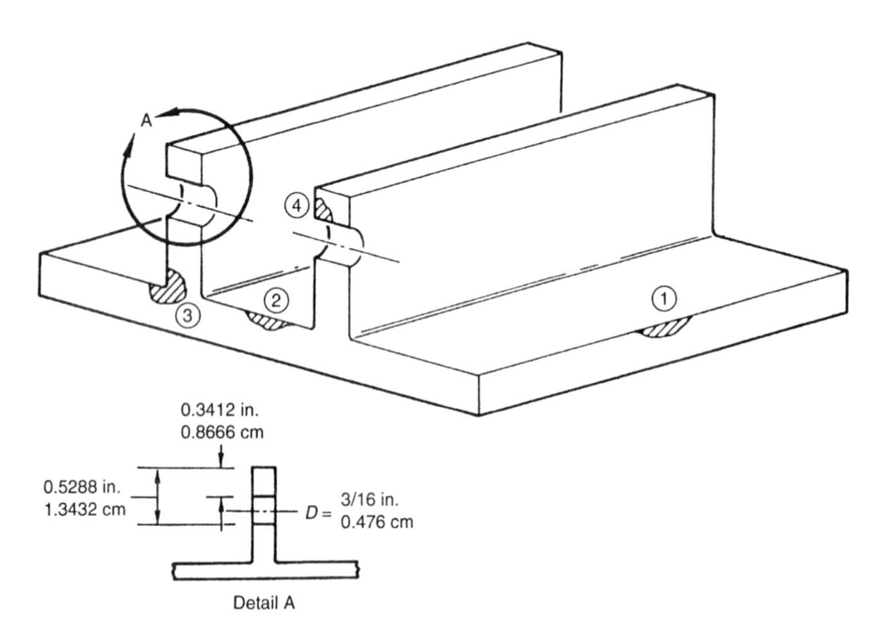

Fig. 8-25a Crack geometries (part through cracks)

maximum stress level is 49.6 MPa (7.2 ksi). Predict the number of cycles required to grow a crack from 12.7 cm (5 in.) total length to a total length of 22.86 cm (9 in.).

The analysis procedures are summarized in Table 8-3 for four increments of crack length. To keep the problem simple, the panel is assumed to be an infinite sheet containing a through-the-thickness crack. Thus the stress intensity factor would always be $\sigma\sqrt{\pi a}$. The propagation rate over each increment is taken to be the average of the rates at the extremes of the increment. The length of each increment divided by its rate gives the approximate number of cycles required to grow the crack the incremental length. The total number of cycles predicted to grow the crack from 12.7 to 22.86 cm (5 to 9 in.) is the sum of the cycles occurring for all the increments. For this example the total number is 2335 cycles.

Predicting Crack Growth by Interpolation

Sometimes it is necessary to predict fatigue crack growth in a real structure when the geometric correction factor in the K-expression is not available. Interpolation is permissible if the actual crack growth record of a

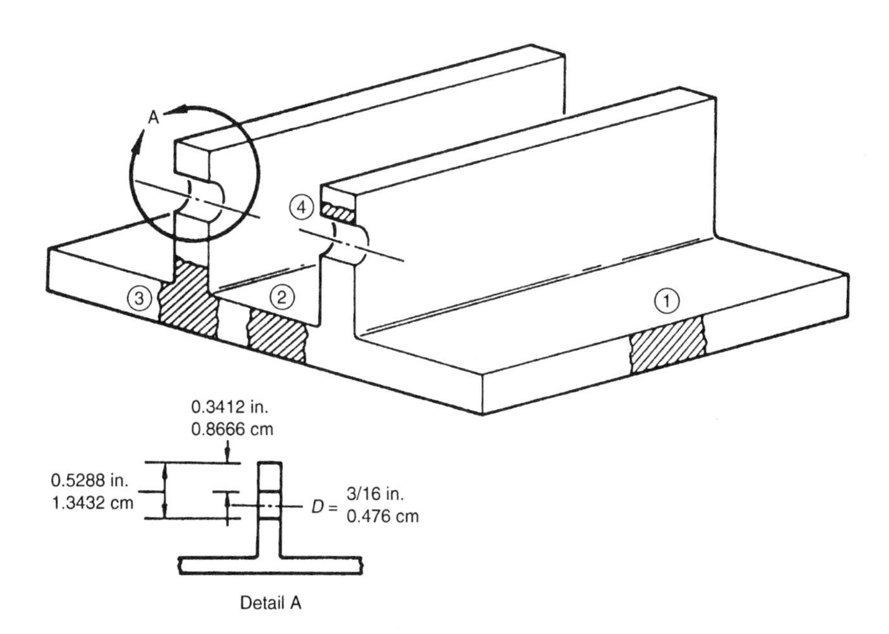

Fig. 8-25b Crack geometries (through cracks)

previously tested panel is available, provided that the panels (the previously tested one and the one under consideration) are otherwise similar. The results of a single test are sufficient if the only difference between the two panels is the material, as illustrated by the example that follows.

Assume that a structural panel (panel A) made of alloy A has been tested. Figure 8-18 shows a smooth plot of crack growth rate versus crack length following the trend of the data points obtained from this test. Now we use alloy B to build a second panel (panel B), which has the same geometry as panel A. The baseline da/dN versus ΔK property for alloy B is given in Fig. 8-17, along with the da/dN plot of alloy A. The objective of this problem is to predict the number of cycles required to grow a crack in panel B from a total initial length of 7.62 cm (3 in.) to a final length of 12.7 cm (5 in.). It is assumed that the stresses are the same in both panels.

In Table 8-4, four increments of crack growth are used. The rate that applies to panel A for each value of average crack length is obtained from Fig. 8-18. This rate is used in Fig. 8-17 (for alloy A) to locate the proper value of ΔK. Because ΔK is invariant from material to material, the rate that applies to the same crack growth increment for alloy B is then found on Fig. 8-17. The increment Δl, i.e., $\Delta(2a)$, divided by the rate for alloy B gives the number of cycles predicted for the increment. The sum over the four increments is a prediction of the number of cycles required for the crack to grow 5.08 cm (2 in.) in panel B.

Variable-Amplitude Crack Growth

During routine inspection of an aircraft, a fatigue crack is found at a rivet hole of a wing skin. The geometric dimensions and material properties are:

- The thickness of the wing skin $t = 6.35$ mm (0.25 in.).
- The hole diameter $2r = 7.9375$ mm (5/16 in.).
- The size of the corner crack $a = c = 1.524$ mm (0.06 in.).
- The wing is made of an aluminum alloy (alloy C). Its da/dN properties are shown in Fig. 8-19. The K_{Ic} and K_c fracture toughness properties for this alloy are, respectively, 33 MPa$\sqrt{m}$ (30 ksi$\sqrt{in.}$) and 55 MPa$\sqrt{m}$ (50 ksi$\sqrt{in.}$).

For the subject aircraft, the design limit stress level is 193 MPa (28 ksi) at the location where the crack occurred. We will seek the number of flights necessary to grow a 1.524 mm (0.06 in.) corner crack to a 12.7 mm (0.5 in.)

through-the-thickness crack. To simplify the problem, crack growth retardation will not be considered. The procedure of life prediction is described in the following steps:

1. Tabulate the operating stress spectrum, as shown in Table 8-5.
2. Construct a table for accommodating all the necessary computational steps (Table 8-6).
3. For a design limit stress level of 193 MPa, determine the critical crack depth for the corner crack by letting $K_{Ic} = S_{limit}\sqrt{\pi a} \cdot Y$, where Y is the geometry factor. In this example we conveniently use a one-dimensional corner crack equation to keep the calculations simple. That is (Ref 8-6):

$$K = 0.712 \cdot S\sqrt{\pi c} \cdot \beta \cdot \alpha_b \cdot \phi_w \quad \text{(Eq 8-16)}$$

where S is the applied far-field stress and β is the modified Bowie factor accounting for the geometry of a corner crack emanating from a hole (Table 8-7, Fig. 8-20). The back face correction factor α_b (adopted from Ref 8-7) is given in Fig. 8-21. The width correction factor is set to unity for this example. In using Eq 8-16, it is required that the crack start with an initial shape of a quarter circle. It is also assumed that the crack will remain a quarter-circular corner crack until it breaks through the plate thickness. In this exercise, we also assume that there is no transitional time during break-through. In other words, the crack is allowed to transform from a quarter-cir-

Table 8-3 Demonstration of constant-amplitude crack growth prediction procedure

l, in.	$\sqrt{\pi l/2}$ $\sqrt{\text{in.}}$	ΔK, ksi$\sqrt{\text{in.}}$	dl/dN in./cycle	Average dl/dN (a), in./cycle	Δl, in.	ΔN, cycles
	From ①	7.2 × ②	Fig. 8-17	From ④	From ①	⑥ ÷ ⑤
①	②	③	④	⑤	⑥	⑦
5	2.80	20.16	0.76×10^{-3}			
				1.005×10^{-3}	1.0	995
6	3.07	22.05	1.25×10^{-3}			
				1.565×10^{-3}	1.0	639
7	3.32	23.85	1.88×10^{-3}			
				2.39×10^{-3}	1.0	419
8	3.54	25.47	2.90×10^{-3}			
				3.55×10^{-3}	1.0	282
9	3.76	27.0	4.20×10^{-3}			
				Total cycles		2335

The number inside a circle identify a given column. (a) Average from two values in column ④.

cular crack to a through-the-thickness crack without going through any transition period.

4. Because the computed critical crack depth is greater than the plate thickness, we will proceed to determine the critical crack length for the through-the-thickness crack. This is accomplished by letting $K_c = S_{limit}\sqrt{\pi c} \cdot \varphi_n \cdot \phi_w$, where φ_n is the Bowie factor given by Eq. 5-9. The computed critical crack length is much greater than 12.7 mm (0.5 in.). In other words, the crack will not become critical during the course of this analysis.
5. To keep the calculations simple, we will not use any load interaction model at this time. We just calculate the equivalent number of cycles in a block spectrum (fractional cycle permissible). Note that the GAG cycles in Table 8-5 are flight cycles. Therefore, the number of cycles per flight in each loading step is equal to cycles per unit block (plus additional cycles indicated) divided by 153 flights per unit block. This calculation is tabulated in Table 8-6.
6. Calculate $\Delta S_{eff} = S_{max} \cdot (1 - R)^m$. The values for the Walker m are $m^+ = 0.5$, $m^- = 0$.
7. Calculate effective ΔK ($\Delta S_{eff}\sqrt{\pi c} \cdot Y$) for several arbitrarily selected crack lengths, e.g., for $c = 0.06, 0.1, 0.15, 0.2, 0.3, 0.4$, and 0.5 in.
8. Using Fig. 8-19, find out what crack growth rate per cycle corresponds to each computed effective ΔK value.
9. Compute Δc per flight, using the results of step 8 times the result of step 5.

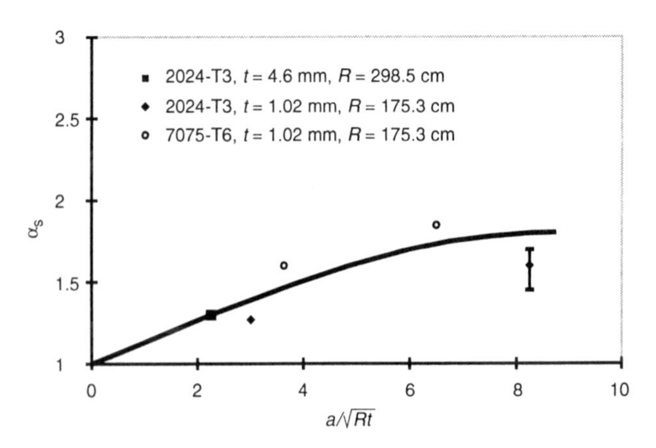

Fig. 8-26 Curvature correction factors for a longitudinal crack in a pressurized cylinder. Trendline: $Y = 1 + 0.1329x + 0.0015x^2 - 0.0007x^3$. Source: Ref 8-8, 8-9

10. Add the total Δc for each stress level in the bottom of Table 8-6.
11. Average the total for each pair of consecutive crack lengths, as shown in Table 8-6. For example, the average for growing a 0.06 in. crack to a 0.1 in. crack is 0.000324 in. per flight.
12. Follow the example given in step 11. The increment between 0.06 in. and 0.1 in. is 0.04 in., and the number of flights needed to grow the crack 0.04 in., based on a crack growth rate of 0.000324 in. per flight, would be 123.
13. Complete the calculations for the other crack lengths and add them up. The final answer is that a 1.524 mm (0.06 in.) corner crack would grow to become a 12.7 mm (0.5 in.) through-the-thickness crack within 498 flights.

Example 6: Determine the Fail-Safety of a Pressurized Fuselage

Description of the Problem

Consider a section of an aircraft fuselage having a long crack arrested by two flat straps, as shown in Fig. 8-22. The fuselage skin is subjected to a pR/t stress of 41.35 MPa (6 ksi) and a shear stress of 34.46 MPa (5 ksi). The axial stress ($pR/2t$) will be ignored in the calculation because it does not contribute to the crack-opening (mode 1) or crack-sliding (mode 2) displacements. The skin and the flat straps are made of 7075-T76 aluminum, $F_{tys} = 413.5$ MPa (60 ksi), $F_{tus} = 482.4$ MPa (70 ksi), $K_c = 62$ MPa$\sqrt{\text{in.}}$ (56.25 ksi$\sqrt{\text{in.}}$) and Ti-6Al-4V annealed, $F_{tyf} = 930.4$ MPa (135 ksi), respectively. Determine the fail-safety of the fuselage.

Table 8-4 Prediction of constant-amplitude crack growth in a structural member based on interpolation of previous test results

l, in.	Average l, (a) in.	dl/dN for alloy A, in./cycle	ΔK, ksi$\sqrt{\text{in.}}$	dl/dN for alloy B, in./cycle	Δl, in.	ΔN, cycles
	From ①	Fig. 8-18	Fig. 8-17	Fig. 8-17	From ①	⑥ ÷ ⑤
①	②	③	④	⑤	⑥	⑦
3.0						
	3.25	0.27×10^{-3}	16	0.043×10^{-3}	0.5	11,600
3.5						
	3.75	0.55×10^{-3}	18.7	0.084×10^{-3}	0.5	5,950
4.0						
	4.25	1.05×10^{-3}	21.3	0.150×10^{-3}	0.5	3,330
4.5						
	4.75	2.00×10^{-3}	23.9	0.275×10^{-3}	0.5	1,820
5.0						
				Total cycles		22,700

Numbers inside circles identify columns. (a) Average from two values in column ①.

The Approach

First, determine the allowable hoop stress by using Eq 4-35. Following the outlines given in Chapter 4, we can determine the values for the parametric variables C_e, A_e, and $2\overline{W}_e$:

$$C_e = 1.048 \text{ for } K_c = 56.25 \text{ ksi}\sqrt{\text{in.}}$$

$$\Sigma A_e/t = (2A/t) \cdot (F_{tyf}/F_{tus}) = (2 \cdot 0.02 \cdot 3.0/0.23) \cdot (135/60) = 1.176$$

$$2\overline{W}_e = 1.231 \text{ for } K_c = 56.25 \text{ ksi}\sqrt{\text{in.}} \text{ and 25 in. stringer spacing}$$

Substituting these values into Eq 4-35, we have:

$$F_{pg} = 1.2F_{tus}\left[\frac{2\overline{W}_e + \Sigma A_e/t}{C_e \cdot l + 2\overline{W}_e}\right] = \frac{1.2 \cdot 70 \cdot (1.231 + 1.176)}{(1.048 \cdot 20) + 1.231} = 9.11 \text{ ksi (62.8 MPa)}$$

Table 8-5 Hypothetical load spectrum (not data)

Sequence	Smax		Cycles per	Cycles, additional every:		Total cycles
No.	ksi	R	unit block	10th block	50th block	(50 blocks)
1	9.996	0.57	2,040	...	...	102,000
Positive	12.852	0.44	1,140	...	...	57,000
maneuver	15.708	0.36	780	...	...	39,000
	18.564	0.31	540	...	...	27,000
	21.42	0.27	300	...	...	15,000
2	5.712	0.00	60	...	...	3,000
Negative	5.712	−0.20	24	...	...	1,200
maneuver	5.712	−0.40	12	...	...	600
	5.712	−0.60	7	2	...	360
	5.712	−0.80	4	2	...	210
	5.712	−1.00	3	6	...	180
	5.712	−1.20	3	...	...	150
	5.712	−1.40	2	4	...	120
	5.712	−1.60	1	8	...	90
	5.712	−1.80	1	2	...	60
	5.712	−2.00	...	6	...	30
	5.712	−2.20	...	3	3	18
3	10.464	−0.04	303	7	...	15,185
Gust	11.6848	−0.14	32	5	...	1,625
	12.7312	−0.21	11	8	...	590
	13.952	−0.28	2	6	...	130
	15.1728	−0.34	...	7	3	38
	16.568	−0.40	...	2	...	10
	17.44	−0.43	...	...	1	1
4 GAG	5.712	−0.54	153	...	...	7,650

Table 8-6 Fatigue crack propagation analysis results

Sequence No.	Cycles per flight	S_{eff}, ksi	$c = 0.06$ in. dK_e	dc	$c = 0.10$ in. dK_e	dc	$c = 0.15$ in. dK_e	dc	$c = 0.20$ in. dK_e	dc	$c = 0.30$ in. dK_e	dc	$c = 0.40$ in. dK_e	dc	$c = 0.50$ in. dK_e	dc
1	13.333	6.54	4.10	20	4.7	25	4.9	33	5.45	48	6.95	100	7.32	120	7.66	133
Positive	7.451	9.58	5.98	35	6.8	52	7.16	61	7.88	75	10.1	180	10.67	205	11.08	231
maneuver	5.098	12.53	7.84	50	8.9	76	9.39	92	10.4	127	13.28	300	14	380	14.5	435
	3.529	15.45	9.66	71	10.98	106	11.58	127	12.75	180	16.37	495	17.26	650	17.9	760
	1.961	18.34	11.48	69	13.05	108	13.75	130	15.2	206	19.45	627	20.5	842	21.25	980
2	0.392	5.71	3.58	0	4.06	0	4.28	0	4.74	0	6.06	2	6.38	2	6.61	2
Negative	0.157	5.71	3.58	0	4.06	0	4.28	0	4.74	0	6.06	1	6.38	1	6.61	1
maneuver	0.078	5.71	3.58	0	4.06	0	4.28	0	4.74	0	6.06	0	6.38	0	6.61	0
	0.047	5.71	3.58	0	4.06	0	4.28	0	4.74	0	6.06	0	6.38	0	6.61	0
	0.027	5.71	3.58	0	4.06	0	4.28	0	4.74	0	6.06	0	6.38	0	6.61	0
	0.024	5.71	3.58	0	4.06	0	4.28	0	4.74	0	6.06	0	6.38	0	6.61	0
	0.020	5.71	3.58	0	4.06	0	4.28	0	4.74	0	6.06	0	6.38	0	6.61	0
	0.016	5.71	3.58	0	4.06	0	4.28	0	4.74	0	6.06	0	6.38	0	6.61	0
	0.012	5.71	3.58	0	4.06	0	4.28	0	4.74	0	6.06	0	6.38	0	6.61	0
	0.008	5.71	3.58	0	4.06	0	4.28	0	4.74	0	6.06	0	6.38	0	6.61	0
	0.004	5.71	3.58	0	4.06	0	4.28	0	4.74	0	6.06	0	6.38	0	6.61	0
	0.002	5.71	3.58	0	4.06	0	4.28	0	4.74	0	6.06	0	6.38	0	6.61	0
3	1.985	10.46	6.55	12	7.45	18	7.85	20	8.68	28	11.1	64	11.7	74	12.14	87
Gust	0.212	11.68	7.3	2	8.30	2	8.75	3	9.68	4	12.38	9	13.03	11	13.51	14
	0.077	12.73	7.95	1	9.05	1	9.54	2	10.55	2	13.48	5	14.2	6	14.73	7
	0.017	13.95	8.725	0	9.93	0	10.45	0	11.55	0	14.8	1	15.6	2	16.18	2
	0.005	15.17	9.975	0	10.78	0	11.34	0	12.55	0	16.08	0	16.91	1	17.55	1
	0.001	16.57	10.35	0	11.75	0	12.4	0	13.72	0	17.55	0	18.5	0	19.18	0
	0.0001	17.44	10.9	0	12.40	0	13.05	0	14.45	0	18.48	0	19.45	0	20.2	0
4 GAG	1.000	5.71	3.57	0	4.06	0	4.28	0	4.74	1	6.06	4	6.38	5	6.61	6
Total dc per flight, μ · in.				260		388		468		671		1788		2299		2658
Average dc per flight, μ · in.				324		428		570		1230		2044		2479		
Number of flights				123		117		88		81		49		40		
Total number of flights = 498																

Note: All dK_c are ksi$\sqrt{\text{in.}}$. All dc are μ · in. per flight.

Next, determine the load interaction using Eq 7-18, which states that if:

$$\left(\frac{K_1}{K_{1c}}\right)^{\xi} + \left(\frac{K_2}{K_{2c}}\right)^{\zeta} = 1$$

then the structure is marginally safe. If the left-side component is <1, the structure is fail-safe. If the left-side component is >1, the structure is not fail-safe. For this problem we choose $\xi = \zeta = 2$. Thus, the above load interaction equation becomes:

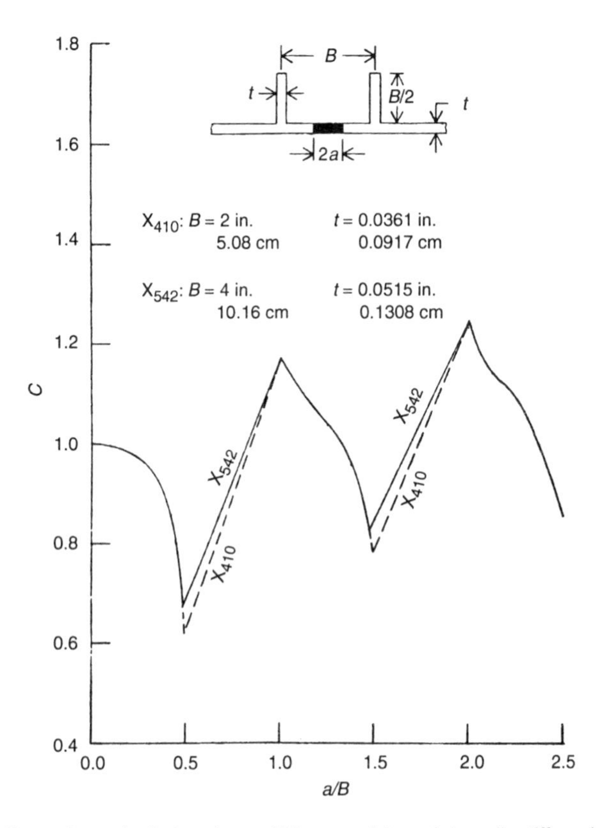

Fig. 8-27 Stress intensity factors for a mid-bay crack in an integrally stiffened skin

$$\left(\frac{K_1}{K_{1c}}\right)^2 + \left(\frac{K_2}{K_{2c}}\right)^2 = 1$$

The general K-expression for all these terms are the same, that is, there is a common element $\sqrt{\pi l/2}$ in each of these terms. Therefore all the K-terms can be replaced by stress terms. Where K_1 corresponds to the applied hoop stress (i.e., 6 ksi), K_2 corresponds to the applied shear stress (5 ksi). Here K_{1c} is the critical stress intensity factor that corresponds to the 9.11 ksi allowable hoop stress, obtained by using Eq 4-35. This allowable hoop stress can be considered an adjusted residual strength value for a given K_c attributed to the influence of curvature, internal pressure, reinforcements, etc. However, we assume that there is no change of residual strength in the shear mode, so that the K_{2c} value for the flat plate is applicable here. From Table 7-1 we obtain $K_{2c} = K_c$ for 7075-T76 aluminum, which leads to an allowable shear stress of 69 MPa (10.06 ksi). Inserting all these stress components into the load interaction equation, we have:

$$\left(\frac{6}{9.11}\right)^2 + \left(\frac{5}{10.06}\right)^2 = 0.433 + 0.247 = 0.68$$

which is <1. Thus, it can be concluded that the structure is fail-safe.

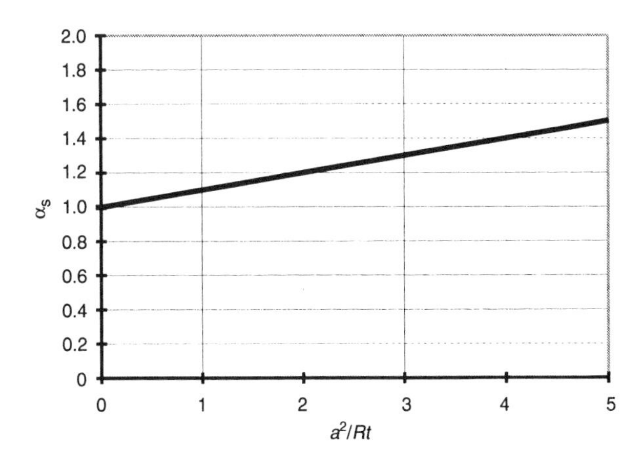

Fig. 8-28 Curvature correction factors for a circumferential crack in a pressurized cylinder. Trendline: $Y = 1 + 0.1x$. Source: Ref 8-10

Example 7: Preliminary Sizing of an Aircraft Crew Compartment

This example illustrates how a life assessment analysis is conducted. The structural assembly under consideration is the crew compartment of an aircraft. To assess the safe-crack-growth life for this structural component, one must first examine Fig. 8-23, the general structural arrangement at the location under consideration. To limit the scope of this illustrative example, the cabin skin opposite the cutout is analyzed. The potential fracture-critical areas have been identified. Cracks would most likely form on the skin and at the fastener hole of the integral stiffener, at locations X_{410} and X_{542} (as shown in Fig. 8-23).

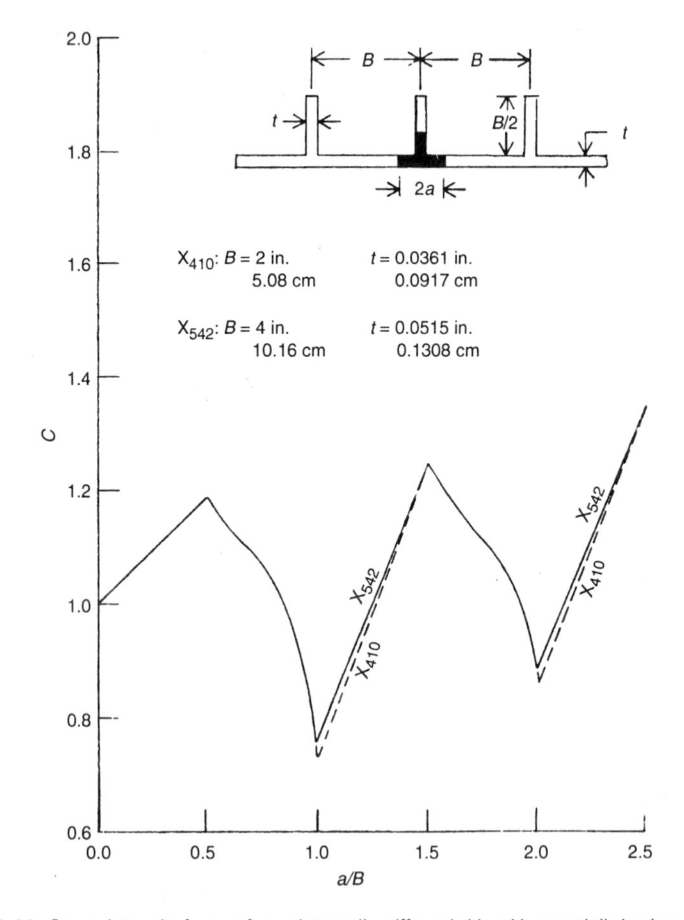

Fig. 8-29 Stress intensity factors for an integrally stiffened skin with a partially broken stiffener

The kind of information needed to conduct life assessment analysis is the next consideration. These variables are discussed in the following paragraphs. However, keep in mind that this example is just a paper study. Hypothetical structural configurations and loading condition are used. They may not represent the actual structural configuration and loading condition of a particular aircraft. Material properties may not represent the latest published allowables. In some cases alternate stress intensity factors are used for the convenience of making an illustration.

Design Stress Level and Structural Dimensions

The primary stresses acting on the cabin skin are the hoop stress pR/t and the axial stress $pR/2t$ (from the internal pressure), combined with some axial

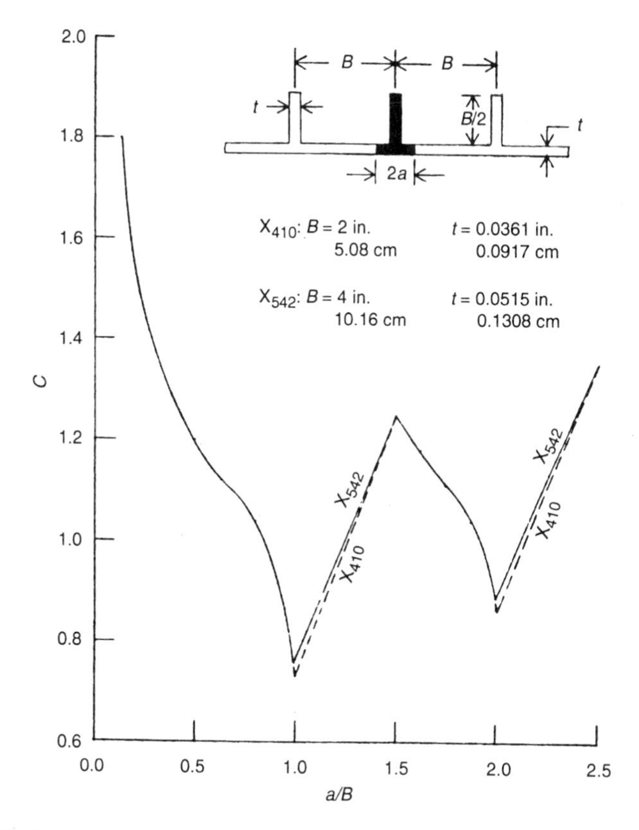

Fig. 8-30 Stress intensity factors for an integrally stiffened skin with a fully broken stiffener

stress coming from the flight load. The expected maximum difference in cabin pressure during flight, ΔP, is 0.11 MPa (16 psi) and the design limit stress is set at 213.7 MPa (31 ksi).

Two locations on the cabin are considered. One location is near the front of the cabin, at X_{410}; the other is located close to the rear of the cabin, at X_{542}. The radii of the curvature at these two locations are 1.78 and 2.54 m (70 and 100 in.), respectively. The required skin thicknesses are $t_{410} = 0.917$ mm (0.0361 in.) and $t_{542} = 1.31$ mm (0.0515 in.), based on static strength and fatigue considerations.

It is anticipated that the aircraft will be flying on two types of missions. There will be six type B missions after each type A mission. The operating ΔP for the type A mission is 0.101 MPa (14.7 psi) and the operating ΔP for the type B mission is 0.069 MPa (10 psi). The corresponding pR/t stresses are 196.5 MPa (28.5 ksi) and 133.75 MPa (19.4 ksi), respectively. Therefore, the load spectrum for the longitudinal crack will be 196.5 MPa (28.5 ksi) for one cycle plus 133.75 MPa (19.4 ksi) for six cycles.

Cracks can occur in both the longitudinal and circumferential directions, and separate load spectra are needed for each of these cracks. The load spectrum for the circumferential crack is the $pR/2t$ stress, i.e., 98.25 MPa (14.25 ksi) for the type A mission and 66.87 MPa (9.7 ksi) for the type B mission. In addition, flight loads, which usually come from axial bending, are superimposed on the internal pressure. The magnitude of these additional loads are 68.9 MPa (10 ksi) of axial stress for the type A mission and an additional 34.47 MPa (5 ksi) of axial stress for the type B mission. Consequently, the final load spectra for the circumferential crack will be 167.19 MPa (24.25 ksi) for one cycle and 101.35 MPa (14.7 ksi) for six cycles.

Material

The subject crew compartment is built with 2219-T851 aluminum alloy. The mechanical properties for this alloy are: $F_{ty} = 317$ MPa (46 ksi), $K_c = 68.1$ MPa$\sqrt{\text{m}}$ (62 ksi$\sqrt{\text{in.}}$), $K_{Ic} = 38.5$ MPa$\sqrt{\text{m}}$ (35 ksi$\sqrt{\text{in.}}$), and $\Delta K_0 = 3.85$ MPa$\sqrt{\text{m}}$ (3.5 ksi$\sqrt{\text{in.}}$). The da/dN curve for this alloy is plotted in Fig. 8-24.

Crack Geometry

Figure 8-25(a) shows the four most probable types of damage and their locations. These part-through cracks would eventually become through-the-thickness cracks. Because the sheet is thin, it would take only a short time for these cracks to grow through the thickness. Therefore, for the purpose of simplification, it has been assumed that all these cracks are initially through-the-thickness cracks, as shown in Fig. 8-25(b).

Particular reference is made to cases 3 and 4. In case 3, it is assumed that the crack is located at the root of the stiffener. The crack lengths in all three directions are the same, and the crack growth rates in all three directions are also the same. In case 4, it is assumed that initially only one crack exists at the edge of the rivet hole, as shown. When the crack breaks through the small ligament of the stiffener, another crack will shortly form on the other side of the rivet hole. In other words, it becomes an edge crack, and it will propagate toward the sheet. When the edge crack propagates through the entire stiffener and the thickness of the sheet, the stiffener is completely broken, and the crack will be a circumferential through-the-thickness crack on the skin. In this case, the criterion for breakthrough of the small ligament due to the crack emanating from the rivet hole and the criterion for the edge crack breaking through the wall are that the total plastic zone ($2r_y$) in front of the crack has penetrated through the free boundaries.

Stress Intensity Factors

Consider a design in which the spacing of the integral stiffener is 10.16 cm (4 in.) for the X_{542} location and 5.08 cm (2 in.) for the X_{410} location. The thickness of the stiffeners is the same as the thickness of the skin at each

Table 8-7 Beta factor (Bowie factors modified for one-dimensional corner cracks subjected to uniaxial tension)

Single crack		Double cracks	
c/r	Beta	*c/r*	Beta
0	3.39	0	3.39
0.1414	2.73	0.1414	2.73
0.2828	2.3	0.2828	2.41
0.4242	2.04	0.4242	2.15
0.5656	1.86	0.5656	1.96
0.707	1.73	0.707	1.83
0.8484	1.64	0.8484	1.71
1.1312	1.47	1.1312	1.58
1.414	1.37	1.414	1.45
2.121	1.18	2.121	1.29
2.828	1.06	2.828	1.21
4.242	0.94	4.242	1.14
7.07	0.81	7.07	1.07
14.14	0.75	14.14	1.03
28.28	0.707	28.28	1

Source: Ref 8-6

location. The height of the stiffeners is set equal to $B/2$ at each location (where B is the stiffener spacing). The size and the location of the rivet hole at the stiffener are presumed to be as shown in detail A of Fig. 8-25. To develop an appropriate stress intensity expression for each of these damage cases, it is necessary to know the dimensions of the horizontal stiffener and the frame. Because the frame is not directly attached to the sheet, its crack arrest capability is negligible. Therefore, its size and spacing are not of concern.

Case 1: Longitudinal Crack. The geometric factors involved in a longitudinal crack are the stress intensity multiplication factors that account for the presence of the frame (perpendicular to the crack) and the horizontal stiffener (parallel to the crack). The handbook-type stress intensity multipli-

Table 8-8 Empirical curvature correction for aluminum cylinders with longitudinal crack

a, in.	α_s
0	1.000
0.1	1.008
0.2	1.017
0.3	1.025
0.4	1.034
0.5	1.042
0.6	1.050
0.7	1.059
0.8	1.067
0.9	1.076
1	1.084
1.2	1.101
1.4	1.118
1.6	1.135
1.8	1.151
2	1.168
2.2	1.185
2.4	1.202
2.6	1.218
2.8	1.235
3	1.251
3.2	1.268
3.4	1.284
3.6	1.301
3.8	1.317
4	1.333
4.2	1.349
4.4	1.365
4.6	1.380
4.8	1.396
5	1.411

Note: $R = 70$ in., $t = 0.0361$ in. Source: Ref 8-8, 8-9

cation factor for the present configuration is not available. Considering past experience, it is reasonable to assume that the total reduction of crack tip stress intensity attributed to the combined effects of the horizontal stiffener and the frame is approximately 15%. In other words, the geometric factor in Eq 5-5 will be $\Pi\alpha = 0.85 \cdot \phi_w \cdot \alpha_s$. Here $\phi_w = 1.0$ for an infinitely wide sheet and α_s is a correction factor for the curved panel. Instead of using the analytical stress intensity solution for the pressurized cylinder given in Chapter 5, we will use an empirically developed α_s value, which is given in Fig. 8-26. To perform a crack growth prediction by using an integrated computed program, it is necessary to simplify the input data format. Tabulated values (having α_s tabulated against crack length) are often required as

Table 8-9 Combined curvature and stiffener correction factors for aluminum cylinders with a circumferential crack

a	C	α_S (a)	$C \cdot \alpha_S$
0	1	1.000	1.000
0.1	1	1.000	1.000
0.2	0.996	1.002	0.998
0.3	0.991	1.004	0.995
0.4	0.986	1.006	0.992
0.5	0.971	1.010	0.981
0.6	0.955	1.014	0.969
0.7	0.928	1.019	0.946
0.8	0.885	1.025	0.907
0.9	0.8	1.032	0.826
1	0.62	1.040	0.645
1.2	0.731	1.057	0.773
1.4	0.842	1.078	0.907
1.6	0.953	1.101	1,050
1.8	1.064	1.128	1.200
2	1.175	1.158	1.361
2.2	1.11	1.192	1.323
2.4	1.064	1.228	1.306
2.6	1.02	1.268	1.293
2.8	0.942	1.310	1.234
3	0.787	1.356	1.067
3.2	0.88	1.405	1.237
3.4	0.973	1.458	1.418
3.6	1.066	1.513	1.613
3.8	1.159	1.571	1.821
4	1.232	1.633	2.012
4.2	1.16	1.698	1.970
4.4	1.12	1.766	1.978
4.6	1.064	1.837	1.955
4.8	0.97	1.912	1.854
5	0.856	1.983	1.698

Note: $R = 70$ in., $t = 0.0361$ in., stiffener spacing = 2 in. (a) α_s values are computed from Fig 8-8. Source: Ref 8-10

input to a computer code. A set of such tabulated factors for location X_{410} is provided in Table 8-8.

Case 2: Circumferential Crack. A through-the-thickness crack symmetrically located at the middle of the bay as shown in Fig. 3-25(b) will be considered. As the crack propagates past the stiffener, the stiffener is also cracked, and the stiffener crack propagates at the same rate as the skin crack. Using the technique described in Chapter 4, a set of curves that describes the variations of K at positions relative to the stiffener location was constructed and is presented in Fig. 8-27. Thus, the geometric factor for this case will be $\Pi\alpha = C \cdot \phi_w \cdot \alpha_s$ with $\phi_w = 1.0$ for the infinitely wide sheet. Here, C is the stress intensity modification factor for the integrally stiffened panel and α_s is a numerical solution for the circumferential crack in a pressurized cylinder (Ref 8-10), given in Fig. 8-28. As mentioned above, tabulated geometric correction factors are required as input to a computer code. This time, tabulation of both α_s and C (against crack length) is required. A set of such combined factors for location X_{410} is provided in Table 8-9.

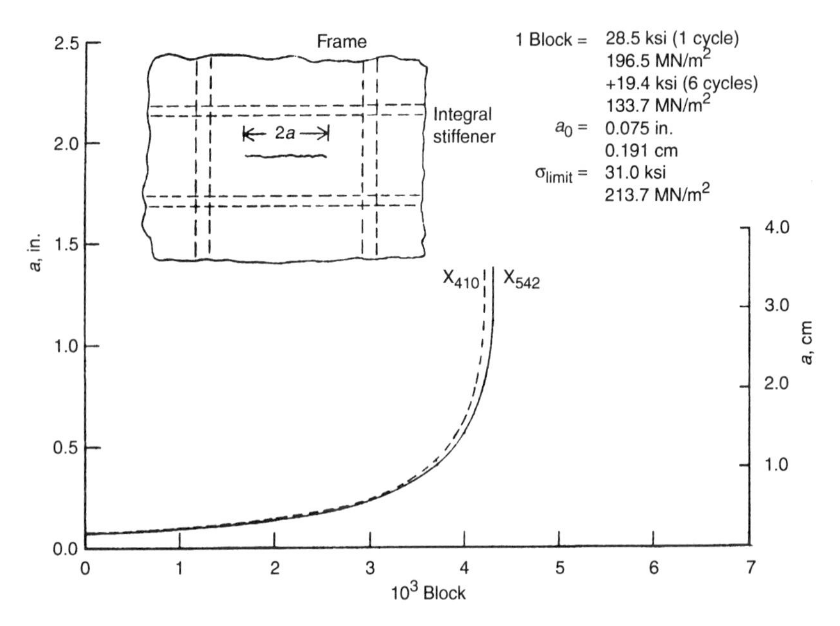

Fig. 8-31 Predicted crack growth history for a longitudinal crack in the crew compartment skin

Case 3: Circumferential Crack at the Root of Integral Stiffener. This case is essentially the same as case 2, except that the crack starts at the bottom of the integral stiffener and simultaneously grows in the stiffener and in the skin. The expression for $\Pi\alpha$ is the same as for case 2, but a different set of C-curves (given in Fig. 8-29) is used.

Case 4: Crack Emanating from a Rivet Hole. As discussed above, this case could result in three separate steps of crack propagation:

1. The crack propagates from the short-edge side of the rivet hole.
2. After breakthrough, an edge crack starts from the opposite side of the rivet hole, propagating toward the skin.
3. The skin crack propagates away from a broken stiffener.

The geometric factors for these cracks are $\Pi\alpha = \phi_B$ for the crack from a rivet hole, $\Pi\alpha = 1.122$ for the edge crack, and $\Pi\alpha = C \cdot \alpha_s$ for the skin crack. Here ϕ_B is the Bowie's factor, α_s is the curvature correction factor for the circumferential crack, and C is the integral stiffener influence factor given in Fig. 8-30. The angle and the frame (see Fig. 8-23) connect with the integral stiffener and provide load paths adjacent to the riveted stiffener;

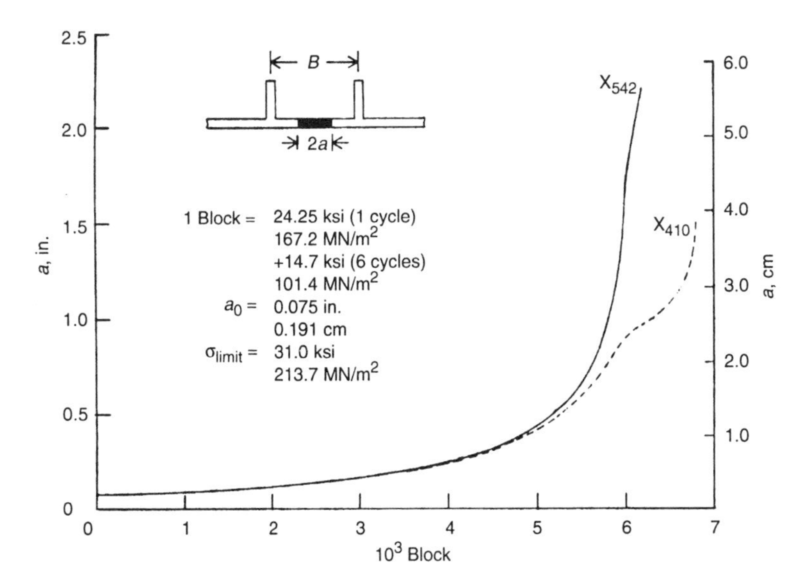

Fig. 8-32 Predicted crack growth history for a circumferential crack in the crew compartment skin

similarly, the sheet skin provides load-path continuity on the other side of the stiffener. Therefore, finite width-correction factors are not applied to any of these crack geometries.

Initial Crack Lengths

On the basis of nondestructive evaluation, it was agreed in the procurement that the following assumed initial crack lengths would be used:

- The through-the-thickness cracks: $2a_0 = 0.38$ cm (0.15 in.)
- A through-crack emanating from a rivet hole: $a_0 = 0.127$ cm (0.05 in.)

The initial crack length for cases 1, 2, and 3, therefore, is $a_0 = 0.19$ cm (0.075 in.). For case 4, the initial crack length at the edge of the fastener hole is 0.127 cm (0.05 in.).

For case 4, the crack growth calculations will stop at a crack length equal to $(0.3412 - 2r_y)$. This crack length is estimated to be 0.698 cm (0.275 in.). The second step is to assume a 1.34 cm (0.5288 in.) edge crack growing toward the skin. The calculations for the second step will stop at a crack length equal to $(B/2 + t - 2r_y)$. This crack length is estimated to be 1.94 cm

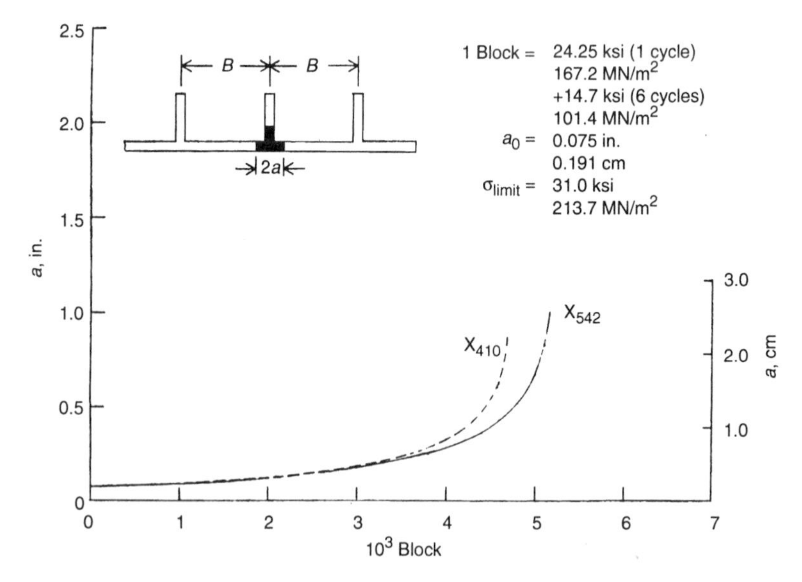

Fig. 8-33 Predicted crack growth history for a circumferential crack at the bottom of an integral stiffener in the crew compartment skin

(0.767 in.) for X_{410} and 3.86 cm (1.52 in.) for X_{542}. The initial crack length for the skin crack in the third step, after breaking the integral stiffener, is uncertain. In this example, it is assumed that this crack length equals the plastic zone size at breakthrough, i.e., $2a = 2r_y$, where $2r_y$ is 0.68 cm (0.268 in.) for X_{410} and 1.35 cm (0.5315 in.) for X_{542}.

Crack Growth Predictions

The current crew compartment configuration consists only of welded integrally stiffened skins, so the residual strength or fail-safe capability for the crew compartment skin cannot be increased by taking advantage of load transfer in any form (e.g., divided planks or attached stiffener). Therefore, the residual strength analysis in this case is combined with the fatigue crack propagation analysis. This is done by computing the K-value at each crack length using the design limit stress of 213.7 MPa (31 ksi). When the K-value at any crack length reaches K_c, final failure of the panel is assumed.

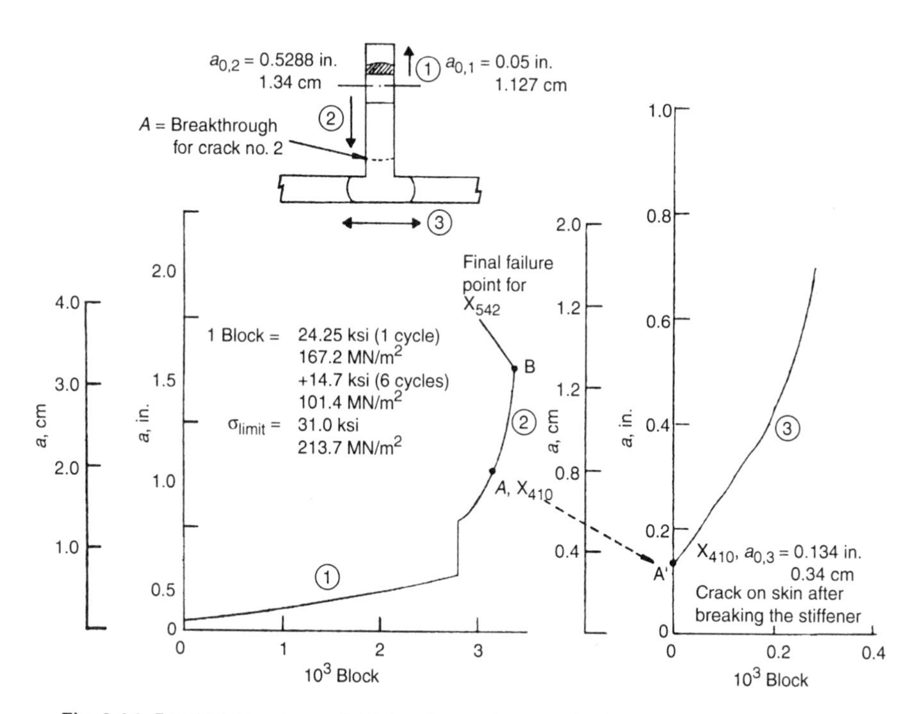

Fig. 8-34 Predicted crack growth history for crack emanating from a rivet hole

The predicted crack growth histories of cases 1 to 3 are presented in Fig. 8-31 to 8-33. The crack growth curves in these figures show that the safe-crack-growth life in either location, X_{410} or X_{542}, is approximately the same for a given crack geometry. Case 1, longitudinal crack, exhibits the shortest life (approximately 4000 loading blocks). Case 2, a circumferential crack between two integral stiffeners, exhibits the longest life (approximately 6000 loading blocks).

The predicted results for case 4 are presented in Fig. 8-34. It will take 2800 loading blocks to grow the 1.27 mm (0.05 in.) crack at one side of the hole to break through the small ligament adjacent to the hole, and then it will take another 350 loading blocks for the crack at X_{410} to break up the entire stiffener. After that, it takes an additional 290 loading blocks to completely crack the skin. As for the crack at X_{542}, complete failure of the whole panel occurs while the edge crack (step 2) is propagating toward the skin. Nevertheless, the total safe-crack-growth life at either location for the case 4 crack is approximately 3400 loading blocks, regardless of where the final failure points occur. This is the shortest life among all four assumed configurations.

REFERENCES

8-1. J.C. Ekvall, T.R. Brussat, A.F. Liu, and M. Creager, Preliminary Design of Aircraft Structures to Meet Structural Integrity Requirements, *J. Aircr.*, Vol 11, 1974, p 136–143

8-2. D. Shows, A.F. Liu, and J.H. FitzGerald, Application of Resistance Curves to Crack at a Hole, *Fracture Mechanics: Fourteenth Symposium*, Vol II, *Testing and Applications*, STP 791, American Society for Testing and Materials, 1983, p II-87 to II-100

8-3. A.F. Liu and D.F. Dittmer, "Effect of Multiaxial Loading on Crack Growth," Report AFFDL-TR-78-175 (in three volumes), Air Force Flight Dynamics Laboratory, Wright-Patterson Air Force Base, Dayton, OH, Dec 1978

8-4. A.F. Liu, J.E. Allision, D.F. Dittmer, and J.R. Yamane, Effect of Biaxial Stresses on Crack Growth, *Fracture Mechanics*, STP 677, American Society for Testing and Materials, 1979, p 5–22

8-5. A.F. Liu and J.R. Yamane, Crack Growth under Equibiaxial Tension, *Res Mech.*, Vol 5, 1982, p 1–11

8-6. A.F. Liu, Stress Intensity Factor for a Corner Flaw, *Eng. Fract. Mech.*, Vol 4, 1972, p 175–179

8-7. A.S. Kobayashi and W.L. Moss, Stress Magnification Factors for Surface-Flawed Tension Plate and Notched Round Bar, *Fracture 1969*, Chapman & Hall, London, 1969

8-8. W.J. Crichlow, The Ultimate Strength of Damaged Structure—Analysis Methods with Correlating Test Data, *Full Scale Fatigue Testing of Aircraft Structures*, Pergamon Press, 1960, p 149–209

8-9. W.J. Crichlow, "A Systems Approach to Material Selection and Design for Structural Integrity," paper presented at Symposium on Crack Propagation of the 7000-Series Aluminum Alloys (McClelland Air Force Base, CA), 29–30 April 1969

8-10. F. Erdogan and M.M. Ratwani, "Fracture of Cylindrical and Spherical Shells Containing a Crack," paper presented at First Int. Conf. on Structural Mechanics in Reactor Technology (Berlin, Germany), 20–24 Sept 1971

SELECTED REFERENCES

- W.E. Anderson, Fatigue of Aircraft Structures, *Int. Metall. Rev.*, Vol 17, 1972, p 240–263
- W.E. Anderson and L.A. James, Estimating Cracking Behavior of Metallic Structures, *Journal of the Structural Division*, ASCE, Vol 96 (No. ST4), April 1970, p 773–790
- R.G. Baggerly, Hydrogen-Assisted Stress Cracking of High-Strength Wheel Bolts, *Engineering Failure Analysis*, Vol 3, 1996, p 231–240
- J.P. Gallagher, F.J. Giessler, A.P. Berens, and R.M. Engle, Jr., "USAF Damage Tolerant Design Handbook: Guidelines for the Analysis and Design of Damage Tolerant Aircraft Structures," Report AFWAL-TR-82–3073, Flight Dynamics Laboratory, Air Force Wright Aeronautical Laboratories, Wright-Patterson Air Force Base, Dayton, OH, May 1984
- R.M. Gamble and P.C. Paris, Cyclic Crack Growth Analysis for Notched Structures at Elevated Temperature, *Mechanics of Crack Growth*, STP 590, American Society for Testing and Materials, 1976, p 345–367
- U.G. Goranson, "Damage Tolerance—Facts and Fiction," paper presented at 17th Symposium of the International Committee on Aeronautical Fatigue (Stockholm, Sweden), 9–11 June 1993
- U.G. Goranson, J. Hall, J.R. Maclin, and R.T. Watanabe, Long-Life Damage Tolerant Jet Transport Structures, *Design of Fatigue and Fracture Resistant Structures*, STP 761, American Society for Testing and Materials, 1982, p 47–90
- U.G. Goranson and J.T. Rogers, "Elements of Damage Tolerance Verification," paper presented at 12th Symposium of the International Committee on Aeronautical Fatigue (Toulouse, France), 25–31 May 1983
- D.I. Nwosu and D.O. Olowokere, Evaluation of Stress Intensity Factors for Steel Tubular T-Joints Using Line Spring and Shell Elements, *Eng. Failure Analysis*, Vol 2, 1995, p 31–44
- L. Riedinger and J.M. Waraniak, "Application and Integration of Design Allowables to Meet Structural-Life Requirements," Paper ICAS-84–3.8.1, presented at 14th Congress of the International Council of the Aeronautical Sciences (Toulouse, France), 9–14 Sept 1984
- A.I. Rifani and A.F. Grandt, Jr., A Fracture Mechanics Analysis of Fatigue Crack Growth in a Complex Cross Section, *Eng. Failure Analysis*, Vol 3, 1996, p 249–265
- S.H. Smith and N.D. Ghadiali, Fatigue Crack Growth Life Evaluation of the Turbine Blades in a Low-Pressure Steam Turbine, *Fracture Mechanics: Fourteenth Symposium,* Vol II, *Testing and Applications*, STP 791, American Society for Testing and Materials, 1983, p II-120 to II-139

- S.H. Smith, N.D. Ghadiali, A. Zahoor, and M.R. Wilson, Fracture Tolerance Analysis of the Solid Rocket Booster Servo-Actuator for the Space Shuttle, *Design of Fatigue and Fracture Resistant Structures*, STP 761, American Society for Testing and Materials, 1982, p 445–476

Conversion Table

To convert from:	To:	Multiply by:
1 in.	mm	25.4
1 $in.^2$	mm^2	645.16
1 kip	N	4448.22
1 lb	kg	0.4536
1 ksi	MPa	6.892
1 ksi $\sqrt{in.}$	MPa$\sqrt{m}$	1.0988
1 lbf · in.	N · m	0.113
1 lbf · ft	N · m	1.356
1 radian	degree	57.295
°F (fahrenheit)	°C (centigrade)	5/9 · (°F – 32)
°C (centigrade)	°T (Kelvin)	1 · (273.16 · °C)

Index

A

D

F

M

Q

R

T

U

V

W

Y

Z